Lecture Notes in Electrical Engineering

Volume 160

For further volumes:
http://www.springer.com/series/7818

Jiadong Sun · Jingnan Liu · Yuanxi Yang
Shiwei Fan
Editors

China Satellite Navigation Conference (CSNC) 2012 Proceedings

The 3rd China Satellite Navigation Conference (CSNC 2012) Guangzhou, China, May 15–19, 2012 Revised Selected Papers

Springer

Jiadong Sun
China Aerospace Science and Technology Corporation
Chinese Academy of Sciences
Beijing
People's Republic of China

Jingnan Liu
GPS Engineering Research Center
Wuhan University
Wuhan 430079
People's Republic of China

Yuanxi Yang
China National Administration of GNSS and Applications
Beijing
People's Republic of China

Shiwei Fan
Beijing
China

ISSN 1876-1100 e-ISSN 1876-1119
ISBN 978-3-642-29174-6 e-ISBN 978-3-642-29175-3
DOI 10.1007/978-3-642-29175-3
Springer Heidelberg New York Dordrecht London

Library of Congress Control Number: 2012935670

Printed on acid-free paper

Springer is part of Springer Science+Business Media (www.springer.com)

Scientific Committee

Chairman:
Jiadong Sun, China Aerospace Science and Technology Corporation

Vice-Chairman:
Rongjun Shen, China
Jisheng Li, China
Qisheng Sui, China
Zuhong Li, China Academy of Space Technology
Shusen Tan, Beijing Global Information Center of Application and Development

Executive Chairman:
Jingnan Liu, Wuhan University
Yuanxi Yang, China National Administration of GNSS and Applications
Shiwei Fan, China

Committee Members:
Qingjun Bu, China
Liheng Wang, China Aerospace Science and Technology Corporation
Yuzhu Wang, Shanghai Institute of Optics and Fine Mechanics, Chinese Academy of Sciences
Guoxiang Ai, National Astronomical Observatories, Chinese Academy of Sciences
Shuhua Ye, Shanghai Astronomical Observatories, Chinese Academy of Sciences
Zhaowen Zhuang, National University of Defense Technology
Qifeng Xu, PLA Information Engineering University
Houze Xu, Institute of Geodesy and Geophysics, Chinese Academy of Sciences
Guirong Min, China Academy of Space Technology
Xixiang Zhang, China Electronics Technology Group Corporation
Lvqian Zhang, China Aerospace Science and Technology Corporation
Junyong Chen, National Administration of Surveying, Mapping and Geoinformation
Benyao Fan, China Academy of Space Technology

Organizing Committee

Secretary General:
Haitao Wu, Navigation Headquarter Office, Chinese Academy of Sciences

Vice-Secretary General:
Guangzhou Ouyang, Navigation Headquarter Office, Chinese Academy of Sciences
Wenhai Jiao, China Satellite Navigation Office
Dawei Wang, BeiDou Office, The Economic and Information Commission of Guangdong Province
Xiuwan Chen, Institute of Digital China, Peking University
Chuang Shi, GNSS Research Center of Wuhan University
Miao Yu, Beijing Institute of Space Science and Technology Information

Committee Members:
Xinyu Wang, Navigation Satellite Systems Engineering Center, China Academy of Aerospace Electronics Technology
Jun Lu, China Satellite Navigation Office
Yan Liu, Guangzhou HAIGE Communications Group Incorporated Company
Xilai Li, Beijing Global Information Center of Application and Development
Shancong Zhang, Beijing UCAS Space Technology Co., Ltd
Xuetian Shao, Academy of Opto-Electronics, Chinese Academy of Sciences
Gang Hu, BDStar Navigation
Xiang'an Zhao, National Defense Science and Technology Information Center
Cantao Zhong, Satellite Navigation Joint Research Center, Ministry of education of PRC
Hong Yuan, Navigation Headquarter Office, Chinese Academy of Sciences
Zhong Dou, National Time Service Center, Chinese Academy of Sciences

Editorial Board

Preface

COMPASS (Beidou) satellite navigation system is China's own satellite navigation system, independently developed and compatible with the rest of the global satellite navigation systems. It provides highly reliable positioning, navigation, and timing services, as well as short-message communication for all users with all-weather, all-time, and worldwide. Currently, COMPASS (Beidou) satellite navigation system has launched 10 satellites, and a basic system has been set up. On Dec 27, 2011, a press conference on COMPASS (Beidou) navigation system was held in Beijing to announce its main performance during the test period, such as the service area, positioning accuracy, velocity accuracy, and timing accuracy. The public release of a "beta" or test version of the COMPASS (Beidou) Interface Control Document (ICD) was also announced. Retaining the active positioning service and short message communication service, COMPASS (Beidou) from that day officially started providing positioning, navigation, and timing services to China and surrounding areas.

COMPASS (Beidou) system encourages domestic and foreign enterprises to participate in its R & D and application. In addition to COMPASS (Beidou) 10 satellites launched so far, China plans to launch six more into orbit in 2012 to bolster its accuracy and expand its service to cover most of the Asia-Pacific area.

China Satellite Navigation Conference (CSNC) is an open platform for academic exchanges in the field of satellite navigation. Its aim is to encourage technological innovation, accelerate GNSS engineering, and boost the development of the satellite navigation industry in China.

The third China Satellite Navigation Conference (CSNC 2012) will be held on May 15–19, 2012, in Guangzhou city, China, sponsored by China Satellite Navigation Office, Department of High and New Technology Development and Industrialization, Ministry of Science and Technology, P.R.C, China National Space Administration, State Administration of Science, Technology and Industry for National Defense, Department of Comprehensive Planning, Ministry of Transport, P.R.C, Department of Science and Technology, Ministry of Education, P.R.C and others. The CSNC 2012 will cover a wide range of activities, including technical seminars, academic exchange, forum, exhibition, as well as CSNC-ION joint panel.

The conference topics are:

1. COMPASS (Beidou)/GNSS Navigation Application;
2. Satellite Navigation Model and Method;
3. Integrated Navigation and New Methods;
4. Satellite Navigation Signal System, Compatibility & Interoperability;
5. Precise Orbit Determination and Positioning;
6. Satellite Navigation Augmentation and Integrity Monitoring;
7. Atomic Clock Technique and Time-Frequency System;
8. COMPASS (Beidou)/GNSS Test and Evaluation Technology; and
9. COMPASS (Beidou)/GNSS User Terminal Technology.

The proceedings include 189 papers selected from 597 technical papers through a strict peer-review process, to be presented at the CSNC 2012. All the 189 papers are divided into nine chapters following the nine topics of the conference. In addition, 24 papers were recommended for publication in national and international journals such as SCIENCE CHINA Physics, Mechanics & Astronomy, and Advances in Space Research. More than 300 papers are included in the CSNC 2012 Electronic Proceedings and posted on the conference.

All the 33 session chairs (see the name list of Editorial Board) and over 100 reviewers are gratefully acknowledged for their time and effort in the review process.

Jiadong Sun
Chair of CSNC 2012

Contents

Part I
Satellite Navigation Signal System, Compatibility & Interoperability

Research on Radio Frequency Compatibility of Beidou User Equipment

Shusen Tan, Lin Li and Chi Xie

Abstract This paper researches the radio frequency compatibility (RFC) of Beidou user equipment systemically. Firstly, the interference margin of the Beidou user equipment is analyzed. Secondly, the protection demand for the satellite navigation signal at the user equipment level is raised. Lastly, the RFC index of user equipment is studied out. Based on the RFC index, alter the Beidou user equipment rationally, and then the RFC issue can be resolved preferably.

1 Introduction

Beidou satellite navigation system is allocated in the L-band and S-band which component has the advantage of high technology maturity and high performance price ratio. Then, a number of radio services are competing fiercely for this resource. Despite very crowded, there are still many new military and civilian application systems to seek the resources available in recent years.

Beidou system has been applied to many fields such as information acquisition, command and control, precision strike, battlefield rescue, readiness training, security of after installation, manage and control of border, etc. With broad application of the Beidou system, the radio frequency compatibility (RFC) issues between the Beidou user equipment and the application system of adjacent band become increasing prominent and more complex.

S. Tan · L. Li (✉) · C. Xie
Beijing Global Information Center, Beijing 100094, China
e-mail: lindda@126.comchina.lilin@gmail.com

J. Sun et al. (eds.), *China Satellite Navigation Conference (CSNC) 2012 Proceedings*,
Lecture Notes in Electrical Engineering 160, DOI: 10.1007/978-3-642-29175-3_1,

The application systems adjacent to the Beidou system are of many kinds and great number. Both high power radar, and a large number of widely distributed low-power broadcast and communications equipment are included. Their frequency bands adjacent to the Beidou system band, partially overlapping or even completely overlapping; their requirements of out-of-band suppression and spurious suppression are not enough fitness; and their transmit power are higher even up to hundreds of kilowatts.

Even though the level of out-of-band suppression of Beidou user equipment is so strong as to infinity, the out-of-band component, spurious component and intermodulation component of transmitted signal of the adjacent channel systems may fall into the bandwidth of Beidou user equipment. The RF compatibility issue then induced. Moreover, the out-of-band suppression of the Beidou user equipment is limited. The out-of-band signal with strong power also leads to the RFC issue. There is no overall and thorough solution to the frequent RFC issue of late.

It is unrealistic to expect severe RFC test to be dealt with Beidou equipment alone. More reasonable solution is as following. First, the explicit protection demand of each system is proposed respectively. Secondly, the protection demands of various systems are carefully weighed, multilateral protection agreement is then reached. Finally, the proper deployment and orderly adjustment of various systems are made according to the multilateral protection agreement.

This paper analyses the RFC ability of Beidou user equipment, dissertates protection demand of Beidou navigation signal at the user equipment level, and develops the RFC index of Beidou user equipment. The RFC issue is then resolved rather thoroughly.

2 Analysis of RFC

2.1 Anti-Interference Principle

In Beidou navigation system, there are BPSK-R, binary offset carrier (BOC) and alternate binary offset carrier (AltBOC) modulations, etc. BPSK-R denotes direct sequence spread spectrum (DSSS) signals generated using BPSK signaling with rectangular chips. BOC and AltBOC signals are generated using DSSS techniques but employ portions of a square wave for the chips.

The received signal is the following

$$\begin{aligned} r(t) &= s(t) + j(t) + n(t) \\ &= \sqrt{2P_S} d(t) s_{p_n}(t) \cos(2\pi f_0 t) + j(t) + n(t) \end{aligned} \tag{1}$$

where $s(t)$ is the spread spectrum signal carrying data information $d(t)$, $j(t)$ is the interference, $n(t)$ is the thermal noise, their power are P_S, P_J and P_N, respectively; f_0 is the carrier frequency; $s_{p_n}(t)$ is the spreading waveform.

In the user equipment, an identical and precisely time-synchronized replica spreading symbol $S_{Pn}(t)$ is generated and correlated (multiply and filter) with the received signal. Since the processing of correlating and dispreading in time domain equates the processing of convolving of power spectral density (PSD) in frequency domain. Then, the spread spectrum signal is compressed to its original narrow bandwidth with only the data modulation remaining. The noise spectra density is still white Gaussian noise. The interference has now been spread, and its spreading extent is determined by the PSDs of spreading waveform and interference.

After the convolution of PSD, the convolved power spectral density is as follows

$$S(f) = P_J \int_{-\infty}^{\infty} |H_R(v)|^2 S_{p_n}(v) S_J(f - v) dv \tag{2}$$

where $S_{p_n}(v)$ is the PSD of the spreading waveform normalized to unit area over infinite bandwidth, $S_J(f)$ is the PSD of the interference normalized to unit area over infinite bandwidth. Suppose that the receiver filter is very wide, so that the $H_R(f)$ can be treated as approximate unity at frequencies where the desired signal has appreciable power, so that the above formula can be simplified to be

$$S(f) = P_J \int_{-\infty}^{\infty} S_{p_n}(v) S_J(f - v) dv \tag{3}$$

Then, the interference power within bandwidth B_d is $P_{J,eff} = \int_{-B_d}^{B_d} S(f) df$.

Note that the PSD of spreading waveform with short period is discrete lines, and the PSD of spreading waveform with long period is continuous spectrum. That interference spectrum convolve with the discrete lines of spreading waveform is more complex than continuous spectrum. To simplify analysis, the discrete lines spectrum is approximated by continuous spectrum in this paper.

Since simplifying the above, the convolved spectrum is flat over the bandwidth B_d in the vicinity of the carrier frequency, and its amplitude is $S(0) = P_J \kappa_{JS}$, where the spectral overlap coefficient (SOC) is $\kappa_{JS} = \int_{-\infty}^{\infty} S_{p_n}(v) S_J(v) dv$. Then, the effective interference power is $P_{J,eff} = S(0)(2B_d)$.

The effective interference power reduce the un-jammed signal-to-noise-power ratio (SNR) of the received signal, P_S/P_N, to a lower value, $(S/N)_{eff}$ as follows

$$(S/N)_{eff} = \frac{P_S}{P_N + P_{J,eff}} \tag{4}$$

The interference margin is the interference-to-signal-power ratio (ISR) when $(S/N)_{eff}$ reduced to $(S/N)_{th}$, where $(S/N)_{th}$ is threshold SNR corresponding to the desired performance.

Substituting $(S/N)_{th}$ and $P_{J,eff}$ into (4) yields

$$M_J = \frac{(NSR)_{diff}}{2\kappa_{JS} B_d} \tag{5}$$

where $(NSR)_{diff} = 1/(S/N)_{th} - 1/(P_S/P_N)$.

According to (5), the SOC, un-jammed SNR, threshold SNR and related bandwidth codetermine the in-band interference margin of user equipment.

2.2 Anti-Interference Capability

2.2.1 Spectral Overlap Coefficient

Observe that the SOC depends on the spectrum of the spreading waveform as well as the spectrum of the interference. Examples of different types of interference can now be evaluated.

Narrowband Interference (NBI)

For single tone interference centered at f_J', the power spectrum can be modeled as $S_J(f) = (1/2)\delta(f + f_J') + (1/2)\delta(f - f_J')$, where δ (•) is the Dirac delta function having infinite amplitude, vanishing width, and unit area. Substituting for this interference power spectral density in SOC formula yields $\kappa_{JS} = S_{p_n}(f_J')$.

For narrowband interference having certain bandwidth, the SOC satisfies $\kappa_{JS} \leq S_{p_n}(f_J')$.

Matched Spectrum Interference (MSI)

Consider now when the interference has the same PSD as the spreading waveform. This situation could arise from multiple access interference or from an interference whose spectrum is matched to that of the spreading waveform. Substituting the PSD of spreading waveform into SOC formula yields $\kappa_{JS} = \int_{-\infty}^{\infty} [S_{p_n}(f)]^2 df$.

Bandlimited White Noise Interference (BWNI)

When the interference has flat spectrum centered at f_J' and extending from $f_J' - B_J/2 \leq f \leq f_J' + B_J/2$, its amplitude is$1/(2B_J)$. Substituting these into SOC formula yields $\kappa_{JS} = (1/B_J) \int_{f_J'-B_J/2}^{f_J'+B_J/2} S_{p_n}(f) df$.

If B_J is large enough so that almost all of the signal power is included within $[f_J - B_J/2, f_J + B_J/2]$, then $\kappa_{JS} = 1/B_J$. When the interference spectrum is centered on the spreading waveform spectrum so that $f_J' = 0$, the SOC becomes $\kappa_{JS} = (1/B_J)\int_{-B_J/2}^{B_J/2} S_{p_n}(f)df$.

BPSK-R(n) is a modulation synthesized by BPSK with rectangular chips at a chip rate of $f_c = nf_{nom}$, where f_{nom} is the nominal reference frequency 1.023 MHz. The BPSK-R(n) spreading waveform can be described as [1]

$$s_{BPSK-R}(t) = \sum_{k=-\infty}^{\infty} a_k g_{BPSK-R}(t - kT_c) \tag{6}$$

where a_k is spreading code; T_c is the chip period, and the reciprocal of the chip period is known as the chip rate; $g_{BPSK-R}(t) = \begin{cases} 1/\sqrt{T_c} & 0 \leq t \leq T_c \\ 0 & else \end{cases}$ is rectangular chip waveform.

The corresponding power spectral density is [1–3]

$$S_{BPSK-R}(f) = T_c sinc^2[\pi f/(nf_{nom})] \tag{7}$$

where $sinc(x) = \sin x/x$.

Substituting this PSD into SOC formula yields the following.

Narrowband Interference

If the narrowband interference is placed at the spectral maximum of a BPSK-R spreading waveform, then $\kappa_{JS} = S_{p_n}(0) = T_c$. If instead the interference is placed at a frequency other than the spectral peak, κ_{JS} is smaller than T_c, meaning that the interference has less effect.

Matched Spectrum Interference

Consider now when the interference has the same power spectral density as BPSK-R spreading waveform. Then, $\kappa_{JS} = \int_{-\infty}^{\infty} \{T_c sinc^2[\pi f/(nf_{nom})]\}^2 df = 2T_c/3$.

Bandlimited White Noise Interference

When the interference spectrum is centered on the spectrum of the BPSK-R spreading waveform so that $f_J' = 0$, the spectral overlap coefficient becomes $\kappa_{JS} = (1/B_J)\int_{-B_J/2}^{B_J/2} T_c sinc^2[\pi f/(nf_{nom})]df$. When in addition $B_J = 2f_c$, so that the interference covers the null-to-null main lobe of the signal spectrum, the SOC is presented as $\kappa_{JS} = (1/2.22)T_c$.

Table 1 Summary of spectral overlap coefficients (SOCs)

Parameter	NBI	MSI	BWNI
(1, 1)	$T_c/1.9$	$T_c/3$	$T_c/4.6$
(6, 1)	$T_c/2.4$	$T_c/4.4$	$T_c/5.4$
(14, 2)	$T_c/2.5$	$T_c/4.5$	$T_c/5.4$
(15, 10)	$T_c/2.2$	$T_c/3.7$	$T_c/5.1$
(m, n)	$[T_c/2.5, T_c/1.9]$	$[T_c/4.5, T_c/3]$	$[T_c/5.5, T_c/4.6]$

The notation BOC(m,n) is shorthand for a BOC modulation generated using an mf_{nom} square wave subcarrier frequency and an nf_{nom} chip rate. When m/n is an integer, a BOC spreading waveform can be described as [1]

$$s_{BOC}(t) = \sum_{k=-\infty}^{\infty} a_k g_{BOC}(t - kT_c) \tag{8}$$

where $g_{BOC}(t) = g_{BPSK-R}(t)sgn[\sin(2\pi f_s t + \varphi)]$, sgn is the signum function (1 if the argument is positive, -1 if the argument is negative), the relationship between square wave subcarrier frequency and chip rate is $mf_c = nf_s$, φ is a selectable phase angle.

Two common values of φ are 0 and 90, for which the resultant BOC signals are referred to as sine phased or cosine phased, respectively. The BOC subscripts s and c refer to sine-phased and cosine-phased, respectively.

The power spectral density for a sine-phased BOC(m,n) is [1]

$$S_{BOCs}(f) = \begin{cases} \frac{1}{nf_{nom}}\{h_{BOCs}(f)sinc[\pi f/(nf_{nom})]\}^2 & 2m/n even \\ \frac{1}{nf_{nom}}\left\{h_{BOCs}(f)\frac{\cos[\pi f/(nf_{nom})]}{\pi f/(nf_{nom})}\right\}^2 & 2m/n odd \end{cases}$$

and the power spectral density for a cosine-phased BOC(m,n) is

$$S_{BOCc}(f) = \begin{cases} \frac{4}{nf_{nom}}\{h_{BOCc}(f)sinc[\pi f/(nf_{nom})]\}^2 & 2m/n even \\ \frac{4}{nf_{nom}}\left\{h_{BOCc}(f)\frac{\cos[\pi f/(nf_{nom})]}{\pi f/(nf_{nom})}\right\}^2 & 2m/n odd \end{cases}$$

where $h_{BOCs}(f) = \tan[\pi f/(2mf_{nom})]$, $h_{BOCc}(f) = \frac{\sin^2[\pi f/(4mf_{nom})]}{\cos[\pi f/(2mf_{nom})]}$.

Substituting for this spreading waveform PSD in SOC formula yields the associated SOC results. Table 1 summarizes these SOCs for four types of BOCs modulation for the three classes of interference types analyzed earlier.

Similarly, the SOC of constant envelope AltBOC modulation can yield according to the following PSD [5, 6]

Table 2 In-band interference margin of COMPASS-S equipment

Satellite signal power (dBm)	NBI M_J (dB)	MSI M_J (dB)	BWNI M_J (dB)
−127	13.64	15.41	17.11
−124	17.15	18.91	20.61

$$S_{AltBOC}(f) = \begin{cases} \frac{4nf_{nom}}{(\pi f)^2}\left[\frac{\sin[\pi f/(nf_{nom})]}{\cos[\pi f/(2mf_{nom})]}\right]^2 h_{AltBOC}(f) & 2m/n even \\ \frac{4nf_{nom}}{(\pi f)^2}\left[\frac{\cos[\pi f/(nf_{nom})]}{\cos[\pi f/(2mf_{nom})]}\right]^2 h_{AltBOC}(f) & 2m/n odd \end{cases}$$

where $h_{AltBOC}(f) = \cos[\pi f/(2mf_{nom})]^2 - \cos[\pi f/(2mf_{nom})] - 2\cos[\pi f/(2mf_{nom})]\cos[\pi f/(4mf_{nom})] + 2$.

2.2.2 In-band Interference Margin of COMPASS-S Equipment

According to Sect. 2.1, the interference margin is apparently related to the desired performance. The desired performance; bit error rate (BER) performance, for example. The threshold SNR is the required minimum SNR of the data demodulation, accordingly, the bandwidth concerned is the information rate.

According to the encoded mode and decoded mode, the threshold SNR is 5 dB corresponding to the bit error rate smaller than 10^{-5}. The unilateral power spectral density of noise is −202.5 dBW/Hz [4], as a result of the antenna noise temperature, host noise figure, insertion loss of feeders and band-pass filter, etc. The un-jammed SNR is calculated according to the received satellite signal power. The SOC of spreading waveform with the three classes of interference types is determined by the associated parameters. Substituting these into (5), the interference margin is acquired as shown in Table 2.

2.2.3 In-band Interference Margin of COMPASS-B2 Equipment

Table 3 shows the COMPASS-B2 navigation signal parameters.

According to the encoded mode and decoded mode, the threshold SNR is 6 dB corresponding to the bit error rate smaller than 10^{-6}. The unilateral power spectral density of noise is −202.5 dBW/Hz, as a result of the antenna noise temperature, host noise figure, insertion loss of feeders and band-pass filter, etc. The un-jammed SNR is calculated according to the received satellite signal power. The SOC of AltBOC spreading waveform with the three classes of interference types is determined by the associated parameters. Substituting these into (5), the interference margin is acquired as shown in Table 4.

Table 3 Parameter of COMPASS-B2 navigation signal

Navigation signal	Carrier frequency (MHz)	Chip rate (MHz)	Bit/symbol rate (bps/sps)	Secondary code length (bits)	Modulation type
B2ax data(I)	1191.795	10.23	25/50	10	AltBOC (15, 10)
B2ay pilot(Q)			No	200	
B2x data(I)			50/100	10	
B2y pilot(Q)			No	200	

Table 4 In-band interference margin of COMPASS-B2 equipment

Satellite signal power (dBm)	NBI M_J (dB)	MSI M_J (dB)	BWNI M_J (dB)
−133	38.89	32.42	42.26
−123	38.98	32.51	42.34

2.3 Radio Frequency Compatibility

The design for out-of-band suppression makes the interference suffered attenuation to a certain extent at the RF front-end. Then, the out-of-band interference margin is an improvement on the in-band one. The improvement depends on the level of out-of-band suppression. Constraining to the level of domestic machining technology and components, there is much room for improvement of out-of-band suppression of Beidou user equipment.

As a computation example, assume that satellite signal power is −124 dBm, in-band interference margin is 17 dB, and the level of out-of-band suppression is 40 dB. The allowable interference power is only −67 dBm. If the level of out-of-band suppression increases to 60 dB, then the allowable interference power is up to −47 dBm.

3 Verification of RFC

Figure 1 depicts the tolerable interference power performance as a function of the interference frequency for single tone interference. The received satellite signal power is −124 dBm in this test.

As shown in Fig. 1, the tolerable power of in-band single tone interference is −107.5 dBm, that is, 16.5 dB in interference-to-signal-ratio. This result approaches the interference margin 17.15 dB, as analyzed in Sect. 2.2.2. Besides, the equipment has an out-of-band suppression level 40 dB at carrier frequency plus minus 30 MHz. All this demonstrates that testing result is consistent with the interference margin analysis in Sect. 2.2.

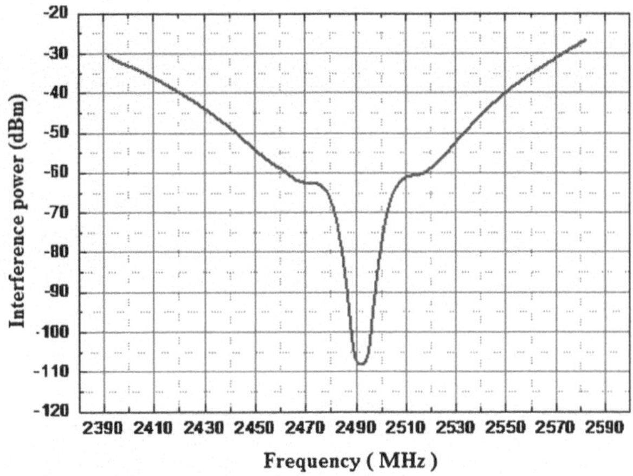

Fig. 1 Testing result of interference margin of S user equipment

4 Protection Demand for the COMPASS-B2 Navigation Signal at User Equipment Level

In order to assure the user equipment work properly, the protection demand of Beidou navigation signal is proposed in this section. Good compromise between harsh (namely, difficult to achieve) and lenient (only the nominal one) is reached. The protection demand for in-band application would have a 3 dB lower ISR performance than least in-band interference margin for the sake of caution. For out-of-band application, its interference margin is the sum of rejection ability of in-band interference and suppression ability of out-of-band interference.

Figure 2 depicts the protection demand for the COMPASS-B2 navigation signal. The associated parameters are given by

- in-band frequency range: carrier frequency ± 27.5 MHz
- tolerable power of in-band interference: –103.58 dBm
- level of out-of-band interference suppression: ≥35 dB (@carrier frequency ± 55 MHz, linear increase with frequency offset)
- saturation power: ≥–67.94 dBm

5 RFI Index of User Equipment

To restrict the Beidou user equipment go beyond the capability associated with the navigation signal protection demand, this section studies out the RFC index consisting of dynamic range, in-band interference margin and out-of-band interference suppression. The RFC index of user equipment is determined as follows

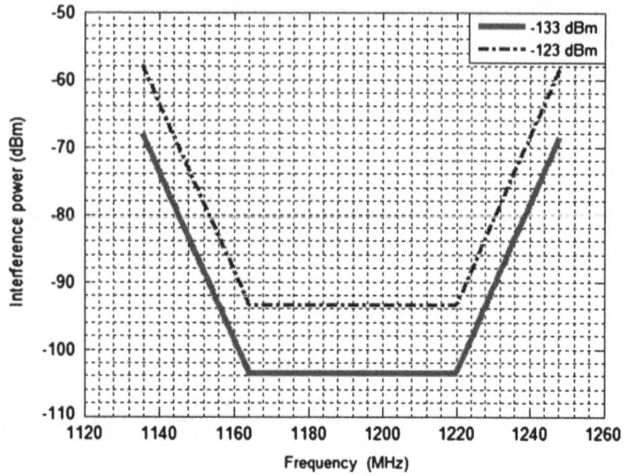

Fig. 2 Protection demand for the COMPASS-B2 navigation signal

- Saturation dynamic range: saturation input power $P_{Si} \geq -57\text{dBm}$
- Linear dynamic range: 1 dB compression point $P_{1dB} \geq -67$ dBm
- Spurious-free dynamic range: three order intermodulation intercept point $IPIP_3 \geq -47$ dBm
- In-band interference margin: ≥32 dB (@satellite signal power −133 dBm)
- Out-of-band interference suppression: ≥40 dB (@carrier frequency ± 55 MHz)

Note that the level of out-of-band interference suppression (in decibel) increases linearly with the frequency offset at least.

6 Conclusion

This paper makes systemic researches on the radio frequency compatibility issue of Beidou user equipment. Firstly, the in-band interference margin, which is based on the SOC of interference with spreading waveform, is analyzed. Secondly, the protection demand, which is a good compromise between harsh and lenient, is reached. Finally, the RFC index of Beidou user equipment is developed via the in-band interference rejection and out-of-band interference suppression. Ground on these three aspects, the RFC issue can be resolved rather thoroughly.

References

1. Kaplan, E. D., & Hegarty, C. J. (2006). *Understanding GPS: Principles and applications* (2nd ed.). Boston: Artech House.
2. Pratap, M., & Per, E. (2006). *Global positioning system: Signals, measurements, and performance* (2nd ed.). Lincoln: Ganga-Jamuna Press.

3. Parkinson, B. W., & Spilker, J. J., Jr. (1996). *Global positioning system: Theory and application*. Washington, DC: American Institute of Aeronautics and Astronautics.
4. Richard, A. P. (2004). *Modern communications jamming principles and techniques*. Boston: Artech House.
5. Shivaramaiah, N. C., & Dempster, A. G. (2009). The Galileo E5 AltBOC: Understanding the signal structure. *International Global Navigation Satellite Systems Society IGNSS Symposium, December 1–3*.
6. Lee, Y., Lee, Y., Yoon, T., Song, C., Kim, S., & Yoon, S. (2009). AltBOC and CBOC correlation function for GNSS signal synchronization. In O. Gervasi et al. (Eds.), *ICCSA 2009, Part II, LNCS* (Vol. 5593, pp. 325–334). Heidelberg: Springer.

Interoperability Feasibility Analysis Between Beidou and GPS

Xiaochun Lu, Jun Lu, Xue Wang, Yan Bai and Tao Han

Abstract Interoperability has become a focus of Global Navigation Satellite System (GNSS) and a development aspect, and incurs much focus among the world. For the purpose of maximum benefit, a series of interoperability researches and cooperation are putting forward. In this paper, according to the development of Beidou and GPS, we analyzed the interoperability feasibility between them as three phases. First of all, we introduced the signal plane of Beidou and GPS, separated their interoperability into three phases. After that, we analyzed the signal performance and service performance during each phases. Finally, based on the analyzing results, we offer a reference for Beidou signal design.

1 Introduction

Interoperability has become a focus of Global Navigation Satellite System (GNSS) and a development aspect, and incurs much focus among the world [1]. For the purpose of maximum benefit, a series of interoperability researches and

X. Lu · X. Wang · Y. Bai · T. Han (✉)
National Time Service Center, 710600 Xi'an, People's Republic of China
e-mail: hantao@ntsc.ac.cn

X. Lu · X. Wang · Y. Bai · T. Han
Key Laboratory of Precision Navigation and Timing Technology,
Chinese Academy of Sciences, 710600 Xi'an, People's Republic of China

J. Lu
Beijing Institute of Tracking and Telecommunications Technology,
No. 26 Beiqing Road, Haidian District, Beijing, China

Y. Bai
Graduate School of the Chinese Academy of Science, 100040 Beijing,
People's Republic of China

J. Sun et al. (eds.), *China Satellite Navigation Conference (CSNC) 2012 Proceedings*,
Lecture Notes in Electrical Engineering 160, DOI: 10.1007/978-3-642-29175-3_2,

Table 1 Signal in the second and third phase of Beidou

Phase	Beidou second	Beidou third
Signal	B1(I)	B1x/y
Center frequency (MHz)	1561.098	1575.42
Code rate (MHz)	2.046	1.023
Modulation	BPSK	TMBOC(6,1,4/33) + BOC(1,1)
Services	Open	Open

Table 2 Signal of current GPS and future GPS

Phase	Current GPS	Future GPS
Signal	L1C/A	L1 C
Center frequency (MHz)	1575.42	1575.42
Code rate (MHz)	1.023	1.023
Modulation	BPSK	TMBOC(6,1,4/33) + BOC(1,1)
Services	Open	Open

cooperation are putting forward. Research of interoperability includes both technical factors, such as signal design, satellite payload, user terminal; and non-technical factors, like market and industry. It should consider not only combining with other systems, but also vindicate own benefits and keep some independence. Thus, research of interoperability should act premeditated, arranged, approached with consider of both technical factors and non-technical factors [2–5].

In this article, we first introduced in development of navigation signal in Beidou and GPS; and separated the interoperability process between then into three steps. Then we analyzed the feasibility of their interoperability in each step. At last, suggestions of Beidou construction and signal system design are given.

2 Discussion of Interoperability Phases Between Beidou and GPS

2.1 Open Signals

Civil signal in GPS L1 band is C/A code-BPSK(1); future civil signals in L1 band is MBOC (TMBOC in pilot channel and BOC(1,1) in data channel).

Civil signal in the second phase of Beidou is BPSK(2), and provides the service in the important area (30E–180E, 70S–70N); civil signal in the third phase is MBOC (TMBOC in pilot channel and BOC(1,1) in data channel) (Tables 1, 2).

Now Beidou system has launched eight satellites of phase-2 and primary formed the positioning ability. Domestic industries have designed CMOS chip in L1 band (1561.098 and 1575.42 MHz); and they planed to produce receivers that

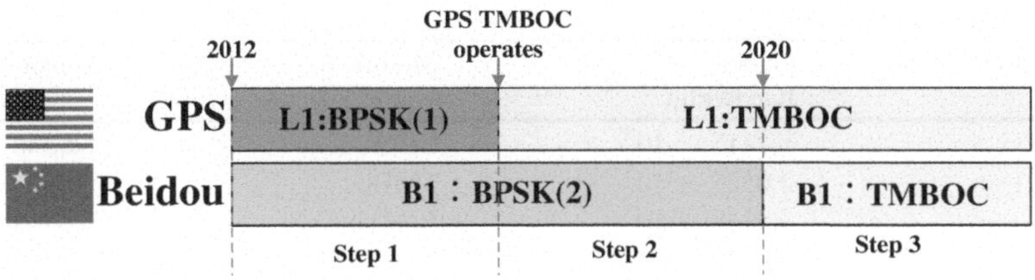

Fig. 1 Phases of interoperability between Beidou and GPS

able to receive GPS L1 C/A and Beidou B1 BPSK(2), which make base of interoperability.

2.2 *Suggestion of Interoperability Between Beidou and GPS*

According to current resources, interoperability between Beidou B1 and GPS L1 should be operated in three phases (Fig. 1).

- Phase 1: interoperability between BPSK(1) in GPS L1 band (1575.42 MHz) and BPSK(2)-I in Beidou regional system B1 band (1561.098 MHz);
- Phase 2: interoperability between TMBOC in GPS L1 band (1575.42 MHz) and BPSK(2)-I in Beidou regional system B1 band (1561.098 MHz);
- Phase 3: interoperability between TMBOC in GPS L1 band (1575.42 MHz) and TMBOC in Beidou global system B1 band (1575.42 MHz).

3 Feasibility Analysis of Interoperability in Phase 1

In phase 1, Beidou uses B1 BPSK(2)-I signal with a regional constellation; GPS uses BPSK(1) signal with the current constellation including 30 satellites.

3.1 *Signal Performance Analysis*

3.1.1 Gabor Bandwidth

Gabor bandwidth is the best index to estimate the infection from receive-bandwidth to tracking accuracy; the more Gabor bandwidth is, the better tracking accuracy a signal has. Under the same code loop bandwidth and same receive carrier to noise ratio, the Root Mean Square (RMS) code tracking accuracy depends on RMS bandwidth:

Table 3 Gabor bandwidth

Modulation	Gabor bandwidth	Modulation	Gabor bandwidth
BPSK(1)	1.4415×10^5	CBOC−	3.5348×10^5
BPSK(2)	2.0525×10^5	TMBoc	3.552×10^5
CBOC+	3.5348×10^5	BOCsin(2,2)	3.5558×10^5

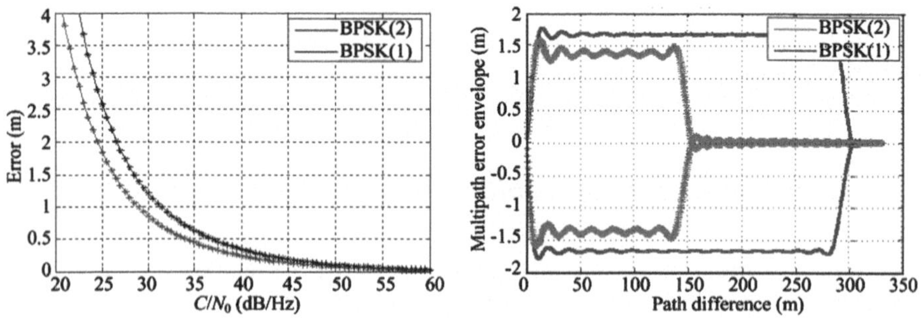

Fig. 2 Tracking error (*left*), multipath error (*right*) of BPSK(1) and BPSK(2)

$$\beta_{rms} = \sqrt{\int_{-\beta_r/2}^{\beta_r/2} f^2 G_S(f) df}. \quad (1)$$

β_{rms} is RMS bandwidth (namely Gabor bandwidth), β_r is bilateral receive bandwidth, $G_S(f)$ is PSD. The RMS bandwidth of different modulations is shown in Table 3.

3.1.2 Tracking Error

$$\sigma^2_{NELP} = \frac{B_L(1 - 0.25 B_L T) \int_{-\beta_r/2}^{\beta_r/2} G_S(f) \sin^2(\pi f \Delta) df}{\frac{C}{N_0}\left(2\pi \int_{-\beta_r/2}^{\beta_r/2} f G_S(f) \sin(\pi f \Delta) df\right)^2} \times \left[1 + \frac{\int_{-\beta_r/2}^{\beta_r/2} G_S(f) \cos^2(\pi f \Delta) df}{T \frac{C}{N_0}\left(\int_{-\beta_r/2}^{\beta_r/2} G_S(f) \cos(\pi f \Delta) df\right)^2}\right], \quad (2)$$

$G_S(f)$ is signal power spectrum, C/N_0 denotes carrier wave-noise ratio, β_r is forward bandwidth, B_L is circle bandwidth, T is integral time, Δs correlator pace (unit: second). From Fig. 2, we know that BPSK(2) has better tracking error than BPSK(1).

3.1.3 Multipath

Receive signal with multipath can be equal as:

$$r(t) = a_0 e^{j\varphi_0} x(t - \tau_0) + \sum_{n=1}^{N} a_n e^{j\varphi_n} x(t - \tau_n). \tag{3}$$

where a_0 is the extent of firsthand signal; φ_0 is the phase of firsthand signal; $x(t)$ is the complex envelope of sending signal; τ_0 is the time delay of firsthand signal; N is the number of path of multipath signal; a_n is the extent of multipath signal; φ_n is the phase of multipath signal; τ_n is the time delay of multipath signal. Multipath error can be depicted as:

$$\varepsilon_\tau \approx \frac{\pm\tilde{a}_1 \int_{-\beta_r/2}^{\beta_r/2} S(f)\sin(2\pi f\tilde{\tau}_1)\sin(\pi f d)df}{2\pi \int_{-\beta_r/2}^{\beta_r/2} fS(f)\sin(\pi f d)[1 \pm \tilde{a}_1 \cos(2\pi f\tilde{\tau}_1)]df}. \tag{4}$$

where, $\tilde{a}_1 = a_1/a_0$ is the extent ratio from multipath signal to firsthand signal, d is correlator space.

The mean multipath error $A(\tau)$ can be calculated by:

$$A(\tau) = \frac{1}{\tau}\int_0^\tau |\max(E(x)) - \min(E(x))|dx, \tag{5}$$

where $E(x)$ is the curve function of multipath error envelope, τ is code time delay. Then the even multipath error is:

$$\varepsilon_a(\tau_1) = \frac{1}{\tau_1}\int_0^{\tau_1}\left[\frac{\left|\varepsilon(\tau)|_{\varphi=0}\right| + \left|\varepsilon(\tau)|_{\varphi=180}\right|}{2}\right]d\tau, \tag{6}$$

where $\varepsilon(\tau)$ is the function of multipath error envelope, $\varepsilon_a(\tau_1)$ is the function of even multipath error, τ_1 and τ are multipath signal time delay. Suppose the receiver front bandwidth is $B = 30$ MHz; correlator space is $d = 1/20$ chip; extent ratio from multipath signal to firsthand signal is $a_1 = -5$ dB [see Fig. 2 (right)].

3.1.4 Code Cross-Correlation

Code cross-correlation can validate the cross-correlation between Beidou ranging code and GPS L1 C/A code. Simulate the maximum value after uniformed correlation between 32 GPS codes to 32 GPS codes, 13 Beidou codes to 13 Beidou codes and 32 GPS codes to13 Beidou codes (Fig. 3).

Beidou B1 BPSK(2) signal has a better tracking performance, better anti-jamming performance and better anti-multipath performance than GPS C/A signal.

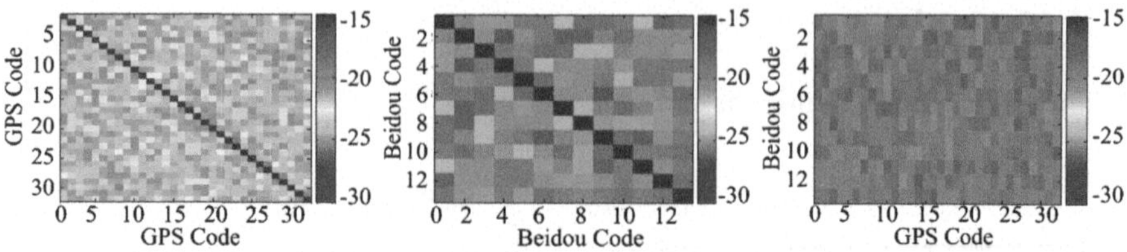

Fig. 3 Cross-correlation of GPS ranging code (*left*), Beidou ranging code (*middle*), GPS ranging code and Beidou ranging code (*right*)

Interoperability of the two signals makes a higher receiver ranging accuracy. Interoperability among the two signals does not rises to much receiver complexity, while enhances the service performances for users.

3.2 Analysis of Service Performance

3.2.1 Accuracy

Service accuracy is employed to depict the difference between the real value and measurement value of positioning, timing and velocity measure [6]. Space signal accuracy includes User Range Error (URE), User Range Ratio Error (URRE), User Range Acceleration Error (URAE) and User Timing Error. It is an important index in assessment the contribution of satellite ephemeris and forecasted clock error to users' positioning error, velocity error and timing error. Assessment of space signal accuracy requires the exactitude orbit and clock error.

Theoretically, the positioning error is depends on pseudo range measurement noise, satellite location error and Positioning Dilution of Precision (PDOP).

$$\sigma_u = \text{PDOP} \times \sigma \tag{7}$$

where PDOP is a non-randomly factor which determines by the location of user and satellites; σ is related to URE and is a randomly factor.

Service accuracy is the difference between users' real position, velocity, timing value and their measurements. Accuracy = UERE × PDOP, where UERE is make up of URE and UEE ($UERE = \sqrt{(URE)^2 + (UEE)^2}$); UERE is related with satellite clock error, ephemeris error, atmosphere model error, multipath error and receiver clock error. In a given system, with the same URE and same UEE, the positioning accuracy is determined by PDOP.

3.2.2 Availability

Service availability is the time ratio of congruous threshold value of positioning, timing, velocity measurement in a prescriptive time period (usually a recursive period) and a given region.

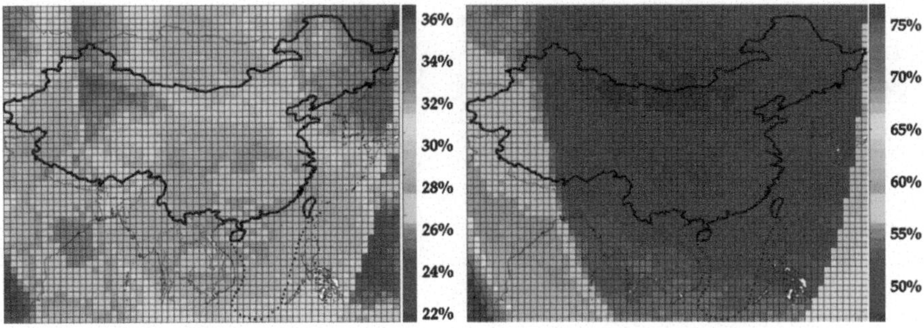

Fig. 4 Average PDOP reducing in China [compare with GPS (*left*), Beidou (*right*)]

Suppose the threshold of positioning accuracy and timing accuracy are σ_p, σ_t. The URE is related to positioning accuracy:

$$\sigma_p = \sqrt{g_{11} + g_{22} + g_{33}} \times \text{URE} = \text{PDOP} \times \text{URE}. \tag{8}$$

Timing accuracy is (GPS SPS 2008):

$$\sigma_t = \sqrt{g_{44}} \times \Delta t = \text{TDOP} \times \Delta t = \frac{\text{TDOP} \times \text{URE}}{c}, \quad (c \text{ is velocity of light}). \tag{9}$$

Thus, the threshold of positioning accuracy and timing accuracy in a given system is:

$$\begin{pmatrix} \sigma_P \\ \sigma_t \end{pmatrix} = \begin{pmatrix} \text{PDOP} \\ \text{TDOP} \div c \end{pmatrix} \times \text{URE}. \tag{10}$$

For the same threshold of positioning accuracy and timing accuracy, debasement of PDOP and TDOP will help to add the threshold. Because the range of URE with randomicity will be enlarged, which leads the enhancement of GNSS availability, this extension is related with the distribution of URE.

3.2.3 Simulation

Calculate average PDOP, TDOP of interoperability constellation, GPS constellation, Beidou constellation in 7 days (a return period) point by point both in the region of China and U.S. Then compare their average to solely system, as follows:

From the simulation results:

- In China, the average PDOP reducing from interoperability constellation to GPS is 29.01%, the average TDOP reducing is 25.77%; so positioning accuracy and availability will add 29.01%, timing accuracy and availability will add 25.77% (Fig. 4).
- In China, the average PDOP reducing from interoperability constellation to Beidou is 53.33%, the average TDOP reducing is 61.23%; so positioning

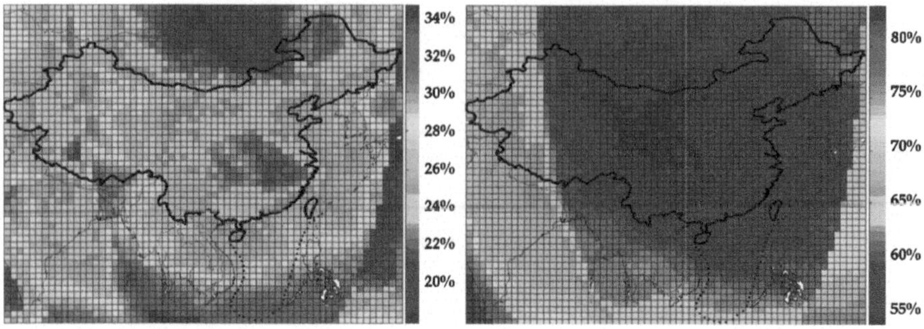

Fig. 5 Average TDOP reducing in China [compare with GPS (*left*), Beidou (*right*)]

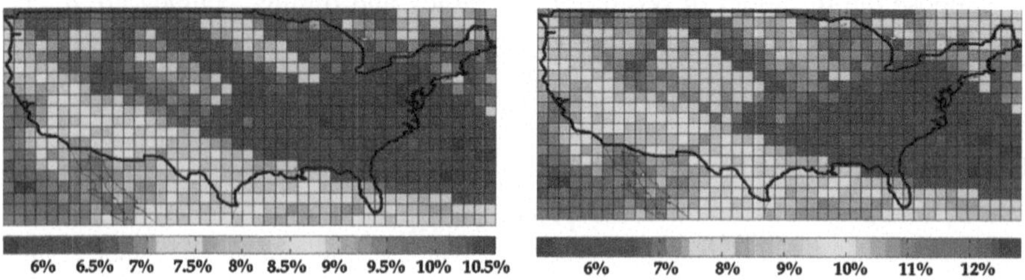

Fig. 6 Average PDOP (*left*), TDOP (*right*) reducing in the U.S. (compare with GPS)

accuracy and availability will add 53.33%, timing accuracy and availability will add 61.23% (Fig. 5).

- In the U.S., the average PDOP reducing from interoperability constellation to GPS is 7.26%, the average TDOP reducing is 8.04%; so positioning accuracy and availability will add 7.26%, timing accuracy and availability will add 8.04% (Fig. 6).

4 Feasibility Analysis of Interoperability in Phase 2

In phase 2, Beidou uses B1 BPSK(2)-I signal with a regional constellation; GPS uses TMBOC signal with the current constellation including 30 satellites.

4.1 Signal Performance Analysis

Figure 7 (left) shows the correlation peak curve of TMBOC and BPSK(2); where TMBOC has the most sharp correlation peak; and from Fig. 7 (right), it shows that TMBOC has a less tracking error than BPSK(2).

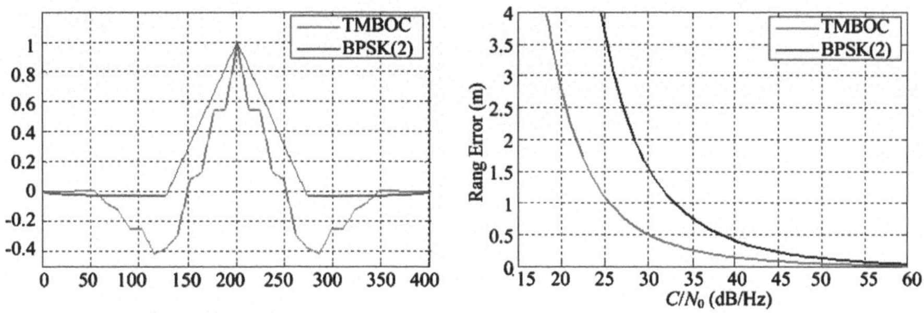

Fig. 7 Correlation peak (*left*), tracking error (*right*) of BPSK(2) and TMBOC

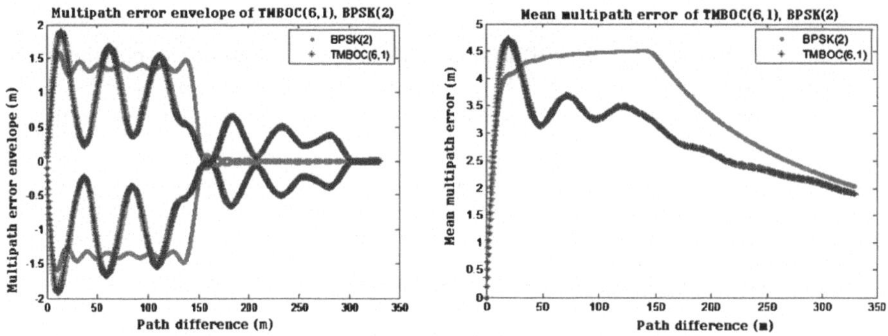

Fig. 8 Multipath error envelope (*left*), multipath error (*right*) of BPSK(2) and TMBOC

GPS TMBOC has better ranging accuracy, anti-jamming and anti-multipath than Beidou B1 BPSK(2), but the difference is not big. Interoperability at different frequency improves the anti-jamming and availability of system (Fig. 8).

4.2 Analysis of Service Performance

Satellite constellation in phase 1 and 2 are the same, so the service performance change will be the same:

- In China, the average PDOP reducing from interoperability constellation to GPS is 29.01%, the average TDOP reducing is 25.77%; so positioning accuracy and availability will add 29.01%, timing accuracy and availability will add 25.77%.
- In China, the average PDOP reducing from interoperability constellation to Beidou is 53.33%, the average TDOP reducing is 61.23%; so positioning accuracy and availability will add 53.33%, timing accuracy and availability will add 61.23%.

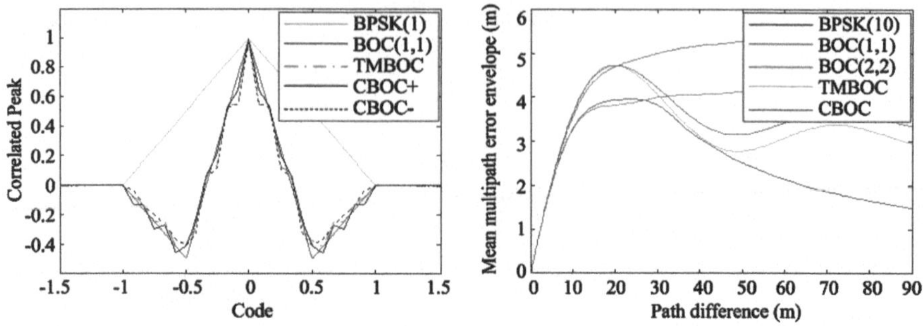

Fig. 9 Correlation peaks (*left*), multipath running average error (*right*) of BPSK, BOC, TMBOC and CBOC

- In the U.S., the average PDOP reducing from interoperability constellation to GPS is 7.26%, the average TDOP reducing is 8.04%; so positioning accuracy and availability will add 7.26%, timing accuracy and availability will add 8.04%.

5 Feasibility Analysis of Interoperability in Phase 3

Navigation signal both of Beidou and GPS in phase 3 is TMBOC; constellation of Beidou is its global constellation while GPS will keep its current constellation.

5.1 Signal Performance Analysis

Correlation peak curve has a direct relation with ranging accuracy, anti-multipath and anti-jamming performance. The sharper the peak, the better performance it has. Figure 9 (left) shows the correlation peak curves of BPSK(1), BOC(1,1), TMBOC and CBOC. Generally, we can get different multipath performance for each kind of navigation signal based on its multipath error and multipath running average error. Assume that receiver front bandwidth is 30 MHz, with correlator space of 1/20 chip and the ratio of multipath to direct path of −6 dB. Figure 9 (right) shows the envelop curves of multipath average errors for BPSK(10), BOC(1,1), BOC(2,2), TMBOC Pilot and CBOC Pilot.

If we put the modulated signal with best performance in the first place, and that with worst performance in the end, then we can easily get the following results: BPSK(10), BOC(2,2), TMBOC(6,1,4/33), CBOC-(6,1,1/11), BOC(1,1), BPSK(1); where the difference between TMBOC(6,1,4/33) and CBOC-(6,1,1/11) is very small.

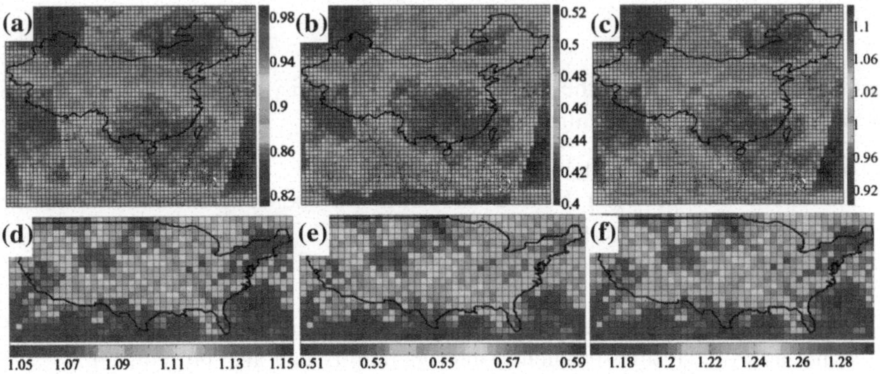

Fig. 10 PDOP in China (**a**), the U.S. (**b**); TDOP in China (**c**), the U.S. (**d**); GDOP in China (**e**), the U.S. (**f**)

Thus, the particular technologies of time division in sub-carrier wave, second coding for ranging codes, channel separation and message coding, make TMBOC signal has a better acquisition performance, better tracking performance, better demodulation performance, better anti-jamming performance and better anti-multipath performance.

5.2 Analysis of Service Performance

In phase 3 of interoperability between Beidou and GPS, not only did the signals of Beidou system change, but also did its satellite constellation.

From the simulation results:

- In China, the average PDOP reducing from interoperability constellation to GPS is 30.14%, the average TDOP reducing is 28.76%; so positioning accuracy and availability will add 30.14%, timing accuracy and availability will add 28.76%.
- In China, the average PDOP reducing from interoperability constellation to Beidou is 17.75%, the average TDOP reducing is 20.14%; so positioning accuracy and availability will add 17.75%, timing accuracy and availability will add 20.14%.
- In the U.S., the average PDOP reducing from interoperability constellation to GPS is 35.10%, the average TDOP reducing is 36.66%; so positioning accuracy and availability will add 35.10%, timing accuracy and availability will add 36.66%.
- In the U.S., the average PDOP reducing from interoperability constellation to Beidou is 35.86%, the average TDOP reducing is 37.97%; so positioning accuracy and availability will add 35.86%, timing accuracy and availability will add 37.97% (Fig. 10).

6 Summary

In this paper, we analyzed the interoperability feasibility among Beidou and GPS in three phases; and we obtain:

1. Beidou and GPS could achieve interoperability;
2. interoperability between Beidou and GPS will be of great benefits to the improvement in service performance;
3. with the development of each system, interoperability between the two systems will deepen;
4. interoperability could be achieved between Beidou B2 signal and GPS L5 signal as well, so dual-frequency interoperability can also be realized.

Acknowledgments This paper is supported by the National Natural Science Foundation of China (Grant No. 11073022), CAS "Western Lights" (29O905YR0001).

References

1. ICG (2007). Report of the First Meeting of the International Committee on Global Navigation Satellite Systems (ICG) Providers Forum. In the first meeting of ICG, Bangalore, September 4, 2007.
2. Grigory, S. (2008). The Russian Federation view on GNSS compatibility and interoperability. In the third meeting of the International Committee on Global Navigation Satellite Systems (ICG), Pasadena, USA, December 8–12.
3. Han, T., Lu, X. C., et al. (2011). Differential equation dynamical system based assessment model in GNSS interoperability. *Science China: Physics, Mechanics & Astronomy, 54*(6), 996–1003.
4. Lu, J., & Yang, Q. W. (2010). Study on GNSS interoperability. *Science China: Physics, Mechanics & Astronomy, 40*(5), 1–8.
5. Li, J. W., Li, Z. H., Hao, J. M., et al. (2009). A preliminary study on compatibility and interoperability of GNSS. *Journal of Geometrics Science and Technology, 26*(3), 177–180.
6. Han, T., Lu, X. C., Rao, Y. N., et al. (2011). Saturated GDOP value analysis in satellite navigation constellation. CSNC 2011, Shanghai.

Cycle Structure Analysis of QC-IRA-B Codes Based on Circulant Permutation Matrices

Jianhui Wang, Kai Zhang, Xiaomei Tang and Gang Ou

Abstract Based on analyzing the cycle structure of QC-IRA-B codes, QC-IRA-*d* code was presented in another paper without analysis process in detail and this kind of code is suitable for channel coding scheme of navigation data of COMPASS. This paper analyzes the cycle structure of QC-IRA-B codes based on circulant permutation matrices. The analysis of this paper shows that some special cycle structures result in lots of low-weight code words which worsen the error-correcting performance of QC-IRA-B codes. This paper also shows why QC-IRA-*d* codes have good error-correcting performance.

1 Introduction

After the discovery of turbo codes [1], LDPC codes were rediscovered in 1996 [2]. Irregular-repeat-accumulate codes (IRA codes) can be seen as the intersection of LDPC codes and turbo codes. IRA codes have linear encoding complexity and good error-correcting performance [3]. Aline et al. studied the optimization of the degree distribution of IRA codes [4], and their results showed that the error-correcting performance of IRA codes with good degree distribution is not worse than good irregular LDPC codes. To improve the encoding speed of IRA codes, a partially parallel encoder was presented [5]. Fossorier presented a kind of quasi-cyclic LDPC codes (QC-LDPC codes) based on circulant permutation matrices [6]. And partion-and-shift LDPC codes [7, 8] can be seen as another kind of

J. Wang (✉) · K. Zhang · X. Tang · G. Ou
Satellite Navigation Research and Development Center,
National University of Defense Technology, Changsha, China
e-mail: wjh0369@163.com

J. Sun et al. (eds.), *China Satellite Navigation Conference (CSNC) 2012 Proceedings*,
Lecture Notes in Electrical Engineering 160, DOI: 10.1007/978-3-642-29175-3_3,

QC-LDPC codes based on circulant permutation matrices. Compared with random LDPC codes, QC-LDPC codes have lower encoding complexity [9]. And what's more, QC-LDPC codes can be decoded in a partially parallel decoder, which is valuable for engineering applications [10–14].

In [15], we propose a new class of codes, quasi-cyclic IRA-*d* codes (QC-IRA-*d* codes), which can combine the advantages of IRA codes and QC-LDPC codes. QC-IRA-*d* codes have lower encoding complexity than IRA codes and can be decoded in a partially parallel way just like QC-LDPC codes [15], so this kind of codes can be used in the ranging and communication system of COMPASS. Reference [15] has compared two kinds of QC-IRA codes, QC-IRA-B codes (base QC-IRA codes) and QC-IRA-*d* codes. The analysis process about cycle structure of QC-IRA-B codes is not described in detail in [15], which is the basis to design QC-IRA-*d* codes. In this paper, the cycle structures resulting in lots of low-weight code words is analyzed in detail.

2 IRA Codes and QC-LDPC Codes Based on Permutation Matrices

2.1 IRA Codes

The parity-check matrix of an IRA code is as follows.

$$H = [H_1 \quad H_2] \tag{2.1}$$

H_1 is a sparse matrix constructed randomly whose rows have the same number of '1's, and H_2 is a structured matrix as follows.

$$H_2 = \begin{bmatrix} 1 & 0 & 0 & \cdots & 0 \\ 1 & 1 & 0 & \cdots & 0 \\ 0 & 1 & 1 & \ddots & \vdots \\ \vdots & \ddots & \ddots & \ddots & 0 \\ 0 & \cdots & 0 & 1 & 1 \end{bmatrix} \tag{2.2}$$

Figure 1 is the Tanner graph of an IRA code. Information nodes correspond to columns of H_1, and parity nodes correspond to columns of H_2. Information nodes and parity nodes all can be called variable nodes (circles) corresponding to all columns of the parity check matrix. Check nodes (squares) correspond to all rows of H.

A code word of an IRA code can be written as

$$w = [s \quad c] \tag{2.3}$$

where s represents information bits, c represents parity bits. And the following equation holds.

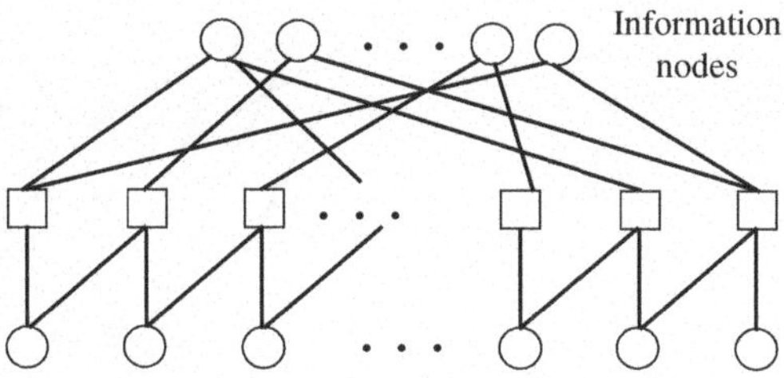

Fig. 1 Tanner graph of an IRA code

$$Hw^T = 0 \tag{2.4}$$

Substituting (2.3) for w, the encoding equation is

$$H_1 s^T + H_2 c^T = 0 \tag{2.5}$$

From (2.5), the parity bits can be written as

$$c^T = H_2^{-1} H_1 s^T \tag{2.6}$$

2.2 QC-LDPC Codes Based on Permutation Matrices

The parity-check matrix of a QC-LDPC code based on circulant permutation matrices can be written as follows.

$$H_{QC} = \begin{bmatrix} I(s_{1,1}) & I(s_{1,2}) & \cdots & I(s_{1,k}) \\ I(s_{2,1}) & I(s_{2,2}) & \cdots & I(s_{2,k}) \\ \vdots & \vdots & \ddots & \vdots \\ I(s_{c,1}) & I(s_{c,2}) & \cdots & I(s_{c,k}) \end{bmatrix} \tag{2.7}$$

$I(s_{i,j})$ is either a circulant permutation matrix or a zero matrix. $s_{i,j}$ is the right shift number of $I(s_{i,j})$ over a identity matrix with the same dimension. If $s_{i,j}$ is negative, $I(s_{i,j})$ represents a zero matrix. Reference [16] analyzed the cycle structure characteristics of the Tanner graph corresponding to H_{QC}. QC-LDPC codes have low encoding complexity [9]. And what's more, they can be decoded in a partially parallel way easily, which is valuable for engineering [10–14].

The Tanner graph corresponding to H_{QC} is illustrated in Fig. 2. For simplicity, the dimension of $I(s_{i,j})$ is 3×3. The variable nodes (circles) in the same dashed circle correspond to the columns of some $I(s_{i,j})$. And the check nodes (squares) in the same dashed circle correspond to the rows of some $I(s_{i,j})$. We call the nodes in the same dashed circle a group. From Fig. 2, we can learn that variable nodes in the same group will not connect with the same check node and check nodes in the same group will not connect with the same variable node. This kind of codes can be decoded in a partially parallel decoder [10].

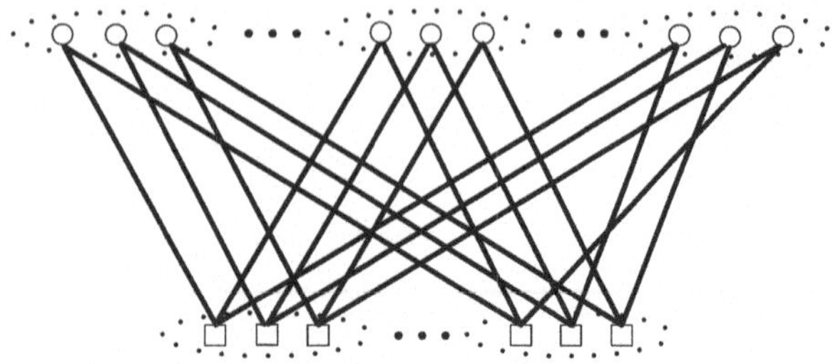

Fig. 2 Tanner graph of a QC-LDPC code based on circulant permutation matrices

3 QC-IRA-B Codes

If the submatrix H_1 of a IRA code has quasi-cyclic structure, we call this code quasi-cyclic IRA code (QC-IRA code). To combine the characteristics of IRA codes and QC-LDPC codes, a simple method is that let H_1 have similar form to H_{QC}. This kind of IRA codes are called QC-IRA-B codes (base QC-IRA codes) [15]. We will show that QC-IRA-B codes have a lot of code words with low weight (number of '1's in a code word). And the exist of many low weight code words results in the poor performance of QC-IRA-B codes. Based on the analysis of QC-IRA-B codes, we present another kind of codes, called QC-IRA-*d* codes, which have good performance.

QC-IRA-B codes have similar parity-check matrix with IRA codes. And the H_1 matrix of a QC-IRA-B code is as follows

$$H_1 = \begin{bmatrix} I_{1,1} & \cdots & I_{1,N} \\ \vdots & \ddots & \vdots \\ I_{M,1} & \cdots & I_{M,N} \end{bmatrix} \tag{3.1}$$

$I_{i,j}$ is either a circulant permutation matrix or a zero matrix with dimension $p \times p$ (p is a prime number generally). The Tanner graph of a QC-IRA-B code is showed in Fig. 3. The information nodes in the same dashed circle correspond to the columns of some $I_{i,j}$, and the check nodes in the same dashed circle correspond to the rows of some $I_{i,j}$.

The thick edges contain two kinds of cycle structures corresponding to degree-2 and degree-3 information nodes, which result in low weight code words. The two kinds of cycle structures are showed in Fig. 4a, b. Similar cycle structures corresponding to degree-4 and degree-5 information nodes are illustrated in Fig. 4c and d. The short cycles inside the Tanner graph corresponding to H_1 can be prolonged by setting proper right shift number to each $I_{i,j}$ [6–8, 16].

In Fig. 4a–d, every check node has connections only with two variable nodes inside the cycle structure (not necessary to have no connection with variable nodes outside the cycle structure), and every variable node has no connection with check nodes outside the cycle structure. That is to say, if the bits of the zero code word corresponding to the variable nodes in Fig. 4a are flipped from '0' to '1', the new vector is also a code word whose weight is 4 (because every parity equation

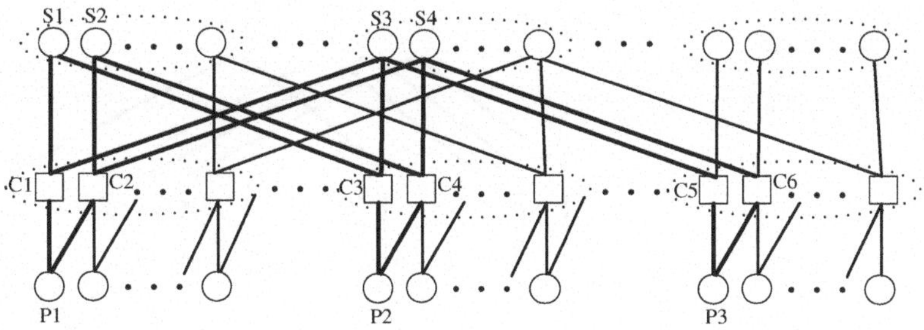

Fig. 3 Tanner graph of a QC-IRA-B code

Fig. 4 Cycle structures corresponding to low weight code words of QC-IRA-B codes

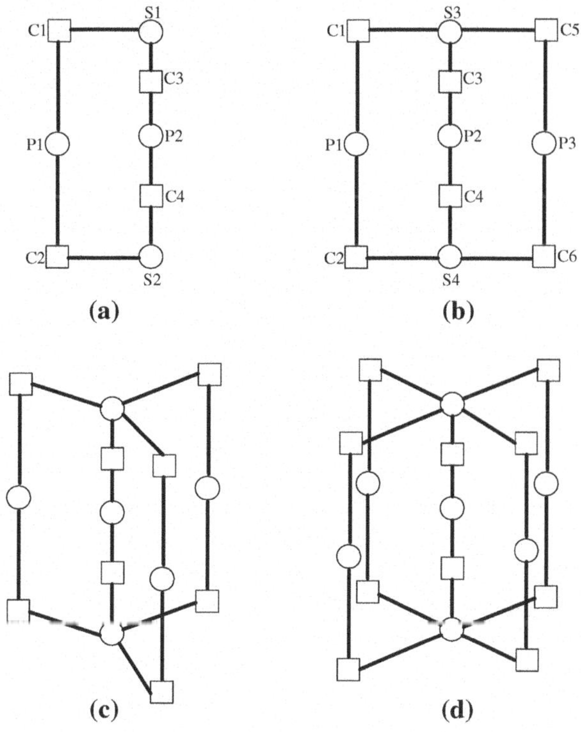

corresponding to the check nodes in Fig. 4a is equal to the sum of two '1's, all parity check equations corresponding to the check nodes in Fig. 4a hold). So the cycle structure of Fig. 4a corresponds to a weight-4 code word. Similarly, the cycle structures of Fig. 4b–d correspond to a weight-5, weight-6 and weight-7 code word respectively. From Fig. 4, we can conclude that degree-x information nodes of QC-IRA-B codes result in cycle structure corresponding to a weight-$x + 2$ code word.

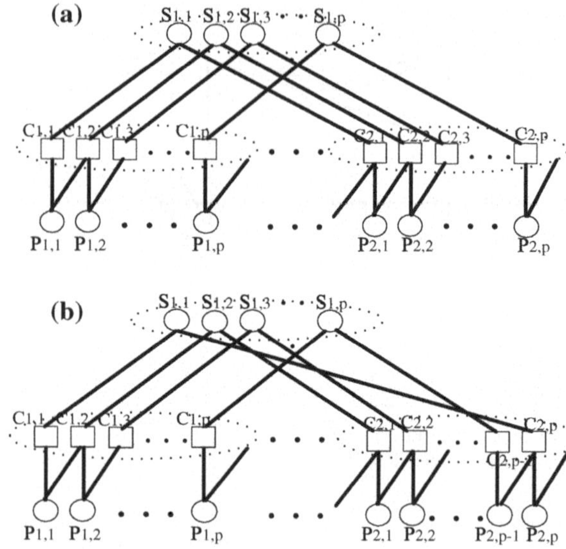

Fig. 5 Tanner graph of degree-2 information nodes group

Now we will analyze the weight distribution of code words corresponding to degree-2 information nodes. In Fig. 5a, $S_{1,1}$, $S_{1,2}$,..., $S_{1,\mathrm{p}}$ denote the p information nodes in the same information nodes group (information nodes in the same dashed circle). $C_{1,1}$, $C_{1,2}$,..., $C_{1,\mathrm{p}}$ and $C_{2,1}$, $C_{2,2}$,..., $C_{2,\mathrm{p}}$ denote the two check nodes group (check nodes in the same dashed circle) in connection with the degree-2 information nodes group. And $P_{1,1}$, $P_{1,2}$,..., $P_{1,\mathrm{p}}$ and $P_{2,1}$, $P_{2,2}$,..., $P_{2,\mathrm{p}}$ denote the parity nodes in connection with the two check nodes groups.

When the information nodes group connects with the two check nodes groups with the same right shift number (not necessary to be zero), as illustrated in Fig. 5a, the weight distribution of code words corresponding to the degree-2 information nodes group will be analyzed firstly. For simplicity, the right shift number in Fig. 5a is zero. The closed cycle ($S_{1,1}$, $C_{1,1}$, $P_{1,1}$, $C_{1,2}$, $S_{1,2}$, $C_{2,2}$, $P_{2,1}$, $C_{2,1}$, $S_{1,1}$) corresponds to a weight-4 code word. The set of all closed cycles corresponding to weight-4 code words in Fig. 5a is

$$S_{cycle,4} = \left\{ \begin{pmatrix} S_{1,k}, C_{1,k}, P_{1,k}, C_{1,k+1}, \\ S_{1,k+1}, C_{2,k+1}, P_{2,k}, \\ C_{2,k}, S_{1,k} \end{pmatrix} \middle| 1 \le k \le p-1 \right\} \tag{3.2}$$

This set contains $p-1$ elements, corresponding to $p-1$ weight-4 code words. The set of all closed cycles corresponding to weight-6 code words in Fig. 5a is

$$S_{cycle,6} = \left\{ \begin{pmatrix} S_{1,k}, C_{1,k}, P_{1,k}, C_{1,k+1}, \\ P_{1,k+1}, C_{1,k+2}, S_{1,k+2}, \\ C_{2,k+2}, P_{2,k+1}, C_{2,k+1}, \\ P_{2,k}, C_{2,k}, S_{1,k} \end{pmatrix} \middle| 1 \le k \le p-2 \right\} \tag{3.3}$$

Table 1 Weight distribution of code words corresponding to the information nodes group in connection with two check nodes groups with different right shift numbers

Weight	$p+2$	4	6	...	$2k_1$
Number of code words	k_1k_2	$p-2$	$p-4$	...	$p-2(k_1-1)$
Weight	$2(k_1+1)$	$2(k_1+2)$	...	$2(k_2-1)$	$2k_2$
Number of code words	k_2-k_1	k_2-k_1-1	...	2	1

This set contains $p-2$ elements, corresponding to $p-2$ weight-6 code words.

Similarly, the set of all closed cycles corresponding to weight-$2p$ code words in Fig. 5a is

$$S_{cycle,2p} = \left\{ \left(\begin{array}{l} S_{1,k}, C_{1,k}, P_{1,k}, C_{1,k+1}, \\ P_{1,k+1}, C_{1,k+2}, P_{1,k+2}, \ldots, \\ P_{1,p-1}, C_{1,p}, S_{1,p}, C_{2,p}, \\ P_{2,p-1}, C_{2,p-1}, P_{2,p-2}, \ldots, \\ C_{2,k+2}, P_{2,k+1}, C_{2,k+1}, \\ P_{2,k}, C_{2,k}, S_{1,k} \end{array} \right) \middle| k = 1 \right\} \tag{3.4}$$

This set contains one element, corresponding to one weight-$2p$ code word.

When the information nodes group in connection with the two check nodes groups with the same right shift number, from the analysis above, we can infer that the number of weight-$2a$ code words is $p+1-a$.

When the information nodes group in connection with the two check nodes groups with different right shift numbers, we can adjust the locations of information nodes to ensure that one of the right shift numbers is zero, as illustrated in Fig. 5b. For simplicity, the two right shift numbers are zero and $p-1$ in Fig. 5b. The analysis of weight distribution corresponding to $S_{1,2}$, $S_{1,3}$,…, $S_{1,p}$ is similar to the situation illustrated in Fig. 5a. We just need to analyze the weight distribution of code words corresponding to $S_{1,1}$. The closed cycle passing through $S_{1,1}$ and $S_{1,2}$ is ($S_{1,1}$, $C_{1,1}$, $P_{1,1}$, $C_{1,2}$, $S_{1,2}$, $C_{2,1}$, $P_{2,1}$, $C_{2,2}$, $P_{2,2}$,…, $C_{2,p-1}$, $P_{2,p-1}$, $C_{2,p}$, $S_{1,1}$). This cycle contains $p+2$ variable nodes and corresponds to a weight-$p+2$ code word. Similarly, the closed cycle passing through $S_{1,1}$ and $S_{1,k}$ ($3 \le k \le p$) corresponds to a weight-$p+2$ code word. So, passing through $S_{1,1}$, there are $p-1$ weight-$p+2$ code words.

When the right shift numbers are 0 and k_1 ($2 \le k_1 \le p-2$), the weight distribution is different. Let k_2 be equal to $p-k_1$. When $k_1 \le k_2$, the weight distribution is listed in Table 1. When $k_2 < k_1$, we can get the weight distribution through interchanging k_1 and k_2.

The analysis above indicates that low weight code words have a large percent in the weight distribution of degree-2 information nodes group. And changing the right shift numbers can not improve the weight distribution.

The weight distribution of information nodes group with higher degree than 2 can be analyzed similarly, and we can get similar conclusions. For simplicity, we just give the weight distribution of a degree-x ($x \ge 2$) information nodes group

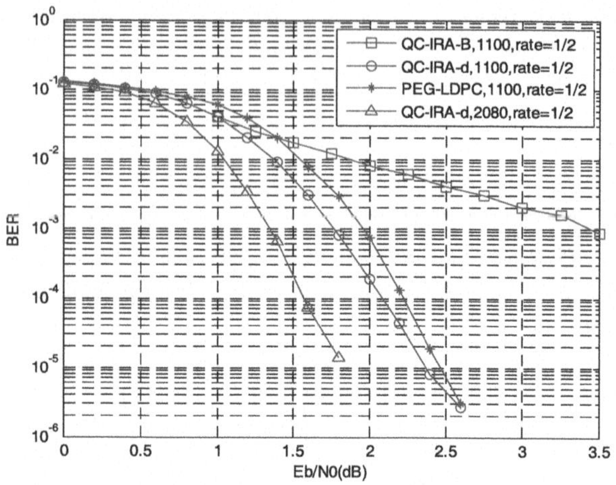

Fig. 6 Performance of a QC-IRA-B code, two QC-IRA-*d* codes and a PEG-LDPC code

which connects with x check nodes groups with the same right shift number. The number of weight-$(ax + 2)$ code words is $p - a$ $(1 \le a \le p - 1)$. When the information nodes group in connection with x check nodes groups with different right shift numbers, the weight distribution analysis is more complicated, but the low-weight code words still have the largest percent.

4 Simulations

To verify the performance of QC-IRA-*d* codes, we construct a QC-IRA-B code, a QC-IRA-*d* code and a PEG-LDPC code (constructed by PEG algorithm) [17] with the same code length (1100), code rate (1/2) and degree distribution. Besides we construct a QC-IRA-*d* code with code length 2080 and code rate 1/2. As illustrated in Fig. 6, QC-IRA-*d* code has better performance than QC-IRA-B code and is as good as the PEG-LDPC code. With the same code rate, the QC-IRA-*d* code with longer code length has better performance. All the codes are decoded with BP algorithm in AWGN channel, and the modulation is BPSK.

5 Conclusion

In this paper, the cycle structures resulting in lots of low-weight code words of QC-IRA-B codes are analyzed through Tanner graph. The analysis results are valuable for designing QC-IRA-*d* codes [15]. QC-IRA-*d* codes have lower encoding complexity than IRA codes and can be decoded in a partially parallel decoder like QC-LDPC codes. Simulation results show that QC-IRA-*d* codes have better performance than QC-IRA-B codes and the performance of QC-IRA-*d* codes are as good as PEG-LDPC codes. Because of low encoding complexity and partially parallel decoder, QC-IRA-*d* codes are valuable for engineering applications, and it is suitable for the channel coding scheme of navigation data of COMPASS.

References

1. Berrou, C., Glavieux, A., & Thitimajshima, P. (1993). Near Shannon limit error-correcting and decoding: Turbo codes. *IEEE International Conference on Communications*, Geneva, Switzerland, May 1993 (pp. 1064–1070).
2. Mackay, D. J. C., & Neal, R. M. (1996). Near Shannon-limit performance of low-density parity-check codes. *Electronics Letters*, *32*, 1645–1646.
3. Jin, H., Khandekar, A., & McEliece, R. (2000). Irregular repeat-accumulate codes. *The Second International Conference on Turbo Codes*, Brest, France, September 2000.
4. Roumy, A., & Guemghar, S. (2004). Design methods for irregular repeat–accumulate codes. *IEEE Transactions on Information Theory, 50*(8), 1711–1727.
5. Hwang, S. O., Myung, S., Lee, H., Park, S.-I., & Lee, J. Y. (2010). Partial parallel encoder for IRA codes. *IEEE Electronics Letters, 46*(2), 135–137.
6. Fossorier, M. (2004). Quasi-cyclic low-density parity-check codes from circulant permutation matrices. *IEEE Transactions on Information Theory, 50*(8), 1788–1793.
7. Lu, J., & Moura, M. F. (2005). Partition-and-shift LDPC codes. *IEEE Transactions on Magnetics, 41*(10), 2977–2979.
8. Lu, J., & Moura, J. M. F. (2006). Structured LDPC codes for high-density recording: large girth and low error floor. *IEEE Transactions on Magnetics, 42*(2), 208–213.
9. Li, Z., Chen, L., & Zeng, L. (2006). Efficient encoding of quasi-cyclic LDPC codes. *IEEE Transactions on Communications, 54*(1), 71–81.
10. Wang, Z., & Cui, Z. (2007). A memory efficient partially parallel decoder architecture for quasi-cyclic LDPC codes. *IEEE Transactions on VLSI Systems, 15*(4), 483–488.
11. Dai, Y., Chen, N., & Yan, Z. (2008). Memory efficient decoder architectures for quasi-cyclic LDPC codes. *IEEE Transactions on Circuits and Systems, 55*(9), 3697–3710.
12. Jiang, N., Peng, K., & Song, J. (2009). High-throughput QC-LDPC decoders. IEEE Transactions on Broadcasting, 55(2), 251–259.
13. Wu, T.-C., Hu, Y.-W., & Lee, C.-M. (2010). Optimization of memory utilization for partially parallel QC-LDPC decoder. *ISITA2010*, Taichung, Taiwan October 2010.
14. Dai, Y., & Yan, Z. (2008). Optimal overlapped message passing decoding of quasi-cyclic LDPC codes. *IEEE Transactions on VLSI Systems I: Regular Papers, 16*(5), 565–578.
15. Wang, J., Lei, J., Ni, S., Tang, X., & Ou, G. (2011). QC-IRA-d codes based on circulant permutation matrices. *IEEE Communications Letters, 15*(11), 1224–1227.
16. Lei, J., Wang, J., & Gao, Y. (2008). The analysis of cycle structure for a class of quasi-cyclic LDPC codes. *IEEE International Conference on Communication Systems*, November 2008.
17. Hu, X. Y., Eleftheriou, E., & Amold D. M. (2001). Progressive edge-growth tanner graphs. *IEEE Global Telecommunication Conference*, February 2001 (pp. 995–1001).

Interference Analysis and Simulation Between GPS and Galileo in China

Yao Wang, Bo Zhang, Xianzhi Luo and Jian Xie

Abstract With update of GPS modern and building of Galileo, the interference between the two systems was avoidable. The international traditional interference methodology was given. The more accurate signal model was derived, considering true pseudo random code and data rate. The effects of different signal parameters onto Spectral Separation Coefficient (SSC) were not the same. By the way of simulation, it was concluded that the effect of signal with short code and low data rate onto SSC was obvious, and the effects of signals from different satellites with same modulation onto SSC were almost identical. Finally, the interference in L1 between GPS and Galileo in china region was simulated. And the results showed that the major interference was GPS intrasystem interference in the case of receiving GPS C/A signals, meanwhile maximum quantity was 2.637 dB; the major interference was the one from GPS C/A signals in the case of receiving Galileo E1 OS signals, meanwhile maximum quantity was 0.348 dB.

1 Introduction

On October 21, 2011 the two European Galileo formally satellite were launched by the Russian launch vehicle of the "Union", marking the long-stalled Galileo plans to start again. And in 2012, two Galileo satellites will be launched to constitute the

Y. Wang (✉) · B. Zhang · X. Luo · J. Xie
The 54th Research Institute of CECT, Shijiazhuang, China
e-mail: wy_hrbeu@163.com

Y. Wang · B. Zhang · X. Luo · J. Xie
Satellite Navigation Technology and Equipment Engineering Technology Research Center of Hebei Province, Shijiazhuang, Hebei, China

J. Sun et al. (eds.), *China Satellite Navigation Conference (CSNC) 2012 Proceedings*, Lecture Notes in Electrical Engineering 160, DOI: 10.1007/978-3-642-29175-3_4,

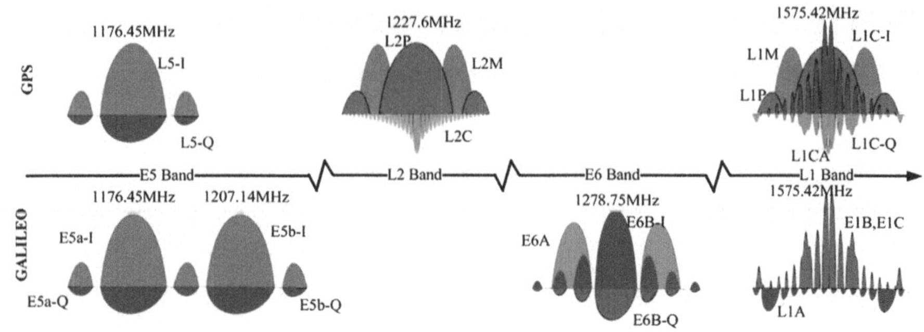

Fig. 1 The final GPS/Galileo signal plan

Galileo IOV simplest system. The Beidou system was announced to run initially on December 27, 2011. Within the next few years, satellite navigation will break the monopoly of the current GPS. The global navigation satellite system (GNSS) will be constituted by GPS, GLONASS, Galileo, Compass and Augmentation system. Various navigation systems interfere with each other is inevitable with the limited radio spectrum of satellite navigation. In order to achieve compatibility with the GPS signal, the signal of Galileo's development has experienced a tortuous process. The Working Group of the European Union and the United States finally reached agreement on the Galileo signal design in 2007 [1] as shown in Fig. 1.

Abroad, some literatures have analyzed the interference of the GPS and Galileo [2–4], However, most of them focused on old signal baseline before final one, not considering new modulation. On the basis of interference analysis method proposed by the United States and Europe, this paper detailed the signal model, analyzed the influence of signal parameters on the spectral separation factor. In the L1-band, the simulation analysis of the compatibility of GPS/Galileo in China was done according to Galileo final signal baseline.

2 Interference Analysis Methodology

2.1 Methodology Description

The degradation of effective carrier-to-noise-power relation-ship was as evaluation criterion of interference between systems [5]. It was because that acquisition, carrier tracking and data demodulation were related to output signal-to-noise-plus-interference ratio (SNIR) of correlator. SNIR was defined as the squared mean of correlator output divided by its variance, for more details in [6]. In addition, the effect of Doppler frequency offset should be taken into account. With the assumption that the band-pass filter was ideal, the effective C/N_0 was expressed by

$$\left(\frac{C}{N_0}\right)_{eff} = \frac{C_s}{N_0 + I_{\text{intra}} + I_{\text{inter}} + I_{ext}} = \frac{C_s}{N_0 + \sum_{j=1}^{N_{\text{intra}}} \sum_{i=1}^{M_{\text{intra}}} C_{ij}\kappa_{ijs} + \sum_{j=1}^{N_{\text{inter}}} \sum_{i=1}^{M_{\text{inter}}} C_{ij}\kappa_{ijs} + I_{ext}} \tag{2.1}$$

where

$$\kappa_{ijs} = \int_{-\beta_R/2}^{\beta_R/2} G_{ij}(f + f_{dopij}) G_s(f + f_{dops}) df \tag{2.2}$$

C_s was received desired signal power. C_{ij} was received interfering power of the i-th interfering signal on the j-th satellite. N_0 was thermal noise PSD. I_{intra} was equivalent white-noise PSD due to aggregate interference from all signals in reference constellation. I_{inter} was equivalent white-noise PSD due to aggregate interference from all signals in other constellation. I_{ext} was equivalent white-noise PSD due to aggregate interference from other systems. M_{intra} was the number of interfering signals by a satellite in reference constellation. M_{inter} was the number of interfering signals by a satellite in other constellation. N_{intra} was the number of visible satellite in reference constellation. N_{inter} was the number of visible satellite in other constellation. $G_s(f)$ was normalized PSD of desired signal. $G_j(f)$ was normalized PSD of interfering signal. β_R was receiver front-end two-sided-width. f_{dop} was signal Doppler frequency shift. κ_{ijs} was Spectral Separation Coefficient (SSC) between interfering signal ij and desired signal s. The interference of signals was evaluated by degradation of $(C/N_0)_{eff}$, and it was as following equations, causing by intrasystem and intersystem.

$$\Delta\left(\frac{C}{N_0}\right)_{\text{intra}} = \frac{\dfrac{C}{N_0 + I_{ext}}}{\dfrac{C}{N_0 + I_{\text{intra}} + I_{ext}}} = 1 + \frac{I_{\text{intra}}}{N_0 + I_{ext}} \tag{2.3}$$

$$\Delta\left(\frac{C}{N_0}\right)_{\text{inter}} = \frac{\dfrac{C}{N_0 + I_{ext} + I_{\text{intra}}}}{\dfrac{C}{N_0 + I_{\text{intra}} + I_{ext} + I_{\text{inter}}}} = 1 + \frac{I_{\text{inter}}}{N_0 + I_{\text{intra}} + I_{ext}} \tag{2.4}$$

In dBs as follow

$$\Theta_{\text{intra}} = 10\log\left(1 + \frac{I_{\text{intra}}}{N_0}\right) \tag{2.5}$$

$$\Theta_{\text{inter}} = 10\log\left(1 + \frac{I_{\text{inter}}}{N_0 + I_{\text{intra}}}\right) \tag{2.6}$$

2.2 Signal Model

The more accurate signal model was established, which considered the effect of true pseudo-random noise code and data. It made the compatibility methodology more accurate. The receiver baseband signal was composed of a desired signal and a finite number of disturbing navigation signals. The signal model can be expressed by

$$\begin{aligned} s(t) &= \sqrt{P_0} d_0(t) g_0(t) \\ &+ \sum_{i=1}^{M} \sqrt{2P_i} d_i(t-\tau_{i,0}) g_i(t-\tau_{i,0}) \cos(\Delta\omega_{i,0} t + \varphi_{i,0}) + n(t) \end{aligned} \tag{2.7}$$

where, P_i represented received power. $d_i(t)$ was the data stream with data duration T_d and data rate $f_d = 1/T_d$. M was number of interference signals. $\Delta\omega_{i,0}$ and $\varphi_{i,0}$ are residual phase and Doppler after down-conversion. $n(t)$ was thermal noise. $\tau_{i,0}$ was the code phase difference the ith signal and the desired signal with index 0. $g_i(t)$ was the pulse-modulation code whose expression was

$$g(t) = w(t) \sum_{l=-\infty}^{\infty} c_l \phi(t - lT_c) \tag{2.8}$$

where, $w(t)$ was time-window of length T_d. c_l was the pseudo-random code sequence, with period N, code length T_c and pulse shape $\phi(t)$. The PSD of desired signal can be expressed by the Ref. [7]

$$G_s(f) = \frac{1}{T_d} G_d(f) |G(f)|^2 \tag{2.9}$$

In this expression, $G_d(f)$ was the PSD of data. Since the data were independent, then $G_d(f) = 1$. $G(f)$ was the Fourier transform of the pulse $g(t)$

$$G(f) = FT(w(t)) * FT\left(\sum_{l=-\infty}^{\infty} c_l \phi(t - lT_c)\right) \tag{2.10}$$

Though the previous calculations, it could be got

$$G_s(f) = T_d \left(\frac{1}{NT_c}\right)^2 \Phi^2(f) |X_{code}(f)|^2 \sum_{k=-\infty}^{\infty} \sin c\left(\pi T_d \left(f - \frac{k}{NT_c}\right)\right)^2 \tag{2.11}$$

Where, $\Phi(f)$ was the Fourier transform of code pulse shape. $X_{code}(f)$ was the code transform, which can be got by FFT. According to different signal structures, two forms can be derivated

$$G_s(f) = \frac{1}{NT_c} \Phi^2(f) |X_{code}(f)|^2 \sum_{k=-\infty}^{\infty} \delta\left(\pi NT_c \left(f - \frac{k}{NT_c}\right)\right)^2 \tag{2.12}$$

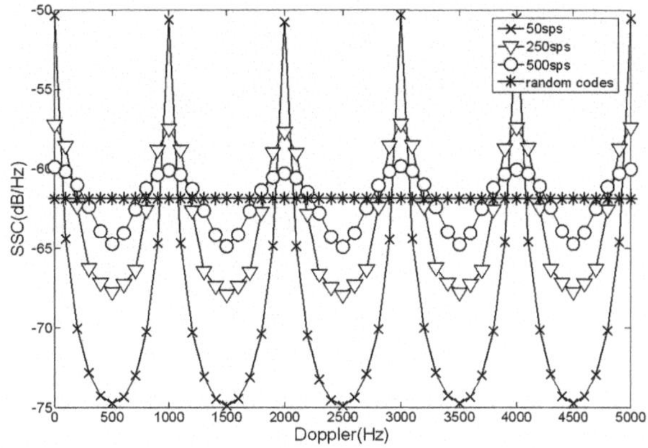

Fig. 2 SSC variations for different data rates (GPS C/A)

$$G_s(f) = T_d\left(\frac{1}{NT_c}\right)^2 \Phi^2(f) \quad (2.13)$$

The signal shown in Fig. 1 could be classed into three types: one type was civil data channel signal, with low code rate and short code period, which can be expressed by Eq. (2.11). The other type civil pilot channel signal, without navigation modulation, could be expressed by Eq. (2.12). such as L1C-I, E5a-Q, E5b-Q etc. The third type was military or public regulated service signal, with high code rate and complex code structure, could be expressed by Eq. (2.13).

3 Characteristic Analysis of the Spectral Separation Coefficient

3.1 Effect of Data Rate onto SSC

From Eq. (2.1), SSC was defined by sum of dot product between desired signal and interfering signal. And it reflected the level of two signals' spectrum overlap. So the SSC was related to signal PSD mostly. See Eq. (2.11), code and data worked together to create line-spectrum of PSD. The data rate can smooth signal PSD, so it also can influence the SSC. The true GPS C/A signal was considered, with the data rates of 50, 250 and 500sps. Figure 2 showed that the SSC was changed with the different Doppler frequency offset. It was also got that the oscillation of SSC was reduced with the increasing of data rate and finally approached to the case of random code (SSC = −61.9 dB/Hz). Oscillation period was 1000 Hz. It can be concluded that data rate was an indispensable factor for analyzing interference between signals with short code.

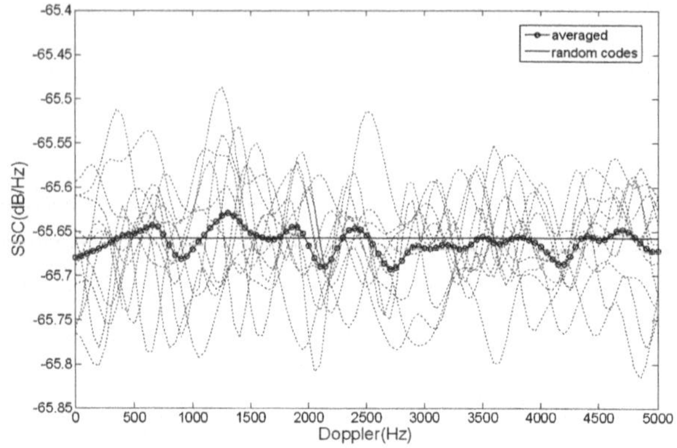

Fig. 3 SSC variations for 12 date channels of E1 OS signal

3.2 Effect of Code onto SSC

The period of code can lead to line-spectrum in signal PSD that can make interference coefficient was changed with the different Doppler frequency offset. When period of code was determinate, different code sequences can affect the interference coefficient. The SSC variations for 12 Galileo E1 OS data channels with the different Doppler frequencies were showed in Fig. 3. For each Doppler, the standard deviation for the variations of the SSC was 0.0524 dB/Hz, which can be considered as negligible when considering magnitude for the SSC itself (about −65.6577 dB/Hz). It can be concluded that the effect to SSC of different code sequences from same group was small when the period of code was determinate.

3.3 The Calculation of SSC

Through analysis of previous sections, the factors, including Doppler frequency offset, data rate and code rate etc., can affect the SSC, especially for signals with short code. Take GPS C/A signals as desired signal, with 24 MHz receiver front-end bandwidth, the SSCs between different signals and GPS C/A were showed in Fig. 4. It can be seen when C/A signal self-interference was obvious. This was the result that the two GPS C/A code had obvious line-spectrum. Figure 5 showed the situation of E1B as the desired signal. Finally, the SSCs between different signals were given in Table 1.

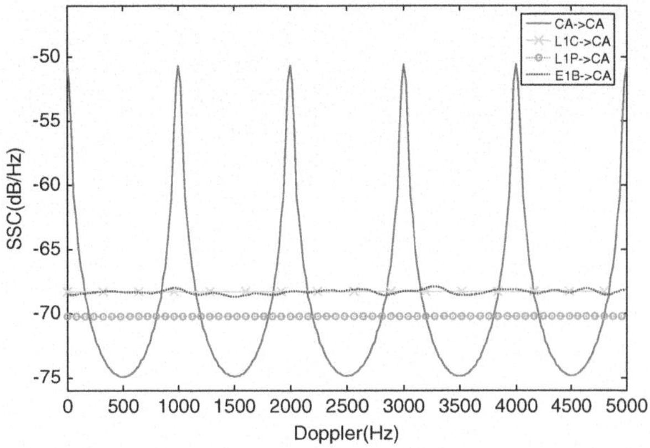

Fig. 4 SSC for different signals and C/A

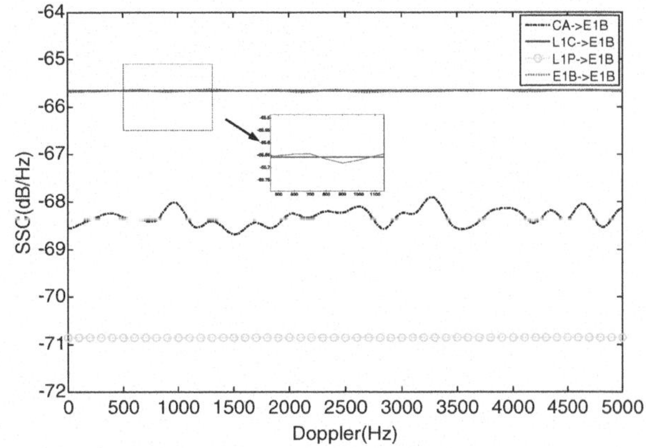

Fig. 5 SSC for different signals and E1B

Table 1 The SSC for different signals(dB/Hz)

	L1 CA	L1C	L1P	L1 M	E1B
L1 CA	See Fig. 4	−68.28	−70.25	−88.02	See Fig. 4
L1C	−68.28	−65.65	−70.86	−82.90	−65.66
L1P	−70.25	−70.86	−71.87	−81.00	−80.29
L1M	−88.02	−82.90	−81.00	−73.24	−83.00
E1B	See Fig. 5	−65.66	−80.29	−83.00	See Fig. 5

4 Interference Simulation of GPS/Galileo

4.1 Simulation Model

In order to simulate the interference between GPS and Galileo in China, the space constellation parameters were shown in Table 2.

Table 2 Space constellation parameters

Parameter	GPS	Galileo
Constellation	24/6	Walker (27/3/1)
Eccentricity	0	0
Inclination	55°	56°
Semi-major axis	26562 km	29600.3 km

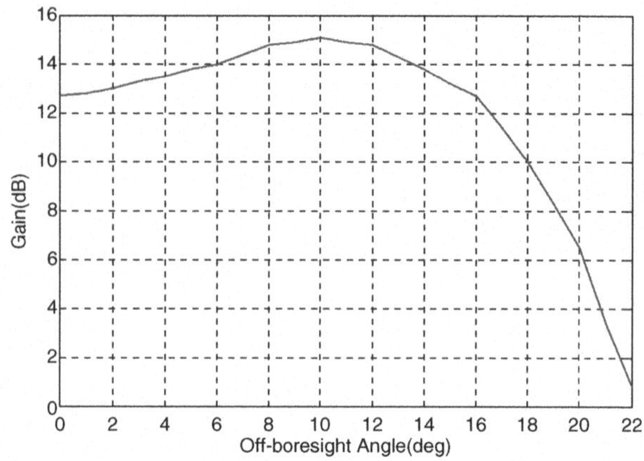

Fig. 6 Typical satellite antenna gain

There were lots of influencing factors when signal transmitting from satellite to receiver, such as satellite antenna gain, free space loss, atmospheric loss, polarization mismatch loss and receiver antenna gain. Where the transmit antenna gain was given in Fig. 6.

The signal Doppler frequency offset can be expressed by

$$\Delta f = \frac{f_c}{c}\left[\left(\vec{v} - \vec{u}\right) \cdot \vec{a}\right] \tag{4.1}$$

where, f_c was emission frequency. $\vec{v}$ was satellite speed vector. $\vec{u}$ was receiver speed vector. $\vec{a}$ was line of sight unit spatial vector from receiver to transmitter. c was speed of light.

As shown in Fig. 1, signals in L1 were the most complex. So two different cases were simulated as follow.

1. GPS C/A ← Galileo E1B
2. Galileo E1B ← GPS C/A

Signal parameter referred to Table 3. Simulation configuration referred to Table 4. We should know that Galileo E1 OS signal includes two parts Galileo E1B (date channel) and E1C (pilot frequency channel). Here we ignored the influence from E1C.

Table 3 GPS/Galileo signal parameters in L1

System	Signal	Modulation	PRN rate (Mcps)	PRN code length	Data rate (sps)
	L1 CA	BPSK(1)	1.023	1023	50
	L1C-I	BOC(1,1)	1.023	10230	100
GPS	L1C-Q	TMBOC	1.023	1800×10230	N/A
	L1P	BPSK(10)	10.23	6.19×10^{12}	50
	L1 M	BOC(10,5)	5.115	N/A	N/A
Galileo	E1 OS	CBOC	1.023	4092	250

Table 4 Simulation parameters

Parameter	GPS ← Galileo	Galileo ← GPS
Simulation time	1 day	1 day
Time resolution	5×60 s	5×60 s
Grid resolution	$0.25° \times 0.25°$	$0.25° \times 0.25°$
Elevation angle	10°	10°
Emission power	GPS: 14.5 dBW	Galileo: 15 dBW
Front end bandwidth	24 MHz	24 MHz

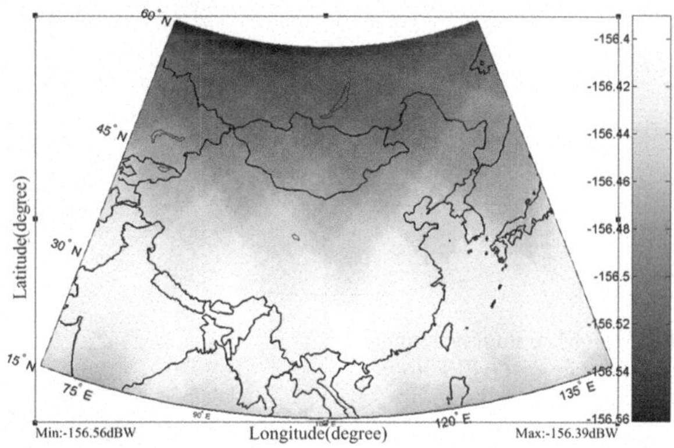

Fig. 7 Distribution of GPS mean received power

4.2 Results

Based on previous models and configuration parameters, intra and intersystem interference of GPS and Galileo were simulated. The signal model was given by Eq. (2.7). Considering real PRN code, results showed maximum degradation of effective C/N_0 and mean power of every longitude-latitude grid in simulation period. Mean received power of GPS and Galileo all over the world were shown in Figs. 7 and 8 separately.

GPS intra-interference in L1 were shown in Fig. 9, where maximum value was 2.637 dB and minimum value was 1.118 dB, GPS L1 C/A signal was desired signal and other GPS signals in L1 (GPS C/A, L1P, L1M and L1C) were

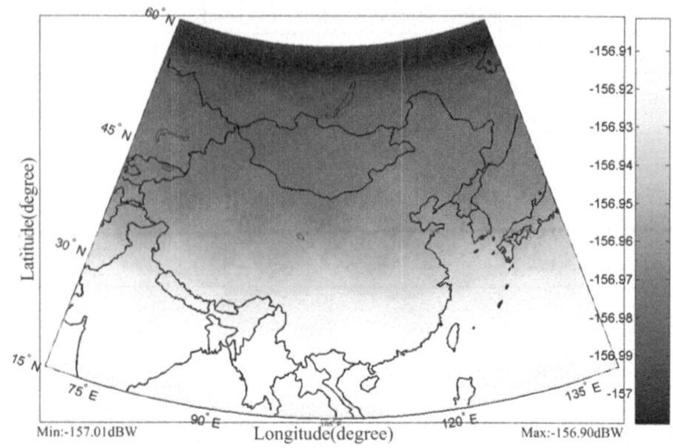

Fig. 8 Distribution of Galileo mean received power

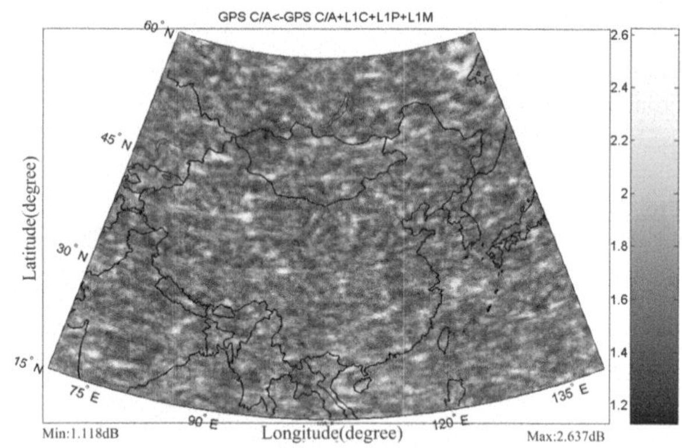

Fig. 9 Max intrasystem interference of GPS

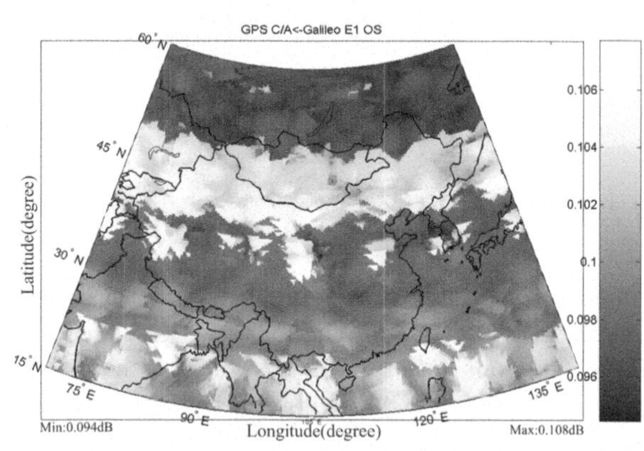

Fig. 10 Max intersystem interference of Galileo to GPS

interfering ones. Inter-interference of Galileo to GPS was shown in Fig. 10, where maximum value was 0.108 dB and minimum value was 0.094 dB, GPS C/A was desired signal and Galileo E1B was interfering one. Galileo intra-interference

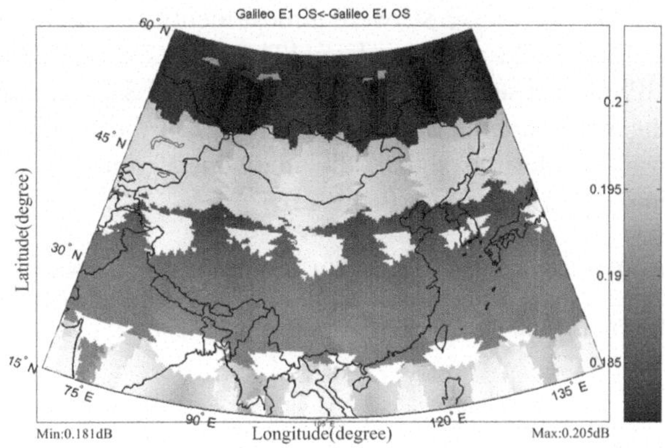

Fig. 11 Max intrasystem interference of Galileo

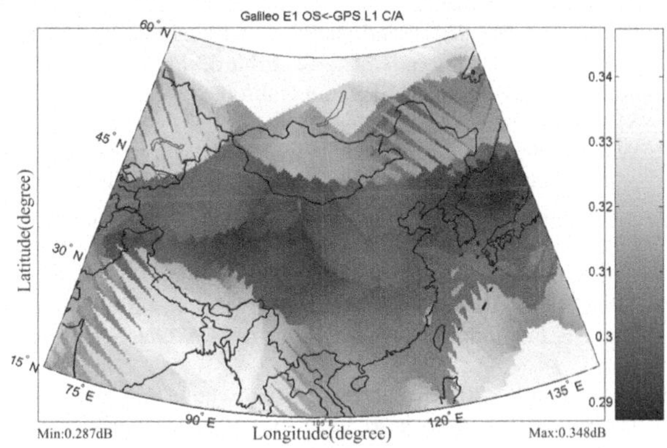

Fig. 12 Max intersystem interference of GPS to Galileo

was shown in Fig. 11, where maximum value was 0.205 dB and minimum value was 0.181 dB. The inter-interference of GPS to Galileo was shown in Fig. 12, where maximum value was 0.348 dB and minimum value was 0.287 dB, Galileo E1B signal was desired signal and GPS L1 C/A was interfering signal.

5 Conclusion

The general interference analysis methodology was presented in this paper. On the base of this methodology, signal model was deduced more accurately with real PRN code and data rate. And according to the structure of signal, all kinds of satellite navigation signals were divided into three types. Besides, the effects of data rate and pseudo random code onto SSC were analyzed. It can be concluded that effect of signal with short code and low data rate onto SSC was the most obvious, moreover, the effects of signals from different satellites with same modulation onto SSC were almost identical. Finally, the compatibility of GPS and Galileo in L1 was simulated. The results showed that the main interference was GPS intrasystem interference when the GPS C/A signal was the desired signal, and the maximum interference value was 2.637 dB, and when the Galileo E1 OS signal was the desired signal, the main interference was from GPS C/A signals, the maximum value was 0.348 dB.

Acknowledgments This work was supported by a grant to "Intelligent navigation and rescue terminals and their regional application" (2011AA120502) from the National Key Technology R&D Program during the twelfth five-year plan.

References

1. Avila-Rodriguez, J. A., Hein, G. W., et al. (2008). The MBOC modulation: the final touch to the Galileo frequency and signal plan. *Navigation, 55*(1), 15–28.
2. Godet, J., de Mateo, J. C., & Erhard, P., et al. (2002). Assessing the radio frequency compatibility between GPS and Galileo. *Proceedings of ION GPS 15th ITM* (pp. 1260–1269). Manassas: U.S. ION.
3. Wallner, S., Hein, G. W., & Pany, T., et al. (2005). Interference computations between GPS and Galileo. *Proceedings of ION GNSS 18th ITM* (pp. 861–876). Manassas: U.S. ION.
4. Avila-Rodriguez, J. A., Hein, G. W., & Wallner, S., et al. (2005). Revised combined Galileo/GPS frequency and signal performance analysis. *Proceedings of ION GNSS 18th ITM* (pp. 846–860). Manassas: U.S. ION.
5. Titus, B. M., Betz, J. W., & Hegarty, J., et al. (2003). Intersystem and intrasystem interference analysis methodology. *Proceedings of ION GNSS 16th ITM* (pp. 2061–2069). Manassas: U.S. ION.
6. Betz, J. W. (2000). Effect of narrowband interference on GPS code tracking accuracy. *Proceedings of ION 2000 NTM* (pp. 16–27). Manassas: U.S. ION.
7. Meyr, H., Moeneclaey, M., & Fechtel, S. A. (1998). *Digital communication receivers*. USA: Wiley.

A New Method of Multipath Error Analysis for Band-Limited BOC Signal

Bin Tang, Wei Wang and Lei Wang

Abstract Based on the shortage of the multipath error analysis for GNSS signal at present, the effect on DLL because of multipath signal was analyzed, the expression of autocorrelation function for band-limited BOC signal is introduced, and then a new method of multipath error analysis for band-limited BOC signal is presented. Using of multipath parameters on some typical environment, multipath error envelope and mean multipath error are evaluated by the new method. Simulation results show that the new method is feasible and convenient.

1 Introduction

Binary Offset Carrier (BOC) modulation is the main modulation mode for modern GPS and Galileo. BOC modulation has obvious advantages compared to ordinary pseudo-code modulation in multipath mitigation, but the multipath error can not be completely eliminated. At present, there are three main methods for navigation signal multipath error analysis and evaluation. First, the mean multipath error is obtained by much of simulation based on a statistical model of multipath channel [1]. Second, the multipath error envelope is calculated by the implicit expression of the multipath error based on non-band-limited assumptions [2]. Third, based on the band-limited conditions, the linear equivalent of the code loop discriminator is given, and the explicit expression of the multipath error envelope is obtained [3]. It takes very long time for the first method do a large number of simulation based on random multi-path delay. The second method is only fit for the receiver whose pre-correlation

B. Tang (✉) · W. Wang · L. Wang
Beijing Global Information Center of Application and Exploration, 100094 Beijing, China
e-mail: tangbinfriend@yahoo.com.cn

J. Sun et al. (eds.), *China Satellite Navigation Conference (CSNC) 2012 Proceedings*, Lecture Notes in Electrical Engineering 160, DOI: 10.1007/978-3-642-29175-3_5,

bandwidth is much larger than the code rate. The third method has a good application in the signal structure design and multipath performance assessment, but it need assume a small error condition and introduce an approximate linear model when deducing the multipath error expression. In addition, the third method need do numerical integral based on pre-correlation bandwidth when using explicit expression to calculate multipath error, and the process of calculation is relatively complex. In this paper, based on the second method, a new multipath error envelope analysis method in the band-limited condition is introduced. The new method takes into account the pre-correlation bandwidth without any approximation, and the process of calculation is relatively simple without any numerical integral.

In this paper, the influence of multipath interference on the code tracking loop is first analyzed, the expression of the delay lock loop (DLL) discriminator output is given, then the autocorrelation function of BOC modulation in band-limited condition is introduced. Based on the explicitly expression of the autocorrelation function, a new multipath error envelope analysis method is introduced. Finally, the proposed method is validated by simulation.

2 Multipath Error Introduced to the Code Tracking Loop

Multipath signal gets to the receiver antenna through a longer path than the direct signal, and its carrier phase and code phase is different from the direct signal. When the mixed-signal received by the receiver, the error will be introduced in the processing of acquiring and tracking. Especially, the error of code tracking will lead to the pseudo-range error.

Without regard to the navigation data and noise, the multipath signal and the direct signal received by a GNSS receiver can be expressed [4]:

$$r(t) = \sum_{i=0}^{M} a_i A \cdot g(t - \tau_i) \cos(\omega_m (t - \tau_i) + \theta(t)) \tag{1}$$

where,

A is the amplitude of the direct signal
a_i is the amplitude scale of the multipath signal. (a_0 is the amplitude scale of the direct signal, $a_0 = 1$)
i is the index of the multipath signal
τ_j is the delay of the multipath signal. (τ_0 is the amplitude of the direct signal, and $\tau_0 = 0$)
ω_m is the intermediate frequency of carrier wave
$g(\cdot)$ is for the modulation pseudo-code

In order to facilitate the process of multipath error analysis, only one-way multipath signal is assumed, then Eq. (1) becomes:

$$r(t) = A \cdot g(t) \cos(\omega_m t + \theta(t)) + a_1 A g(t - \tau_1) \cos(\omega_m (t - \tau_1) + \theta(t)) \tag{2}$$

The influence on the pseudo-range measurements from multipath signal embodies the curve of DLL discriminator. Commonly, there are three kinds of DLL discriminator used in the GNSS receiver code loop, which are coherent early minus late discriminator (EML), non-coherent early minus late power discriminator (EMLP), and non-coherent dot product power discriminator (DP). It can prove that the multipath error of EMLP and DP is the same as the coherent EML when carrier phase error is 0° (i.e., the carrier loop is locked) [3]. Therefore, under the assumption that the carrier phase error is 0°, the multipath error of coherent EML will be analyzed in the following. The results of multipath error analysis apply equally to EMLP and DP. The output of EML discriminator is (amplitude normalized) [3]:

$$\begin{aligned} D_{EML} &= R(\Delta\tau_k - d \cdot T_c) - R(\Delta\tau_k + d \cdot T_c) \\ &\quad \pm a_1[R(\Delta\tau_k - \tau_1 - d \cdot T_c) - R(\Delta\tau_k - \tau_1 + d \cdot T_c)] \end{aligned} \tag{3}$$

where,

I_E is the output of in-phase early correlator
I_L is the output of in-phase early correlator
$R(\cdot)$ is the autocorrelation function of pseudo-code
d is the correlator spacing
$\Delta\tau_k$ is estimation error of code phase
T_c is width of one chip

The multipath signal will make the code phase deviate from zero when the output of discriminator is zero, and this offset is the code multipath error. Let the Eq. (3) to be 0 (i.e., the code loop is locked), code tracking error caused by multipath signal can be calculated using an iterative approach. The key to get multipath error is to determine the explicit expression of the autocorrelation function in the band-limited condition.

3 Analysis of the Autocorrelation Function for Band-Limited BOC

A BOC signal is obtained through the product of an PRN code with a synchronized square wave sub carrier, which can be expressed as BOC(n,m). The code frequency f_c is $m * 1.023$ MHz, the sub carrier frequency f_s is $n * 1.023$ MHz. The sine sub-carrier can be modeled as [5]:

$$SC(t) = \mathrm{sgn}[\sin(2\pi f_s t)] \tag{4}$$

Assuming $l = n/m$, the time domain waveform of BOC(n,m) can be written as:

$$s(t) = \begin{cases} 1 & t \in [(2k - l)t_c, (2k - l + 1)t_c] \\ -1 & t \in [(2k - l + 1)t_c, (2k - l + 2)t_c] \end{cases} \tag{5}$$

Fig. 1 Modulating square wave for BOC(10,5)

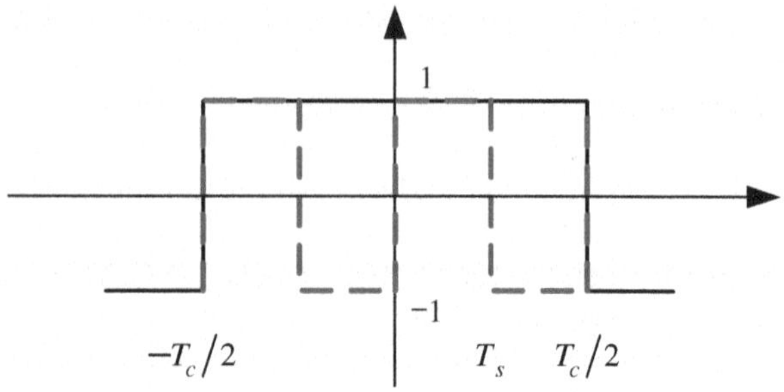

where,

$k \in [0, \ldots, l-1]$,
t_c is $T_c/(2l)$,
T_c is $1/f_c$.
T_s is $1/f_s$.

For example, the time domain waveform of BOC(10,5) is shown in Fig. 1. Then the Fourier transform corresponding to Eq. (5) can be written as:

$$S_{BOC}(\omega) = \sum_{k=0}^{l-1} \left[\int_{(2k-l)t_c}^{(2k-l+1)t_c} e^{-j\omega t} dt - \int_{(2k-l+1)t_c}^{(2k-l+2)t_c} e^{-j\omega t} dt \right] \tag{6}$$

The power spectrum $P_{BOC}(\omega)$ for the BOC modulation is obtained:

$$P_{BOC}(\omega) = S^*_{BOC}(\omega) S_{BOC}(\omega) \tag{7}$$

The power spectrum envelope of BOC(n,m) is [6]:

$$P_{BOC}(f) = \begin{cases} f_c \left[\dfrac{\tan\left(\frac{\pi f}{2f_s}\right) \sin\left(\frac{\pi f}{f_c}\right)}{\pi f} \right]^2, n = \dfrac{2f_s}{f_c}, even \\ f_c \left[\dfrac{\tan\left(\frac{\pi f}{2f_s}\right) \cos\left(\frac{\pi f}{f_c}\right)}{\pi f} \right]^2, n = \dfrac{2f_s}{f_c}, odd \end{cases} \tag{8}$$

The autocorrelation function is the inverse Fourier transform of the power spectral density. Therefore, the autocorrelation function of BOC(n,m) can be expressed [7]:

$$\begin{aligned} R_{BOC}(t) = \sum_{k=-l+1}^{l-1} (l - |k|)[&2\mathrm{Tri}(t/t_c - 2k) \\ &- 2\mathrm{Tri}(t/t_c - 2k - 1) - 2\mathrm{Tri}(t/t_c - 2k + 1)] \end{aligned} \tag{9}$$

The front-end filter of the receiver has an important effect on the signal power spectrum. The band-limited trigonometric function Tri can be expressed [7]:

$$R_{BL}(\tau)=\frac{1}{\pi}(\tau+1)Si[2\pi b(\tau+1)]+\frac{1}{2\pi^2 b}\cos[2\pi b(\tau+1)]+\frac{1}{\pi}(\tau-1)Si[2\pi b(\tau-1)]$$
$$+\frac{1}{2\pi^2 b}\cos[2\pi b(\tau-1)]-\frac{2\tau}{\pi}Si[2\pi b\tau]-\frac{1}{\pi^2 b}\cos(2\pi b\tau) \qquad (10)$$

where, $Si(x)=\int_0^x \mathrm{sinc}(y)dy$, the pre-correlation bandwidth B is $4l\cdot bf_c$. According to equation (9), the autocorrelation function of BOC modulation in the band-limited condition can be expressed:

$$R_{BOC}^{BL}(t)=\sum_{k=-l+1}^{l-1}(l-|k|)[2R_{BL}(t/t_c-2k)$$
$$-R_{BL}(t/t_c-2k-1)-R_{BL}(t/t_c-2k+1)] \qquad (11)$$

Taking into account the normalization of the autocorrelation function and the change of delay step, the normalized autocorrelation function of BOC(n,m) in the band-limited condition can be expressed:

$$R_{BOC}^{BL}(t)=\sum_{k=-l+1}^{l-1}(l-|k|)[R_{BL}(2l\cdot t/t_c-2k)-0.5R_{BL}(2l\cdot t/t_c-2k-1)$$
$$-0.5R_{BL}(2l\cdot t/t_c-2k+1)]/l \qquad (12)$$

4 The Simulation Validation for the New Method

BOC(10,5), BOC(15,2.5) and BOC(14,2) are usually used in the GNSS signal, whose multipath error are analyzed in the following.

The first step, the autocorrelation function of BOC modulation is calculated with Eq. (12). For BOC(10,5), n is 10, m is 5, and l is 2. The normalized autocorrelation function of BOC(10,5) in the band-limited condition is shown in Fig. 2. The normalized autocorrelation function of BOC(15,2.5) and BOC(14,2) are shown in Figs. 3 and 4. For clarity, only the right side of the autocorrelation function is shown. When b is 1, the pre-correlation bandwidth are 40, 60 and 56 MHz for BOC(10,5), BOC(15,2.5) and BOC(14,2). When b is 0.5, the pre-correlation bandwidth are 20, 30 and 28 MHz for BOC(10,5), BOC(15,2.5) and BOC(14,2).

It can be seen from Figs. 2, 3 and 4:

(1) The autocorrelation function curve of BOC modulation is made up of a group of interconnected segments. There are many positive peaks and negative peaks in the curve. The sum of positive peaks and negative peaks are $4n/m-1$, and

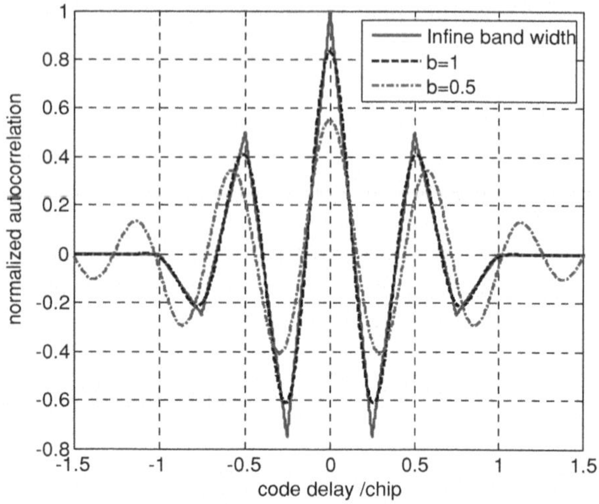

Fig. 2 Normalized autocorrelation of band-limited BOC(10,5)

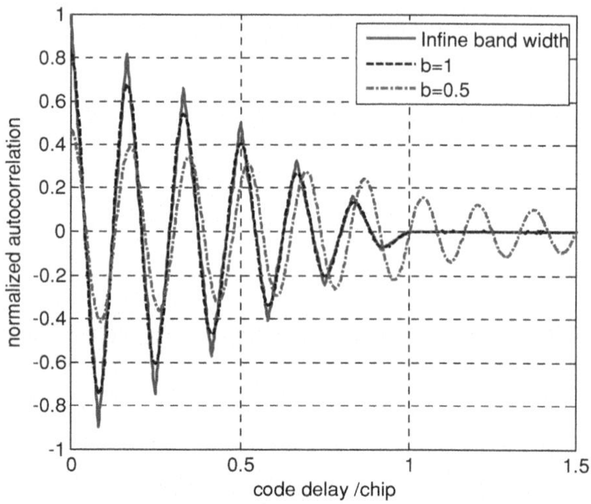

Fig. 3 Normalized autocorrelation of band-limited BOC(15,2.5)

the number of positive peaks is one less than the negative peaks. The delay of adjacent positive peak and negative peak is $m/(2n)$ chip.

(2) The main lobe of the BOC spectrum can not be completely covered when the pre-correlation bandwidth is narrow, and the autocorrelation curve will be deformed. So the following will analyze multipath error with the band-limited condition b = 1.

The second step, substitute the value of normalized autocorrelation function into Eq. (3), the implicit expression of the code phase tracking error $\Delta\tau_k$ will be obtained. Let the output of discriminator D_{EML} to be 0 $\Delta\tau_k$ is the error caused by multipath signal. The signal amplitude ratios a_1 are 0.5, 0.2 and 0.05 in typical

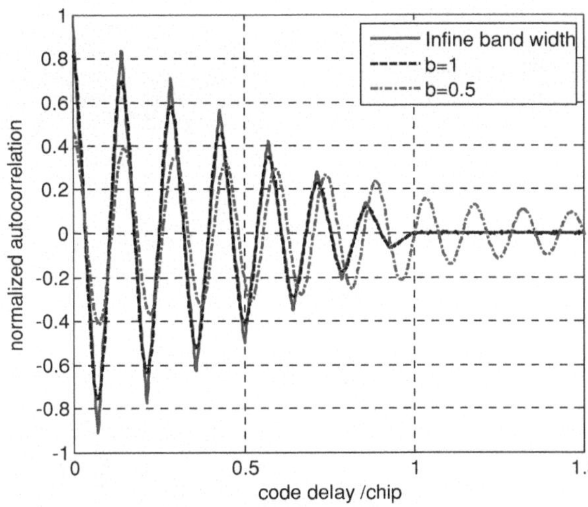

Fig. 4 Normalized autocorrelation of band-limited BOC(14,2)

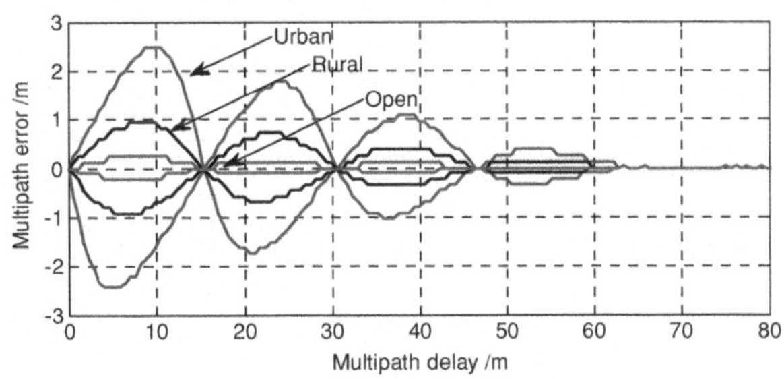

Fig. 5 BOC(10,5) multipath error envelopes for different multipath environments

application environment, which are respectively the environment of urban, rural and open [1].

The third step, let the Eq. (3) to be 0, the iterative calculation will be done for different multipath delay τ_1, and the relationship between multipath error envelope and multipath delay can be obtained. The multipath error envelope of BOC(10,5), BOC(15,2.5) and BOC(14,2) are shown on Figs. 5, 6 and 7.

The forth step, multipath error envelope is the maximum deviation caused by different multipath delay, which reflects the worst multipath error. A more reliable way for multipath performance assessment is to calculate the mean multipath error. For this purpose, only the absolute envelope values are considerer and their cumulative sum is used to compute mean multipath error. The formula is:

$$\bar{E}(\tau) = \frac{1}{\tau}\int_0^{\tau} \{[abs(E^+(\tau)) + abs(E^-(\tau))]/2\} d\tau \tag{13}$$

where, $E^+(\tau)$ is positive multipath error envelope, $E^-(\tau)$ is negative multipath error envelope. The curve of mean multipath error for different application environment is shown in Fig. 8.

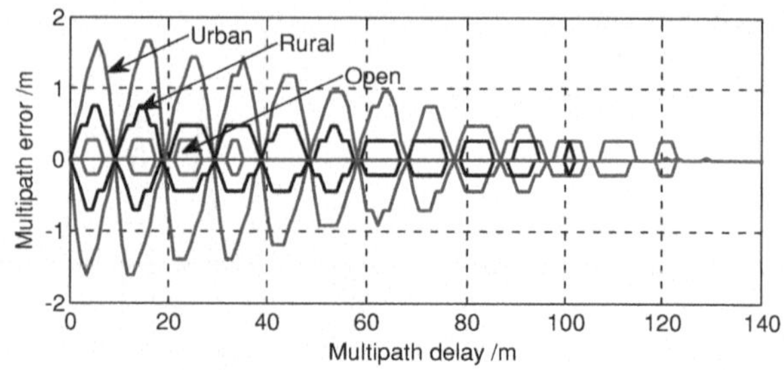

Fig. 6 BOC(15,2.5) multipath error envelopes for different multipath environments

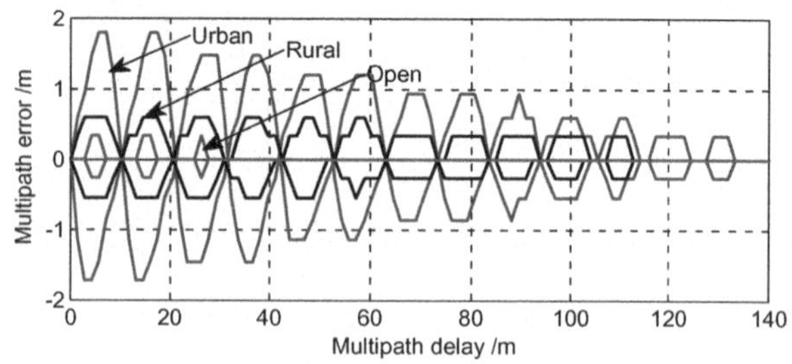

Fig. 7 BOC(14,2) multipath error envelopes for different multipath environments

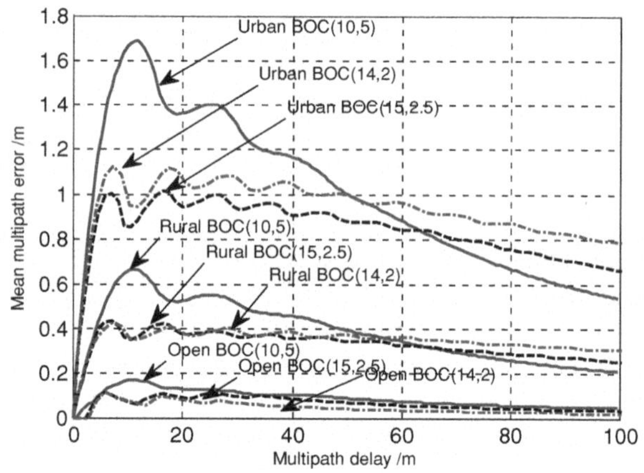

Fig. 8 BOC mean multipath errors for different multipath environments

5 Conclusion

In this paper, the relation between the output of the DLL discriminator and the multipath error is obtained based on the study of multipath error introduced to the code tracking loop. Then the autocorrelation function of BOC modulation in band-limited condition is deduced, and a new multipath error envelope analysis method is introduced. The simulation results show that the new method can be used to calculate multipath error envelope and mean multipath error for band-limited BOC modulation signal, the process of calculation is more simple, which does not require complex numerical integral calculation.

References

1. Markus, I., Rodriguez, J., & Hein, G. W. (2005). Criteria for GNSS multipath performance assessment. *ION GNSS 18th International Technical Meeting of the Satellite Division*, (pp. 2166–2177).
2. Zhao, W., & Li, C. (2009). Analysis of the multipath error of BOC modulated signal. *Journal of Geomatics Science and Technology, 26*(4), 258–260.
3. Tang, Z., Hu, X., & Huang, X. (2009). Analysis of multipath rejection performance in GNSS signal design. *Huazhong University of Science and Technology, 37*(5), 1–4.
4. El-Sayed Abdel-Salam Gadallah (1998). *Global positioning system receiver design for multipaths mitigation*. Dissertation, Air Force Institute of Technology Air University, USA, 55–57
5. Strodl, K., Naddeo, G., & Samson, J., et al. (2001). System verification approach, methods, and tools for Galileo. ION GNSS 2003, Portland, USA (pp. 2446–2456).
6. Betz, J. W. (2001). Binary offset carrier modulations for radio navigation. *Journal of the Institute of Navigation, 48*(4), 227–246.
7. Winkel, J. (2003). *Modeling and simulating GNSS signal structures and receivers*. Dissertation, Universitat der Bundeswehr Munchen, Munich, 50–57

Optimum Waveform Design for GNSS Signals Based on PSWF

Chengeng Su, Shuren Guo, Hongwei Zhou, Sihui Liu and Wei Wang

Abstract Signal waveform is an important factor to the navigation system such as tracking precision, anti-jamming and anti-multipath. Prolate Spheroidal Wave Functions (PSWF) are widely used in Ultra Wide Band (UWB) communication area. They are complete orthogonal set in the time domain that maximizes energy in the specified bandwidth. This paper presents PSWF bases method and orthogonal decomposition of waveforms. By optimizing coefficients, we can obtain optimum waveforms to minimize target functions with constrains. It presents a new method for waveform design of the navigation signals.

Introduction

Signal waveform is key to the GNSS signal design, for it decides the autocorrelation function and power spectrum properties.

The waveform of navigation signal can be divided into two types: rectangular waveform and continuous function waveform. The former one is easy to utilize and has been applied in conventional GNSS signals, but the modulated signal adopting rectangular waveform is not continuous in phase. In recent years, the research on continuous waveform is on its way. In ION 2009, Felix Antreich enhanced chip pulse shape design with the help of Prolate Spheroidal Wave Functions (PSWF) in bound limits [1, 2]. These design methods will promote the progress of GNSS signal design, which make it more flexible and optimum.

C. Su (✉) · S. Guo · H. Zhou · S. Liu · W. Wang
Beijing Institute of Tracking and Telecommunication Technology,
100094 Beijing, China
e-mail: hancecilia@126.com

J. Sun et al. (eds.), *China Satellite Navigation Conference (CSNC) 2012 Proceedings*, Lecture Notes in Electrical Engineering 160, DOI: 10.1007/978-3-642-29175-3_6,

This paper studied the basic theories of PSWF, solved the orthogonal decomposition based on which we could express any waveform approximately. The followed example demonstrated its feasibility. In this way, this paper provided a referenced method for GNSS signal design.

1 Prolate Spheroidal Wave Functions

1.1 Basic Theory of PSWF

Based on Heisenberg's uncertainty principle, band-limited and time-limited are paradoxical, for band-limited signal used to be continuous in infinite time range [3]. Does the signal exist that is both band-limited and time-limited? In the 1960 s, Bell Laboratories proposed one optimum waveform—prolate spheroidal wave functions, which are band-limited without inter-symbol interference and last for one chip.

Mathematically, "concentrated in the interval of [−Tc/2, Tc/2] with most of its energy in the frequency band of [−B, B]" means that this normalized waveform function $\psi(t)$ which is restricted to (1.1) has maximum energy J:

$$\int_{-Tc/2}^{Tc/2} \psi^2(t)dt = 1 \tag{1.1}$$

$$J = \int_{-B}^{B} |G(f)|^2 df \tag{1.2}$$

where G(f) is the normalized spectrum corresponding to $\psi(t)$. So that

$$J = \int_{-Tc/2}^{Tc/2} \int_{-Tc/2}^{Tc/2} \psi(t)\psi(\tau)\frac{\sin B(t-\tau)}{\pi(t-\tau)} dtd\tau \tag{1.3}$$

After building the augmented functional with Lagrange multipliers, we can infer that, to maximize its energy in [−B, B], $\psi(t)$, which is continuous in the interval of [−Tc/2, Tc/2], should satisfy (1.4), [4].

$$\int_{-Tc/2}^{Tc/2} \psi(\tau)\frac{\sin 2\pi B(t-\tau)}{\pi(t-\tau)} d\tau = \lambda\psi(t) \tag{1.4}$$

where λ denotes the concentration degree of $\psi(t)'s$ energy in [−B, B]. Equation (1.4) also implies that $\psi(t)$ is the function of time t, and depends on bandwidth. $C = T_cB$, and C denotes the product of interval and bandwidth, which represents the freedom of design.

The functions set $\{\psi_1(t), \psi_2(t)..., \psi_n(t)\}$ which satisfy (1.4) are PSWF's orthogonal decompositions, corresponding to $\{\lambda 1, \lambda 2..., \lambda n\}$. PSWF have these properties:

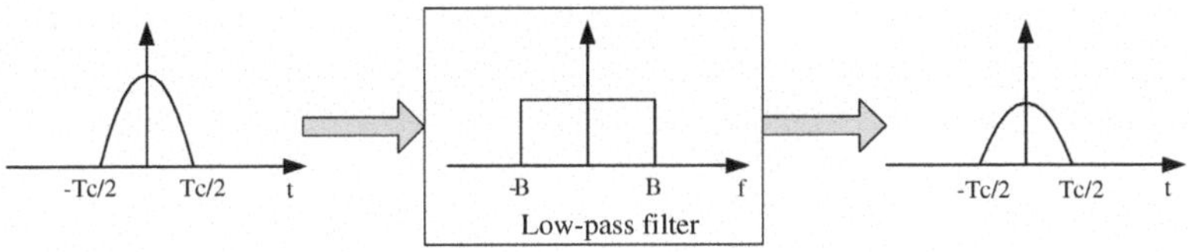

Fig. 1 PSWF sketch map

$$\int_{-\infty}^{\infty} \psi_m(t)\psi_n(t)dt = \begin{cases} 1 & m = n \\ 0 & m \neq n \end{cases} \tag{1.5}$$

$$\int_{-Tc/2}^{Tc/2} \psi_m(t)\psi_n(t)dt = \begin{cases} \lambda_m & m = n \\ 0 & m \neq n \end{cases} \tag{1.6}$$

In fact, PSWF denote the waveforms whose shapes in time domain nearly keeps constant except the amplitude when filtered by the ideal low-pass filter with single bandwidth B (Fig. 1).

1.2 Solve PSWF

Since it's difficult to solve the PSWF in close form, we usually adopt numerical method. After discretization, (1.4) turns into:

$$\lambda\psi[n] = \sum_{m=-N/2}^{N/2} \psi[m] \cdot h[n-m] \tag{1.7}$$

where

$$h[n] = \frac{\sin 2\pi B(nTc/N)}{\pi nTc/N} \tag{1.8}$$

Then

$$\lambda\psi = H\psi \tag{1.9}$$

where

$$H = \left\{ \begin{array}{cccc} h[0] & h[-1] & \cdots & h[-N] \\ h[1] & h[0] & \cdots & h[-N+1] \\ \vdots & \vdots & \vdots & \vdots \\ h[N/2] & h[N/2-1] & \cdots & h[-N/2] \\ \vdots & \vdots & \vdots & \vdots \\ h[N] & h[N-1] & \cdots & h[0] \end{array} \right\} \tag{1.10}$$

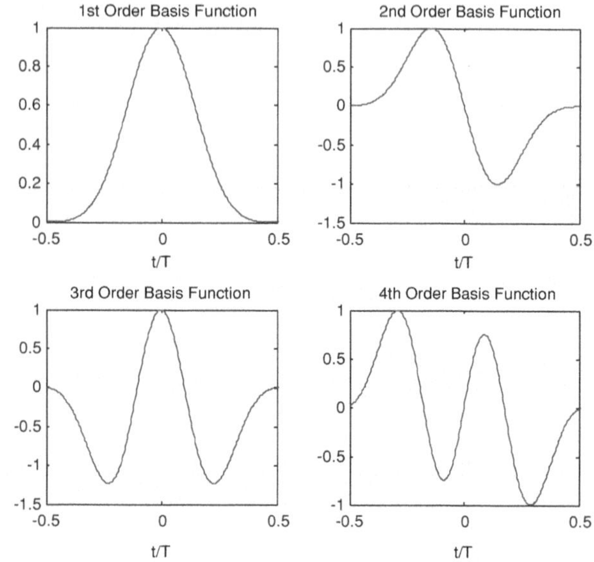

Fig. 2 PSWF bases in time domain

Then λ is the eigenvalue of matrix H and denotes the extent of energy concentration, and $\lambda = [\lambda_1, \lambda_2, \cdots \lambda_n]$; ψ is the eigenvector for λ, and $\psi = [\psi_1, \psi_2, \cdots \psi_n]$ are set of complete orthogonal basis. Thus, the complicated problem has been transformed into a tractable problem of solving eigenvalues and eigenvectors of matrix H.

After sizing down the eigenvalues where λ_1 represents that the energy is the most concentrated, we simulate the basis function ψ_n in time domain and frequency domain as show in Figs. 2 and 3. The spectrum of 1st-order basis function is similar to BPSK signal, 2nd-order to BOC(1,1), 3rd-order to BOC(1.5,1), 4th-order to BOC(2,1), etc.

2 Signal Expressions by PSWF

PSWF are set of complete orthogonal basis. Thus, any signal waveform could be expressed by PSWF basis in unique form. Assuming that the PSWF set is $\psi = [\psi_1, \psi_2, \cdots \psi_n]$, then the signal waveform $p(t)$ can be expressed to be:

$$p(t) = \sum_{k=1}^{\infty} \alpha_k \psi_k(t) \tag{2.1}$$

Considering

$$\int_{-\infty}^{\infty} \psi_m(t)\psi_n(t)dt = \begin{cases} 1 & m = n \\ 0 & m \neq n \end{cases} \tag{2.2}$$

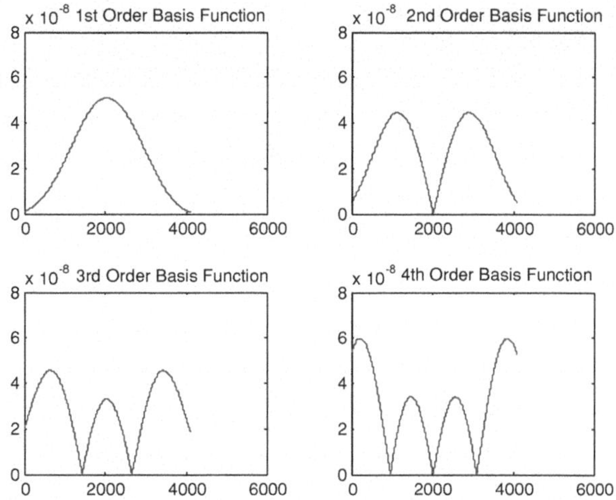

Fig. 3 PSWF bases in frequency domain

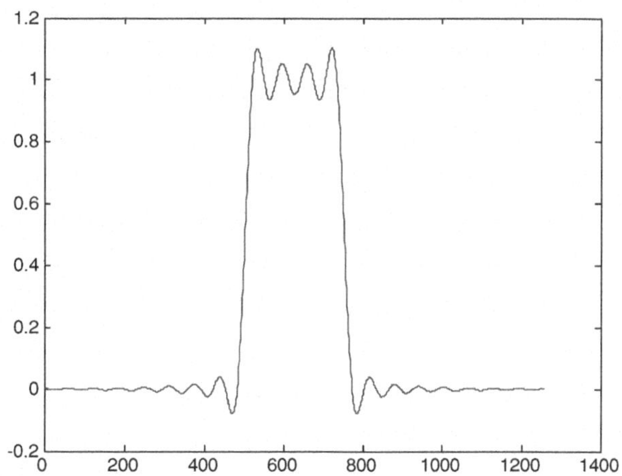

Fig. 4 Waveform of BPSK represented by PSWF

Then

$$\alpha_k = \int_{-\infty}^{\infty} p(t)\psi_k(t)dt \tag{2.3}$$

Thus, the signal can be expressed by the former M basis functions approximately as:

$$p(t) \approx \sum_{k=1}^{M} \alpha_k \psi_k(t) \tag{2.4}$$

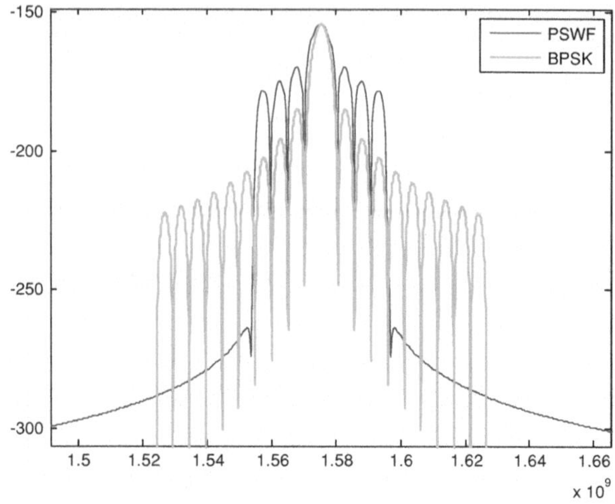

Fig. 5 Spectrum of BPSK represented by PSWF

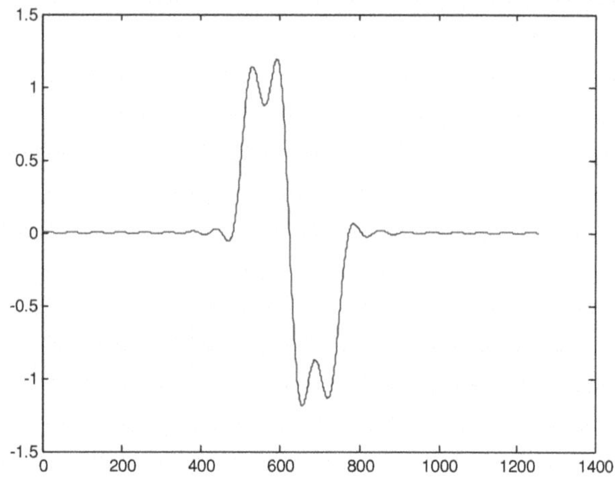

Fig. 6 Waveform of BOC represented by PSWF

(1) Expressing BPSK signal with PSWF

The approximate waveform and spectrum density of BPSK signal with 20 orders basis function is shown as Figs. 4 and 5.

(2) Expressing BOC signal with PSWF

The approximate waveform and spectrum density of BOC signal with 20 orders basis function is shown as Figs. 6 and 7.

Despite of the difference that introduced by the insufficient orthogonal series, the shapes of both signals in time domain and frequency domain are quite similar. If the orthogonal series are sufficient, the shapes will be the same.

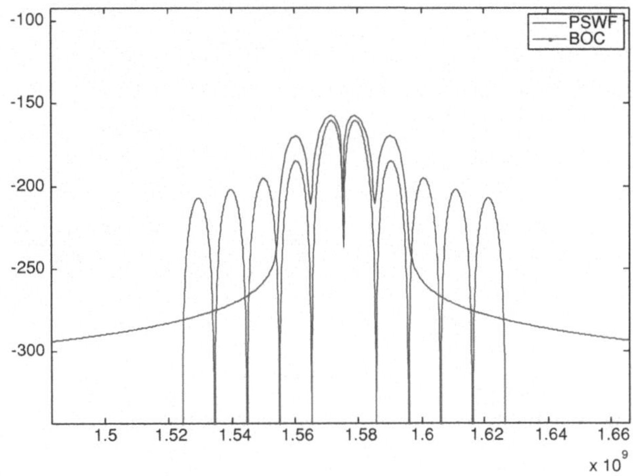

Fig. 7 Spectrum of BOC represented by PSWF

3 Optimization of Waveform Design with PSWF

3.1 Optimization Routine

Since that any signal can be uniquely expressed by PSWF, we could set the restraint and target functions. We could get the suitable signal waveform by optimizing the weighting coefficients, and the problem can be transformed into optimization problem under nonlinear conditions.

The routine is as follows:

(1) Expressing the waveform by basis functions

Express the waveform into the linear combine of orthogonal PSWF basis functions.

$$p(t) = \sum_{m=1}^{\infty} \alpha_m \psi_m(c, t) \quad \text{where} \quad c = \pi T_c B \tag{3.1}$$

(2) Selecting the target functions

Choose suitable target functions that we most concern, i.e. the target function could be the Gabor bandwidth or code tracking error function when the precision is key to the design.

(3) Setting the constraints

Set the constraints which the waveform should be restrained to, mainly including the side lobes of correlating function, the external inhibition, the separation of spectrum, anti-jamming, anti-multipath, etc.

(4) Parameters optimization

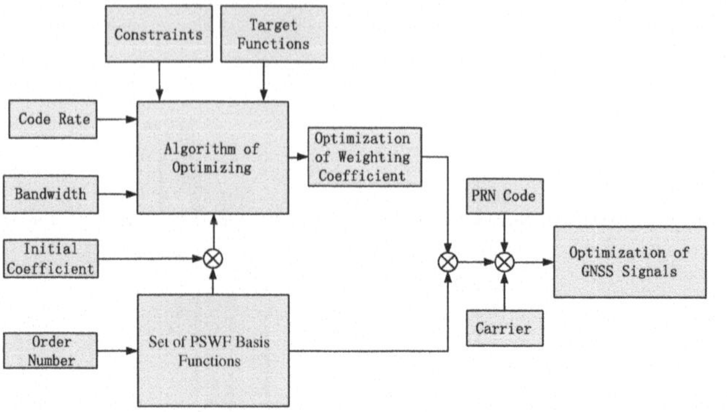

Fig. 8 Flow chart of optimization based on PSWF

Select the weighting coefficient $\{\alpha_m\}$ to optimize the target functions under the constraints.

The routine is shown in Fig. 8.

3.2 Simulation and Analysis

1. Parameter settings

- Bandwidth: B = 20 MHz;
- Code Rate: f = 1 MHz;
- Basis function order: 20
- Target Function: $SSC_{BPSK,\ BPSK}$–$SSC_{waveform,\ BPSK}$;
  $SSC_{BPSK,\ BPSK}$ denotes the spectrum separation coefficient between BPSK(1) and BPSK(1); $SSC_{waveform,\ BPSK}$ denotes the spectrum separation coefficient between BPSK(1) and the designed waveform.
- Constraint: normalization of sum of squares for weighting coefficient.

2. Optimizing results

The optimized coefficients are:

No.	Coefficient	No.	Coefficient
1	9.216057608555903E−4	11	−0.04072240078632876
2	−8.761790373487091E−4	12	8.804206986606916E−4
3	−0.0805206274116532	13	−0.027120060134857904
4	2.5028931073465275E−4	14	−0.31916368815525276
5	−0.06357737750737735	15	−0.23172621715926695
6	2.420491397225373E−4	16	0.1960348157221141
7	−0.05779551816395052	17	0.14574288818325803

(continued)

(continued)

No.	Coefficient	No.	Coefficient
8	1.6432170147098648E−5	18	−0.04682155484821568
9	0.03086068575540897	19	−0.3795299299196015
10	2.9760562828911864E−4	20	0.7896126893249475

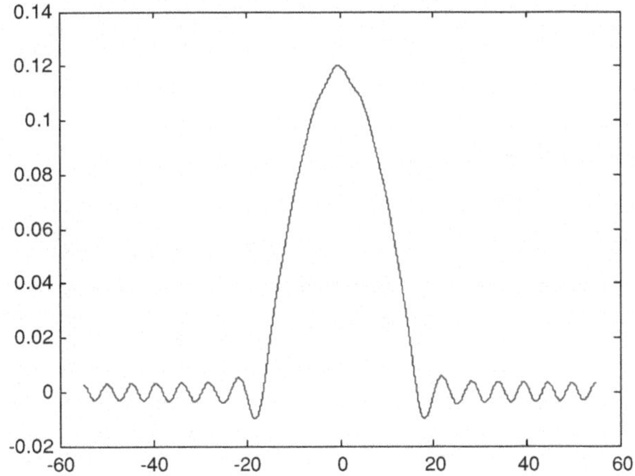

Fig. 9 Optimum waveform in time domain

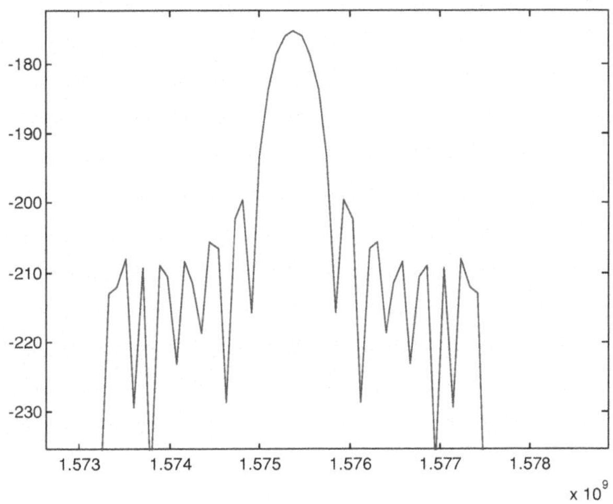

Fig. 10 Spectrum of optimum waveform

The shape of waveform is shown as Figs. 9 and 10, corresponding to time domain and frequency domain.

Figures 9 and 10 show that the designed waveform is similar to BPSK both in time domain and frequency domain. When the basis function order comes to infinite, we will get the ideal BPSK signal.

4 Conclusion

In this work we studied the basic theories of Prolate Spheroidal Wave Functions, proposed methodology in order to solve the PSWF mathematically with orthogonal decomposition. Since any waveform can be expressed with basis functions of PSWF uniquely, we could set the target functions and constraints to optimize the coefficients. The demonstration in the end proved its feasibility. Optimization routine presented in this paper could be a reference for future GNSS signal design.

References

1. Antreich, F. (2009). *Joint optimization of spectral separation and navigation performance in GNSS signal design. 22nd International Meeting of the Satellite Division of the Institute of Navigation* (22–25). Savannah: The Institute of Navigation.
2. Antreich, F., & Nossek, J. A. (2007). *On chip pulse shape design for global navigation satellite systems. Proceedings of the 2nd CNES-ESA Workshop on GNSS Signlas and Signal Processing, GNSS SIGNALS*. Noordwijk: ESA-ESTEC.
3. Zhao, Z., Wang H., & Li H. (2010). Waveform design method based on prolate spheroidal wave functions. *Radio Communications Technology, 36*, 26–28.
4. Halpern, P. (1979). Optimum finite duration nyquist signals. *IEEE Transactions on Communications, 27*(6), 884–888.

Inter-satellite Traffic Data Modeling for GNSS

Feihong Dong, Jing Lv, Yong Yu, Qingqing Wang and Caiwu Wang

Abstract The high precision positioning and Auto Navigation (AutoNav) ability achievement of next generation Global Navigation Satellite System (GNSS) largely depend on the navigation satellite constellation network buildup via Inter-Satellite Links (ISLs). The ephemeris data, TT&C data, integrity monitoring data, AutoNav data etc. (hereafter called traffic data) transmitting in constellation network is one of the important researches. According to the characteristic of GNSS traffic data, building a suitable Inter-Satellite Traffic Data Model (ISTDM) is precondition to design ISLs communication protocols and increase the efficiency of traffic data transmission. The paper analyzes the traffic data requirement in GNSS ISLs, points out the distribution character of traffic data and abstracts out three basic classes. A general hybrid ISTDM is given. To highlight the ISTDM feature, comparisons of traffic data volume change rule are made with that of Ground Communication Network (GCN) and Satellite Communication System (SCS). The main points are proposed when to design ISLs communication protocols. The ISTDM presented in this paper could be a reference model for GNSS construction.

F. Dong (✉) · J. Lv · Y. Yu · C. Wang
Institute of Communication Engineering, PLA University of Science and Technology, 210007 Nanjing, China
e-mail: feihong070@qq.com

Q. Wang
School of Earth Sciences and Engineering, Hohai University, 210098 Nanjing, China

J. Sun et al. (eds.), *China Satellite Navigation Conference (CSNC) 2012 Proceedings*, Lecture Notes in Electrical Engineering 160, DOI: 10.1007/978-3-642-29175-3_7,

1 Introduction

The GPS III, GALILEO, COMPASS and GLONASS make high precision positioning and AutoNav ability as their development goals [1–3]. The AutoNav is the work mode that GNSS fulfills user's high precision positioning requirement without the support of ground TT&C system in a long time. This mode need inter-satellite ranging, navigation satellites data exchange and processing onboard etc. Those will enable the GNSS constellation system continuously corrects the ground upload station long-term forecasting ephemeris and clock parameters, and creates the navigation massage and maintains basic constellation configuration autonomously [4]. The fundamental of the AutoNav is to realize the GNSS traffic data creation, transmission and maintenance autonomously.

There are numbers of references about GCN and SCS traffic models [5–7]. Those models have made a great contribution to the GCN and SCS construction and running. In GCN and SCS, the traffic is always divided into three types: voice, data and video. The establishment of model is with a view to achieve integrated transmission of the three types, and the design of communication protocols is also around how to satisfy three types traffic requirement. However, the GNSS traffic data has its own special compared with the open GCN and SCS. In GNSS constellation, the traffic can be group in data type of above, so the previous models' accuracy is not sufficient and that can not meet the system requirement. A Traffic Data Model (TDM) which is specifically applicable in GNSS ISLs needs to be researched. In Ref. [8], the GPS ISLs traffic data is simulated based on precise ephemeris which is provided by International GNSS Service (IGS). That simulation just contains the satellite broadcast data, and the data model configuration is single, which can not estimate traffic data comprehensively. So that can not simulate the real process of ISLs traffic data exactly. In this paper, we focus on the GNSS inter-satellite traffic data model analysis.

The rest of this paper is organized as follows, Sect. 2 analyzes the requirement of all kinds of traffic inner GNSS constellation, Sect. 3 tells the characteristics of different traffic and establishes an ISTDM, Sect. 4 simulates the ISTDM, analyzes the result in detail, and compares the traffic volume with GCN, SCS. The main points are proposed when to design ISLs communication protocols in Sect. 5, and finally Sect. 6 sums up the whole paper.

2 Traffic Data Requirement

The GNSS ISLs traffic data requirement is studied in this section to provide basis for Sect. 3. Traffic data include two main aspects: one is navigation service data, such as essential ephemeris, integrity data etc. and the other is system support information, such as TT&C, ISL ranging, etc.

2.1 Uploaded Data Relay Requirement

Data uploading means that navigation satellites ephemeris, integrity data etc. are updated by the uplink stations, such as GPS III three High Speed Ground Antennas (HSGA) which will be establish in Continental United States. The traditional method is to upload data when the satellite is visible for station. However, for the next generation GNSS, things may be different. ISLs will be use to relay the upload data when the satellite is invisible. Generally speaking, upload relay data include periodic data, such as basic navigation data, integrity data, regional augment service data etc. and nonperiodic or random data, such as onboard software update data, TT&C backup data, other payloads management data.

2.2 TT&C Data Relay Requirement

Before GPS Block IIR, there is no ISL, the satellite working conditions information, attitude information etc. telemetering data are collected by TT&C stations around the word, and then transmitted to Control Center. Each satellite is monitored 45–90 times everyday. And the telecontrol data is upload 1–2 times and 45–90 min everyday when the satellite crosses over the TT&C station. However, GPS III will change that status. The full-connected network and HSGA will give the Main Control Station (MCS) an uninterrupted TT&C link; provide a continuous and near real-time telemetering and telecontrol for all on orbit satellites 24 h a day. The dependence on TT&C stations will reduce obviously. In the constellation with ISLs, the continuous data back of telemetering data, ISL ranging data, time synchronization data etc. for invisible satellites will depend on ISL relay. The operational state information of whole constellation network is needed back to MCS also to observe the abnormal system operation. The telemetering data is usually periodic, includes key payloads working conditions data, whole constellation telemetering data, ISL ranging data, constellation network operation data, satellite attitude data etc. The telecontrol data is usually nonperiodic, includes payload command data, whole constellation telecontrol data, satellite attitude control data etc.

2.3 Constellation Network Operation Data Requirement

After the full-connected constellation network completed, it can autonomously manage and control the on–off of ISLs, maintain the network topology, distribute failure nodes information to whole system, collect and process the ISL ranging data. When there is no failure nodes, the routing information update periodic based on known network topology and predictable satellite movement law.

When any node is failure, the network will trigger routing update. Constellation network operation traffic data usually includes nonperiodic, such as network topology data, ISLs rebuild data, nodes failure distribution data; and periodic data, such as routing distribution data, constellation autonomic operation data, and ISL ranging data [9].

Of course, the analyzed requirements above are just based on the understanding of GNSS primary traffic. The potential traffic data requirement may continue pour out belongs with the next generation GNSS design, construction and application.

3 Traffic Data Characteristic Description and Modeling

In this section, the paper will describe the characteristic of GNSS traffic data, pick-up traffic data parameter. Then, a GNSS network ISTDM is provided which consists of three primary models.

3.1 Period Traffic Data

From Sect. 2, we have known that Periodic Traffic Data (PTD) is a very large proportion in the constellation network. The characteristics describing PTD are period (T), data transmission time (λ), data volume (M), delay (τ). Thus, a PTD can be expressed as Eq. 1.

$$P(t-T-\tau) = (\underbrace{p_1,\ p_2,\ p_3,\cdots p_{M-1},\ p_M,}_{\lambda}\ \underbrace{\cdots\ \cdots}_{T-\lambda}) \tag{1}$$

where, p_i is information bit, $T-\lambda$ is interspace period of time, in which gap, other data can be transmitted. We suppose that there are N periodic traffic data streams. Then, the total PTD can be expressed as Eq. 2.

$$PTD(t) = \sum_{i=1}^{N} P_i(t-T_i-\tau_i) = \sum_{i=1}^{N} (\underbrace{p_{i,1}, p_{i,2}, p_{i,3}, \cdots, p_{i,M_i-1}, p_{M_i}}_{\lambda_i}\underbrace{,\cdots\cdots}_{T_i-\lambda_i}) \tag{2}$$

For PTD, an appropriate transmission scheduling timetable can be designed which shows the beginning and ending time of each stream depending on its parameters, T, λ, M, τ. Overall, this timetable can make the transmission of streams continuous and smooth. Finally, the timetable can make the optimization transmission come true.

3.2 Poisson Distribution Traffic Data

In communication system, the arrival time probability of users is considered obeying Negative Exponential Distribution (NED) [10]. However, in GNSS, the nonperiodic traffic data can be divided into two classes: one is general random traffic, such as satellite payload failure data. At the beginning of research, its arrival time probability can be considered obeying NED also. The other is the traffic data which has the characteristic of long range dependence, such as network topology information, ISLs rebuild data etc. using the self-similarity model to simulate the arrival time probability of this data has higher accuracy.

The general random traffic is very much like traffic of SCS, which has strong bursts. Its time distribution is very non-stationary. In some bursts cases, the network may be congestion. This traffic is usually Short Range Dependence (SRD). For GNSS inter-satellite traffic data, they are mainly onboard software update data, TT&C backup data, nodes failure distribution data etc. We suppose the traffic data model is $PDTD(t)$, then the $PDTD(t)$ is stationary independent increment Poisson distribution. For any $t_0, t \geq 0$, increment $PDTD(t + t_0) - PDTD(t_0)$ is nonnegative, and meets Eq. 3.

$$P\{PDTD(t + t_0) - PDTD(t_0) = k\} = \frac{\lambda_p t^k}{k!} e^{-\lambda_p k}, \quad \lambda_p > 0, k = 0, 1, 2, \cdots \tag{3}$$

3.3 Long Range Dependence Traffic Data

When analyzing the traffic data of GNSS, we will find some traffic data whose arrival time probability is random in certain extent, but it is different from Poisson distribution. Its time distribution has Long Range Dependence (LRD). The LRD process is the random process which has statistical self-similarity in long time scale. The process meets the Eqs. 4 and 5.

$$LRD(t) \stackrel{d}{=} a^{-H} LRD(at), \ \forall t \geq 0, \forall a > 0, 0.5 < H < 1 \tag{4}$$

$$\sum_{k=0}^{\infty} r(k) = \infty \tag{5}$$

where, $\stackrel{d}{=}$ means equal of distribution, H is self-similarity parameter which is also called Hust parameter. H means the degree of self-similarity. When the H increases, the long range dependence is augment. $r(k)$ is $LRD(t)$ autocorrelation function.

3.4 Inter-satellite Traffic Data Modeling

Three basic traffic data has been analyzed above, but they are always hybrid together and constantly changing with time. We suppose that traffic data volume is a function of time, then the GNSS inter-satellite hybrid traffic data model called ISTDM can be expressed by Eq. 6.

$$ISTDM(t) = \alpha_1(t)PTD(t) + \alpha_2(t)PDTD(t) + \alpha_3(t)LRD(t) \tag{6}$$

where, $\alpha_1(t)$, $\alpha_2(t)$, $\alpha_3(t)$ are the functions of time, represent different traffic weight at the moment t. In addition, the expectation always meets in Eq. 7.

$$E[\alpha_1(t)] > E[a_2(t)],\ E[\alpha_1(t)] > E[\alpha_3(t)] \tag{7}$$

From Eqs. 6 and 7, we known that the inter-satellite traffic data distribution character is quite complex, which not only contains PTD, PDTD but also contains LRD. And the weight of traffic is varied with time. However, the good news is, the PTD takes majority of all traffic data, which provides favors to transmission.

4 Simulation, Analysis and Comparison

4.1 ISTDM Simulation

When simulating, distribution character of three kinds of traffic data can be considered mutually independent events. Each traffic data distribution multiplies by its weight which is varied with time. And then, the total distribution character gets.

Simulation parameters set: for periodic traffic data, n data sources are produced, which is different by rate, T, λ, M, τ. For Poisson distribution traffic data, parameter λ_p = 10 kbps. For long range dependence traffic data, k random data sources are produced. The simulation time is 24 h and sample rate is 1,000/h. So 24,000 test results are get as Fig. 1. From Fig. 1a, we can see that the total time distribution of periodic traffic data also has some regularity, whose rate is between 0 and 30 kbps. At certain points, there is some burrs. That was because numbers of streams are sending data at the same moment, which will make speed increased suddenly. This problem further illustrates the importance of appropriate transmission scheduling timetable. Figure 1b shows the Poisson distribution traffic data volume statistic, and Fig. 1c shows long range dependence traffic data volume statistic. We find that the volume of LRD is steadier than that of Poisson distribution. This can reduce the data stream surge of network. Figure 1d is the total ISTDM volume statistic 24 h a day. The expectation is 19.35 kbps. The peak volume is 56.31 kbps, and the Peak to Mean Ratio (PMR) is 2.91.

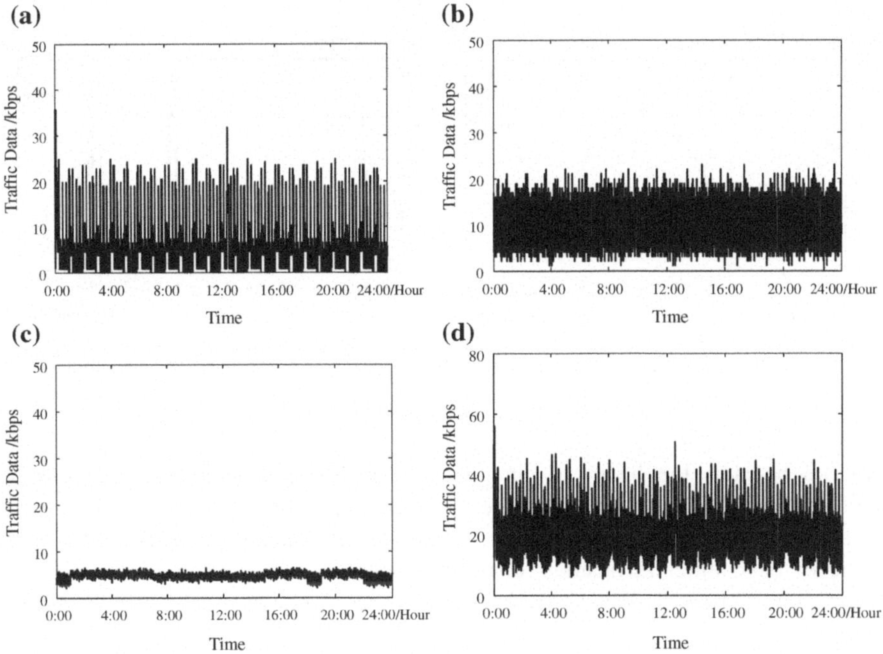

Fig. 1 Three kinds of traffic data volumes and total volume simulation. **a** the volume statistic of PTD; **b** the volume statistic of PDTD; **c** the volume statistic of LRD; **d** the total time distribution of ISTDM

4.2 Traffic Volume Comparison with GCN and SCS

4.2.1 GCN Traffic Data Volume

The Fig. 2a shows the telephone traffic volume statistic in one network of somewhere during a workday. We can lean that the volume is changing with time, and has obvious bi-peak phenomenon: one is during 12:00–13:00 pm; and the other is during 22:00–23:00 pm. Therefore, for GCN, its network steam has distinct regularity. The main factor which affects the volume is the users' work and rest time. In festival and holiday, Fig. 2a telephone traffic volume statistic may change.

4.2.2 SCS Traffic Data Volume

For GSO satellite communication system, the beam of antenna covers the same area of earth surface all along, whose traffic source is invariable and traffic volume statistic is like GNC [11]. However, for non-GSO system, the traffic source is

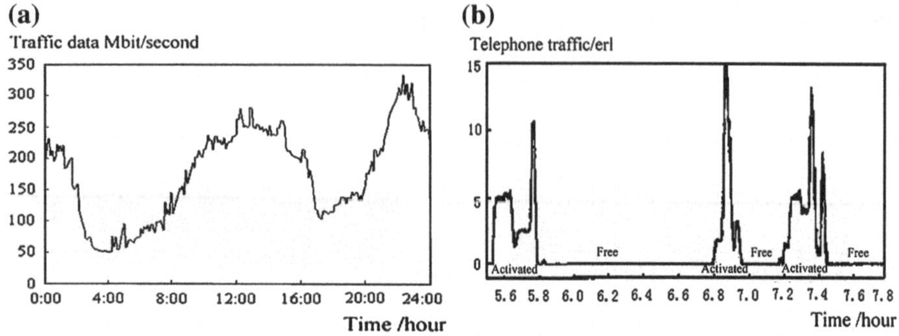

Fig. 2 **a** GCN traffic data volume simulation. **b** LEO satellite traffic data volume simulation

variable because the beam of antenna covers different areas with the mobile satellite. The Fig. 2b shows LEO satellite traffic data volume in one cycle of earth about 2.4 h. As is shown in Fig. 2b, when the LEO satellite crosses over land, traffic of each antenna beam is changing with city and country, mountain and desert, day and night. When the satellite crosses over ocean, there is very little traffic.

4.2.3 Comparison

- The traffic data volume in GCN and SCS is changing with user's behavior and time etc. as is shown in Fig. 2. But that of GNSS is relatively stable, mainly related with transmission strategy, which is shown in Fig. 1d.
- The GCN or SCS achieve communication between user and user, or user and server. Most of the traffic is end-to-end and broadcasting. The non-user segment is being service state, which produces a little management data, routing data etc. but in GNSS, every satellite is both traffic source and sink, participates the maintenance of whole network at the same time. When without regarding to Search and Rescue (SAR) and RDSS, the user of GNSS can even be ignored.
- In GCN and SCS, the integrated transmission of voice, data and video is needed. The total transmission rate is always high, and delay is low. For example when the delay is greater than 0.1 s, the ear will be aware of that. So the voice is demanding for delay. However, in GNSS, its traffic is low rate data, which is lower delay requirement, but higher reliability demand except some special steams such as telecontrol data, which is both low delay and high reliability demand.

5 Recommendations and Concerns

In this section, some recommendations and concerns are given for designing inter-satellite communication protocol in GNSS for example COMPASS. We think some recommendations and concerns below are important items based on the analysis and comparison with GCN and SCS.

- Design an appropriate transmission scheduling timetable for periodic traffic data which says the beginning and ending time of each stream. This timetable can make the transmission of streams continuous and smooth, and can make the optimization transmission come true. It is similar with Vehicle Support Requirements (VSR) which is told in Ref. [1] via pre-establishing a suitable transmission schedule.
- An exclusive ID should be used to mark each traffic data stream based on its requirement of Bit Error Rate (BER), priority level, etc. when designing inter-satellite communication protocol [12].
- A seasonable ISL capability should be designed to achieve a certain reliability demanding based on PMR.
- Because the GNSS constellation network transmits low rate data and the BER need is low, a high accuracy is needed and certain delay can be permitted.
- The GNSS has two operating mode: during peacetime, the system provides precise ephemeris to achieve high-performance positioning and navigation service based on ground uplink stations and TT&C stations; during wartime, the system will work in AutoNav mode which is independent of ground segment. So the GNSS is a "ground-based and space-based" network. In different mode, the traffic data requirement may change. This is a concern when designing inter-satellite communication protocol.

6 Conclusions

In this paper, GNSS inter-satellite traffic data model is researched. The ISTDM is given which consists of three basic models. They are period traffic data model, Poisson distribution traffic data model and long range dependence traffic data model. From the ISTDM simulation result, we know that the inter-satellite traffic data are different in some significant characteristics from communication system. Through traffic data volume comparison with that of GCN and SCS, the characteristics of GNSS inter-satellite traffic data are highlighted. Paper argues that the factors such as traffic data scheduling, data volume, PMR, delay, AutoNav mode etc. should be considered when design GNSS constellation communication protocol.

Acknowledgments Foundation Item: This work is supported by National Natural Science Foundation of China (60972062 and 61032004).

References

1. Luba, O., Boyd, L., & Gower, A, ect. (2005). GPS III system operations concepts. *IEEE. Aerospace and Electronic Systems, 20*, 10–18.
2. Fernandez, A., Sanchez, M., Beck, T., & Amarillo, F. (2010). Future satellite navigation system architecture inter-satellite ranging and orbit determination, *ION ITM*, 872–879.
3. Lin, Y., He, S., Zheng, J., & Chun, H. (2010). Development recommendation of inter-satellite links in GNSS. *Spacecraft Engineering, 19*(6), 1–7.
4. Shuai, P., & Qu, G. (2005). Autonomous navigation techniques of navigation constellation autonomously updated ephemerides. *Journal of Spacecraft TT&C Technology, 24*(5), 8–13.
5. Krunz, M., & Tripathi, S. K. (1997). On the characterization of VBR MPEG steam. *Performance Evaluation Review, 25*, 192–202.
6. Heyman, D. P., & Lakshman, T. V. (1996). Source model for VBR broadcast-vide traffic. *IEEE Transactions Networking, 4*(1), 40–48.
7. Crovella, M. E., & Bestavros, A. (1997). Self-similarity in world wide web traffic: Evidence and possible causes. *IEEE/ACM Transactions Networking, 5*(6), 835–846.
8. Liu, Y., & Yang, X. (2010). GPS Inter-satellite link and simulation of ISL data. *Journal of Time and Frequency, 33*(1), 39–47.
9. Li, Z., Zheng, J., & Zhang, L. (2011). Enlightenment from evolution of the GPS payload for Beidou system. CSCN2011, 16–17.
10. Zhang, G., & Zhang, H. (2001). *Satellite mobile communication system* (pp. 219–222). Beijing: People's post and Telecommunication Publishing House.
11. Feng, S., Jie, X., Li, Y., & Li, G. (2010). Traffic modeling and simulation in low earth orbit constellation system. *Journal of Astronautics, 31*(1), 179–184.
12. Dong, F., Lv, J., Li, H., & Wu, C. (2011). The link layer protocol design for GNSS inter-satellite link based on AOS. ICITIS2011, 02, 997–1000.

Disclaimer

The views expressed in this article are those of the authors and do not reflect the official policy or position of the Chinese Government.

Analysis of Signal Distortion Impact on Code Tracking Bias for High-Order BOC Modulation

Zhengwen Zhang, Yanhong Kou, Jiansheng Liu and Jingtao Sui

Abstract The high-order binary offset carrier (BOC) modulation can make full use of the allocated frequency spectrum resource to improve the compatibility with other GNSS signals in the same band, the anti-multipath capability and the code tracking accuracy. High-order BOC signals have been adopted by Galileo and COMPASS system. The distortions of high-order BOC signals and their impacts on signal performance need to be further explored. Based on the 2nd-order threat model of Evil Waveform (2OS EWF) proposed by ICAO, three types of threat mode—TMA, TMB, and TMC for BOC signals are modeled in this paper. The impacts of various models and parameters on the correlation function are analyzed. Next, the code tracking biases are calculated under different combinations of threat modes, distortion parameters, and correlator spacing. Finally, the optimum selections of correlator spacing for receivers are proposed based on simulation results.

1 Introduction

The high-order BOC modulation yields narrower main-peak in autocorrelation function (ACF) and more deviated split spectrum than BPSK and low-order BOC modulation. It can make full use of the allocated frequency spectrum resource to improve the compatibility with other GNSS signals in the same band, the

Z. Zhang (✉) · Y. Kou · J. Liu · J. Sui
School of Electronic and Information Engineering,
Beihang University, Beijing, People's Republic of China
e-mail: zzw0816@126.com

J. Sun et al. (eds.), *China Satellite Navigation Conference (CSNC) 2012 Proceedings*,
Lecture Notes in Electrical Engineering 160, DOI: 10.1007/978-3-642-29175-3_8,

anti-multipath capability and the code tracking accuracy. High-order BOC modulation signals have been adopted by Galileo and COMPASS system. By far most studies on high-order BOC signals, however, focused on ideal signals without comprehensive consideration of signal distortions induced by payload signal generation, channel propagation, and receiver processing, and their impacts on signal acquisition, tracking, and positioning.

The 2OS EWF proposed by ICAO is established for anomalies of GPS SVN19 satellite in 1993. It is used to study the signal deformations caused by hardware failure or non-ideal characteristics of the satellite signal generation payload. It models three failure modes: digital (TMA), analog (TMB), and combination (TMC), which respectively account for the impacts of non-ideal characteristics of digital components, analog components and the overall generation chain of the satellite payload on navigation signal [1]. The GPS lab in Stanford University has proposed a set of evil waveform monitoring and analysis methods [1, 2], which have been applied to satellite integrity monitoring, WAAS and GBAS [3–5]. In recent years, many studies on signal quality analysis have been carried out in China, including navigation signal quality monitoring and analysis [6, 7], navigation satellite signal quality autonomous monitoring (SQM) [8].

The 2OS EWF model originally aimed at BPSK modulation. References [9, 10] proposed the application of the EWF model in modernized GNSS Signals. This paper analyzes the characteristics of BOC signal distortions based on the EWF model in Sect. 2, and investigates the impacts of different distortion modes and parameters on three unambiguous tracking methods for BOC signals in Sect. 3. The code tracking biases induced by the distortions are simulated in Sect. 4. The correlator spacing selections under signal distortions are summarized in Sect. 5 and conclusions are drawn in Sect. 6.

2 EWF Model for BOC Modulation

The baseband BOC signal can be expressed as [9]:

$$c_{m,n}(t) = c_m(t)s_n(t) \tag{1}$$

where $s_n(t)$ is the primary code sequence with a chip rate of n*1.023 MHz, $c_m(t)$ is the square-wave sub-carrier with a frequency of m*1.023 MHz.

The ideal autocorrelation function (ACF) of a BOC signal is given by:

$$R_{m,n}(\tau) = \int_{-\infty}^{+\infty} c_{m,n}(t)c_{m,n}^*(t-\tau)dt \tag{2}$$

where $c_{m,n}^*(t)$ is the complex conjugate of $c_{m,n}(t)$.

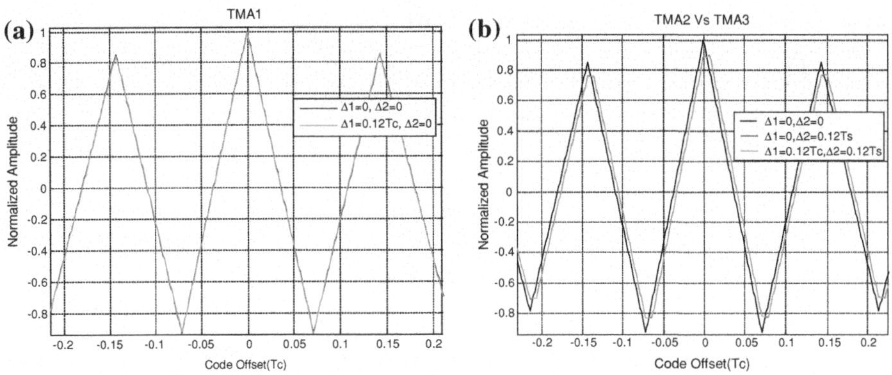

Fig. 1 CCF of BOC(14,2) signal under TMA distortions

2.1 TMA Distortion

For BOC signal, TMA distortion may occur in the primary code sequence generation or the sub-carrier sequence generation. We define the TMA distortion in the primary code sequence as TMA1, in the sub-carrier sequence as TMA2, and in both sequences as TMA3. Thus the three TMA distortion forms for BOC modulation are defined by:

$$c_{m,n,\Delta}(t) = \begin{cases} c_m(t)s_{n,\Delta 1}(t) & \text{TMA1} \\ c_{m,\Delta 2}(t)s_n(t) & \text{TMA2} \\ c_{m,\Delta 2}(t)s_{n,\Delta 1}(t) & \text{TMA3} \end{cases} \tag{3}$$

where $s_{n,\Delta 1}(t)$ is the primary code signal with a falling-edge delay of $\Delta_1 \mathrm{T_c}$, $c_{m,\Delta 2}(t)$ is the sub-carrier signal with a falling-edge delay of $\Delta_2 \mathrm{T_s}$, $\mathrm{T_c}$ is the primary code chip period, and $\mathrm{T_s}$ is a half of the sub-carrier period. In a BOC receiver with the optimal correlator, the cross-correlation function (CCF) between the input baseband signal and local replica baseband signal is given by:

$$R_{TMA}(\tau, \Delta) = R_{m,n,\Delta}(\tau) = \int_{-\infty}^{+\infty} c_{m,n,\Delta}(t) c^*_{m,n}(t-\tau) dt \tag{4}$$

Figure 1 shows the CCF (zoomed in view near the main peak) of BOC(14,2) under TMA distortions. The falling-edge delay for a GPS C/A code is typically within ±12% of a code chip [9]. We use $\Delta_1 = 0.12\mathrm{T_c}$, $\Delta_2 = 0.12\mathrm{T_s}$ in the simulation. It can be seen that TMA1 has few influences on the CCF. The CCF of TMA3 is very close to TMA2, which shows a code phase offset of Δ_2, and a dead-zone atop the correlation peak.

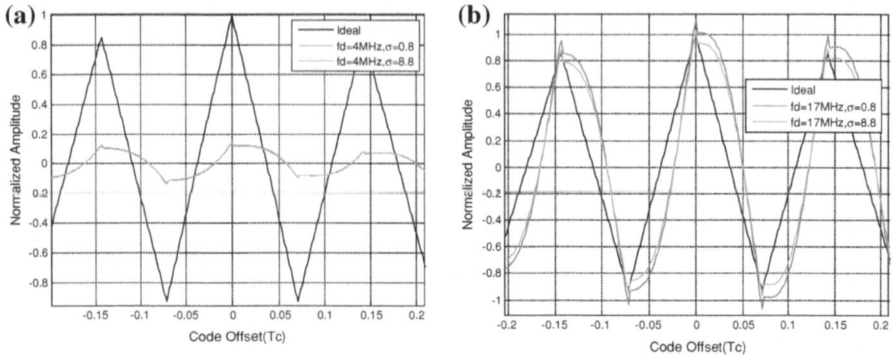

Fig. 2 CCF of BOC(14,2) signal under TMB distortions

2.2 TMB Distortion

Unlike the digital signal generation circuits in satellite payloads, the analog components for BOC signal generation have no much difference with that of the BPSK signal. So the same TMB model of BPSK modulation can be applied to BOC modulation, which can be described by the unit step response of a second-order system [1]

$$e(t) = \begin{cases} 0 \\ 1 - \exp(-\sigma t)\left[\cos 2\pi f_d t + \dfrac{\sigma}{2\pi f_d} \sin 2\pi f_d t\right] \end{cases} \tag{5}$$

where f_d is the damped frequency of oscillation, σ is the damping factor. The ranges of the two TMB parameters proposed by ICAO are: $4 \leq f_d \leq 17$ MHz, $0.8 \leq \sigma \leq 8.8$ Mnepers/s [1]. For the high-order BOC signals discussed in this paper the ranges are: $1 \leq f_d \leq 50$ MHz, $0.1 \leq \sigma \leq 10$ Mnepers/s [9].

TMB distortion can cause correlation peak asymmetry. Figure 2 shows the CCF (zoomed in view near the main peak) of BOC(14,2) under TMB distortions. It can be seen that the amplitude and asymmetry of the CCF are mainly affected by f_d: The higher the f_d is, the closer the CCF is to ideal ACF, while the peak values and the shape of correlation peaks are affected by σ.

2.3 TMC Distortion

TMC distortion is the combination of TMA and TMB distortion [1], and accordingly there are three modes, defined as TMC1, TMC2, and TMC3.

The CCF in a matched receiver can be generalized as follow

$$R_{TMC}(\tau, \sigma, f_d, \Delta) = h_{2nd}(\tau, \sigma, f_d) * R_{m,n,\Delta}(\tau) \tag{6}$$

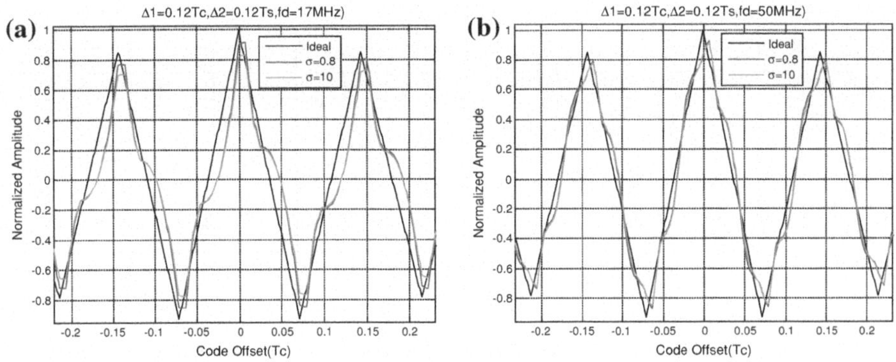

Fig. 3 CCF of BOC(14,2) signal under TMC3 distortions

Obviously, the CCF under TMC1 distortion is similar to TMB, and the CCF between TMC2 and TMC3 distortions are similar. Figure 3 simulates the CCF of BOC(14,2) under TMC3 distortions.

The figure shows that the CCF under TMC distortion is mainly affected by the TMB parameter f_d: as the f_d increases, the symmetry of correlation peaks is improved, while the TMA parameters Δ and the TMB parameter σ affect the shape and peak value of correlation peaks.

3 Unambiguous Tracking Algorithms for BOC EWF Signal

The multi-peaks in the ACF of high-order BOC signals might cause false acquisition or lock onto the side-peaks in the presence of noise and multipath. This section analyzes the impact of BOC EWF signals on three unambiguous tracking algorithms, including BPSK-like (BL), Bump-Jump (BJ) and Dual Estimate (DE).

1. BPSK-like Algorithm (BL)

The BL algorithm regards the two main lobes of the BOC signal as two separate BPSK signals and tracks them independently [11]. Only the parameters Δ_1, f_d and σ can change the CCF in receivers, just like BPSK modulation, while the parameter Δ_2 will not impact the BL algorithm.

2. Bump-Jump Algorithm (BJ)

The BJ algorithm compares the currently locked peak value with adjacent side-peaks to determine the main peak. Two local replica codes named "very early" (VE) and "very late" (VL) with a spacing of ±1 sub-carrier chip are added [12]. The impact of distortion modes and parameters on CCF are the same as the simulation results in Sect. 2 for the BOC signal.

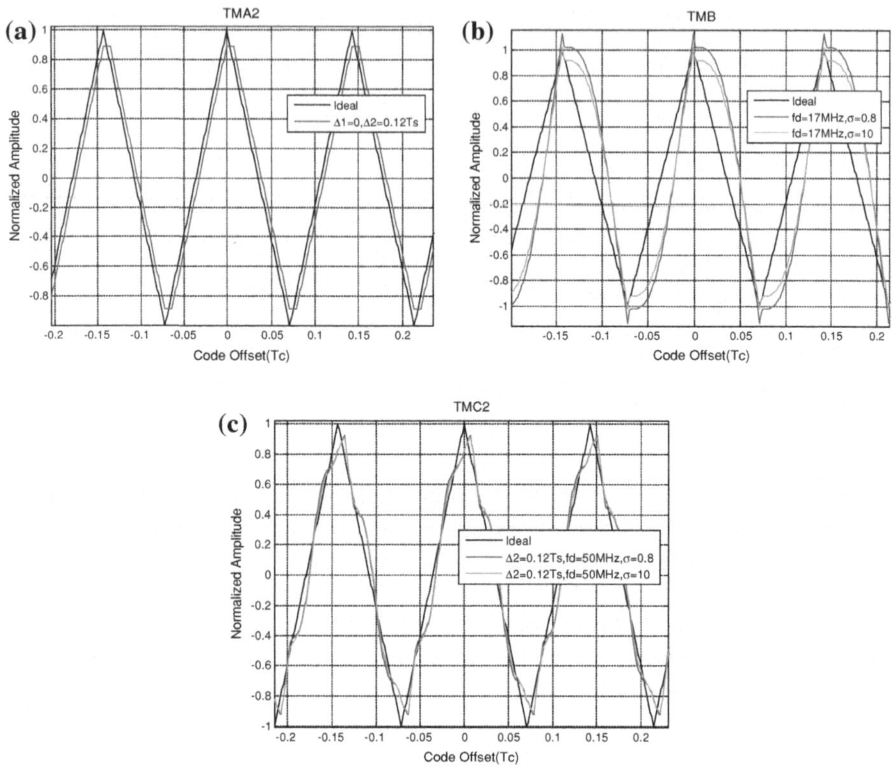

Fig. 4 CCF of BOC (14,2) signal under EWF distortions for DE algorithm

3. Dual Estimate Algorithm (DE)

The DE method considers the primary code and sub-carrier sequence as two non-synchronized components that can be tracked independently using a code locked loop (CLL) and a subcarrier locked loop (SLL) respectively [13]. The CCF for CLL is similar to the BL method, and the ideal CCF for SLL is composed of a set of correlation peaks with the same amplitude and slope, as the black curve shown in Fig. 4. The DE method can reach both the accuracy of the SLL and the robustness of the CLL.

Obviously, the parameters Δ_1, f_d and σ influence the CCF for CLL, while Δ_2, f_d and σ change the CCF for SLL. Figure 4 shows the CCF of BOC(14,2) for SLL under TMA2, TMB and TMC2 distortions respectively. It can be seen that the parameter Δ_2 causes code phase offset and a dead-zone atop the correlation peaks, which is similar to that of BOC modulation under TMA2 distortion in Fig. 1b, while the deformation of the CCF under TMB and TMC2 distortions is similar to that of BOC modulation in Figs. 2b and 3b.

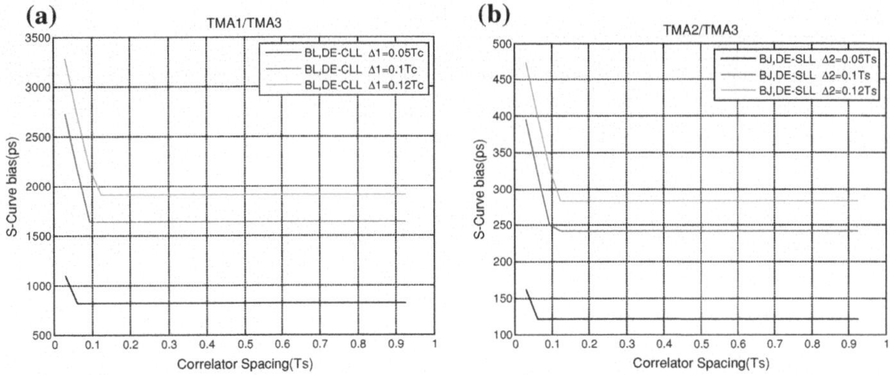

Fig. 5 S-curve bias with TMA distortions

4 Code Tracking Bias of High-Order BOC EWF Signal

Based on the analysis of the EWF model for BOC modulation and the three unambiguous tracking algorithms, it can be inferred that the code chip distortions will lead to code tracking biases depending on tracking algorithms especially for high-order BOC signals. In this section, influences of distortion modes, parameters and tracking algorithms on high-order BOC code tracking biases are compared in terms of S-curve bias. The simulation is performed using the typical noncoherent early-minuse-late power discriminator (NC-EMLP), with the E-L-spacing set to less than $1T_s$.

4.1 Code Tracking Bias with TMA Distortion

Figure 5 simulates the S-curve bias under different TMA distortions and parameters using different algorithms. It can be seen that the code tracking bias under TMA distortion becomes a constant when the E-L-spacing is larger than Δ_1 or Δ_2, and for the same E-L-spacing, the code tracking bias increases with Δ_1 or Δ_2.

4.2 Code Tracking Bias with TMB Distortion

Figure 6 simulates the influence of TMB distortions on code tracking bias with different combinations of the distortion parameters and tracking methods. It can be seen that with large E-L-spacing ($> 0.2T_s$), the code tracking bias increases with f_d decreases. The biases for the BJ and DE algorithms are close together. More simulations show that the code tracking bias increases with increased E-L-spacing

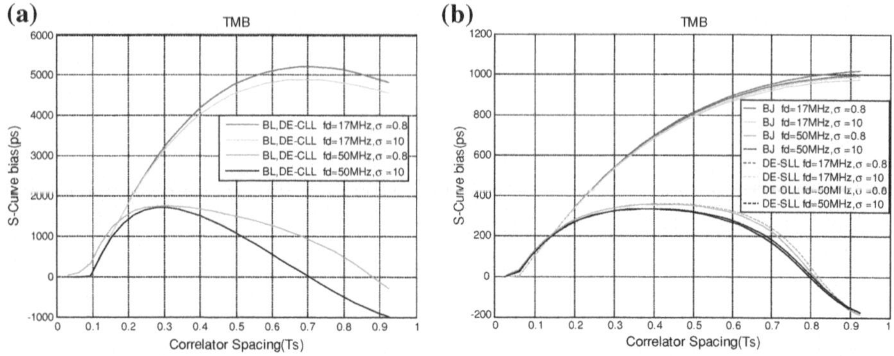

Fig. 6 S-Curve bias with TMB distortions

when f_d is less than $2f_s$ (sub-carrier frequency), otherwise the code tracking bias decreases with increased E-L-spacing when f_d is higher than $2f_s$.

4.3 Code Tracking Bias with TMC Distortion

The analysis of the three unambiguous tracking algorithms implies that: (1) for the BL method, the S-curve bias under TMC3 distortion is the same as TMC1, and TMC2 is equivalent to TMB; (2) for the BJ and DE methods, TMC1 is equivalent to TMB in terms of S-curve bias, and TMC3 is equivalent to TMC2. Figure 7 simulates the S-curve bias under TMC3 distortion with different parameters and algorithms respectively. Figure 7a for the BL method shows that: (1) the code tracking bias is dominated by Δ_1-induced bias when the E-L-spacing is less than $0.4T_s$; (2) with low f_d, the parameter σ will impact the code tracking bias obviously. Figure 7b for the BJ and DE methods shows that: (1) the biases under TMC distortions for the two methods are close together; (2) with low f_d, the code tracking bias is dominated by Δ_2-induced bias when the E-L-spacing is less than $0.4T_s$; (3) with high f_d, the code tracking bias shows combined effects of TMA and TMB distortion.

5 Optimization of E-L-Spacing for High-Order BOC Modulation

Based on the above analysis of code tracking bias for high-order BOC signals with different distortion modes, parameters and tracking methods, we can draw some optimization criteria for E-L-spacing selection in receivers: (1) If there is only TMA distortion, the E-L-spacing should be larger than Δ_1 or Δ_2. (2) If there is only

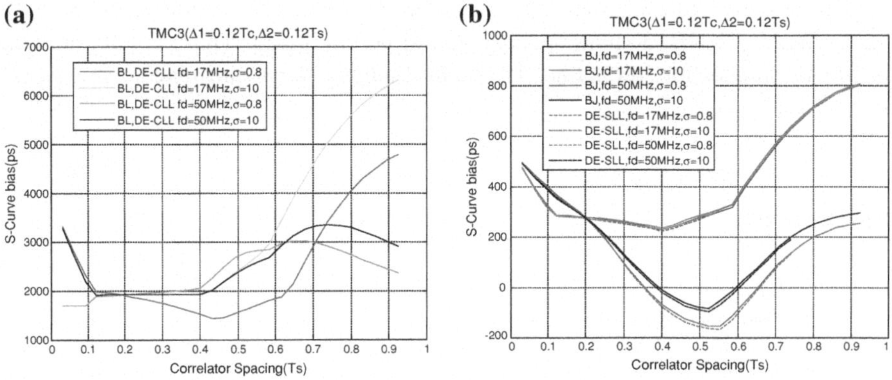

Fig. 7 S-curve bias with TMC3 distortions

TMB distortion, a small E-L-spacing can ensure a high code tracking accuracy. When f_d is higher than $2f_s$, the same code tracking accuracy can be achieved with a larger E-L-spacing. (3) The situation of TMC distortions are relatively complicated due to various combinations of parameters and tracking methods. For the BL method, the selection of E-L-spacing is similar to TMB distortion when $\Delta_1 = 0$, otherwise $[\Delta_1 T_s, 0.4T_s]$ is chosen. For the BJ and DE methods, the selection of E-L-spacing is also similar to TMB distortion when $\Delta_2 = 0$. If $\Delta_2 \neq 0$, the E-L-spacing is chosen in $[\Delta_2 T_s, 0.4T_s]$ with smaller f_d.

6 Conclusions

Based on the analysis of the EWF model for BOC modulation signal, the paper has discussed the impacts of different distortion modes and parameters on correlation functions by grouping the TMA, TMB and TMC distortions into several situations. The simulation shows that TMA in sub-carrier generation will cause the offset and top-chopping of the correlation function and TMB can lead to a wavelike fluctuation of the correlation curve, while TMC combines the two effects.

Then the impacts of signal distortion on three unambiguous tracking methods for high-order BOC signals are analyzed. The simulation of the S-curve biases under different distortion modes, parameters and tracking methods show that TMA introduces constant code tracking bias when the E-L-spacing is larger than the code chip falling-edge delay (Δ_1 or Δ_2), whereas TMB introduces small code tracking bias when the correlator spacing is small enough, or a larger E-L-spacing with f_d higher than $2f_s$.

Finally, the paper summarizes the optimization criteria of E-L-spacing selection for different high-order BOC signal distortions and different unambiguous tracking methods. When TMA dominates the distortion, the correlator spacing larger than the falling-edge delay should be considered. When the distortion is mainly contributed by TMB, a small spacing should be employed. These results provide useful reference for reducing the code tracking bias introduced by signal distortions.

References

1. Phelts, R. E. (2001). *Multicorrelator techniques for robust mitigation of threats to GPS signal quality*. Palo Alto: Stanford University.
2. Mitelman, A. M. (2004). *Signal quality monitoring for GPS augmentation systems*. Palo Alto: Stanford University.
3. Wong, G., Phelts, R. E., Walter, T., & Enge, P. (2010). Characterization of signal deformations for GPS and WAAS satellites. ION GNSS.
4. Shloss, P., Phelts, R. E., Walter, T., & Enge, P. (2002). A simple method of signal quality monitoring for WAAS LNAV/VNAV, ION GPS.
5. Macabiau, C., & Chatre, E. (2000). Signal quality monitoring for protection of GBAS users against evil waveforms. ION GPS.
6. Lu, X., & Zhou, H. (2010). Methods of analysis for GNSS signal quality. *Scientia Sinica Physics, Mechanics and Astronomy, 40*, 528–533
7. Han, L. (2011). *Research and software design verification on GNSS signal quality monitoring*. Beijing: Beihang University.
8. Gan, X., Yang, Y., & Guo, X. (2011). The study of the techniques for multi-Correlators GPS signal monitoring. Assisi Perugia: CSNC.
9. Phelts, R. E., & Akos, D. M. (2006). Effects of signal deformations on modernized GNSS signals. *Journal of Global Positioning Systems, 5*(1–2), 2–10.
10. Phelts, R. E. (2007). Range biases on modernized GNSS codes. *European navigation conference GNSS/TimeNav*
11. Martin, N., Leblond, V., Guillotel, G., & Heiries, V. (2003). BOC (x,y) signal acquisition techniques and performance, ION GPS.
12. Fine, P., & Wilson, W. (1999). Tracking algorithm for GPS offset carrier signals. ION GPS.
13. Hodgart, M. S., Blunt, P. D., & Unwin, M. (2007). The optimal dual estimate solution for robust tracking of binary offset carrier (BOC) modulation. ION GNSS.

Analog Distortion of Wideband Signal in Satellite Navigation Payload

Yibo Chen, Yanhong Kou and Zhengwen Zhang

Abstract The non-ideal components in the navigation signal generation chain of a GNSS satellite payload can induce various signal distortions, which lead to ranging error and impact positioning accuracy. This paper investigates the non-ideal characteristics of analog components in the payload including the up-convertor, filter, HPA, and OMUX. We establish the distortion model of the analog signal generation chain, and propose the simulation methods for phase noise and HPA memory effect, as well as the performance evaluation formulation. Then the distortions of the most demanding AltBOC(15,10) signal after each component are analyzed in the frequency, modulation, and correlation domains. The code phase and carrier phase tracking errors are further assessed. The simulation results quantify the impact of each non-ideal component on the signal performance. The study provides a useful reference for the design and evaluation of navigation signal and payload as well as the specification of satellite signal quality.

1 Introduction

The non-ideal components in the signal generation chain of a GNSS satellite payload can induce signal distortions, which lead to ranging error and impact positioning accuracy. The imperfect analog components such as up-convertor, filter, high power amplifier (HPA) and output multiplexer (OMUX) may cause amplitude and phase distortions, which degrade the correlation peak, introduce

---

Y. Chen · Y. Kou (✉) · Z. Zhang
School of Electronic and Information Engineering, Beihang University, Beijing, People's Republic of China
e-mail: buaakyh@126.com

J. Sun et al. (eds.), *China Satellite Navigation Conference (CSNC) 2012 Proceedings*, Lecture Notes in Electrical Engineering 160, DOI: 10.1007/978-3-642-29175-3_9,

extra noise to the carrier phase, and finally increase the code tracking bias and carrier phase tracking error. These influences are more obvious for wideband navigation signals [1, 2].

Emilie et al. [1] investigated the phase noise introduced by satellite clock and HPA in the payload, the simulation method of the phase noise source, however, was not mentioned. Ref. [3] analyzed the effects of the non-ideal filter and HPA on BPSK and BOC signals. As most other similar studies, the memoryless model of the satellite HPA was employed which ignored the memory effect of the HPA on wideband signals. Refs. [4, and 5] explored the non-linear distortion of a memory HPA for wideband communication signals, which can serve as a reference for the analysis of the wideband navigation signal distortion in payload.

This paper studies the non-ideal characteristics of the analog components in the payload, and establishes analytical distortion model and the numerical simulation method. Taking AltBOC(15,10) as an example, the impacts of these non-ideal components on wideband navigation signals are then analyzed and simulated in the frequency, modulation, and correlation domains. Finally the code tracking bias and carrier phase tracking error induced by the signal distortions are assessed.

2 Analog Distortion Modeling and Simulation Method

2.1 *Payload Model*

The satellite payload for modern GPS system is mainly composed of the navigation signal generation unit and the L band transmission unit, with a few differences in function and modularization between Block IIR-M and Block IIF payloads [6]. The Galileo GIOVE-B payload has a similar architecture of function modules as GPS [7]. Based on the analysis of GPS and Galileo satellite payload composition, a payload model can be summarized for signal distortion analysis, as shown in the block diagram of Fig. 1.

We will focus on the impact of the analog subsystem on wideband signal quality in the following discussion.

2.2 *Up-convertor*

As the frequency source driving navigation signal generation, the performance of the atom frequency standard in payload is essential for the signal quality. The phase noise of the atomic clock is introduced into the local oscillator and then into the radio frequency (RF) carrier through the phase-locked frequency-multiplier circuit, which contaminates the signal phase.

Figure 2 shows the linear phase noise model of the phase locked loop (PLL) frequency synthesizer. There are mainly two noise sources: the noise injected into

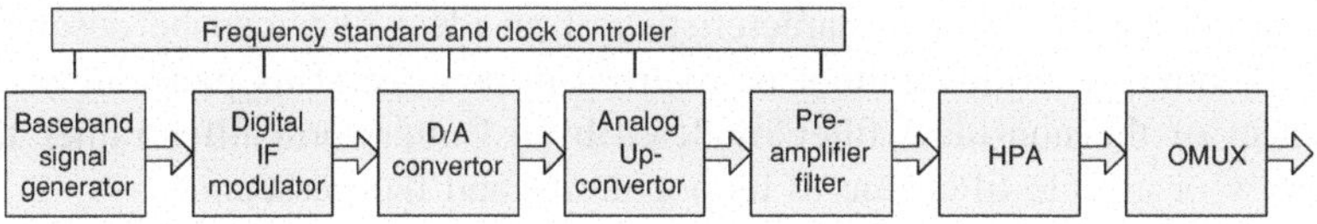

Fig. 1 Model of onboard navigation signal generation payload

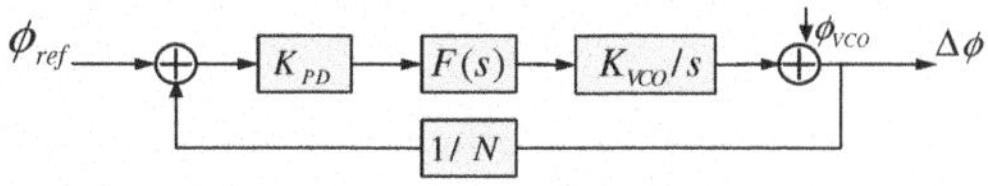

Fig. 2 Phase noise model of PLL synthesizer

the PLL along with the signal, such as the phase noise of the reference clock ϕ_{ref}; the internal noise produced by the PLL components, among which the noise from the voltage-controlled oscillator (VCO) ϕ_{VCO} is the dominating one.

Based on the phase noise model in Fig. 2, the Laplace domain expression of the synthesizer output phase noise can be obtained by [8]

$$\Delta\phi(s) = \phi_{ref}(s) \cdot N \cdot H(s) + \phi_{VCO}(s) \cdot [1 - H(s)] \tag{1}$$

where $H(s)$ is the closed-loop transfer function of the PLL frequency synthesizer. Assuming that the two noise sources are independent of each other, the phase noise power spectral density (PSD) of the synthesizer output can be written as [2]

$$S_{out}(f) = S_{ref}(f) \cdot N^2 \cdot |H(f)|^2 + S_{VCO}(f) \cdot |1 - H(f)|^2 \tag{2}$$

where $S_{ref}(f)$ is the phase noise PSD of the input reference clock, $S_{VCO}(f)$ is the phase noise PSD of the VCO. With the phase noise PSDs of both the input reference clock and the VCO known in most applications, the phase noise spectrum of the PLL output can be calculated by (2). However, when performing numerical simulation and analysis about the impact of the phase noise in the modulation domain and the correlation domain, we need to obtain the phase noise time sequence $\Delta\phi(t)$. The phase noise can be deduced from the power law spectrum model using the method proposed by [9]. We first deduce the time sequences of the input reference clock and the VCO $\phi_{ref}(t)$, $\phi_{VCO}(t)$ from their PSDs respectively, then calculate the frequency spectrum $\Delta\phi(j\omega)$ by (1), and finally obtain the phase noise sequence of the PLL output $\Delta\phi(t)$ by inverse Fourier transform.

2.3 Filter

The filter follows the up-convertor to filter out the harmonic and out-of-band noise/interference before the signal is sent to the HPA. Nevertheless, its band-limiting effect will impair the constant envelope property of the wideband signal.

The amplitude-frequency characteristics of an ideal filter can be described as brick wall filtering, which cannot be realized in practice. This paper investigates the impact of the non-ideal filter modeled by a Butterworth filter rather than a Chebyshev or elliptic filter due to its popularity and flat passband.

2.4 HPA and Multiplexer

To ensure the power efficiency, the HPA usually operates in its saturation zone, which leads to obvious non-linear distortion. In narrow band communication, the non-linear characteristics of the HPA can be described by the AM/AM and AM/PM curves, which mean that the output amplitude and phase change with the input amplitude memorylessly. Besides the non-linear distortion, HPA will present a strong memory effect on a wideband input. Due to this memory effect the input–output relation curves show dynamic characteristics [4].

A common Wiener model is employed here to describe the non-linear characteristics of the memory HPA. Wiener model is a cascade of a linear time invariant (LTI) system and a memoryless non-linear system for which the conventional Saleh model is usually applicable. The AM/AM and AM/PM functions of Saleh model can be expressed as:

$$A(|s(t)|) = \frac{\alpha_A |s(t)|}{1 + \beta_A |s(t)|^2}, \Phi(|s(t)|) = \frac{\alpha_\Phi |s(t)|^2}{1 + \beta_\Phi |s(t)|^2} \quad (3)$$

where $s(t)$ is the output of LTI system, and α_A, β_A, α_Φ, β_Φ are the coefficients.

The multiplexer (OMUX) before the transmission antenna performs the combination of multiple signals, and its filtering effect can help to eliminate the regrown spectrum after HPA and the intermodulation products. Usually the OMUX can be approximated by a set of Butterworth or Chebyshev filters, of which the bandwidth is the transmission bandwidth of a specific signal.

3 Assessment Methods of Analog Distortion

3.1 Constellation Plot and PSD

The modulation constellation will rotate and disperse due to phase noise, filter band-limiting effect and HPA non-linear distortion, which can lead to wrong signal detection and decision as well as phase tracking error in the receiver.

The PSD directly reflects the distorted signal in the frequency domain, where the band-limiting effect and the HPA spectrum regrowth can be observed.

3.2 Correlation Function, S-curve, and Code Tracking Bias

The correlation function is fundamental for navigation signal performance analysis. The normalized cross-correlation function (CCF) of the distorted signal and the locally generated reference signal can be calculated by [10].

$$CCF(\varepsilon) = \frac{\int_0^{T_p} s_{BB-\Pr e \Pr oc}(t) \bullet s^*_{\mathrm{Ref}}(t-\varepsilon)dt}{\sqrt{(\int_0^{T_p} |s_{BB-\Pr e \Pr oc}(t)|^2 dt) \bullet \left(\int_0^{T_p} |s_{\mathrm{Ref}}(t)|^2 dt\right)}} \tag{4}$$

where $s_{BB-\Pr e \Pr oc}(t)$ is the input baseband distorted signal, $s_{\mathrm{Ref}}(t)$ is local reference signal. The signal distortion will cause the correlation loss and the deformation of the correlation function and the phase discriminator function (S-curve) of the code tracking loop (DLL). The zero-crossing point of the S-curve may deviate from the ideal point, which leads to a code phase bias (the so-called lock-point bias) of the DLL.

Taking the coherent DLL as an example, its S-curve can be expressed as [11]

$$SCurve(\varepsilon, \delta) = real[CCF(\varepsilon - \frac{\delta}{2})] - real[CCF(\varepsilon + \frac{\delta}{2})] \tag{5}$$

δ is the early-late-spacing. The lock-point bias $\varepsilon_{bias}(\delta)$ (in chips) is defined as

$$SCurve(\varepsilon_{bias}(\delta), \delta) = 0 \tag{6}$$

The corresponding code tracking bias ξ is (in meters)

$$\xi = c \cdot |\varepsilon_{bias}(\delta)| \cdot T_c \tag{7}$$

where T_c is the code chip length (in seconds), c is the speed of light.

3.3 Carrier Phase Tracking Error

The estimation of the PLL tracking error can be used to assess the impact of the phase noise introduced by the up-convertor and the HPA non-linear distortion on carrier phase tracking [2].

$$\sigma^2 = \int_0^{\infty} S_{N_e}(f) \cdot |1 - H(jf)|^2 df \tag{8}$$

where $S_{N_e}(f)$ is the single side-band PSD of the PLL input phase noise, H is the transfer function of the PLL. A traditional three-order PLL is used to estimate the phase error in this study.

4 Simulation Results Analysis

4.1 Ideal AltBOC(15,10) Signal

The impact of analog distortions on the performance of wideband navigation signals is simulated taking the most demanding AltBOC(15,10) signal as an example. As a modified version of BOC, AltBOC(15,10) can be expressed as an 8PSK signal [12]. Figure 3 shows the modulation constellation and normalized PSD of the constant envelope AltBOC signal.

4.2 Impact of Up-convertor

Taking the phase noise performance of the Clock Monitoring and Control Unit (CMCU) and the VCO in Galileo satellite payload as an example. [2], we can obtain the phase noise time sequence of the PLL output using the method afore-cited. To test its applicability, the PSD of the deduced time sequence is recalculated and compared with the theoretical PSD by (2). Figure 4 shows that the trends of the two PSDs are reasonably consistent with each other.

The carrier phase shows a fluctuation after incorporating the phase noise of the up-convertor into the AltBOC signal. As shown in Fig. 5, the degradation of the constant envelope is not obvious with the dispersion of the constellation points confined in a small range. Accordingly the PSD shows no obvious change.

4.3 Impact of Filter, HPA, and OMUX

The modules after the up-convertor can be regarded as the combination of a pre-amplifier filter, a HPA and a post-filter. The pre-filter and post-filter are modeled by a 6 order Butterworth filter with a 92 MHz bandwidth which is in accordance with the Galileo E5 transmission bandwidth. The HPA is simulated by the Wiener memory model. The transfer function of the LTI system is $H(z) = 0.7692 + 0.1538z^{-1} + 0.0769z^{-2}$, and the parameters of the memoryless non-linear system are $\alpha_A = 2.1587$, $\beta_A = 1.1517$, $\alpha_\phi = 4.0033$, $\beta_\phi = 9.104$ [5]. As shown in Fig. 6, the AM/AM curve of the memory HPA presents a cluster of curves instead of a single curve. With the constant envelope AltBOC signal as the input, the

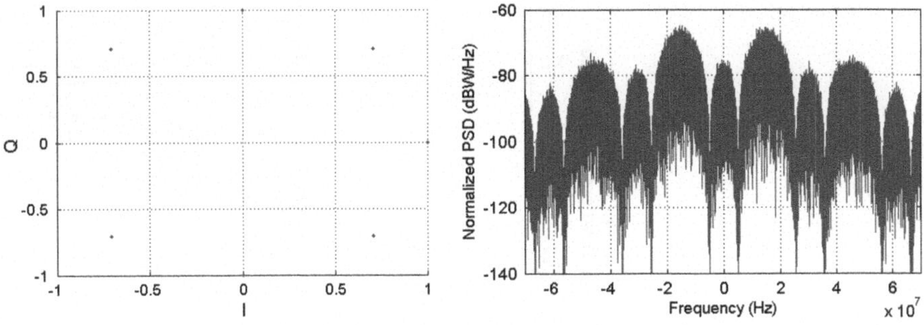

Fig. 3 Constellation and PSD plots of AltBOC signal

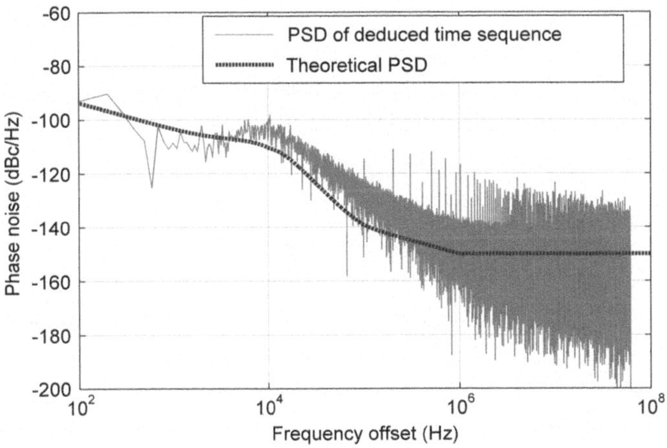

Fig. 4 Comparison between deduced and theoretical PSD of PLL output phase noise

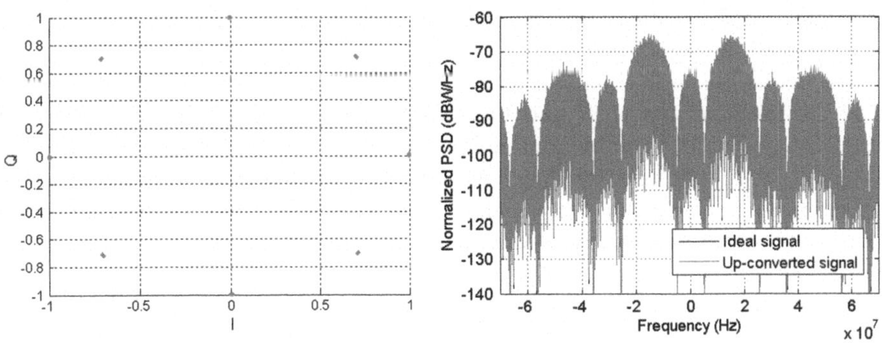

Fig. 5 Constellation and PSD plots after up-conversion

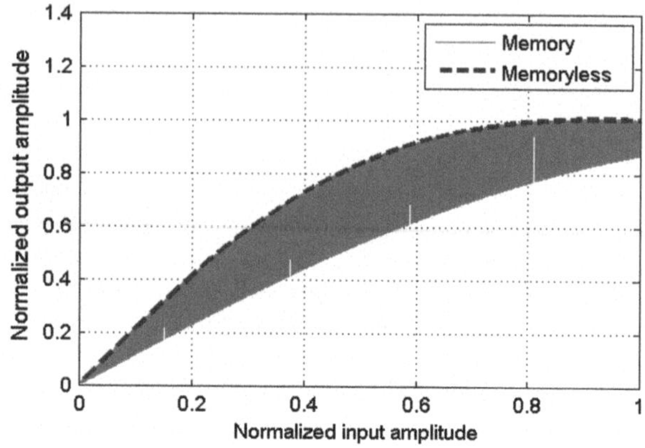

Fig. 6 AM/AM curve

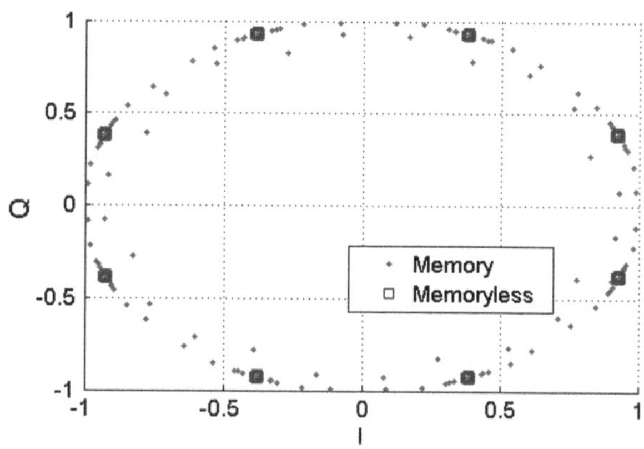

Fig. 7 Constellation of AltBOC signal after HPA

constellation of the HPA output is shown in Fig. 7. For the memoryless HPA, only fixed phase deflection occurs, whereas the memory HPA additionally brings the dispersion of the signal points.

The evolutions of the modulation-domain and frequency-domain characteristics after the prefilter, the HPA and the OMUX are shown in Figs. 8, 9, and 10 respectively. The band-limiting effect can be seen in the PSD evolution. The out-of-band spectrum regrowth induced by the HPA non-linear distortion may cause interferences between adjacent signals or systems. The OMUX filter, however, filters out the regrown spectrum, and finally regulates the output signal to meet the transmission bandwidth requirement.

The ringing effect of the filter can significantly impair the signal constant envelope property. Due to the saturation cut-off effect of the AM/AM response, the fluctuation of the HPA output signal envelope in Fig. 9 becomes smaller than in Fig. 8, whereas the phase shows obvious rotation due to the AM/PM characteristic. Both the envelope fluctuation and the phase rotation are present in the constellation plot of the final output.

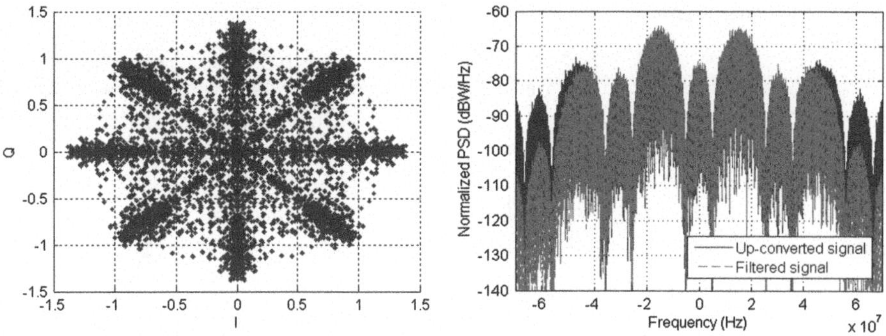

Fig. 8 Constellation and PSD plots after pre-filter

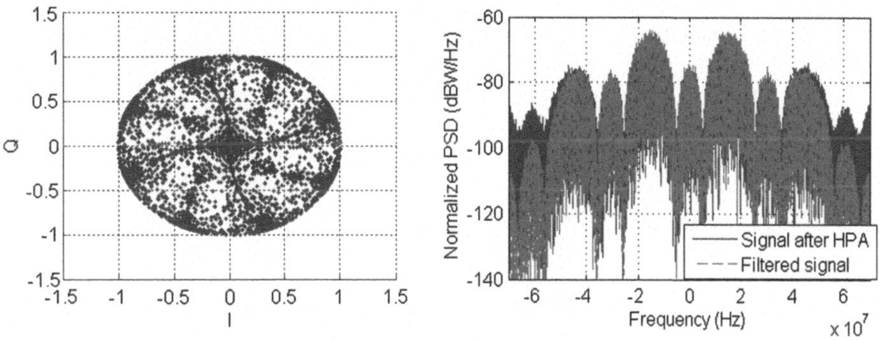

Fig. 9 Constellation and PSD plots after HPA

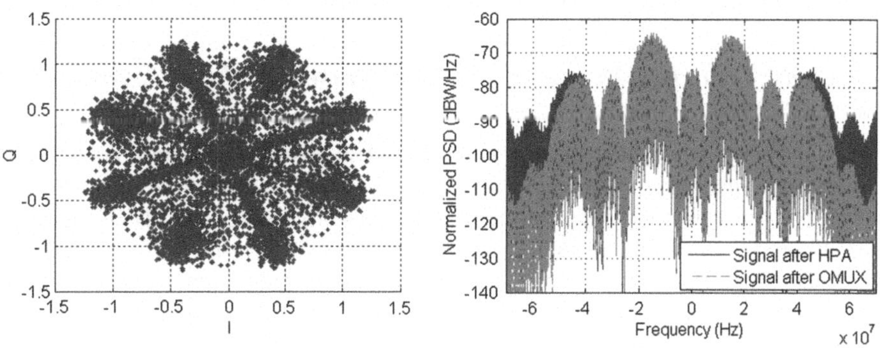

Fig. 10 Constellation and PSD plots after multiplexer

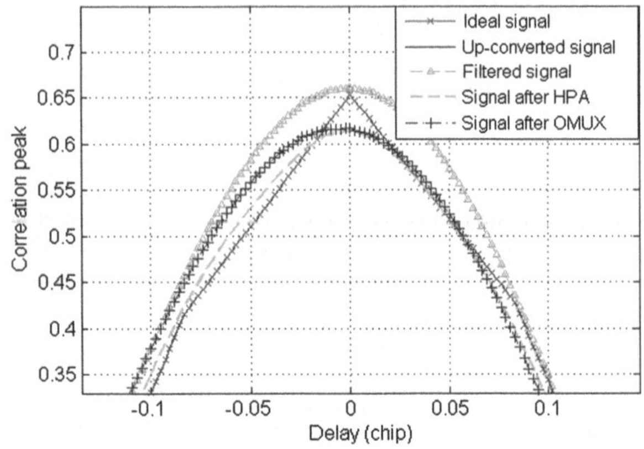

Fig. 11 CCF after each component

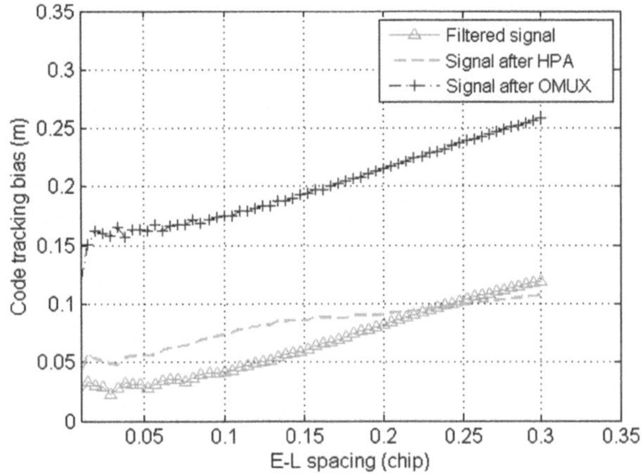

Fig. 12 Code tracking bias after each component

4.4 Code Tracking Bias and Carrier Phase Tracking Error

The CCF for the pilot channel after each non-ideal analog component is analyzed and the main peak is shown in Fig. 11. It can be seen that the phase distortion in the up-convertor has very few influences on the correlation peak. Then the non-ideal filter smoothes the correlation peak, and deforms the shape. Next, the HPA non-linear distortion further degrades the correlation shape, corresponding to the spectrum regrowth. Finally the OMUX modifies the shape and location of the correlation peak.

Accordingly, for the code tracking bias induced by the S-curve distortion, the up-convertor has few influences on the S-curve, whereas the pre-filter, the HPA, and the OMUX leads to increased code tracking bias, as shown in Fig. 12.

The up-convertor and the HPA both introduce phase noise to the receiver carrier loop. The payload output phase noise is the combination of the two parts [2].

Assuming that the loop noise bandwidth is 10 Hz, the RMS carrier phase error of the up-convertor output is estimated to be 0.03° by (8), while the HPA introduced phase error is 3.84° RMS. The result shows that the HPA contributes much more phase noise than the up-convertor. Therefore, appropriate pre-distortion compensations are necessary to ensure fine receiving performance for wideband navigation signals.

5 Conclusion

Based on the non-ideal characteristics of the analog components in GNSS satellite payload, the model for analog signal distortion analysis has been established, and their impacts on wideband navigation signal performances have been assessed taking the AltBOC(15,10) signal as an example. The results show that:

(1) The phase noise in the up-convertor causes a small phase fluctuation;
(2) The non-vertical cut-off effect and non-linear phase response of the pre-amplifier filter with limited bandwidth leads to non-constant envelope, deformed correlation peak, and increased code tracking bias;
(3) The non-linear distortion of the memory HPA introduces extra phase noise to the wideband signal, which degrades the code tracking and carrier phase tracking performance;
(4) The filter in the OMUX can mitigate the regrown spectrum after HPA, on the other hand it may also increase the code tracking bias;
(5) The impact of the payload analog component imperfections on wideband navigation signal performance cannot be ignored in high-end applications.

References

1. Rebeyrol, E., Macabiau, C., Julien, O., et al. (2006). Signal distortion at GNSS payload level. ION GNSS.
2. Rebeyrol, E., Macabiau, C., Ries, L., et al. (2006). Phase noise in GNSS transmission/reception system. ION NTM.
3. Yang, J., Yang, Y., Li, X., & Liu, L. (2010). Design methodology and performance analysis for a novel constant envelope modulation signal of GNSS. ICACIA.
4. Li, M. (2009). *Study on behavioral modeling and predistortion techniques for power amplifiers in broadband communications*. Dissertation, University of Electronic Science and Technology of China.
5. Li, A., Xiao, S., & Zhang, E. (2009). A method of adaptive volterra predistortion based on the discrete wavelet transform. *Signal Processing, 25*(1), 40–43.
6. Rajan, J. A., & Irvine, J. (2005). GPS IIR-M and IIF: Payolad modernization. ION NTM.
7. Benedicto, J., Dinwiddy, S. E., Gatti, G., Lucas, R., & Lugert, M. (2000). *GALILEO: Satellite system design and technology developments*. European Space Agency.
8. Kim, Y. W., & Yu, J. D. (2008). Phase noise model of single loop frequency synthesizer. *IEEE Transactions on Broadcasting, 54*(1), 112–119.

9. Kou, Y., & Zhang, Q. (2004). A method for simulating the crystal oscillator errors in GPS receiver. *Journal of Electronics and Information Technology, 26*(8), 1319–1324.
10. Soellner, M., Kurzhals, C., & Hechenblaikner, G., et al. (2008). GNSS offline signal quality assessment. ION GNSS.
11. Margaria, D. (2007). Galileo AltBOC receivers. *Dissertation, Politecnico di Torino.*
12. Kaplan, E. D., & Hegarty, C. J. (2006). *Understanding GPS principles and applications* (2nd ed.). Washington: Artech House.

The Analysis of Equivalent Power Flux-Density of COMPASS System in New RNSS Band

Chunhai Zhang, Hongtao Li, Wenjun Zhao, Xiaodong Zhao and Siliang Wu

Abstract The 1164–1215 MHz has been allocated to Radionavigation satellite service (RNSS) on a primary. In accordance with the Radio Regulation (RR), the band 960–1215 MHz is also allocated on a primary basis to the aeronautical radionavigation service (ARNS). WRC-03 determined that protection of the ARNS from RNSS can be achieved if the value of the EPFD produced by all the space stations of all RNSS systems in the band 1164–1215 MHz does not exceed the level of -121.5 dBW/m^2/MHz. In this paper, the methodology which used to calculate Spectral Shaping Factor (SAF), Power Flux-density (PFD), Equivalent Power Flux-density (EPFD) is addressed deeply, and the results of COMPASS system is provided.

1 Introduction

The 1164–1215 MHz has been allocated to Radionavigation satellite service (RNSS) on a primary on the World Radio Conference 2000 (WRC-2000). In accordance with the Radio Regulations, the band 960–1215 MHz is also allocated on a primary basis to the aeronautical radionavigation service (ARNS), and the ARNS is safty service. So the RNSS should not produce harmful interference to ARNS. The WRC-03 determined that protection of the ARNS from RNSS can be

C. Zhang (✉) · S. Wu
School of Information and Electronics, Beijing Institute of Technology, 10081 Beijing, China
e-mail: stevenznudt@sina.com, stephen_ch_zhang@sina.com

C. Zhang · H. Li · W. Zhao · X. Zhao
Beijing Global Information Center of Application and Exploitation, 100094 Beijing, China

J. Sun et al. (eds.), *China Satellite Navigation Conference (CSNC) 2012 Proceedings*, Lecture Notes in Electrical Engineering 160, DOI: 10.1007/978-3-642-29175-3_10,

achieved if the value of the equivalent power flux-density (EPFD) produced by all the space stations of all RNSS (space-to-Earth) systems in the band 1164–1215 MHz does not exceed the level of -121.5 dBW/m^2 in any 1 MHz band, and adopted Resolution 609 [1] (WRC-03) in order to ensure that this level is not exceeded.

The Resolution 609 consultation meeting was being held every year since 2003. The COMPASS system has taken part in the meeting 7 times since 2004, and input document which represents the technical characters of COMPASS system has been provided in order to ensure the validity and priority of filling for COMPASS system.

2 Methodology

The ITU-R M.1642 Recommendation [2] has been adopted, which gives a methodology and the reference antenna characteristics for assessing the maximum aggregate equivalent power flux-density level produced at the input of a station of the ARNS (altitude 40,000 feet or 12192 m) by all RNSS systems operating in any portion of the 1164–1215 MHz band.

The relationship between the parameters (including SAF, PFD and EPFD) which need to be provided to 609 consultation meeting and the parameters which used to describe system (including satellite constellation parameters, signal characters) is provided in Fig. 1. We can calculate SAF with the signal configuration, number of signal in the 1164–1215 MHz band, modulation types/central frequency/transmit power of every signal. The maximum EIRP in any 1 MHz can also get from this process. The SAF is only dependent on frequency (f), and independent on space position (longitude and latitude). We can calculate the EPFD from constellation parameters, satellite orbit, signal configuration, maximum EIRP/1 MHz, ARNS antenna pattern and longitude/latitude of the ARNS receive station. The EPFD is different at different position (longitude/latitude/altitude) on the earth. The EPFD results should be provided to the 609 consultation meeting in a Microsoft Excel table, and the Fig. 8 shows the distribution of EPFD on the surface of map.

Power Flux-density (PFD) is a quantity that describes the power density at which energy is transferred by electromagnetic radiation through a real or virtual surface at per unit surface area (m^2). The power flux-density is calculated using the following formula:

$$pfd = 10\log_{10}\left(\frac{10^{\frac{P}{10}}G_t}{4\pi d^2}\right) \quad (1)$$

Equivalent Power Flux-density (EPFD) is defined as the sum of the power flux-densities produced at a receive station, by all the transmit stations within all RNSS systems, taking into account the off-axis discrimination of a reference receiving

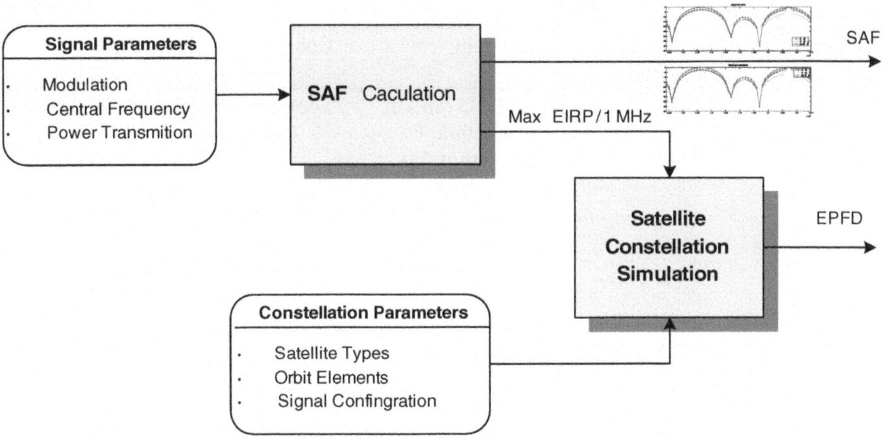

Fig. 1 Diagram of input and output of Res609

antenna assumed to be pointing in its nominal direction. The equivalent power flux-density is calculated using the following formula:

$$epfd = 10\log_{10}\left[\sum_{i=1}^{N_a} 10^{\frac{P_i}{10}} \frac{G_t(\theta_i)}{4\pi d_i^2} \frac{G_r(\varphi_i)}{G_{r.\max}}\right] \quad (2)$$

where N_a is the number of visible satellites from RNSS system to ARNS station, i is the index of satellites, P_i is the RF power at the input of the antenna of the i-th satellite, θ_i is the off-axis angle between the boresight of the transmit station and the direction of the receive station, d_i is the distance between the i-th transmit station and the ARNS receive station, φ_i is the off-axis angle between the boresight of the ARNS receive station and the direction of the i-th transmit station in the RNSS constellation, $G_r(\varphi_i)$ is the gain of receive antenna in the direction of the i-th transmit station, $G_{r,\max}$ is the maximum gain of the receive station antenna.

The EPFD is related to constellation parameters, orbit parameters and signal configuration, the EPFD results of RNSS systems can be obtained by simulation method.

Aggregate Equivalent Power Flux-density (AEPFD) is the summation of EPFD results of some RNSS systems. According with the definition, the EPFD is related with the space position of receive station. The worst case should be considered when the interference from RNSS system to ARNS is calculated. The EPFD of RNSS systems with different signal characteristics should be factored by the appropriate spectral shaping factor (SAF) before summation with other lists or tables.

Spectral Shaping Factor (SAF) is a vector related to frequency, and represents the normalized envelope of all signals in a RNSS system.

Table 1 The constellation of COMPASS system

Satellite types	Number of satellite	Orbit parameters	Constellation characters
GSO	5	Altitude: 35786 km, inclination: 0°, eccentricity: 0	Longitude: 58.75°E, 80°E, 110.5°E, 140°E, 160°E
IGSO	3	Altitude: 35786 km, inclination: 55°, eccentricity: 0	Number of orbit planes: 3, longitude crossing: 118°E
MEO	27	Altitude: 21500 km, inclination: 55°, eccentricity: 0	Number of orbit planes: 3, constellation: Walker 24/3/1, Plus 3 spares

3 Simulation Assumptions

The simulation assumptions include signal characters, constellation parameters of COMPASS system and the antenna pattern of ARNS receive station.

3.1 COMPASS System Characteristics

Two signals are transmitted by COMPASS system, one is a BPSK-R(2) signal at central frequency 1207.42 MHz, the other is a AltBOC (15,10) signal at central frequency 1191.795 MHz. As shown in the Table 1, the constellation of COMPASS system is consisted of 5 GSO, 3 IGSO and 27 MEO. The detail parameters of the constellation of COMPASS can be found in attachment 3 of the Report of the eight Resolution 609 consultation meeting to the ITU Radiocommunication Bureau [3] by the chairman.

3.2 ARNS Station Antenna Parameters

The Fig. 2 provides the antenna gain for elevation values between −90 and 90° of ARNS receive station. The pattern is designed to have a higher gain at elevation angle −30 to 3° in order to receive the signal from TACAN/DEM on the surface of the Earth. The maximum gain is about −1 dB, which appears between −3.5° and 3°. The gain is about −17 dB at +90/−90 degrees elevation angle. The maximum interference will take place when the satellite of RNSS system appears at low elevation angle, not the zenith (where the elevation angle is 90°), according with the antenna pattern of ARNS receiver.

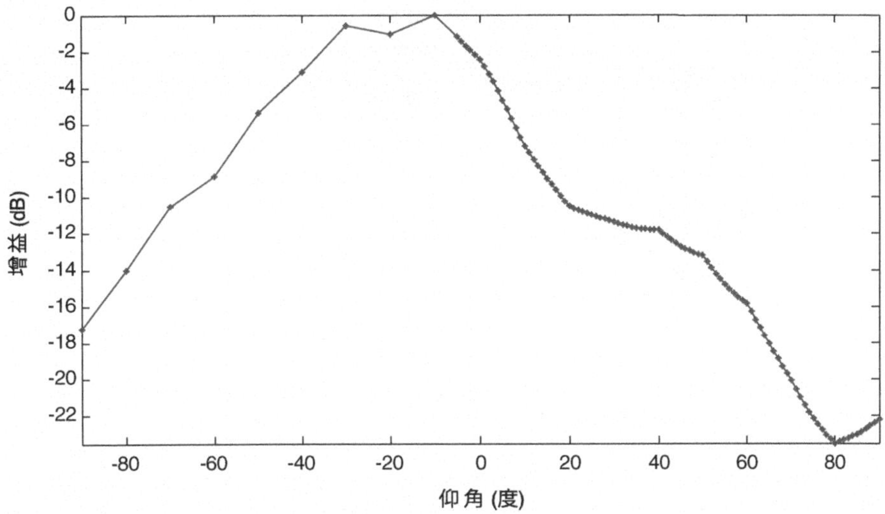

Fig. 2 Antenna gain for ARNS receiver

3.3 Simulation Period and Time Increment

In order to obtain accurate results, the simulation period should be as long as possible, and the time increment should be as short as possible. But on the other hand the total simulation time should be reasonable.

In order to cover the ergodicity of the constellation movement, the simulation period should be multiple of recursive period of the specific constellation. The period of GSO and IGSO constellation is 24 h, and the minimum recursive period of MEO constellation is 8 days (8 × 24 h).

To obtain the accurate required, it is necessary to ensure a sufficient number of EPFD measurements when non-GSP RNSS spacecraft fall into the ARNS antenna pattern of maximum gain visible to a space station. It is important to have 5 hits when the constellation movement is 1°, as shown on Fig. 3. As the COMPASS system is concerned, the time increment should be less than 48 s for IGSO constellation and less than 25 s for MEO constellation.

4 Simulation Results and Discussion

4.1 SAF

The calculation of SAF should consider about the different signal characteristics and EIRP of every signal in RNSS system. The normalized PSD of BPSK-R(2) and AltBOC (15,10) are shown on Fig. 4.

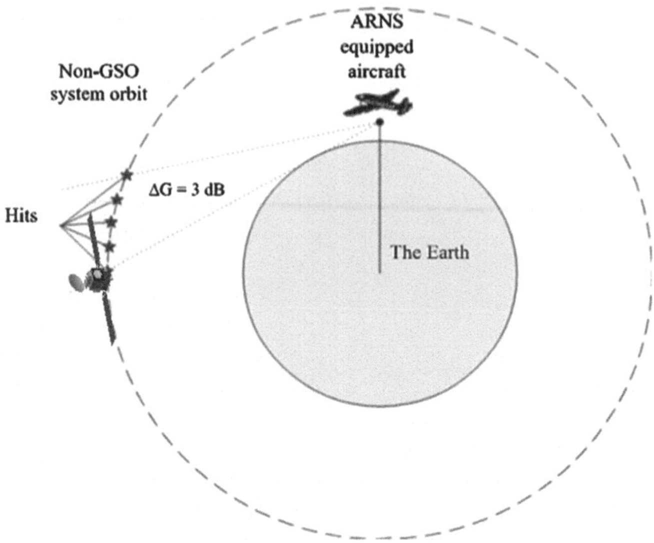

Fig. 3 The relative relationship of satellite and ARNS receiver

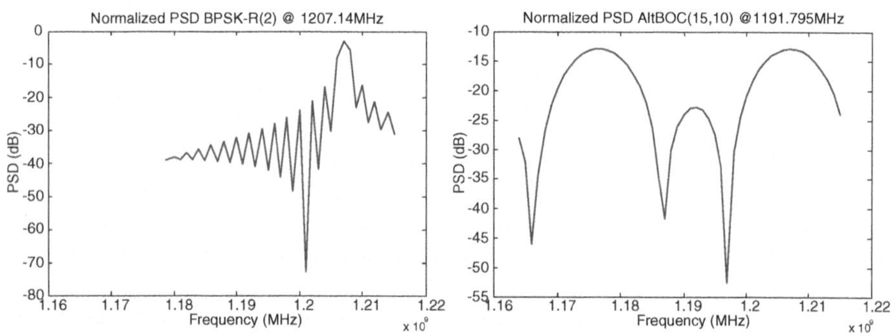

Fig. 4 Normalized power spectrum density

The normalize Power Spectral Density (PSD) represents the distribution of power on different frequency (1 MHz). The normalization is relative to the total power. It can be calculated with following formula:

$$psd_{norm}(k) = \frac{psd(k)}{\sum_k psd(k)} \tag{3}$$

The combination of normalized PSD of different signal should be factored by the power level. The SAF of COMPASS system is provided in Fig. 5.

The SAF needs to be normalized too, but it is normalized relative to the maximum EIRP at any 1 MHz

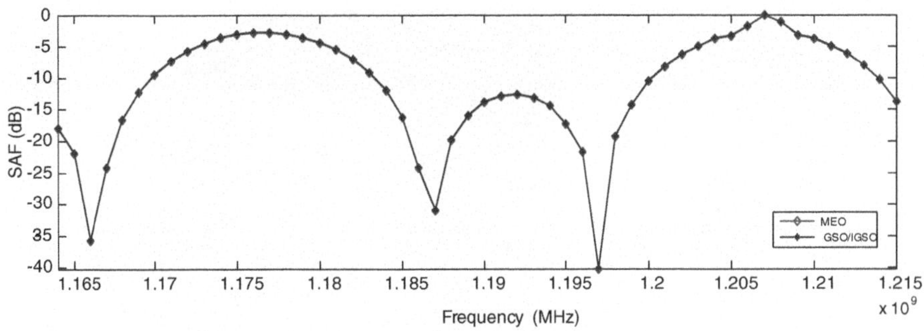

Fig. 5 SAF result

4.2 EPFD

The EPFD should be calculated according to the types of satellites, where:

1) Each GSO satellite will provide a table of maximum EPFD versus latitude and longitude, and a SAF vector. If there are more than one GSO satellites in a system, more EPFD table will be provided.
2) For non-GSO satellites, the non-GSO not having a geosynchronous period will provide a maximum EPFD versus latitude (applicable at all longitudes) and a SAF vector, the non-GSO having a geosynchronous period will provide a list of maximum EPFD versus latitude and longitude, and a SAF vector.

The EPFD results of GSO, IGSO and MEO of COMPASS system is calculated as follow, they are discussed respectively.

4.2.1 GSO

The EPFD of different GSO is only a shift along the longitude if they have the same signal configuration. The EPFD of 80E satellite of COMPASS system is provided in Fig. 6. It is shown that the maximum interference takes place at low elevation angle, not the zenith (where the elevation angle is 90°). The elevation angle is negative inside the red circle, and the EPFD is negative infinite when the elevation angle is less than the mask angle in the central area of the circle (shown as blue).

4.2.2 IGSO

The EPFD results of IGSO constellation is shown as Fig. 7. The crossing longitude of IGSO constellation is 118E, and track of sub satellite point is like an '8' shape. The distribution of EPFD is like a '∞' shape, and is never becoming uniform horizontal stripes like the MEO with the extension of simulation time.

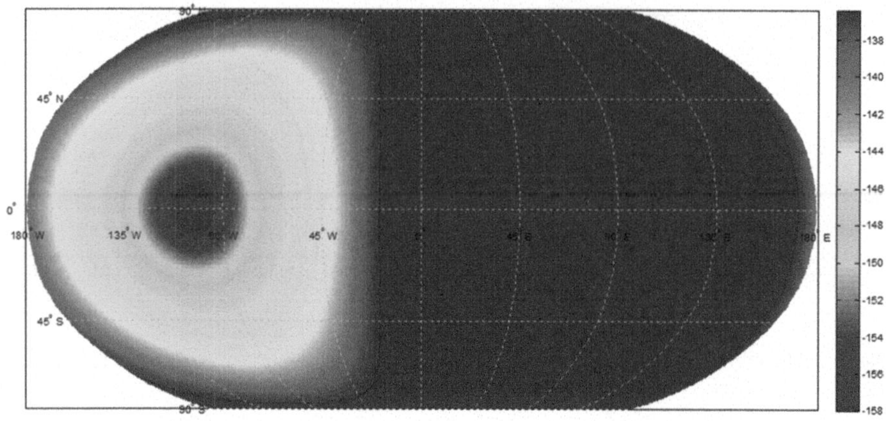

Fig. 6 EPFD values of 80E satellite

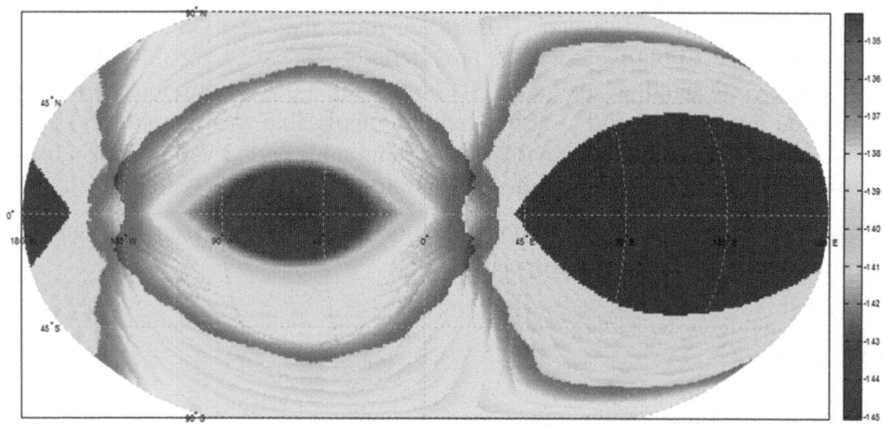

Fig. 7 EPFD values of IGSO constellation

4.2.3 MEO

The EPFD result of MEO is shown as Fig. 8. The uniform horizontal stripes are very obvious in this figure.

For the non-GSO constellation not having a geosynchronous period, the aforementioned effect will result in the maximum EPFD at any longitude becoming the maximum EPFD at all longitudes if the simulation were extended to cover all possible states of the constellation. That is to say the EPFD would spread horizontally to become uniform horizontal strips not dependent on latitude.

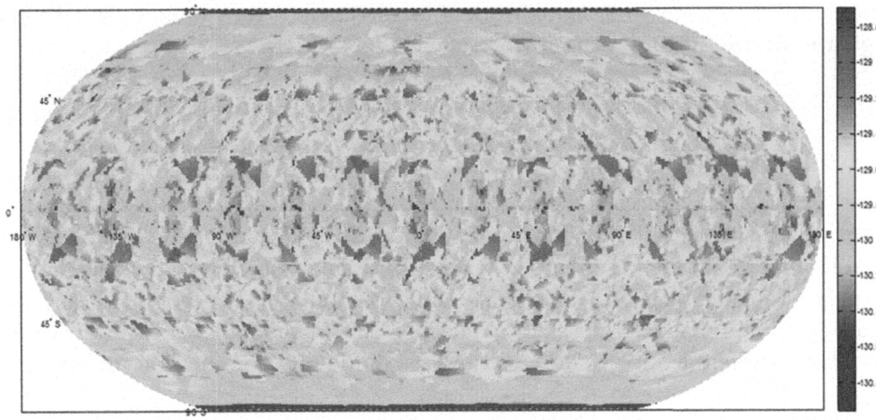

Fig. 8 EPFD values of MEO constellation

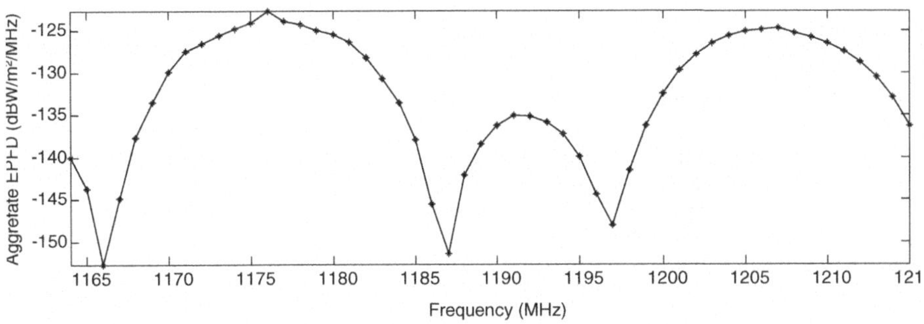

Fig. 9 Aggregate EPFD values of the 8th 609 consultation meeting

4.3 AEPFD

Calculation of the AEPFD by all space station of RNSS in 1164–1215 MHz band need to combinate the EPFD factored with SAF of all systems listed in attachment 1 of Report of the eight Resolution 609 consultation meeting to the ITU Radiocommunication Bureau [3] by the chairman. The AEPFD results of the 8th Resolution 609 Consultation Meeting, which was hold in September 2011, is provided in Fig. 9. Detailed characteristics of these systems, which were used for the aggregate computation, are available on the Resolution 609 Forum page within the ITU web site [4].

The maximum EPFD of all RNSS satellites is determined to be no greater than -122.64 dBW/m^2/MHz, i.e. 1.14 dB below the Resolution 609 limit of -121.5 dB(W/m2/MHz). The maximum AEPFD value is at 1176 MHz, which is introduced by the RNSS systems work at this frequency, for example the GPS L5 signal is BPSK-R(10) at central frequency 1176.45 MHz.

5 Conclusion

This paper analyses the parameter definition and calculation methodology of Resolution 609 on the baseline of COMPASS signal characters and constellation. Simulation results of GSO, IGSO and MEO constellation are provided, and the space distribution pattern are analyzed deeply. The methodology and results could be taken as reference for COMPASS in Resolution 609 consultation meeting.

Acknowledgments We would like to acknowledge the contribution of Dr. Zhijian LIU, Mrs Xia GE and Mrs Xinyan Zhuang. The results of COMPASS system are based on assumptions. It is provided only for technology discussion, not represents any official standpoints.

References

1. ITU. (2008). Radio Regulations (Edition of 2008).
2. ITU-R. (2007). Methodology for assessing the maximum aggregate equivalent power flux-density at an aeronautical radionavigation service station from all radionavigation-satellite service systems operating in the 1164–1215 MHz band (p. 15).
3. ITU. (2011). Report of the Eighth Resolution 609(Rev WRC-07) Consultation meeting to the ITU Radiocommunication Bureau (p. 7).
4. http://www.itu.int/ITU-R/space/res609/.

Analysis of the Effect of Carrier's Spurious Signals on the Code-Tracking Accuracy

Jiaxing Liu, Haibin Chu, Jinjun Zheng, Zhigang Han and Jun Chi

Abstract Carrier's spurious signals are the unideal frequency fractions containing in a single-frequency carrier signal. Spur-to-carrier rejection is a kernel index to characterize the unideal quality of a carrier signal. This paper presents a method for analyzing the effect of carrier's spurious signals on the code-tracking error. First, the time-domain and frequency-domain expressions of the interference signals caused by carrier's spurious signals are deduced strictly, based on code-tracking loops employing coherent early-late processing (CELP) and noncoherent early-late processing (NELP) respectively. In order to describe the degradation of code-tracking accuracy caused by carrier's spurious signals, and to separate it from the total code-tracking error, the upper bounds of additional code-tracking error (UBACE) for CELP and NELP are defined through theoretical deduction. After that three scenes on different kinds of spur-to-carrier rejections are set up, and the UBACEs for CELP and NELP are simulated. Finally the influences of the corresponding parameters on the UBACEs for CELP and NELP are analyzed by using the theoretical formulas and simulation results. The method herein can be applied to evaluate the quality of generated signals, as well as to determine the index demand of the emitter's spur-to-carrier rejection.

J. Liu (✉) · H. Chu · J. Zheng · Z. Han · J. Chi
Institute of Spacecraft System Engineering, CAST, 100094 Beijing, People's Republic of China
e-mail: liujiaxing@gmail.com

J. Sun et al. (eds.), *China Satellite Navigation Conference (CSNC) 2012 Proceedings*, Lecture Notes in Electrical Engineering 160, DOI: 10.1007/978-3-642-29175-3_11,

1 Introduction

Carrier's spurious signals are the unideal frequency fractions containing in the single-frequency carrier signal. Spur-to-carrier rejection, which reflects the relation between the respective spur's power and the carrier's power, is a kernel index to characterize the quality of a carrier signal. The quality of the carrier signal determines the quality of the transmitting signal, consequently spur-to-carrier rejection is an indispensable index in signal system designing. However, in the engineering practice people usually pay attention to the characteristics of base-band or intermediate-frequency signals (such as correlation of spread codes, bandwidth of the major lobe, power spectrum density and modulation) and those of transmitting signals (such as bandwidth and power), while taking less cognizance of the effects of the unidealness of carrier signals on signal receiving and processing.

Actually in the generation of the transmitting signal, the carrier's spurious signals, along with the carrier signal, are modulated by the base-band signal, or up-convert the intermediate-frequency signal. After being amplified and filtered, the interference signals are generated at the emitter end. The interference signals, along with the useful signal, are received by the receivers after propagation in space, and their fractions that fall in the passband of the filter at the receiver's front end interferences the code-tracking loop, so as to reduce the code-tracking accuracy.

This paper presents a method for analyzing the effect of carrier's spurious signals on the code-tracking error. First, the time-domain and frequency-domain expressions of the interference signals caused by carrier's spurs are deduced strictly, based on code-tracking loops employing coherent early-late processing (CELP) and noncoherent early-late processing (NELP) respectively. In order to describe the degradation of code-tracking accuracy caused by carrier's spurs, and to separate it from the total code-tracking error, the upper bounds of additional code-tracking error (UBACE) for CELP and NELP are defined by theoretical deduction. After that three scenes on different kinds of spur-to-carrier rejections are set up, and the UBACEs for CELP and NELP are simulated taking example for the GPS P code. Finally the influences of the corresponding parameters on the UBACEs for CELP and NELP are analyzed by using the theoretical formulas and simulation results. The method herein can be applied to determine the index demand of the emitter's spur-to-carrier rejection.

2 Effect on CELP Code-Tracking Error

Considering the code-tracking loop employing CELP discriminator, the signal's input port of the code-tracking loop is the signal's receiving end in this article. The variance [1] of code-tracking error is

$$\sigma^2_{\text{CELP}} = \frac{B_{\text{L}}(1-0.5B_{\text{L}}T)\int_{-\beta_r/2}^{\beta_r/2} S_s(f)\sin^2(\pi f\Delta)df}{(2\pi)^2 C_s^{\text{r}}/N_0\left(\int_{-\beta_r/2}^{\beta_r/2} fS_s(f)\sin(\pi f\Delta)df\right)^2} + \frac{B_{\text{L}}(1-0.5B_{\text{L}}T)\int_{-\beta_r/2}^{\beta_r/2} S_l(f)S_s(f)\sin^2(\pi f\Delta)df}{(2\pi)^2 C_s^{\text{r}}/C_l^{\text{r}}\left(\int_{-\beta_r/2}^{\beta_r/2} fS_s(f)\sin(\pi f\Delta)df\right)^2} \tag{1}$$

where B_{L} is the single bandwidth for equivalent noise (Hz), T is the integral time (s), β_{r} is the double bandwidth of the receiver's frontend (Hz), Δ is the correlator's spacing (s), $S_s(f)$ is the signal's normalized power spectrum density at the receiving end (/Hz), $S_l(f)$ is the interference signal's normalized power spectrum density at the receiving end (/Hz), N_0 is the white noise's power spectrum density at the receiving end (W/Hz), C_s^{r} is signal's power defined on the infinite bandwidth at the receiving end (W), C_l^{r} is the interference signal's power defined on the infinite bandwidth at the receiving end (W). For a general receiver design, Δ is small enough, so Eq. (1) is simplified by

$$\sigma^2_{\text{CELP},\Delta\to 0} = \sigma^2_{\text{CELP,white},\Delta\to 0} + \frac{B_{\text{L}}(1-0.5B_{\text{L}}T)\chi_{ls}}{(2\pi)^2 C_s^{\text{r}}/C_l^{\text{r}}\alpha_s^2} \tag{2}$$

where $\sigma^2_{\text{CELP,white},\Delta\to 0}$ is the CELP code-tracking error when only white noise exists and Δ is extremely small, which is expressed by

$$\sigma^2_{\text{CELP,white},\Delta\to 0} = \frac{B_{\text{L}}(1-0.5B_{\text{L}}T)}{(2\pi)^2 C_s^{\text{r}}/N_0\alpha_s} \tag{3}$$

$$\alpha_s = \int_{-\beta_r/2}^{\beta_r/2} f^2 S_s(f)df \tag{4}$$

$$\chi_{ls} = \int_{-\beta_r/2}^{\beta_r/2} f^2 S_l(f)S_s(f)df \tag{5}$$

According to Schwartz inequality,

$$\chi_{ls} \le \sqrt{\int_{-\beta_r/2}^{\beta_r/2}\{S_l(f)\}^2 df \int_{-\beta_r/2}^{\beta_r/2} f^4\{S_s(f)\}^2 df} \tag{6}$$

It is given that the Fourier transform of the signal to be modulated namely $x(t)$ is $X(f)$, the carrier's power is C_s, the carrier's frequency is f_0. The useful signal is generated after BPSK modulation by the carrier and the signal to be modulated. Without losing generality, the useful signal's time-domain expression is given by

$$x_s(t) = \sqrt{2C_s}x(t)\cos(2\pi f_0 t) \tag{7}$$

The signal's normalized power spectrum density observed at the receiving end is

$$S_s(f) = \lim_{\tau \to \infty} \mathrm{E}\left\{|X^{\tau}(f)|^2\right\} \Big/ \tau \tag{8}$$

where τ represents the signal's truncated time.

It is assumed that there is only one single-frequency spur in the receiving band, with power of $C_{l,\text{single}}$, frequency of $f_0 + \Delta f$, and $-\beta_r/2 < \Delta f < \beta_r/2$ satisfied. The interference signal is generated after BPSK modulation by the spur and the signal to be modulated. The interference signal's time-domain expression is given by

$$x_{l,\text{single}}(t) = \sqrt{2C_{l,\text{single}}}x(t)\cos(2\pi(f_0 + \Delta f)t + \theta) \tag{9}$$

where θ is the initial phase difference between the single-frequency spur and the carrier. At the transmitting end, the Fourier transform of the interference signal is

$$X_{l,\text{single}}(f) = \frac{\sqrt{2C_{l,\text{single}}}}{2}\begin{bmatrix} e^{-j\theta}X(f + f_0 + \Delta f) \\ +e^{j\theta}X(f - f_0 - \Delta f) \end{bmatrix} \tag{10}$$

At the receiving end, the single interference signal's normalized power spectrum density [2] observed by the receiver is

$$\begin{aligned} S_{l,\text{single}}(f) &= \frac{1}{2}\lim_{\tau \to \infty} \mathrm{E}\left\{|X^{\tau}(f + \Delta f)|^2\right\} \Big/ \tau \\ &\quad + \frac{1}{2}\lim_{\tau \to \infty} \mathrm{E}\left\{|X^{\tau}(f - \Delta f)|^2\right\} \Big/ \tau \\ &= \frac{1}{2}S_s(f + \Delta f) + \frac{1}{2}S_s(f - \Delta f) \end{aligned} \tag{11}$$

Because the signal's power spectrum is mirror symmetric, with main lobe's power density much larger than side lobe, and the receiver's filter band is coincident with the signal's band, consequently

$$\int_{-\beta_r/2}^{\beta_r/2} \{S_s(f \pm \Delta f)\}^2 df \le \int_{-\beta_r/2}^{\beta_r/2} \{S_s(f)\}^2 df \tag{12}$$

Actually, the carrier signal contains plenty of extra frequency fractions. Generally spur-to-carrier rejection in the specific frequency and the total spur-to-carrier rejection are utilized to characterize the carrier's spurs. The spur-to-carrier rejection in the specific frequency is defined by the power ratio of the spur in the corresponding frequency within a extremely narrow band and the carrier. The total spur-to-carrier rejection is the maximum value of the spur-to-carrier rejections in all the specific frequency's. In engineering, the spur-to-carrier rejection can be measured using a spectrum analyzer.

Given an arbitrary carrier signal, it is spitted into extremely narrow frequency lattices in the frequency domain, with one spur in each lattice. For each spur's bandwidth is extremely narrow, it can be regarded as a single-frequency signal. According to the nature of the Fourier transform, in the time domain the actual carrier signal is the composition of all the single-frequency spurs and the pure single-frequency carrier signal. So when the useful signal is generated by the modulation between the pure carrier and the signal to be modulated (or the upconversion), the interference signal is generated by the modulation between the single-frequency spurs and the signal to be modulated (or the upconversion).

Assuming there are $n + 1$ lattices totally in the receiving band, which means there are n single-frequency spurs and one pure carrier. For the respective spur, the power is C_l^i, the frequency is $f_0 + \Delta f^i$, with $-\beta_r/2 < \Delta f^i < \beta_r/2$, and with $\Delta f^u \neq \Delta f^v$ satisfied when $u \neq v$, where $i, u, v = 1, 2, \ldots, n$. The interference signal's time-domain expression is given by

$$x_l(t) = \sum_{i=1}^{n} \sqrt{2C_l^i x(t)} \cos\left(2\pi(f_0 + \Delta f^i)t + \theta^i\right) \tag{13}$$

where θ^i is the initial phase difference between the respective single-frequency spur and the carrier. At the transmitting end, the Fourier transform of the interference signal is

$$X_l(f) = \sum_{i=1}^{n} \frac{\sqrt{2C_l^i}}{2} \begin{bmatrix} e^{-j\theta^i} X(f + f_0 + \Delta f^i) \\ + e^{j\theta^i} X(f - f_0 - \Delta f^i) \end{bmatrix} \tag{14}$$

At the transmitting end, the interference signal's power is

$$\begin{aligned}
C_l &= \lim_{\tau \to \infty} \int_0^{\tau} (x_l(t))^2 dt/\tau \\
&= \lim_{\tau \to \infty} \int_0^{\tau} \left[\sum_{i=1}^{n} \sqrt{2C_l^i x(t)} \cos(\omega^i t + \theta^i) \right]^2 dt/\tau \\
&= \sum_{i=1}^{n} C_l^i + \lim_{\tau \to \infty} \int_0^{\tau} \sum_{\substack{u=1 \\ u \neq v}}^{n} \sum_{v=1}^{n} \begin{bmatrix} 4\sqrt{C_l^u C_l^v} \cos(\omega^u t + \theta^u) \\ \cdot \cos(\omega^v t + \theta^v) \end{bmatrix} dt/\tau \\
&= \sum_{i=1}^{n} C_l^i + \sum_{\substack{u=1 \\ u \neq v}}^{n} \sum_{v=1}^{n} \begin{bmatrix} 2\sqrt{C_l^u C_l^v} \\ \cdot \lim_{\tau \to \infty} \int_0^{\tau} \left\{ \begin{array}{l} \cos[(\omega^u + \omega^v)t + (\theta^u + \theta^v)] \\ + \cos[(\omega^u - \omega^v)t + (\theta^u - \theta^v)] \end{array} \right\} dt/\tau \end{bmatrix} \\
&= \sum_{i=1}^{n} C_l^i + \sum_{\substack{u=1 \\ u \neq v}}^{n} \sum_{v=1}^{n} 2\sqrt{C_l^u C_l^v} \cdot 0 \\
&= \sum_{i=1}^{n} C_l^i
\end{aligned} \tag{15}$$

where $\omega^i = 2\pi(f_0 + \Delta f^i)$. At the receiving end, the interference signal's normalized power spectrum density observed by the receiver is

$$S_l(f) = \frac{1}{C_l} \lim_{\tau \to \infty} \mathrm{E}\left\{ \sum_{i=1}^{n} \frac{C_l^i}{2} \begin{bmatrix} |X^\tau(f + \Delta f^i)|^2 \\ + |X^\tau(f - \Delta f^i)|^2 \end{bmatrix} \right\} / \tau$$
$$= \frac{1}{2C_l} \sum_{i=1}^{n} C_l^i [S_s(f + \Delta f^i) + S_s(f - \Delta f^i)] \tag{16}$$

According to formula (12),

$$\int_{-\beta_r/2}^{\beta_r/2} \begin{bmatrix} S_s(f + \Delta f^u) \\ \cdot S_s(f + \Delta f^v) \end{bmatrix} df$$
$$\leq \sqrt{\int_{-\beta_r/2}^{\beta_r/2} \{S_s(f + \Delta f^u)\}^2 df} \cdot \sqrt{\int_{-\beta_r/2}^{\beta_r/2} \{S_s(f + \Delta f^v)\}^2 df} \tag{17}$$
$$\leq \int_{-\beta_r/2}^{\beta_r/2} \{S_s(f)\}^2 df$$

According to formula (12), (16) and (17),

$$\int_{-\beta_r/2}^{\beta_r/2} \{S_l(f)\}^2 df$$
$$= \int_{-\beta_r/2}^{\beta_r/2} \left\{ \frac{1}{2C_l} \sum_{i=1}^{n} C_l^i [S_s(f + \Delta f^i) + S_s(f - \Delta f^i)] \right\}^2 df$$
$$\leq \int_{-\beta_r/2}^{\beta_r/2} \left\{ \begin{array}{l} \frac{1}{4C_l^2} \sum_{i=1}^{n} 4(C_l^i)^2 \{S_s(f)\}^2 \\ + \frac{2}{4C_l^2} \sum_{\substack{u=1 \\ u \neq v}}^{n} \sum_{v=1}^{n} 4C_l^u C_l^v \{S_s(f)\}^2 \end{array} \right\} df \tag{18}$$
$$= \int_{-\beta_r/2}^{\beta_r/2} \{S_s(f)\}^2 df$$

After substituting formula (18) into (6), getting

$$\chi_{ls} \leq \sqrt{\int_{-\beta_r/2}^{\beta_r/2} \{S_s(f)\}^2 df \int_{-\beta_r/2}^{\beta_r/2} f^4 \{S_s(f)\}^2 df} \tag{19}$$

After substituting formula (19) into (2), it is obtained that

$$\sigma_{\mathrm{CELP},\Delta\to 0} \leq \sqrt{\sigma^2_{\mathrm{CELP,white},\Delta\to 0} + \sigma^2_{\mathrm{CELP,spurUB},\Delta\to 0}}$$
$$\leq \sigma_{\mathrm{CELP,white},\Delta\to 0} + \sigma_{\mathrm{CELP,spurUB},\Delta\to 0} \tag{20}$$

where $\sigma_{\text{CELP,spurUB},\Delta\to0}$ is the upper bound of the additional code-tracking error (UBACE) for CELP caused by carrier's spurs, with expression of

$$\begin{aligned}\sigma_{\text{CELP,spurUB},\Delta\to0} &= \sqrt{\frac{B_{\text{L}}(1-0.5B_{\text{L}}T)\sqrt{\beta_s\gamma_s}}{(2\pi)^2 C_s^{\text{r}}/C_l^{\text{r}}\alpha_s^2}} \\ &= \sqrt{\frac{B_{\text{L}}(1-0.5B_{\text{L}}T)\sqrt{\beta_s\gamma_s}}{(2\pi)^2 C_s/C_l\alpha_s^2}}\end{aligned} \tag{21}$$

where

$$\beta_s = \int_{-\beta_{\text{r}}/2}^{\beta_{\text{r}}/2} \{S_s(f)\}^2 df \tag{22}$$

$$\gamma_s = \int_{-\beta_{\text{r}}/2}^{\beta_{\text{r}}/2} f^4 \{S_s(f)\}^2 df \tag{23}$$

The reason of Eq. (21) is the interference signal generated by the carrier's spurs experiences the same gains and attenuations as the useful signal in the transmitting channel, space transmission and receiving channel. The spur-to-carrier ratio (namely the ratio of the sum of spurious powers and the pure carrier's power) at the receiving end is equal to that at the transmitting end. The paper hereafter distinguishes these two ratios no longer.

3 Effect on NELP Code-Tracking Error

Considering the code-tracking loop employing NELP discriminator, the variance [3] of code-tracking error is

$$\sigma^2_{\text{NELP}} = \sigma^2_{\text{CELP}} \left[\begin{aligned} &1 + \frac{\int_{-\beta_{\text{r}}/2}^{\beta_{\text{r}}/2} S_s(f)\cos^2(\pi f\Delta)df}{TC_s^{\text{r}}/N_0\left(\int_{-\beta_{\text{r}}/2}^{\beta_{\text{r}}/2} S_s(f)\cos(\pi f\Delta)df\right)^2} \\ &+ \frac{\int_{-\beta_{\text{r}}/2}^{\beta_{\text{r}}/2} S_l(f)S_s(f)\cos^2(\pi f\Delta)df}{TC_s/C_l\left(\int_{-\beta_{\text{r}}/2}^{\beta_{\text{r}}/2} S_s(f)\cos(\pi f\Delta)df\right)^2} \end{aligned} \right] \tag{24}$$

When Δ is small enough, Eq. (24) is simplified into

$$\sigma^2_{\text{NELP},\Delta\to0} = \sigma^2_{\text{CELP},\Delta\to0} \left[\begin{aligned} &1 + \frac{1}{TC_s^{\text{r}}/N_0\eta} \\ &+ \frac{\int_{-\beta_{\text{r}}/2}^{\beta_{\text{r}}/2} S_l(f)S_s(f)df}{TC_s/C_l\eta^2} \end{aligned} \right] \tag{25}$$

where

$$\eta = \int_{-\beta_r/2}^{\beta_r/2} S_s(f)df \tag{26}$$

After substituting formula (20) into (25),

$$\begin{aligned}
&\sigma^2_{\mathrm{NELP},\Delta\to 0} \\
&\le \begin{pmatrix} \sigma^2_{\mathrm{CELP,white},\Delta\to 0} \\ + \sigma^2_{\mathrm{CELP,spurUB},\Delta\to 0} \end{pmatrix} \begin{bmatrix} 1 + \dfrac{1}{TC_s^r/N_0\eta} \\ + \dfrac{\int_{-\beta_r/2}^{\beta_r/2} S_l(f)S_s(f)df}{TC_s/C_l\eta^2} \end{bmatrix} \\
&= \sigma^2_{\mathrm{NELP,white},\Delta\to 0} + \sigma^2_{\mathrm{CELP,white},\Delta\to 0} \frac{\int_{-\beta_r/2}^{\beta_r/2} S_l(f)S_s(f)df}{TC_s/C_l\eta^2} \\
&\quad + \sigma^2_{\mathrm{CELP,spurUB},\Delta\to 0} \left[1 + \frac{1}{TC_s^r/N_0\eta} + \frac{\int_{-\beta_r/2}^{\beta_r/2} S_l(f)S_s(f)df}{TC_s/C_l\eta^2} \right]
\end{aligned} \tag{27}$$

where $\sigma^2_{\mathrm{NELP,white},\Delta\to 0}$ is the NELP code-tracking error when only white noise exists and Δ is quite small, expressed by

$$\sigma^2_{\mathrm{NELP,white},\Delta\to 0} = \sigma^2_{\mathrm{CELP,white},\Delta\to 0} \left(1 + \frac{1}{TC_s^r/N_0\eta} \right) \tag{28}$$

According to formula (12) and (16),

$$\begin{aligned}
&\int_{-\beta_r/2}^{\beta_r/2} S_l(f)S_s(f)df \\
&= \int_{-\beta_r/2}^{\beta_r/2} \frac{1}{2C_l} \sum_{i=1}^{n} C_l^i \left[S_s(f+\Delta f^i) + S_s(f-\Delta f^i) \right] S_s(f)df \\
&\le \sum_{i=1}^{n} \frac{C_l^i}{2C_l} \begin{pmatrix} \sqrt{\int_{-\beta_r/2}^{\beta_r/2} \{S_s(f+\Delta f^i)\}^2 df \int_{-\beta_r/2}^{\beta_r/2} \{S_s(f)\}^2 df} \\ + \sqrt{\int_{-\beta_r/2}^{\beta_r/2} \{S_s(f-\Delta f^i)\}^2 df \int_{-\beta_r/2}^{\beta_r/2} \{S_s(f)\}^2 df} \end{pmatrix} \\
&\le \frac{1}{C_l} \sum_{i=1}^{n} C_l^i \int_{-\beta_r/2}^{\beta_r/2} \{S_s(f)\}^2 df \\
&= \int_{-\beta_r/2}^{\beta_r/2} \{S_s(f)\}^2 df
\end{aligned} \tag{29}$$

After substituting formula (29) into (27), it is obtained that

$$\begin{aligned}\sigma_{\text{NELP},\Delta\to 0} &\le \sqrt{\sigma^2_{\text{NELP,white},\Delta\to 0}+\sigma^2_{\text{NELP,spurUB},\Delta\to 0}}\\ &\le \sigma_{\text{NELP,white},\Delta\to 0}+\sigma_{\text{NELP,spurUB},\Delta\to 0}\end{aligned} \tag{30}$$

where $\sigma_{\text{NELP,spurUB},\Delta\to 0}$ is the upper bound of the additional code-tracking error (UBACE) for NELP caused by carrier's spurs,

$$\begin{aligned}&\sigma_{\text{NELP,spurUB},\Delta\to 0}\\ &=\sqrt{\sigma^2_{\text{CELP,white},\Delta\to 0}\frac{\beta_s}{TC_s/C_l\eta^2}+\sigma^2_{\text{CELP,spurUB},\Delta\to 0}\left[1+\frac{1}{TC_s^{\text{r}}/N_0\eta}+\frac{\beta_s}{TC_s/C_l\eta^2}\right]}\\ &=\sqrt{\frac{B_{\text{L}}(1-0.5B_{\text{L}}T)}{(2\pi)^2C_s^{\text{r}}/N_0\alpha_s}\frac{\beta_s}{TC_s/C_l\eta^2}+\frac{B_{\text{L}}(1-0.5B_{\text{L}}T)\sqrt{\beta_s\gamma_s}}{(2\pi)^2C_s/C_l\alpha_s^2}\left[1+\frac{1}{TC_s^{\text{r}}/N_0\eta}+\frac{\beta_s}{TC_s/C_l\eta^2}\right]}\end{aligned} \tag{31}$$

4 Simulations of Effects on GPS P Code

The P code's normalized power spectrum density [4] is

$$S_s(f)=T_C\text{sinc}^2(\pi f T_C) \tag{32}$$

where $T_C = 1/(10.23\times 10^6)$s. The receiver's single bandwidth $\beta_r/2$ is set as 10.23×10^6Hz, the single bandwidth for equivalent noise B_{L} as 1 Hz, and the integral time T as 10 ms.

4.1 Extremely Worst Condition

It is assumed that there is one carrier's spur for each 1 Hz frequency lattice in the signal's main lobe bandwidth, with spur-to-carrier rejection of Spur (dBc). Thus the total spur-to-carrier rejection is Spur (dBc). The spur-to-carrier ratio is

$$C_l/C_s = 2\times 10.23\times 10^6\times 10^{\text{Spur}/10} \tag{33}$$

When C_s/N_0 is equal to 35 dB-Hz, the UBACE's variation with the total spur-to-carrier rejection is illustrated in Fig. 1, where red circles represent NELP UBACEs and black pluses represent CELP UBACEs. It can be seen that UBACE is very sensitive to the total spur-to-carrier rejection under the extremely worst condition.

The effects of different received carrier-to-noise ratios on the curves of NELP UBACEs under the extremely worst condition are shown in Fig. 2, where the 21

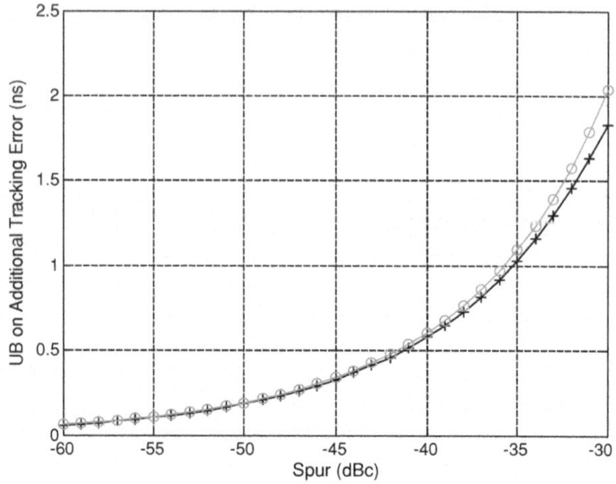

Fig. 1 Variation of UBACE with the total spur rejection under the extremely worst condition

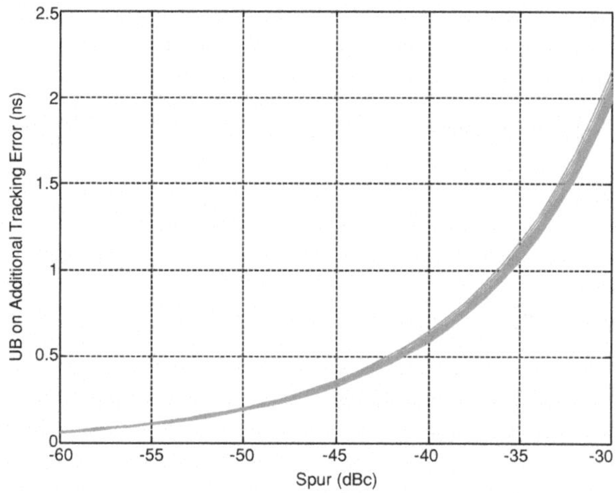

Fig. 2 Effect of C_s/N_0 on NELP UBACE under the extremely worst condition

curves from top to bottom correspond to carrier-to-noise ratios of 30, 31, 32, ..., 50 dB-Hz respectively. It can be observed that the carrier-to-noise ratio effects on NELP UBACE slightly. When the total spur-to-carrier rejection is -30dBc, the differences between the NELP UBACEs corresponding to different carrier-to-noise ratios are all less than 0.2 ns.

4.2 Generally Worse Condition

It is assumed that there is one carrier's spur for each 1 Hz frequency lattice in the signal's main lobe bandwidth, among which 100 spurs have the largest power with spur-to-carrier rejection Spur greater than −60 dBc, and spur-to-carrier rejection

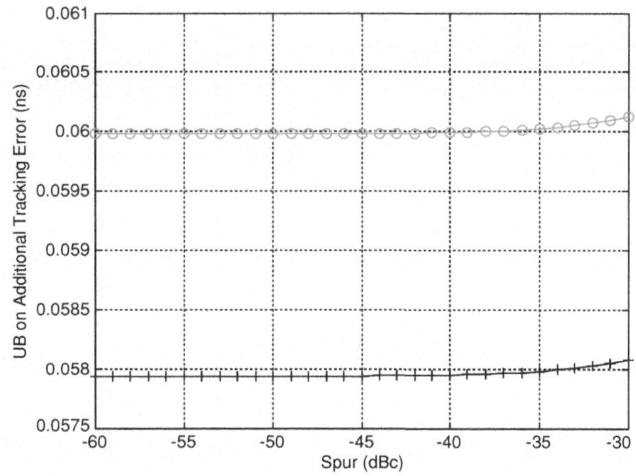

Fig. 3 Variation of UBACE with the total spur rejection under the generally worst condition

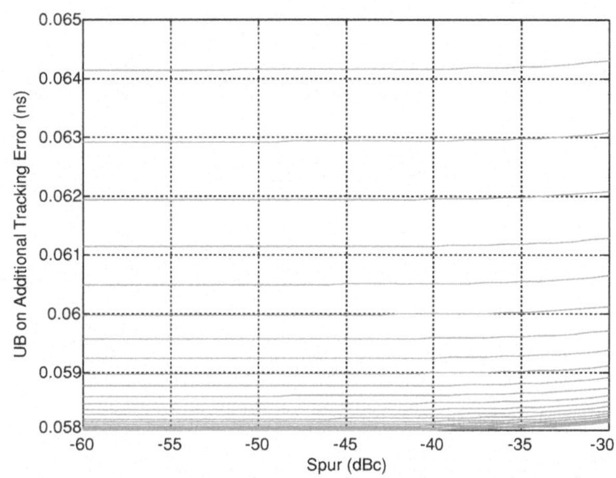

Fig. 4 Effect of C_s/N_0 on NELP UBACE under the generally worst condition

in any other frequency is −60 dBc. So the total spur-to-carrier rejection is Spur (dBc). The spur-to-carrier ratio is

$$C_l/C_s = 100 \times 10^{\mathrm{Spur}/10} + (20.46 \times 10^6 - 100) \times 10^{-60/10} \qquad (34)$$

When C_s/N_0 is equal to 35 dB-Hz, the UBACE's variation with the total spur-to-carrier rejection is illustrated in Fig. 3, where red circles represent NELP UBACEs and black pluses represent CELP UBACEs. It can be seen that the UBACE is not sensitive to the total spur-to-carrier rejection under the generally worse condition.

It is shown in Fig. 4. the effects of different received carrier-to-noise ratios on the NELP UBACE curves under the generally worse condition, where the 21

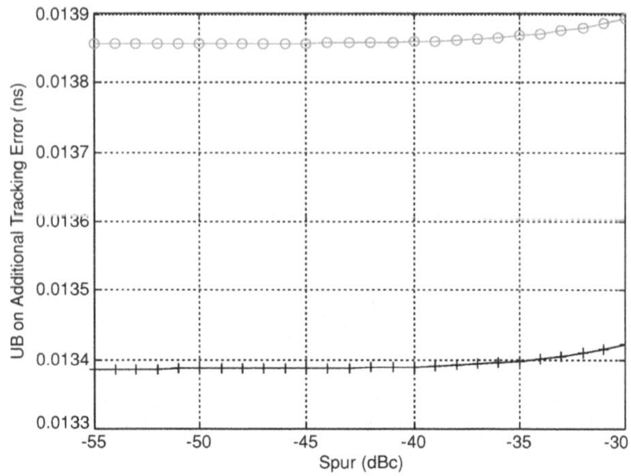

Fig. 5 Variation of UBACE with the total spur rejection under the general condition

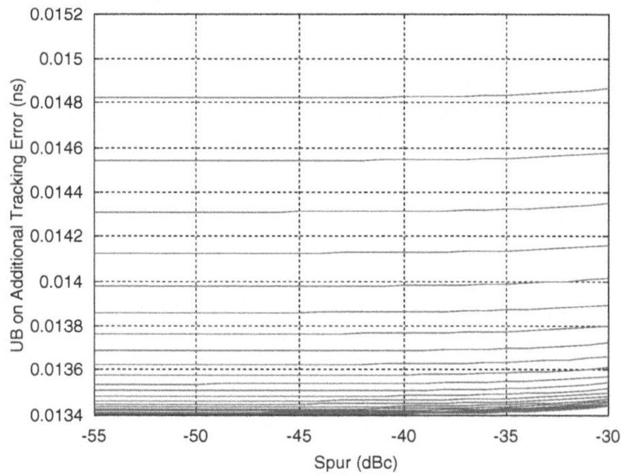

Fig. 6 Effect of C_s/N_0 on NELP UBACE under the general condition

curves from top to bottom correspond to carrier-to-noise ratios of 30, 31, 32, …, 50 dB-Hz respectively. It can be seen that the carrier-to-noise ratio effects on NELP UBACE slightly. For any total spur-to-carrier rejection value, all the differences between the NELP UBACEs under different carrier-to-noise ratios are only less than 0.01 ns.

4.3 General Condition

In this condition, there are very few carrier's spurs with large power in the signal's main lobe bandwidth, and a majority of them keeps less than −70dBc/Hz.

It is assumed that there is one carrier's spur for each 1 Hz frequency lattice in the signal's main lobe bandwidth, among which 6 spurs have the largest power with spur-to-carrier rejection Spur greater than −55 dBc, 10^5 spurs have spur-to-carrier rejection of −55 dBc, 10^5 spurs have that of −60 dBc, 10^5 spurs have that of −65 dBc, 10^5 spurs have that of −70 dBc, and other spurs have that of −75 dBc. So the total spur-to-carrier rejection is Spur (dBc). The spur-to-carrier ratio is

$$\begin{aligned} C_l/C_s = & \ 6 \times 10^{\mathrm{Spur}/10} + 10^5 \times 10^{-55/10} + 10^5 \times 10^{-60/10} \\ & + 10^5 \times 10^{-65/10} + 10^5 \times 10^{-70/10} \\ & + (20.46 \times 10^6 - 4 \times 10^5 - 6) \times 10^{-75/10} \end{aligned} \tag{35}$$

When C_s/N_0 is 35 dB-Hz, the UBACE's variation with the total spur-to-carrier rejection is illustrated in Fig. 5, with red circles for NELP UBACEs and black pluses for CELP UBACEs. It can be seen that UBACE is not sensitive to the total spur-to-carrier rejection under the general condition.

It is shown in Fig. 6 the effects of different received carrier-to-noise ratios on the NELP UBACE curves under the general condition, where the 21 curves from top to bottom correspond to carrier-to-noise ratios of 30, 31, 32, ..., 50 dB-Hz respectively. Obviously the carrier-to-noise ratio effects on NELP UBACE slightly. For any total spur-to-carrier rejection value, all the differences between the NELP UBACEs under different carrier-to-noise ratios are only less than 0.001 ns.

Conclusions can be made below about the influence factors on UBACE, by the theoretical formulas and simulation results above:

1. The spur-to-carrier ratio effects on UBACE greatly, the few spurs with large power have small even tiny effects on UBACE.
2. Unless under the extremely worst condition, UBACE is not sensitive to the total spur-to-carrier rejection.
3. The carrier-to-noise ratio has no effect on CELP UBACE and a tiny effect on NELP UBACE.
4. UBACE is concerned with the receiver's design parameters, among which the single bandwidth for equivalent noise of the code-tracking loop is the main factor.

5 Conclusions

The CELP UBACE and NELP UBACE caused by carrier's spurs are deduced in this paper, in order to analyze the effect on the code-tracking error. Theoretical deduction and simulation indicate that spur-to-carrier ratio is the essential factor to degrade the code-tracking accuracy. Generally the total spur-to-carrier rejection effects on the code-tracking accuracy slightly. Moreover, as one of the receiver's design parameters, the single bandwidth for equivalent noise of the code-tracking loop determines the degradation of the code-tracking accuracy caused by spurs.

References

1. Betz, J. W., & Kolodziejski, K. R. (2009). Generalized theory of code tracking with an early-late discriminator part I: Lower bound and coherent processing. *IEEE Transactions on Aerospace and Electronic Systems, 45*(4), 1538–1550.
2. Holmes, J. K. (1990). *Coherent spread spectrum systems*. Malabar: Krieger Publishing Company.
3. Betz, J. W., & Kolodziejski, K. R. (2009). Generalized theory of code tracking with an early-late discriminator part II: Noncoherent processing and numerical results. *IEEE Transactions on Aerospace and Electronic Systems, 45*(4), 1551–1564.
4. Misra, P., & Enge, P. (2008). *Global positioning system signals, measurements, and performance* (M. Luo, C. Cao & X. Xiao, Trans., 2nd ed.). Beijing: Publishing House of Electronics Industry.

An Evaluation of Multi-GNSS Receiver's Interoperability

Shuangna Zhang, Chao Ren and Tao Ju

Abstract Recently, as a part of multi-GNSS performance test, evaluation of interoperability becomes an challenge in receiver test field. And there is not a standard or a criterion yet. The paper focused on the assessment of Multi-GNSS Receiver's Interoperability. At first, the article analyzed theory of interoperability evaluation, which combined knowledge of statics and rough set., the analysis includes selection method of condition attribute, configuration of test conditions and interoperability evaluation method. After that, we introduced the hardware environment on which evaluation codes operated. At last, taking a multi-GNSS receiver for example, we analyzed interoperability. Moreover, evaluation results are different according to different condition attributes. Experiment records indicate that our interoperability evaluation method is valid, and it can provide different results in response to different service types. In general, our evaluation method could be used in verification of future interoperable receiver. Meanwhile, assessment result could be a guide for optimization of multi-GNSS receiver.

1 Introduction

Evaluation of multi-GNSS receiver's interoperability is necessary during design and research process. The results could be used as references in multi-GNSS receiver's designing and updating. At present, the world has not yet formed a unified standard on assessment of interoperability. The paper raised an evaluation method of multi-GNSS receiver's interoperability which is based on rough set theory. Also, taking a multi-GNSS receiver designed by Space Star Technology as an example, we showed the evaluation flow of interoperability.

S. Zhang (✉) · C. Ren · T. Ju
Space Star Technology Co. Ltd, Beijing, China
e-mail: sindy19821123@163.com

J. Sun et al. (eds.), *China Satellite Navigation Conference (CSNC) 2012 Proceedings*, Lecture Notes in Electrical Engineering 160, DOI: 10.1007/978-3-642-29175-3_12,

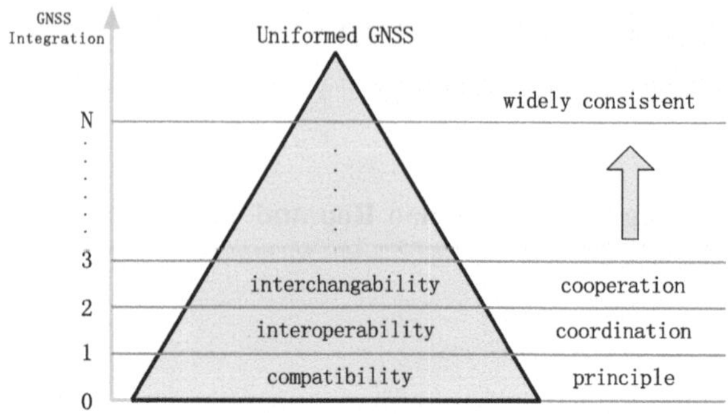

Fig. 1 Levels of GNSS Integration [1]

2 Definition of Interoperability

While some new Global Navigation Satellite System (GNSS) begin to construct, some old GNSS begin to modernize. Every new system should not interfere system in existence in order to achieve compatibility. Meanwhile, everyone wants to utilize all GNSS systems to realize interoperability, by which users can obtain better services. Usually, compatibility and interoperability cannot be separated from each other. As showed in Fig. 1, in different stages of GNSS Integration, compatibility is the basic principle, while interoperability can only be realized by coordination.

Interoperability is the ability of open services of multiple satellite navigation system to be used together to provide better capabilities at the user level than would be achieved by relying solely on one service, without significantly increasing the complexity of receivers. Definition does not illuminate intension of capabilities. Usually, optimization of multi-GNSS receiver's solution included two aspects, consumption reduction and performance improvement. Thereinto, consumption reduction related to cost, complexity, power consumption, etc. Performance improvement included availability, reliability, accuracy, etc. Therefore, when GNSS interoperable signal is designed, some key factors should be considered, such as radio frequency, code pattern, navigation message. The objective is to share information from different GNSS while limiting increase of consumption.

3 Interoperability Evaluation Method Based on Rough Set Theory

In 1980s, mathematician Pawlk from Poland raised Rough set theory, which is used in treatment of incomplete and uncertain information [2]. Since GNSS interoperability is closely related to system structure, operating mechanism, environment and system operation of GNSS, it includes a large number of

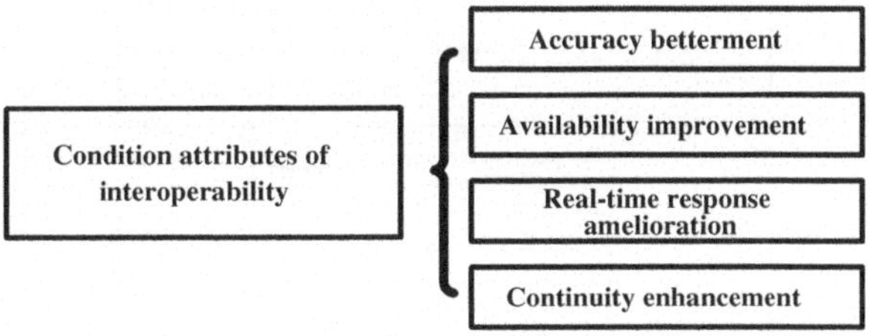

Fig. 2 Interoperability condition attributes

uncertain information. Thus, utilizing tough set theory to evaluate GNSS signal's interoperability is a good choice.

Firstly, we build the elements of condition attributes set and decision attributes, which will be used in integrated assessment of interoperability. Condition attributes of interoperability includes accuracy betterment, availability improvement, real-time response amelioration and continuity enhancement, that is shown in Fig. 2. Decision attributes is integrated assessment result.

Secondly, we collect all the condition attributes data, then we discrete these original data and reduce repetitive condition attributes [3]. After that, we judge the importance of every condition attribute. Finally, every condition attribute's weight can be obtained. According to different signal type and user type, condition attribute set will be different. Thus, even we evaluate the same receiver's interoperability, the result could be varied in response to different service types. Interoperability evaluation flow chart is showed in Fig. 3.

4 Operating Environment of Interoperability Evaluation System

Operating environment of interoperability evaluation system contains five subsystems, start and control subsystem, data collecting subsystem, data storage subsystem, visualization subsystem. The relationship among subsystems is showed in Fig. 4.

Start and control subsystem is used to manage and coordinate operation of the whole system. While data collecting subsystem real-time collects all kinds of data sources, converts data format, rearranges the data and distributes them. The appearance of GPS/GLONASS/Compass/Galileo system simulator designed by Space Star Technology is showed in Fig. 5.

The functions of the data processing subsystem are receiving data, calling program corresponded to the interoperable test evaluation algorithm, pre-processing the measured data, giving the statistics analysis of visible data and tested data, etc. The data storage subsystem is used for real-time collection of the test resources, clarification, storage and query for the assessment data, data playback, and so on. Visualization subsystem is used to set parameters of the system and realize the visualization of the output results.

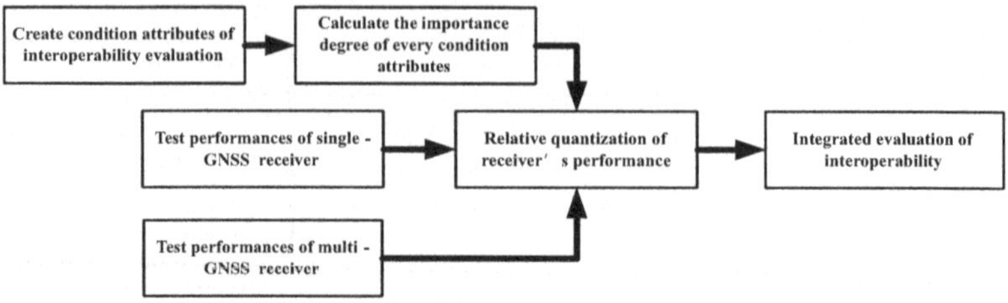

Fig. 3 Flow chart of interoperability evaluation

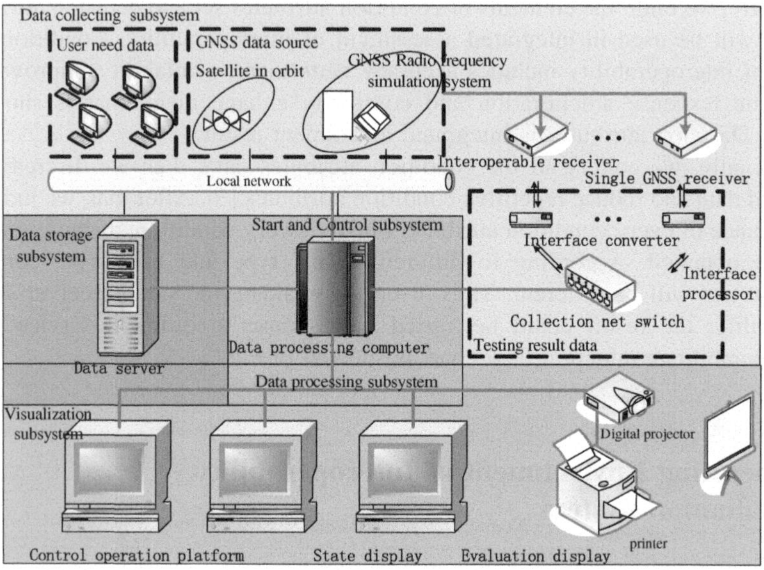

Fig. 4 Structure of interoperability evaluation system

Fig. 5 GPS/GLONASS/Compass/Galileo system simulator

5 An Example of Multi-GNSS Receiver's Interoperability Evaluation

The performance of interoperability is superior to that of the receivers, and its evaluation should show the performance improvement and degrading induced by the interoperability. Meanwhile, interoperability is a composite index, therefore,

Table 1 Performances' weights of receivers based on user need

Service type	Accuracy	Availability	Real-time response	Continuity
Individual safety user	0.2	0.5	0.2	0.1
High-accuracy user	0.5	0.35	0.1	0.05

multi-dimensional analysis for interoperability should be carried out according to the type of the service. Besides, with aspects to the users concerned about, reasonable assessment should be made on the base of different weights.

As known, the performance of receivers is closely related to the design scheme. In this paper, for one type of receivers, quantitative assessment and qualitative assessment have been derived based on the performance comparison between the single receivers and multi-receivers. This method doesn't lose generality, although the assessment is evaluated for special cases.

5.1 Definition of Every Performance's Weights According to User's Need

Performance requirement for receivers varies as the users, for example, high-accuracy receivers, real-time receivers, high-dynamic receivers, are all the kind of receivers which satisfies the special requirements. The performance of the personal users and high-accuracy users is evaluated in this paper. The important parameters for the former one are service availability and continuity, especially for case that the visibility is limited by the environment (jungle, valley, and so on); and for the latter one, like receivers used for measuring, surveying and mapping, positioning accuracy is more important than other parameters. Performance weights of receiver based on user need is listed in Table 1, as it is shown, the weights changed with the users' requirement. It should be noted that, for the purpose of simply demonstrating proposed method, only parameters concerned by most of the users are listed in Table 1 and the corresponding refinements and decomposition have been neglected.

5.2 Performance Measurement of Receivers

Brief introduction for the receiver performance measuring is made on the base of the parameters in Table 1. The relation between the measured parameters and the technical requirements of the users is regarded as the measured result. Meanwhile, some other parameters can be added for commercial purpose, such as volume, weight, price and so on, which are neglected in this paper. Performance test example for individual safety interoperable receivers and high precision

Table 2 Example of individual safety interoperable receiver's performance test

	Performance of receiver		
	Specification	Testing results	Assessment results
Accuracy(m) (positioning)	<5	4.7	(5−4.7)/5 = +0.060
Availability (satellites number from a single GNSS is less than 4)	>60%	65%	(65–60%)/60% = +0.083
Real-time response (first time of first positioning)	<40 s	42 s	(40−42)/40 = −0.050
Continuity (repositioning time after 1 satellite in use is changed)	<30 s	25 s	(30−25)/30 = +0.167

Table 3 Example of high precision interoperable receiver's performance test

	Performance of receiver		
	Specification	Testing results	Assessment results
Accuracy(m) (positioning)	<2	4.7	(2−4.7)/2 = −1.35
Availability (satellites number from a single-GNSS is less than 4)	>70%	65%	(65−70%)/70% = −0.071
Real-time response (first time of first positioning)	<60 s	42 s	(60−42)/60 = +0.3
Continuity (repositioning time after 1 satellite in use is changed)	<40 s	25 s	(40−25)/40 = +0.375

interoperable receivers given in Tables 2 and 3 is served as demonstrating examples, respectively. As it is shown, the testing results varied as the users' type in spite of the same receivers' parameters. Correspondingly, interoperability assessment result will differ.

The parameters in Tables 2, 3, 4, 5 are from one certain type of Space Star Technology Co. Ltd., and the practical tested data is from the corresponding individual safety interoperable receivers.

5.3 Multi-GNSS Receiver's Interoperability Evaluation Based on Service Type

According to the tested results and weights of the receivers' parameters, compressive evaluation for the interoperable receivers is involved in this paper. Based on the tested results in Tables 4 and 5, performance of the receiver is evaluated and the corresponding compressive evaluation results are shown in Table 6 which can be attained by summing the products of receivers' performance weights and tested results.

Table 4 Example of individual safety single receiver's performance test

	Performance of receiver		
	Specification	Testing results	Assessment results
Accuracy(m) (positioning)	<5	4.65	(5−4.65)/5 = +0.07
Availability (satellites number from a single-GNSS is less than 4)	>60%	5%	(5-60%)/60% = −0.917
Real-time response (first time of first positioning)	<40 s	39 s	(40−39)/40 = +0.025
Continuity (repositioning time after 1 satellite in use is changed)	<30 s	35 s	(30−35)/30 = −0.167

Table 5 Example of high precision single receiver's performance test

	Performance of receiver		
	Specification	Testing results	Assessment results
Accuracy(m) (positioning)	<2	4.65	(2−4.65)/2 = −1.325
Availability (satellites number from a single-GNSS is less than 4)	>70%	5%	(5−70%)/70% = −0.929
Real-time response (first time of first positioning)	<60 s	39 s	(60−39)/60 = +0.35
Continuity (repositioning time after 1 satellite in use is changed)	<40 s	35 s	(40−35)/40 = +0.125

$$mark = \sum_{i=1}^{n} para_i \times weight_i \tag{1}$$

where *mark* is the receivers' comprehensive performance evaluation test results, $para_i$ is the test results of the *ith* parameter, $weight_i$ is the weight value for the *ith* parameter which is varied as the user type and $i \in \{1, 2, \ldots n\}$.

The receivers' performance weights for different users' requirements can be derived according to Table 1, and by calculating the results in Tables 2, 3, 4, 5 we can get the test results as shown in Table 6. It can be concluded from Table 6 that for one certain type of interoperable receivers, the compressive evaluation results for different types of users are different from each other. On the other hand, if for the same users' types, the differences of the comprehensive evaluation results between the single receivers and multi-receivers are significant.

As it has been demonstrated above, the evaluation value is minus for those users whose requirement for accuracy is high and positive for those users whose requirement for accuracy is low. Besides, the interoperable receivers has been greatly improved the availability of the system, therefore, for the individual safety users, the comprehensive evaluation results of the interoperable multi-receivers outperform that of the single receiver.

Table 6 Integrated evaluation of receiver performance

Service type	Accuracy	Availability	Real-time response	Continuity	Integrated evaluation
Individual safety user	0.2	0.5	0.2	0.1	0.06
(interoperable receiver)	+0.06	+0.083	−0.05	+0.167	
High precision user	0.5	0.35	0.1	0.05	−0.65
(interoperable receiver)	−1.35	−0.071	+0.3	+0.375	
Individual safety user	0.2	0.5	0.2	0.1	−0.46
(single-GNSS receiver)	+0.07	−0.917	+0.025	−0.167	
High precision user	0.5	0.35	0.1	0.05	−0.65
(single-GNSS receiver)	−1.325	−0.929	+0.35	+0.125	

6 Conclusion

Quantitative analysis and qualitative analysis results of interoperable receivers can be derived by comparing the comprehensive performance evaluation results of the single receiver and interoperable multi-receivers in Table 6 for one kind of service. It can be concluded that for the individual safety users, performance of the system is significantly improved comparing the interoperable multi-GNSS receivers with the single-GNSS receiver. It should be noticed that performance of the receivers is closely related to implementation method of the receivers, no matter what the receivers' type is. For example, the receiver accuracy will increase by choosing satellites according to the DOP, power and pseudo-range of the satellite. And the multi-frequency signal can be helpful to eliminate the errors induced by ionosphere, thus, it can increase the receiver accuracy, however, it will increase the responding time at the same time, that is, the evaluation of the interoperable receivers should not only consider the sources, but also give qualitative comparison according to the type of the receivers and the corresponding service type.

Acknowledgments This research was financially supported by the National High Technology Research and Development (863 Program) Study on simulation of multi-GNSS compatibility and interoperability general technique (No. 2009AA12Z322). The authors are grateful for a grant from 863 Program.

References

1. Stupak, G. (2008). The Russian Federation View on GNSS compatibility and interoperability. In *international committee on GNSS (ICG-3): General designer on GLONASS system.*
2. Jiang, C. (2006). *Study on application of Rough theory used in multi-attributes*. Thesis. Xi Nan Traffic University.
3. Liu, X. (2007). Research on discrete method of ground-to-air missile threat level based on rough set. *Fire Control and Command Control, 32*(11).

Analysis of the Sun Transit Outage Impact on the Inter-satellite Link of the Navigation Satellite

Ying Guo, Meihong Li, Shanbao He, Pan Xin and Liu Tao

Abstract For the global navigation satellite, when the two satellites which constitute an inter-satellite link are approximately in a line with the sun, generally the communication between them will be impacted by the sun transit outage. Based on the character of the native navigation satellite, this paper researched the sun transit outage impact theory, analyzed and emulated the factors which impacted the antenna noise, accomplished the link calculation. By establishing inter-satellite network topology emulate model, forecast the precise happen time of the sun transit outage. It also suggested some feasible methods to avoid the sun transit outage impact to offer reference to constitute the navigation satellite inter-satellite link project.

1 Introduction

The Sun Transit Outage is also known as the Satellite Eclipse. For global navigation satellite, when the two satellites which constitute an inter-satellite link are approximately in a line with the sun, the communication between them will be impacted by the sun transit outage. When the satellite received signal is in a line with the sun and the satellite send signal, the receiving antenna pointed to the send signal satellite, but also to the sun. A strong electromagnetic interference signal from the sun and the satellite signal are in the same direction. The sun's wide electromagnetic spectrum is a huge noise source for inter-satellite link receiving antenna. The interference signal and normal signal focused, that make the

Y. Guo (✉) · M. Li · S. He · P. Xin · L. Tao
China Academy of Space Technology, Beijing, People's Republic of China
e-mail: guoying-tju@163.com

J. Sun et al. (eds.), *China Satellite Navigation Conference (CSNC) 2012 Proceedings*, Lecture Notes in Electrical Engineering 160, DOI: 10.1007/978-3-642-29175-3_13,

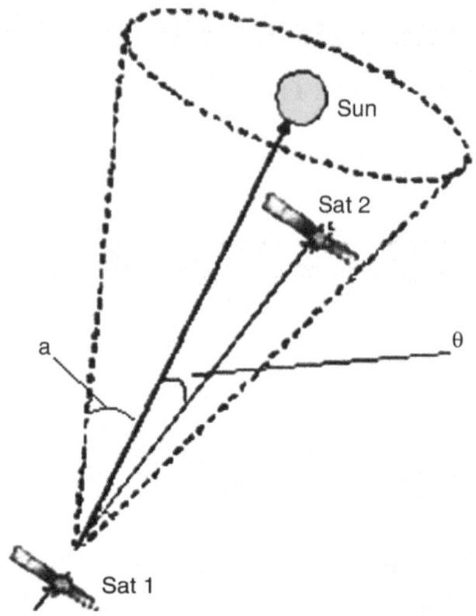

Fig. 1 Satellite sun transit outage sketch map

demodulation system can not distinguish between the satellite signal and the sun's strong interference signal. Accordingly the satellites communicate will be interrupt and the corresponding link may be temporarily interrupt [1]. It is propitious to analyze the sun's noise impact to the antenna noise temperature and forecast the precise happen time of the sun transit outage for reasonably arranging telemetry and telecontrol events during the sun transit outage and offer reference to constitute the navigation satellite inter-satellite link project.

2 Geometric Model of the Sun Transit Outage Analysis

Define the two satellites which established a inter-satellite link are Sat 1 and Sat 2, and the sun transit outage angle of satellite is θ. θ means the angle between "Sat1-Sat2" and "Sat1-Sun" as shown in Fig. 1.

By the spatial geometric relationship in the unified coordinate system, according to the satellite and the sun's ephemeris, forecast the time of the sun transit outage, namely:

$$\begin{cases} \vec{r}_1 = \vec{r}_{sat2} - \vec{r}_{sat1} \\ \vec{r}_2 = \vec{r}_{sun} - \vec{r}_{sat1} \\ \theta = \arccos(\vec{r}_1, \vec{r}_2) \end{cases} \tag{1}$$

$\vec{r}_{sat1}$, $\vec{r}_{sat2}$, $\vec{r}_{sun}$ are vectors of satellite Sat1, Sat2 and the sun, θ is the satellite sun transit outage angle, a is the satellite antenna beam angle. When $0 \leqq \theta \leqq 2a$ or $180\text{-}2a \leqq \theta \leqq 180$, the satellite Sat1 and Sat2's, communication may be subject to the sun transit outage's impact.

NASA's tracking experiments showed that when the antenna point at the sun as the center of the disk from 0 to 1°, the system noise temperature changed significantly with the sun's surface temperature. So this communication link calculations select $\theta = 1°$. In order to forecast more exactly the orbit simulation consider that there is sun transit outage risk when θ is less than 2°.

3 Sun Transit Outage's Impact on Communication

3.1 Principle of Satellite Communication

Satellite communication through the inter-satellite links has the following relationship [2]:

$$P_{EIRP}(dBW) + G_r/T_n(dB/K) - K(dBW) - [L_S + \sum L_i + M] = S/N_0(\text{dB/Hz}) \tag{2}$$

where K = −228.6 dBW/kHz, PEIRP is the Effective Isotropic Radiated Power (dBW,), Tn is the receiving system's total noise temperature (K), Gr is the receiver gain, M is the system margin, Ls is the free space loss, Σli is the other loss, included the atmospheric loss (do not consider in space), loss of antenna polarization, antenna pointing error loss, feeder loss, etc.

Given frequency, Gr/Tn of receiving system determines the impact of sun transit outage. When the receiver gain Gr is Constant, the main consideration is the total noise temperature Tn.

The antenna noise temperature increment due to sun transit outage related factors include the sun surface temperature Ts, the receiving antenna diameter, downlink frequency, the sun diameter, the sun is the main source of spatial noise, the sun surface temperature Ts is [1]:

$$T_s = 120000 f^{-0.75} \tag{3}$$

The unit of Ts is K; f is the signal frequency, unit GHz; when the main lobe width is less than or equal to θ, the antenna noise temperature is equal to the solar noise temperature. Solar noise in the Ka-band spectrum is similar to white noise, has the characteristics of the whole band.

For Ka-band, f is the 22.55–27.5 GHz, Ts is 9992.67–11596.35 K.
For C-band, f = 4 GHz, Ts = 42426 K.
For Ku-band, f = 12 GHz, Ts = 18612 K.

Estimate the antenna equivalent noise increment by the sun surface temperature:

$$\Delta T_s = p \cdot T_s \cdot \eta \cdot r^2 \tag{4}$$

Fig. 2 Antenna and noise source sketch map

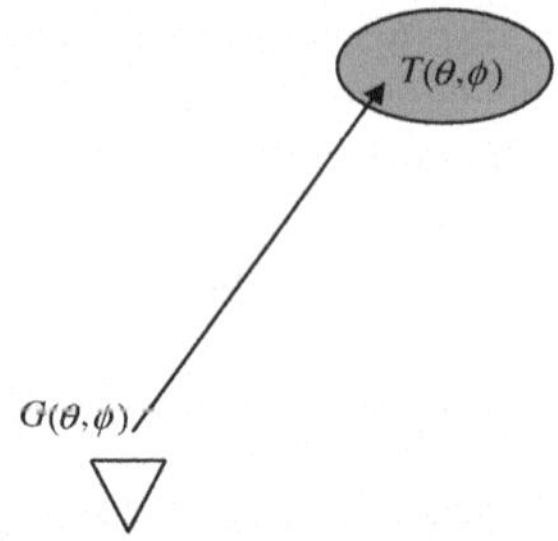

where p is the antenna polarization factor, η is the antenna efficiency, which

$$r = \begin{cases} \mathrm{R}/\lambda & (\lambda > R) \\ 1 & (\lambda \leq R) \end{cases} \tag{5}$$

R is the diameter of the sun's surface, λ is the width of the antenna beam. When the sun's center is on the center of the antenna beam, the impact is strongest. Calculate the Maximum increments, in order to find the link SNR.

3.2 Noise Temperature Calculations

Located in the direction of (θ, Φ), a noise source of the blackbody temperature is T (θ, Φ), by the Plunk blackbody law, the radiation intensity of the noise source (Fig. 2).

$$B(\theta, \phi) = K\frac{1}{\lambda^2}T(\theta, \phi)\Delta f \tag{6}$$

The antenna receives the noise source distributing in the solid angle dΩ, the equivalent area of antenna A_e (θ, Φ) is linked with the antenna gain:

$$G(\theta, \phi) = \frac{4\pi}{\lambda^2}A_e(\theta, \phi) \tag{7}$$

Consider the polarization mismatch factor p (when the noise source is the random polarization orientation of radiation power p = 0.5), namely:

$$A_e(\theta, \phi) = p\frac{\lambda^2}{4\pi}G(\theta, \phi) \tag{8}$$

So the noise power received by the antenna in the direction of (θ, Φ) as follows:

$$\mathrm{d}P_R = K\frac{1}{\lambda^2}T(\theta, \phi)\Delta f pG(\theta, \phi)\mathrm{d}\Omega = \frac{K\Delta f}{4\pi}T(\theta, \phi)pG(\theta, \phi)\mathrm{d}\Omega \tag{9}$$

The noise power P_R is equivalent to the resistance with the noise temperature T_R ($P_R = KT_R\Delta f$), so:

$$\mathrm{d}T_R = p\frac{1}{4\pi}T(\theta,\phi)G(\theta,\phi)\mathrm{d}\Omega \tag{10}$$

The noise temperature caused by noise sources of the whole space:

$$T_R = \frac{1}{4\pi}p\int_{4\pi} T(\theta,\phi)G(\theta,\phi)\mathrm{d}\Omega \tag{11}$$

As mentioned above, T (θ, Φ) is the blackbody temperature of the noise source, but in fact the noise source can not as a blackbody processing. In actual calculation T (θ, Φ) is the brightness temperature of the noise source.

According to the definition of antenna gain pattern,

$$\int_{4\pi} G(\theta,\phi)\mathrm{d}\Omega = 4\pi \tag{12}$$

Therefore, there was

$$T_R = p\frac{\int_{4\pi} T(\theta,\phi)G(\theta,\phi)\mathrm{d}\Omega}{\int_{4\pi} G(\theta,\phi)\mathrm{d}\Omega} \tag{13}$$

The above only consider the noise temperature caused by the pattern gain. The feed network loss is L dB (<0), and the feed network loss is equivalent to the noise temperature [3]

$$T_L = \left(1 - 10^{\frac{L}{10}}\right)T_e \tag{14}$$

Received noise temperature due to the wear and tear is reduced to

$$T'_R = T_R 10^{\frac{L}{10}} \tag{15}$$

The total antenna noise temperature for these two parts is:

$$T_A = T'_R + T_L = 10^{\frac{L}{10}}T_R + \left(1 - 10^{\frac{L}{10}}\right)T_e \tag{16}$$

Figure 3 is the feeder network loss effect simulation. The figure shows, the feed network losses have some impact on the system noise; the system noise deviation can reach 50–100 K.

The receiving system noise temperature changes with the sun perspective diameter, simulation results shown in Fig. 4. Assume that the solar brightness temperature is approximately constant Ts, the sun's perspective diameter $\delta s = 32'$, the solid angle is

$$\Delta\Omega = \int_0^{2\pi}\int_0^{\delta_S/2} \sin\theta\mathrm{d}\theta\mathrm{d}\phi = 2\pi\left(1 - \cos\frac{\delta_S}{2}\right) \approx 1.0831\times 10^{-5}\times 2\pi \tag{17}$$

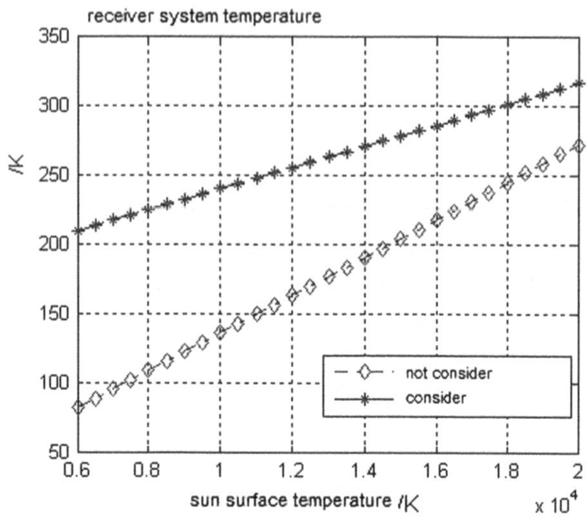

Fig. 3 *Curve* receiver system temperature to the Sun surface temperature

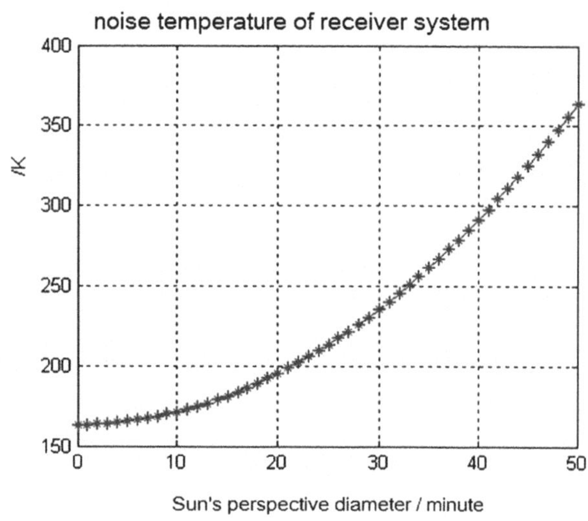

Fig. 4 *Curve* noise temperature of receiver system to the Sun perspective temperature

When the sun transit outage occurred, the sun is near by the receiving antenna beam center (θ_S, Φ_S), the antenna's main beam is wide enough, integral approximation (integral value is approximated as G (θ_S, Φ_S)):

$$\int_{4\pi} T(\theta,\phi)G(\theta,\phi)\mathrm{d}\Omega \approx T_S G(\theta_S,\phi_S)\Delta\Omega \tag{18}$$

Table 1 The inter-satellite link of navigation satellite calculate

Performance parameters	Normal	Sun transit outage
Receiver entrance equivalent noise temperature Te (K)	346.0123375	538.7006468
Antenna equivalent noise temperature Tant (K)	29	245.2
Receiving system G/T (dB/K)	7.870143259	5.947584134
Receive I channel signal despread bit signal to noise ratio Eb/N0 (dB)	1.939155548	0.016596423
Receive the Q channel signal despread bit signal to noise ratio Eb/N0 (dB)	14.93915555	13.01659642
I-way wireless link margin E1 (dB)	0.189768182	−1.732790943
Q-way wireless link margin E2 (dB)	4.439155548	2.516596423

Therefore, the antenna noise temperature as follows:

$$\begin{aligned} T_R &\approx \frac{1}{4\pi} p T_S G(\theta_S, \phi_S) \Delta\Omega = \frac{1}{4\pi}\frac{1}{2} T_S G(\theta_S, \phi_S) 2\pi \left(1 - \cos\frac{\delta_S}{2}\right) \\ &= \frac{1 - \cos\frac{\delta_S}{2}}{4} T_S G(\theta_S, \phi_S) \approx 2.7077 \times 10^{-6} T_S G(\theta_S, \phi_S) \end{aligned} \tag{19}$$

The sun perspective temperature is calculated as follows (frequency f in GHz):

$$T_S \approx 120000 f^{-0.75}$$

At 25 GHz, Ts = 1.0733 × 104, the antenna peak gain is about 37 dBi, so the antenna noise temperature generated by the solar radiation received is:

$$2.7077 \times 10^{-6} \times 1.0733 \times 10^4 \times 10^{3.7} = 145.6538\text{K}$$

The antenna feed network loss is about 2.5 dB, ambient temperature up to about 100°C = 373.15 K, and therefore the total noise temperature is:

$$\begin{aligned} T_A &= 10^{\frac{L}{10}} T_R + (1 - 10^{\frac{L}{10}}) T_e \\ &= 145.65 \times 10^{-0.25} + (1 - 10^{-0.25}) \times 373.15 = 245.2K \end{aligned} \tag{20}$$

3.3 Link Calculations and Conclusions

A Ka-band receiving antenna, for example, the receiving antenna feeder ambient temperature is 290 K. When a link happened the sun transit outage and the ambient temperature up to the highest, the antenna noise temperature is about 245.2 K, link calculation changes as follows due to the antenna equivalent noise temperature increased (Table 1).

By calculation, the sun transit outage will bring about 2 dB interference in Ka-band inter-satellite links, that may cause communication interruption. The next section will forecast the precise happen time of the sun transit outage.

4 Orbit Simulation and Analysis

This paper mainly analysis the sun transit analysis based on the MEO Walker 24/3/2 constellation configuration, including both cases the inter-satellite links within the orbit surface and the different orbit surface link.

4.1 Simulation of Links in the Same Orbit Surface

The simulation object is a Walker 24/3/2 constellation, the simulation time from the epoch January 1, 2011 UTCG zero to 2012 January 1st UTCG zero. Motion model considered J2; antenna beam angle is 1°. The simulation results showed the following conclusions:

(1) The sun transit outage in the same orbit surface is mainly distributed in March 14 to March 22 and September 10 to September 17.
(2) In the fixed link, the sun transit outage happened up to four times in a whole day, and this phenomenon occurs alternately by the two satellite link-building, each time interval was 7 h 23 min approximately, each lasted about 10 min the longest.
(3) Among a variety of topologies, the sun transit outage happened almost at the same time in parallel links.

4.2 Simulation of Different Orbit Surface Link

Different orbit surface fixed link topology will vary according to the number of link-building. The principle that established fixed link satellite is: the front phase satellite (FP), the behind phase satellite (BP), the front plus 4 phase satellite (FP 4), the back plus 4 phase satellite (BP), as shown above (Fig. 5).

The A1 star of the orbit surface, for example, aim at the four kinds of links select 1 species (A1, C8) from the orbit surface C, analyze the existence of the sun transit outage. The same simulation method with the above, results are as follows:

(1) The sun transit outage of navigation satellite link is mainly distributed in March 3 to March 11 and September 4 to September 12. A total of 16 days affected in 1 year. The affected time is less than 40 min in 1 day.
(2) In the fixed link, the sun transit outage happened up to four times in a whole day, and this phenomenon occurs alternately by the two satellite link-building, each time interval was 7 h 22 min approximately, each lasted about 10 min the longest.

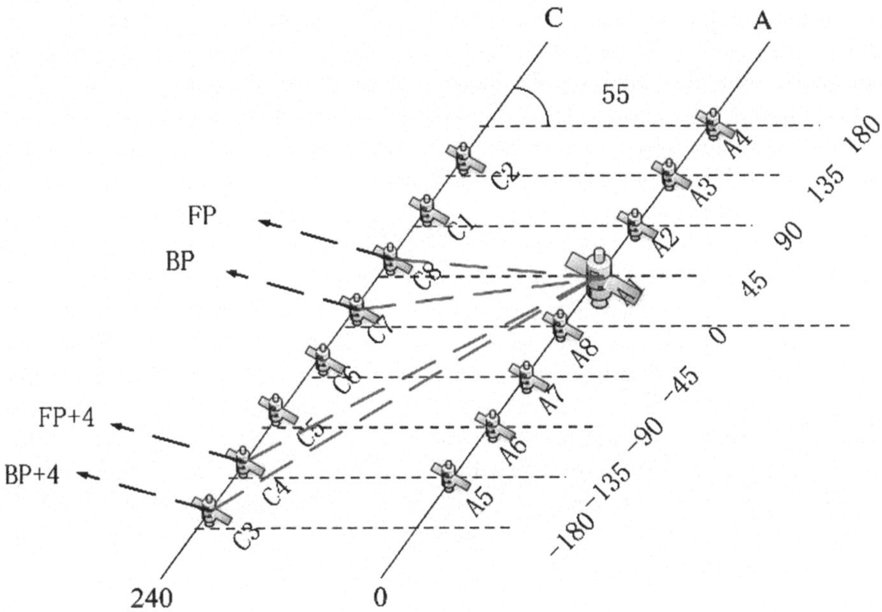

Fig. 5 The 4 fixed inter-satellite links in the different orbit surfaces

4.3 Orbital Analysis

From the space geometry, for the links in the same orbit surface, the sun transit outage happens only when the MEO satellite orbit surface are nearly parallel to the sun's rays. Thus, according to the sun laws, for a certain orbital plane, in each year there are two opportunities to happen sun transit outage, and the fixed link at the same orbital planes, the happen date is approximate the same.

For a fixed link, it is only when the sun reach the beam angle range, to take place the sun transit outage. Therefore, twice happened time interval Δt_s should be equal to the time difference that the sun reaches the beam angle range twice. Analysis shows that if at some point, the sun's rays in the beam angle, when the satellite has been operating half-circle, the fixed link vector and the initial vector parallel, the sun's rays are still in the beam angle. Δt_s equals to the time that MEO satellites turn 180°. Satellite angular velocity $\omega = 0.4063°/\text{min}$, so $\Delta ts = 443.0224$ min.

For different fixed links in the same orbital plane, due to different phase, the moment of the sun transit outage will be different, the time interval is connected with the angle θ between two links:

$$\Delta t_d = \theta/\omega \tag{21}$$

ω is the satellite angular velocity, for the MEO satellites which eight days run 13 laps, $\omega = 0.4063°/\text{min}$. As a result, from the perspective of ensuring communication reliability analysis, the larger θ is, the better links are.

For different orbit surface MEO fixed link, the dates of the sun transit outage are different. The interval ΔD can be solved under the sun and the satellite position, and the two orbit surface phase difference $\Delta\Omega$:

$$\Delta D = \Delta\Omega/\delta\Omega \tag{22}$$

$\delta\Omega$ is the sun precession angular velocity (0.9856°) and orbital plane precession angular velocity difference. The precession angular velocity of the orbital plane is affected by many factors. Due to the non-central gravity of the earth and the gravitational perturbation from the sun and the moon, the MEO satellite ascending node right ascension reduces about 14° a year. But the sun and the moon gravitational perturbation impact are different on different orbit surfaces, each MEO satellite ascending node right ascension changing is not the same, the average annual maximum error is about 0.5° [4]. To the ascending node right ascension every year to reduce 14°, it can be estimated that in adjacent orbital planes, the date of the sun transit outage has a interval of 127 days.

5 Summary

The sun transit outage is inevitable in the navigation satellite link communication, this paper described the phenomenon, explained the sun outage. Through the link calculation and the calculation of the antenna noise, obtained the sun transit outage impact on the inter-satellite link communication.

To avoid the influence of the sun transit outage, the following suggestions:

(1) When the sun transit outage happens, lower communication speed to ensure the normal operation of the link.
(2) Backup link: make use of the different angle links of each satellite to build backup link to achieve flexible switch during sun transit outage.
(3) The phased array antenna can quickly switch at all times; in contrast the reflector antenna must fully consider the impact of the sun transit outage, leaving enough margin in the establishment of the link.

References

1. Mohamadi, F., & Lyon, D. L. (1988). Effects of solar transit in Ku-Band VSAT systems. *IEEE Transactions on Communications, 36*(7), 892–894.
2. Pratt, T., Bostian, C., & Allnutt, J. (2003). *Satellite Communications* (Second ed.). Chichester: Wiley and Sons, Inc.

3. Chang, T., Phillips, T. M., & Acuff, P. A. (1994). A comparative study of solar interference on the Iridium and MSS constellations, *International Journal of Satellite Communications*, vol. 0 000-000.
4. Yuri, U. (2000). Geometric analysis of low-earth-orbit satellite communication system: covering functions. *Journal of Spacecraft and Rockets*, vol. 37, no. 3, pp. 385–391 May–June.

Part II
Precise Orbit Determination and Positioning

Estimation of Crosslink Equipment Delay and Its Contribution to GNSS Orbit Determination and Time Synchronization

Rengui Ruan, Laiping Feng, Xiaolin Jia, Xiaoyong Song, Xianbing Wu and Tao He

Abstract Considering equipment delay, measurement equations combining GNSS pseudorange, carrier phase and crosslink range observations are presented. It is revealed that the equipment delay of crosslink payload appears, in the combined observation equation, in the form of linear combination with the group delay of navigation signal. It was pointed out that the best approach to remove the impact of this combined delay bias on orbit determination and clock solution is to estimate them together with orbit and clock parameters and so on. Method for estimating the combined delay for each satellite was proposed as well as the corresponding data processing strategy for orbit determination and clock solution with simultaneous combined delay estimation. Software for simulation and processing has been developed and the feasibility of the proposed method and data processing strategy is validated by simulation analysis. The result indicates that the accuracy of orbit and clock result would be significantly degraded due to presence of equipment delay; and would be improved to the maximum degree by estimating the combined delay, and the effect is nearly the same as if the combined delay has been exactly corrected.

1 Introduction

Since the early 1980s, the United States has been engaged in research on the technique of crosslink to provide autonomous navigation (AUTONAV) for GPS [1], so that the system would be able to maintain navigation service for a certain period (typically 6 months) in case that the ground operation facility was disabled.

R. Ruan (✉) · L. Feng · X. Jia · X. Song · X. Wu · T. He
Xi'an Research Institute of Surveying and Mapping, 710054 Xi'an, China
e-mail: rrg2002me@163.com

J. Sun et al. (eds.), *China Satellite Navigation Conference (CSNC) 2012 Proceedings*, Lecture Notes in Electrical Engineering 160, DOI: 10.1007/978-3-642-29175-3_14,

Payloads for crosslink and AUTONAV were equipped in the 20 BLOCK IIR Satellites and the sequels. Nowadays, developers of the other global navigation satellite systems, such as GLONASS, GALILEO, are active in developing crosslink for AUTONAV or orbit determination and time synchronization (OD&TS) at Operation Control Segment (OCS) [2]. There are also plans of development of crosslink in BeiDou navigation satellite system programme [3]. Crosslink will not only provide BeiDou with the capability of inter-satellite communication and AUTONAV which will greatly enhance the survivability of the system. Meanwhile, in order to ensure the designed performance index be achieved and crack the dilemma that ground stations can not be build worldwide, OD&TS based on inter-satellite and satellite-ground link observation data will be the basic data processing mode at OCS. It has been shown by many researchers that, by using this mode, the accuracy of orbit determination and clock solution under the conditions of the regionally distributed stations would be greatly improved [4–6]. One of the greatest difficulties researchers are confront with, is how to remove the impact of the equipment delay inherent in crosslink and navigation devices on OD&TS.

Equipment delay, typically including a constant part and a variable part varies with temperature and environment condition, is the propagation time interval of an electrical signal passing through electronic equipment and it is the inherent characteristics of electronic equipment. The stability of the equipment delay is an important indicator to a device. Equipment delay is one of the main factors affect the system performance of GNSS. Not to mention different devices, even the equipment delay of two devices of the same type are different. Therefore the equipment delay of crosslink and navigation payloads are inevitably different with each other. If these equipment delay can not be accurately calibrated, they will appear in the form of systematic errors and will inevitably degrade the accuracy of OD&TS at OCS.

In order to ensure that the desired results of OD&TS with inter-satellite and satellite-ground link data be achieved, It is critical to exactly calibrate or estimate equipment delay. We tried to find out the form that the equipment delay appear in by theoretical analysis and to propose an approach to estimate equipment delay.

2 Observation Equation

Pseudorange and carrier phase are the "natural" observation that GNSS can provide. Taking pseudorange and carrier phase as the main observations for OD&TS, the obtained result of orbit and clock solution will maintain the best consistence with GNSS user algorithms, and at the same time ensure the system simple. In this paper, satellite-ground data is referred to pseudorange and carrier phase.

2.1 Pseudorange and Phase Observation Equation

In order to separate the spectrum of difference navigation signals and eliminate the ionospheric delay effect by method of dual-frequency combination, navigation satellite systems are usually designed with multiple-frequency navigation signals. The Ionospheric Free (IF) combination of two-band is usually the basic observation for OB&TS. Without derivations, the IF combinations of pseudorange and carrier phase were written as follows [7]:

$$P_s^j = \rho_s^j + \delta t_s - \delta t^j + b_s + b^j + M_s^j \cdot d_{zpd,s} + \varepsilon_P \tag{1}$$

$$\phi_s^j = \rho_s^j + \delta t_s - \delta t^j + \tilde{N}_s^j + M_s^j \cdot d_{zpd,s} + \varepsilon_\phi \tag{2}$$

where $\rho_s^j = |\boldsymbol{r}_s - \boldsymbol{r}^j|$: the geometry distance between receiver s and satellite j; $\boldsymbol{r}_s$: position vector of receiver s; $\boldsymbol{r}^j$: position vector of satellite j; P_s^j, ϕ_s^j: IF combination of pseudorange and carrier phase of satellite j observed by receiver s ; ε_P, ε_ϕ: observation noise, include multi-path effect; δt_s: clock offset of receiver s; δt^j: clock offset of satellite j; b_s: group equipment delay (IF combination of pseudorange) of receiver s; b^j: group equipment delay of satellite j; $d_{zpd,s}$: zenith path delay of troposphere; M_s^j: mapping function of troposphere; $\tilde{N}_s^j$: ambiguity parameter, include phase equipment delay and initial phases of both satellite and receiver.

Terms such as antenna phase center offset and the relativistic effect, were omitted in the above equation. In addition, various sources of error need to be very carefully considered in OB&TS of high precision, for example the tidal displacement of stations, the effect of Phase Wind-up and so on [8].

2.2 Observation Equation of Inter-Satellite Pseudorange

Inter-satellite communication and ranging measurement operate in TDMA mode in accordance with the crosslink planning. Taking the crosslink of GPS BLOCK IIR satellites for example [9], each satellite is assigned a 1.5 s time slot within a 36 s TDMA frame. On each frame, the satellite transmits on its assigned slot and listens on the other 23 slots. So each satellite obtains its Inter-Satellite Ranging (ISR) observations with respect to the other satellites (if it can be made link to) on different time slot. This is quite different with the way that a GNSS receiver generates pseudorange and carrier phase observations of all the satellites in view just at the same time. So, for convenience, the ISR observations usually have to be interpolated to the desired time epoch by some algorithm (which is really complex and out the range of this paper) and the derived ISR observations are something like observations observed simultaneously. In this paper, inter-satellite ranging observation is referred to the "derived inter-satellite ranging observation".

To simplify the description, we directly present the formula of the derived ISR observation in a simplified form [10, 11]:

$$L_i^j = \gamma_i^j + \delta t^i - \delta t^j + \Delta^j + \delta^i + \varepsilon_L \tag{3}$$

$$L_j^i = \gamma_j^i + \delta t^j - \delta t^i + \Delta^i + \delta^j + \varepsilon_L \tag{4}$$

where L_i^j: pseudorange observed by satellite i to satellite $j (j \neq i)$; L_j^i: pseudorange observed by satellite j to satellite i $(j \neq i)$; $\gamma_i^j = |\boldsymbol{r}^i - \boldsymbol{r}^j|$: geometry distance for satellite i to j; δt^i: clock offset of satellite i; Δ: equipment delay of the crosslink signal transmitting device; δ: equipment delay of the crosslink signal receiving device; ε_L: observation noise, includes the uncertainty of equipment delay and errors caused by interpolation.

Terms of crosslink phase center offset and the relativistic effect were omitted in Eqs. (3) and (4). In order to separate the information of orbit and clock, there are two combinations of ISR observation usually used in applications [11]. Without consideration of the effect of satellite movement and satellite clock variation which has to be carefully deal with for data processing, they can be written as (not rigorous expression):

Clock-Free combination:

$$L_{i,CF}^j = \frac{L_i^j + L_j^i}{2} = \frac{\gamma_i^j + \gamma_j^i}{2} + \frac{\Delta^j + \delta^j}{2} + \frac{\Delta^i + \delta^i}{2} + \varepsilon_{L,CF} \tag{5}$$

And Geometry Free combination:

$$L_{i,GF}^j = \frac{L_i^j - L_j^i}{2} = \delta t^i - \delta t^j + \frac{\Delta^j - \Delta^i + \delta^i - \delta^j}{2} + \varepsilon_{L,GF} \tag{6}$$

Equation (5), with absence of the clock offset parameters, can only be used for orbit determination; while Eq. (6) can only be used for clock synchronization, for it doesn't contain information of geometry between satellites.

2.3 Combined Observation Equation

In Eqs. (1) and (2), b^j, the group equipment delay of navigation signal, appears explicitly. But the GNSS users with dual-frequency receivers do not need to correct this term. Because it has been included in the clock correction parameters broadcasted in nav-message. Actually, since the group delay parameters are linear related with clock parameters, in OD&TS data processing at OCS, the group delay is accumulated to satellite clock offset. In this way, it's convenient for both data processing and user applications. Considering this point, we had better rewrite Eqs. (1) and (2) in the form which is more consistent with data processing algorithm:

$$P_s^j = \rho_s^j + \Delta t_s - \Delta t^j + M_s^j \cdot d_{zpd,s} + \varepsilon_P \tag{7}$$

$$\phi_s^j = \rho_s^j + \Delta t_s - \Delta t^j + B_s^j + M_s^j \cdot d_{zpd,s} + \varepsilon_\phi \tag{8}$$

where

$$\Delta t_s = \delta t_s + b_s;\ B_s^j = \tilde{N}_s^j - b_s - b^j \tag{9}$$

$$\Delta t^j = \delta t^j - b^j \tag{10}$$

Thus, as mentioned above, the group delay of navigation signal is included by satellite clock offset, while the receiver clock offset absorbs the group delay of receiver. And ambiguity parameters include the linear combination of group equipment delay, phase equipment delay and initial phases of both satellite and receiver.

Satellites with inter-satellite ranging function are equipped with both transmitting and receiving devices of crosslink, and the sum of transmiting equipment delay and receiving equipment delay, $\Delta + \delta$, can be measured via self-loop measurement. This measured value can be used together with Eq. (5) for orbit determination, but offers no help for time synchronization which must be carried out by making use of Eq. (6) or Eqs. (3) and (4). It seems necessary to directly determine the transmitting and receiving equipment delay for each satellite, but this is very difficult and is even considered impossible.

It must be pointed out that Eqs. (7) and (8) are not only the basic observation equations for OD&TS but also the equations for user applications such as positioning, navigation and timing. In order to keep consistent with user algorithms, the satellite clock offsets calculated with ISR observations should also include the group equipment delay of navigation signal. Rather than the combined ISR observation expressed in Eqs. (5) and (6), We adopt the single-way ISR observations. By introducing the clock offset parameters expressed by Eq. (10) into Eqs. (3) and (4), the ISR observation equation for OD&TS can be written as:

$$\begin{aligned} L_i^j &= \gamma_i^j + \Delta t^i - \Delta t^j + \tilde{\Delta}^j + \tilde{\delta}^i + \varepsilon_L \\ L_j^i &= \gamma_j^i + \Delta t^j - \Delta t^i + \tilde{\Delta}^i + \tilde{\delta}^j + \varepsilon_L \end{aligned} \tag{11}$$

where:

$$\begin{aligned} \tilde{\Delta}^j &= \Delta^j - b^j; \tilde{\delta}^i = \delta^i + b^i \\ \tilde{\Delta}^i &= \Delta^i - b^i; \tilde{\delta}^j = \delta^j + b^j \end{aligned} \tag{12}$$

Combining Eq. (11) together with Eqs. (7) and (8) lead to the observation equation for OD&TS:

$$\begin{aligned} P_s^j &= \rho_s^j + \Delta t_s - \Delta t^j + M_s^j \cdot d_{zpd,s} + \varepsilon_P \\ \phi_s^j &= \rho_s^j + \Delta t_s - \Delta t^j + B_s^j + M_s^j \cdot d_{zpd,s} + \varepsilon_\phi \\ L_i^j &= \gamma_i^j + \Delta t^i - \Delta t^j + \tilde{\Delta}^j + \tilde{\delta}^i + \varepsilon_L \\ L_j^i &= \gamma_j^i + \Delta t^j - \Delta t^i + \tilde{\Delta}^i + \tilde{\delta}^j + \varepsilon_L \end{aligned} \tag{13}$$

To note that the equipment delay of crosslink device does not appear alone in the ISR observation equation, but in the form of linear combination together with the group equipment delay of navigation signal. $\tilde{\Delta}$ and $\tilde{\delta}$ denote the combined delay of navigation payload and crosslink payload (noted as "combined delay"). The magnitude of both the transmitting equipment delay Δ and receiving delay $\tilde{\delta}$ of crosslink is up to tens of nanoseconds (one nanosecond equals to about 0.3 m in distance), so is that of group equipment delay of navigation payload. Therefore the magnitude of Δ can be positive or negative tens of nanoseconds, the value of $\tilde{\delta}$ should be (positive) tens of nanoseconds. Since navigation devices and crosslink devices are independent, and operate with different frequency bands, the combined delay $\tilde{\Delta}$ and $\tilde{\delta}$ can not be accurately calibrated neither on ground nor in orbit. It is imperative to estimate these combined delay via data processing.

2.4 Estimation for Combined Equipment Delay

Equipment delay is usually stable during a certain period, which is the pre-condition that the combined delay parameters can be estimated. Equation (13) shows linear relationship among combined delay parameters and satellite clock parameters. Only when satellite clock has been determined or satellite-ground link ranging observation participate in data processing that the correlation among satellite clock and combined delay parameters can be removed. It is easy to think of tow approach for data processing: the first approach is, using only ISR observations expressed by Eq. (11) by fixing satellite clock (or both satellite orbit and clock) in advance, to solve both satellite orbit and combined delay (or only combined delay). Another approach is to use both satellite-ground link observation and the ISR observation to estimate all the unknown parameters including satellite orbit, clock offset and the combined delay at the same time. The former approach requires that, during the whole arc, each satellite should be continuously tracked by monitor stations in order to determine the satellite clock offsets for all epoch. Therefore the monitor stations should be globally distributed. The latter approach only requires that each satellite should be tracked by monitor stations during data processing arc (not requiring continuously tracked), which is quite easy to be fulfilled even if there was only one monitor station. The latter approach takes full advantage of ISR observations so that optimal estimation of satellite orbit, clock offset and combined delay parameters would be achieved. Considering that the amount of combined delay parameters is not large (typically two for each satellite), it will not significantly increase the difficulty and complexity of data processing. We recommended the latter approach.

It is needed to point out that even if satellite clock offsets have been determined, the normal equation formed with observation Eq. (13) is rank deficiency (typically of order one), which means that the absolute values of $\tilde{\delta}$ and $\tilde{\Delta}$ are unobservable. In order to remove the rank one deficiency, rather than eliminate one combined delay parameter directly, we prefer to statistical handling that just choose one combined delay parameter as reference and make prior information constrain on it, that is quite similar to the regular way dealing with clock parameters [12]. In this way, the estimated combined delays are all relative to the constrained reference value. Basic procedure of data processing can be simply described as follows:

(1) Input all pseudorange, carrier phase and ISR observations of one epoch.
(2) Make corrections for various error sources and accumulation of normal equation.
(3) Make constrain to reference clock parameter and eliminate clock parameters epoch by epoch, jump to step (1) until all the observations of the whole arc have been inputted and then continue with step (4).
(4) Make constrain to reference combined delay parameter, make inversion of normal equation and all parameters including combined delay, orbit, ambiguity and tropospheric zenith path delay parameters of stations are solved, except for clock parameters.
(5) Solve clock parameters epoch by epoch.

The estimated combined delay $\tilde{\Delta}$ and $\tilde{\delta}$ can be introduced into Eq. (11) or (13) for real time application or take place of Δ and δ in Eqs. (5) and (6).

3 Simulation and Analysis

3.1 Simulation Condition

Experiment was carried out with the data processing and simulation software for satellite-ground link and inter-satellite link data developed at Xi'an Research Institute of Surveying and Mapping. Basic conditions for simulation are as follows:

Constellation: Walker constellate 24/3/1:55° of MEO satellites;
Monitor stations: 7 stations (located at Beijing, Harbin, Chengdu, Sanya, Shantou, Wurumuqi, and Kashi);
Satellite-Ground link: observations noise 0.3 and 0.002 m for pseudorange and carrier phase respectively;
Crosslink: just like crosslink of GPS BLOCK IIR satellites with signal beam angle in nadir ranged from 15 to 60°.
Clock bias: white noise model with zero-mean and STD. 100 and 1 us for receiver and satellite respectively
Sample interval: 300 s for both satellite-ground link data and derived ISR data
Arc length: 3 days

Table 1 lists the transmitting and receiving equipment delay of crosslink payloads and the group equipment delay of navigation payloads inputted for simulation, which were generated by FORTRAN function RANDOM, as uniform distributed values ranged from 0.0 to 0.3 m.

Table 1 Values of equipment delay input for simulation (unit: m)

PRN	Δ	δ	b	PRN	Δ	δ	b
01	0.180	0.057	0.120	13	0.201	0.283	0.268
02	0.152	0.272	0.098	14	0.271	0.079	0.254
03	0.240	0.268	0.076	15	0.138	0.024	0.298
04	0.120	0.194	0.165	16	0.160	0.121	0.056
05	0.048	0.200	0.293	17	0.283	0.135	0.094
06	0.108	0.062	0.058	18	0.240	0.058	0.214
07	0.171	0.232	0.092	19	0.014	0.186	0.007
08	0.100	0.069	0.285	20	0.258	0.299	0.152
09	0.085	0.150	0.083	21	0.083	0.003	0.155
10	0.228	0.027	0.169	22	0.063	0.282	0.251
11	0.100	0.096	0.184	23	0.009	0.188	0.183
12	0.296	0.183	0.179	24	0.133	0.285	0.135

Based on conditions described above, three sets of date were generated with different inter-satellite ranging precision:

Dataset One: with inter-satellite ranging precision of 1.0 m;
Dataset Two: with inter-satellite ranging precision of 0.3 m;
Dataset Three: with inter-satellite ranging precision of 0.1 m;
Each dataset was used for OD&TS processing with three different data processing schemes:
Scheme One: ignore the equipment delay, just take them as random noise, and the data was directly used for OD&TS without any special process;
Scheme Two: correct the equipment delay with the values in Table 1 and then used for OD&TS;
Scheme Three: estimate the combined delay parameters together with other unknown parameters, the combined delay was treated as constant during the 3 day arc.

Therefore, one group of combined delay estimation result (estimated with Scheme Three) and three groups of orbit and clock result were obtained with respect to each dataset.

3.2 Estimation of Combined Delay

The theoretical (computed with data in Table 1 and Eq. (12)) and estimated values of combined delay and estimation errors with respect to satellites with PRN 1–8 are listed in Table 2. Since the combined receiving delay parameter of satellite with PRN 1, $\tilde{\delta}^1$, was taken as reference in data processing (with prior value of 0.0 m and a covariance constrain of 0.001 m^2), the estimation error of $\tilde{\delta}^1$ is –0.177 m. All estimated values of the combined transmitting and receiving delay have systematical errors with comparative magnitudes and opposite signs.

Table 2 The theoretical and estimated value of combined delay and estimation errors w.r.t. Satellites with PRN 1–8 (unit: m)

PRN	$\tilde{\Delta}$			$\tilde{\delta}$		
	Theoretical	Estimated	Error	Theoretical	Estimated	Error
01	0.060	0.236	0.176	0.177	0.000	−0.177
02	0.054	0.273	0.219	0.370	0.154	−0.217
03	0.164	0.307	0.143	0.344	0.197	−0.147
04	−0.045	0.134	0.179	0.359	0.179	−0.180
05	−0.245	−0.076	0.170	0.493	0.321	−0.172
06	0.050	0.234	0.184	0.120	−0.065	−0.185
07	0.079	0.256	0.177	0.324	0.146	−0.178
08	−0.185	−0.005	0.180	0.354	0.170	−0.184

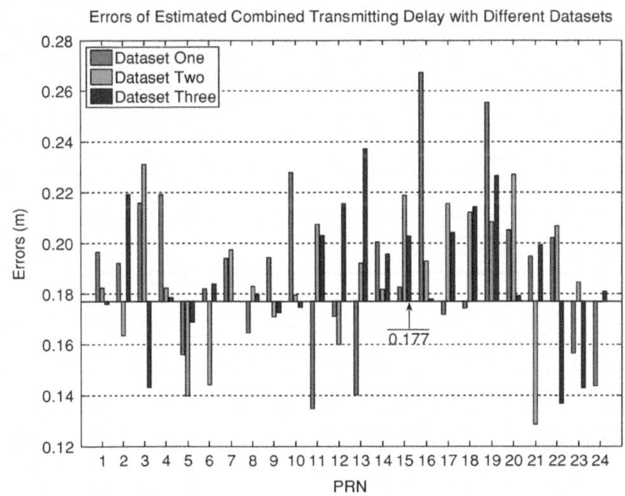

Fig. 1 Errors of combined delay $\tilde{\Delta}$ estimated with different datasets

Figures 1 and 2 show estimation errors of combined delay with respect to each satellite. The tail end of the bar indicates absolute error while the length of the bar represents relative error (compared with the baseline at 0.177 and −0.177 m). It's shown that the estimated results with the three datasets have great difference with each other. The biggest relative error of the combined transmitting delay $\tilde{\Delta}$ is 0.1 m, while 0.12 m for the combined receiving delay $\tilde{\delta}$.

Mean errors (MEAN) and standard deviation (STD) with three datasets were shown in Table 3. The MEANs of $\tilde{\Delta}$ and $\tilde{\delta}$ are nearly equal to the "ideal" value 0.177 m and -0.177 m respectively, with differences less than 0.02 m. In terms of STD, the values decrease from Dataset One to Dataset Three, for the corresponding precision of ISR observations increase. Since the precision of ISR

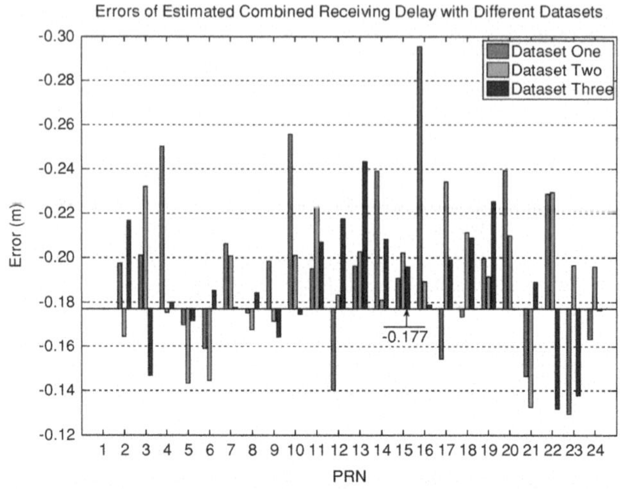

Fig. 2 Errors of the combined delay $\tilde{\delta}$ estimated with different datasets

Table 3 Mean errors and standard deviation of combined delay estimated with different datasets (unit: m)

	Mean		Std	
	$\tilde{\Delta}$	$\tilde{\delta}$	$\tilde{\Delta}$	$\tilde{\delta}$
Dataset One	0.189	−0.195	0.033	0.040
Dataset Two	0.187	−0.190	0.027	0.028
Dataset Three	0.187	−0.186	0.026	0.027

observations is finite, Even if the orbit and clock of satellites were correctly determined, we should not expect too much of the estimation accuracy of combined delay.

3.3 Orbit and Clock Solution

Since the experiment is carried out based on simulation, we had better pay our attention to the difference of the orbit and clock results obtained with different processing schemes. Figure 3 shows the 3D orbit RMS for each satellite obtained from Dataset One with three different processing schemes. It is easy to discover that, compared with the results of Scheme One, Scheme Two and Scheme Three significantly improved the accuracy of satellite orbit, and the difference between the results of Scheme Two and Scheme Three is really small. Mean 3D RMS of all satellites with Scheme One is about 0.7 m, while RMSs for all satellites with Scheme Two and Scheme Three are less than 0.15 m. Scheme Two and Scheme Three enjoy about 75% improvement compared with Scheme One.

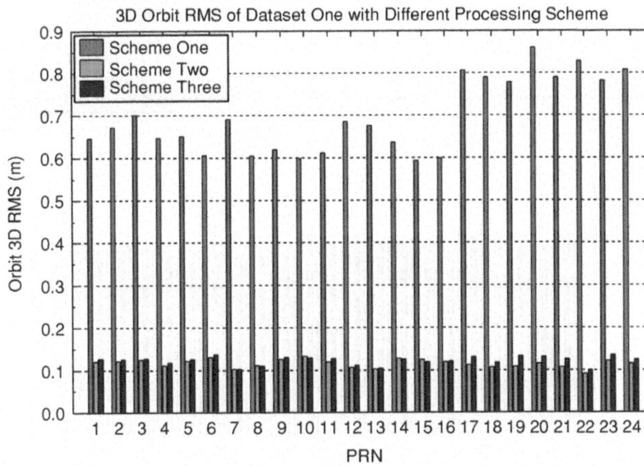

Fig. 3 RMS in 3D of orbit result obtained from dataset one with different schemes

Table 4 RMS in 3D of orbit result with different schemes and datasets (unit : m)

	Dataset One	Dataset Two	Dataset Three
Scheme One	0.695	0.694	0.709
Scheme Two	0.115	0.076	0.041
Scheme Three	0.122	0.076	0.053

Table 4 makes a census of the mean 3D orbit RMS of all satellites with different datasets and schemes. For all of the three Datasets, compared with Scheme One, Scheme Two makes an improvement of 83.5%, 89.0% and 94.2%, while Scheme Three makes an improvement of 82.4%, 89.0% and 92.5%. It is also shown that the effect of Scheme Three is nearly equal to Scheme Two, with an inferiority of less than 2%.

Figure 4 shows the clock RMS for each satellite obtained from Dataset One with three different processing schemes. For most satellites, the RMSs with Scheme Two and Scheme Three are smaller than those with Scheme One. But it's not the case for all satellites, because the ISR noise in Dataset One is bigger than the inputted equipment delay in terms of magnitude and makes dominant impact on the clock results.

Table 5 makes a census of the mean clock RMS of all satellites with different datasets and schemes. For all the three datasets, compared with Scheme One, Scheme Two makes an improvement of 12.1%, 45.9% and 66.6%, while Scheme Three makes an improvement of 10.5%, 40.5% and 56.9%. Once again, It is also observed that the effect of Scheme Three is close to Scheme Two with an inferiority of less than 10%.

It was noticed that, compared with orbit solutions, clock solutions seem less sensitive to the equipment delay. That's reasonable, because the clock parameters

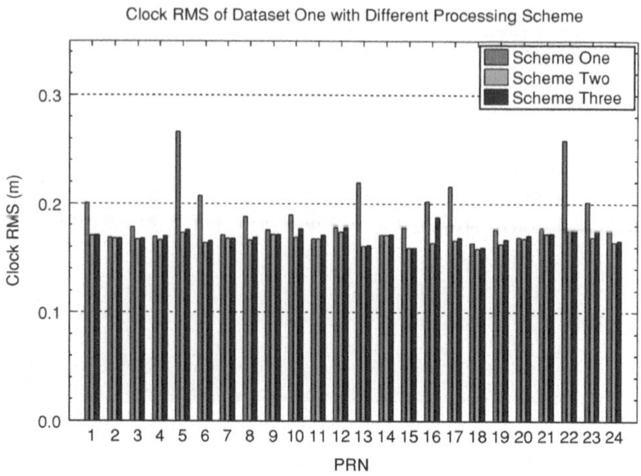

Fig. 4 RMS of clock solutions obtained from dataset one with different schemes

Table 5 RMS of clock solutions with different schemes and different datasets (unit: m)

	Dataset One	Dataset Two	Dataset Three
Scheme One	0.190	0.101	0.087
Scheme Two	0.167	0.055	0.029
Scheme Three	0.170	0.060	0.037

are strongly correlated to the observations of the current epoch (and insensitive to the observations of the other epochs); although, for the whole processing arc, equipment delay acts as systemic errors, however, for a certain epoch, the equipment delay of different satellites exhibit the nature of randomicity. Since the equipment delay values inputted for simulation are generated by RANDOM function (in practice, the real equipment delay values of different satellites are something like non-zero mean random values), and, in our experiments, the magnitude of inter-satellite ranging noise is bigger than or equal to that of the input hardware delay. So the impact of equipment delay on clock solution is not as significant as that on orbit solution. It's necessary to make out that, the actual equipment delay of satellite payloads is ten times larger than the values presented in this paper, and is certainly much bigger than the magnitude of ranging noise, and their impact on both orbit and clock solution is unacceptable.

4 Summary and Conclusion

With consideration of equipment delay, combined observation equation with satellite-ground link range and inter-satellite range observation is present, method for equipment delay estimation and the corresponding strategy for data processing are proposed, experiments with simulation were conducted and the proposed method and data processing strategy were validated, Some conclusions were achieve:

1. In the combined observation equation, equipment delay of crosslink payloads appear in the form of linear combinations with equipment delay of navigation payloads. The combined delay are mathematically independent with satellite clock offsets, therefore, in order to remove the impact on orbit determination and clock synchronization, the combined delay must be estimated in data processing.
2. If being treated as measurement noise, equipment delay would greatly reduce the accuracy of orbit and clock results, and the more precise the inter-satellite ranging observations are, the more sensitive to the equipment delay. Relatively speaking, orbit result is more sensitive than clock result.
3. The result of orbit and clock solution would be improved to the maximum degree by estimating the combined delay, and the effect is nearly equal to the case that the combined delay has been exactly corrected.

Although our work shows the feasibility and great effect to estimate the combined delay in data processing for orbit determination and clock synchronization. It doesn't mean that it's unnecessary to calibrate the crosslink equipment delay on orbit by self-loop measurement. After all, only the directly observed equipment delay (generally not exact) would represents the status of crosslink devices. Actually, equipment delay of crosslink devices would have been determined by manufactures before the satellite are launched into orbit. So the equipment delay appeared in this paper may usually stands for errors of equipment delay determination, and their values would be either positive or negative, however, the methods and conclusions in this paper still work.

References

1. Rajan, J. A., Brodie P., & Rawicz H. (2003). Modernizing GPS autonomous navigation with anchor capability. ION GPS/GNSS, Protland, OR, pp. 1534–1542.
2. Hein, G. (2011). Inter satellite links for GNSS, Xian.
3. China Satellite Navigation Office (2011). *Implementation of BeiDou navigation satellite system programme*, CSNC 2011, Shanghai.
4. Geng, T., Liu, J., Zhao, Q., & Shi, C. (2011). *Compass precise orbit determination based on space-ground monitoring network. GEODAEETICA et CARTOGRAPHICA SINICA, 40*(Sup.), 47–51 (In Chinese).
5. Liu, W., Gong, X., & Zhang, W. (2011). *GPS orbit determination with real ground tracking observations and simulative crosslink ranging observations*, CSNC 2011, Shanghai, pp. 451–455 (In Chinese).
6. Cai, Z., Han, C., Du, Y., & Yang, L. (2011). *Orbit determination method and accuracy analysis utilizing satellite-ground and satellite-satellite links*, CSNC 2011, Shanghai, pp. 472–476 (In Chinese).
7. Kleusberg, A., & Teunissen, P. J. G. (1996). *GPS for geodesy*. Berlin: Springer.

8. Wei, Z., & Ge, M. (1997). *Mathematical models for GPS relative positioning*. Beijing: Surveying and Mapping Press (In Chinese).
9. Parkinson, B. W., & J. Spilker, J. J. (1996). *Global Positioning System: Theory and Applications (Volume II)*. Washington: American Institute of Aeronautics and Astronautics, Inc.
10. Liu, W. (2008). *Research and simulation on autonomous orbit determination and combined orbit determination of navigation satellites*. Wuhan: Wuhan University (In Chinese).
11. Song, X. (2009). *Study on orbit determination of COMPASS navigation satellites*. Xian: Chang'an University (In Chinese).
12. Ge, M. (2010). Real-time precise positioning services. Xi'an.

A New Ambiguity Resolution Method for PPP Using CORS Network and its Real-time Realization

Xuan Zou, Weiming Tang, Chuang Shi and Jingnan Liu

Abstract The aim of this article is to improve the method of network RTK based on undifferenced (UD) observation corrections (abbreviated as URTK hereafter) by modeling the code UD observation residual. Thus rover could obtain more accurate approximate coordinate by code Single Point Positioning (SPP), and then, speeding up the ambiguity convergence time. In addition, based on the improved URTK method, a three-step method for single-differenced (SD) ambiguity resolution of rover is proposed, which could solve the ionosphere-free (LC) ambiguities in several seconds. Lastly, based on the existing network RTK software and the real-time data stream simulator, the efficiency of our new methods have been proved with a real-time test. The results show that rover could gain rapid, cm-level and real-time Precise Point Positioning (PPP) service using CORS network.

X. Zou (✉) · W. Tang · C. Shi · J. Liu
GNSS Research Center, Wuhan University, 129 Luoyu Road, 430079, Wuhan, China
e-mail: supermanzx1982@sina.com

W. Tang
e-mail: wmtang@whu.edu.cn

C. Shi
e-mail: shi@whu.edu.cn

J. Liu
e-mail: jnliu@whu.edu.cn

J. Sun et al. (eds.), *China Satellite Navigation Conference (CSNC) 2012 Proceedings*, Lecture Notes in Electrical Engineering 160, DOI: 10.1007/978-3-642-29175-3_15,

1 Introduction

Network RTK and PPP are two representative techniques for high accuracy Global Navigation Satellite System (GNSS)-based positioning [1]. PPP positioning accuracy can reach cm-level to dm-level for static or dynamic applications [2, 3]. However, due to the fact that PPP-based solutions use real-valued ambiguities, long convergence times are experienced [4]. Network RTK could provide rapid and precise relative positioning service [5]; however, it can only serve within the regional reference network. In order to take the advantages of both PPP and network RTK, network RTK based on undifferenced observation corrections (URTK) method was put forward by Ge Maorong in [6], which allows rovers to gain rapid and precise absolute positioning service comparable to standard network RTK. CORS-based network PPP-RTK is also a very effective method [7], but compared with the URTK method, it is more complex as it introduces the bases on clocks and ionospheric delay.

In this paper, we improved the original URTK method by modeling the code UD observation residual, and then a method for solving the SD LC ambiguities between satellites at a single station is presented. Finally, a real-time URTK software is developed based on the existing network RTK software, and the efficiency of our new methods have been proved with a real-time test.

2 Modeling Errors Using CORS Network

2.1 Network RTK Based on UD Observation Corrections

The concept of URTK method is shown as follows [6]. There are well-known methods to resolve double-differenced (DD) ambiguities and get the corresponding observation residuals [8–10]. Since the mapping function between DD ambiguities and their corresponding UD ambiguities shows a rank defect. If $n - m$ additional parameters are available and chosen as the basis, such as integer-valued UD ambiguities, the DD ambiguities can be converted into UD ambiguities that preserve the integer nature of DD ambiguities.

$$\underset{n\times 1}{N} = \begin{bmatrix} \underset{(n-m)\times n}{G} \\ \underset{m\times n}{D} \end{bmatrix}^{-1} \begin{bmatrix} \underset{(n-m)\times 1}{b} \\ \underset{m\times 1}{B} \end{bmatrix} \tag{1}$$

where m and n are the number of DD ambiguities (B) and UD ambiguities (N); b is the additional integer-valued basis; G and D are the corresponding mapping functions written in matrix form.

In the case of one DD-ambiguity (B_{01}^{ij}) and three additional integer-valued basis (b_0^i, b_0^j, b_1^i), the (1) also can be expressed as (2) in detail.

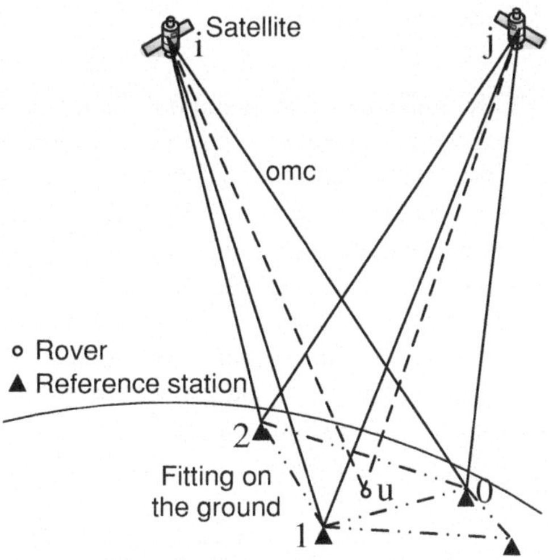

Fig. 1 Modeling *omc* in the direction of each common visible GNSS satellite with surrounding reference stations

$$\begin{bmatrix} N_0^i \\ N_0^j \\ N_1^i \\ N_1^j \end{bmatrix} = \begin{bmatrix} 1 & 0 & 0 & 0 \\ 0 & 1 & 0 & 0 \\ 0 & 0 & 1 & 0 \\ 1 & -1 & -1 & 1 \end{bmatrix}^{-1} \begin{bmatrix} b_0^i \\ b_0^j \\ b_1^i \\ B_{01}^{ij} \end{bmatrix} \tag{2}$$

With the UD ambiguities (N_k^i) gotten by (1), the phase UD observation residual (omc_{Lk}^i) between reference station (k) and satellite (i) can be computed using the following relation.

$$omc_{Lk}^i = \lambda L_k^i + \lambda N_k^i - \rho_k^i - ct_k + ct^i + M_k^i \tag{3}$$

where λ, L_k^i, ρ_k^i, c, t_k, t^i and M_k^i are the wavelength, phase observation, geometric distance, speed of light, clock error of station and satellite, and the modeled corrections (Fig. 1).

Thus the model for correction can be constructed based on these phase *omc* for each satellite [6, 11]. The logic for taking such basis, that is, the assumed UD ambiguities with arbitrarily integer-valued bias on the SD level at reference stations, is that the validity of modeling will not be violated [12].

This model deals with a small area in the direction of each individual satellite, thus is able to effectively model the influences of ionospheric delay, tropospheric delay, as well as satellite hardware delay, orbit and clock error, while the integer nature of ambiguity is maintained at the same time. Rovers inside the reference network could interpolate corrections with their approximate coordinates, and gain rapid PPP service with SD ambiguity resolution.

2.2 Improvements

URTK is originally proposed only to model the phase *omc*. The tropospheric delay between the corrected phase observation and the pure code observation is not consistent. In addition, the approximate coordinate of rover is gotten by SPP, whose accuracy is only 5–10 m. The coordinate error makes the float ambiguity solutions and their variance–covariance matrix inaccurate, thus increasing the difficulty of ambiguity resolution and affecting its success rate and reliability. Therefore, the accuracy of code observation should be improved so as to obtain more accurate approximate coordinate by SPP.

Based on the above analysis, we improved URTK method by modeling the code UD observation residual omc^i_{Pk} at the reference stations. Similar to (3), omc^i_{Pk} can be expressed as follows:

$$omc^i_{Pk} = P^i_k - \rho^i_k - ct_k + ct^i + M^i_k \tag{4}$$

omc^i_{Pk} is mainly composed of the ionospheric delay, tropospheric delay, orbit and clock errors, it can be interpolated using the same method of omc^i_{Lk} by (5). And then, the rover could obtain more accurate approximate coordinate by SPP with its corrected code observations.

3 Three-step Method for SD Ambiguity Resolution of the Rover

A triangulation of all reference stations is defined firstly using the Delaunay method to have the best geometry [13]. The corresponding triangular is selected according to the approximate user position. The model corrections of any rover station inside the reference network can be calculated by means of linear interpolation with horizontal coordinates as variables using *omc* of the same epoch at the three reference stations of the select triangle.

$$omc_u = \frac{(x_2y_u - x_uy_2)(omc_1 - omc_0) + (x_uy_1 - x_1y_u)(omc_2 - omc_0)}{x_2y_1 - x_1y_2} + omc_0 \tag{5}$$

where x and y are the horizontal coordinates of stations 1, 2, u relative to station 0; *omc* is the observation residual at reference station.

The same orbits, clocks and tropospheric delay model for generating *omc* are used to generate the observation equations at the rover side. Since the ionospheric delay, unmodeled tropospheric delay and satellite orbit errors in the directions between a satellite and stations are not strictly linear dependent, the model corrections of rover calculated by (5) also contain some errors. Taking into account that the wavelength of L_1 is only 19 cm, it is difficult to ensure the success rate and reliability of ambiguity resolution if the SD L_1 ambiguities are resolved directly, and even cause errors.

The wavelength of wide lane (WL) observation is 86 cm and the influence on it is relatively small, so that its SD ambiguities can be easily fixed, and then the rover could get much more accurate approximate coordinates than SPP. In addition, the interpolation precision of LC *omc* at the reference stations using (5) will not be influenced by the ionospheric delay, it should be more accurate than the interpolation of L_1/L_2 *omc*. Based on the characteristics mentioned above, this paper presents a new method which resolves the SD-ambiguities of rover in three steps: WL ambiguity resolution - L_1 ambiguity resolution - LC ambiguity fixed solution.

Although the work presented in this contribution only dual-frequency GNSS data, this method is also applicable to multi-frequency SD ambiguity resolution similarly.

3.1 WL Ambiguity Resolution

(1) The float solution of SD WL ambiguity

As the P1/CA contains smaller observation noise than ionosphere-free combination of pseudorange (PC) observation, so the corrected P1/CA and the corrected WL phase observations are used to form the observation equations.

$$\begin{bmatrix} l_{P1} \\ l_{WL} \end{bmatrix} = \begin{bmatrix} B & 0 \\ B & I\lambda_{WL} \end{bmatrix} \begin{bmatrix} a \\ N_{WL} \end{bmatrix} + \begin{bmatrix} V_{P1} \\ V_{WL} \end{bmatrix} \tag{6}$$

where B is the given design matrix; λ_{WL} is the wavelength of WL observation; I the identity matrix; a is the receiver coordinate; N_{WL} is the SD WL ambiguity; V_{P1} and V_{WL} the corrections of P1/CA and WL; l_{P1} and l_{WL} the observed-minus-computed SD residuals for P1/CA and WL, respectively. As a result of solving (6), one obtains the so-called float solution of WL ambiguity, together with its variance–covariance matrix.

(2) Fixing the SD WL ambiguity

The real-valued SD WL ambiguity and its related variance–covariance matrix are input to the LAMBDA method [8]. Since the LAMBDA method always outputs an integer solution, we have to decide whether this integer solution can be accepted as the correct integer solution with a sufficient amount of reliability. For this purpose the ratio test is carried out [14]. The resolved ambiguities are more reliable when the ratio value is greater. The threshold or critical value of the ratio test for deciding whether this integer solution can be accepted is taken as 1/3.

3.2 L_1 Ambiguity Resolution

Similar to (6), with the resolved SD WL ambiguities, these WL phase data start to act as if they were very precise code data, and they are used to form the observation equations with the corrected LC phase data.

$$\begin{bmatrix} l_{WL} \\ l_{LC} \end{bmatrix} = \begin{bmatrix} B & 0 \\ B & I\lambda_{LC} \end{bmatrix} \begin{bmatrix} a \\ N_{LC} \end{bmatrix} + \begin{bmatrix} V_{WL} \\ V_{LC} \end{bmatrix} \tag{7}$$

where B is the given design matrix; λ_{LC} is the wavelength of LC observation; I the identity matrix; a is the receiver coordinate; N_{LC} is the SD LC ambiguity; V_{WL} and V_{LC} the corrections of WL and LC; l_{WL} and l_{LC} the observed-minus-computed SD residuals for WL and LC, respectively.

For L_1 observation, its related SD ambiguity (N_{L1}) can be expressed as the following combination of LC and WL ambiguities:

$$N_{L1} = \frac{f_1 + f_2}{f_1} N_{LC} - \frac{f_2}{f_1 - f_2} N_{WL} \tag{8}$$

Considering the relation of (8), the LC ambiguity in (7) can be eliminated. The remaining parameters are only the L_1 ambiguity and the receiver coordinate. As a result of solving the modified (7), one obtains the so-called float solution of L_1 ambiguity, together with its variance–covariance matrix (Q_{L1}).

$$Q_{L1} = \left(\frac{f_1 + f_2}{f_1} \right)^2 Q_{LC} \tag{9}$$

Similar to the method in Sect. 3.1, the SD L_1 ambiguities can be obtained by LAMBDA method together with ratio test.

3.3 LC Ambiguity Fixed Solution

Once the integer solution of SD WL and L_1 ambiguities are accepted, the third processing step is executed, in which the reliability of ambiguity resolution is checked by their linear relationship. The observation equations of SD WL and L_1 can be formed as:

$$\begin{cases} m_{WL} & = omc_{WL} + N_{WL}\lambda_{WL} + a\Delta_X + b\Delta_Y + c\Delta_Z \\ m_{L1} & = omc_{L1} + N_{L1}\lambda_{L1} + a\Delta_X + b\Delta_Y + c\Delta_Z \end{cases} \tag{10}$$

where omc the SD residual of corrected phase observation; m is the remaining model error of URTK method; a, b and c are the linear coefficients of coordinates (Δ_X, Δ_Y, Δ_Z).

Since the ionospheric delay can be modeled by URTK method accurately [6]. In addition, m is also a small value. So that it is considered to be the same for L_1 and WL observations, and the following relation can be derived.

$$\hat{N}_{L1} = \frac{omc_{WL} - omc_{L1} + N_{WL}\lambda_{WL}}{\lambda_{L1}} \tag{11}$$

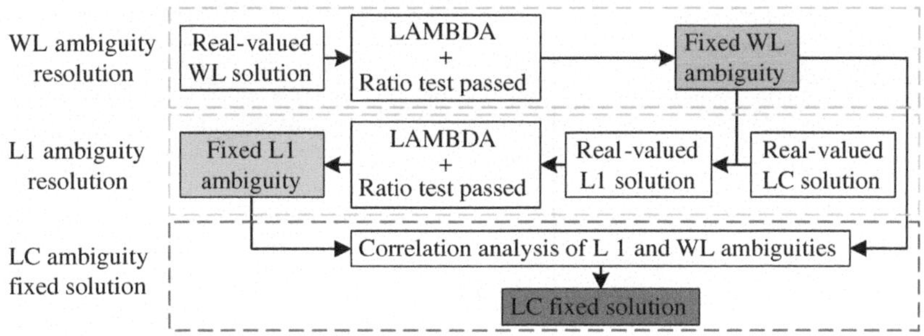

Fig. 2 Three-step method for SD ambiguity resolution of rover

The real-valued SD L_1 ambiguity by (11) is set as $\hat{N}_{L1}$, then the difference between $\hat{N}_{L1}$ and the fixed L_1 ambiguity (N_{L1}) can be used for checking.

$$|N_{L1} - \hat{N}_{L1}| < \delta \tag{12}$$

where δ denotes the threshold or critical value for the checking of the relationship, and it usually can be set as 0.2. When the difference is less than δ, the obtained SD WL and L_1 ambiguities are considered to be reliable. And then, the high-precision PPP based on fixed LC ambiguity resolution can now be realized as (13) (Fig. 2).

$$N_{LC} = \frac{f_1 f_2}{f_1^2 - f_2^2} N_{WL} + \frac{f_1}{f_1 + f_2} N_{L1} \tag{13}$$

4 Analysis of the Real-time Test

4.1 Data Sets and the Real-time Software

(1) Data sets
All the phase and code observations from a CORS network in China have been collected above a cut-off elevation of 10° from 0:10 to 3:00 pm of April 5th 2006, with a sampling interval of 1 s. The data are composed of 3 reference stations and 1 rover station. The checking network is depicted in Fig. 3. The 3 reference stations are at distances of 170 km and all equipped with the same Javad receivers. Since some time is required before the DD ambiguities of reference network can be solved reliably, which is the basis of implementing the URTK method, the first one hour of data has not been taken into account in our checking tests.

(2) Real-time processing software
Based on PANDA [15, 16] and the existing network RTK software Power-NetworkPP [10], our methods mentioned above for real-time and rapid PPP were tested. This real-time processing and checking software is composed of three parts,

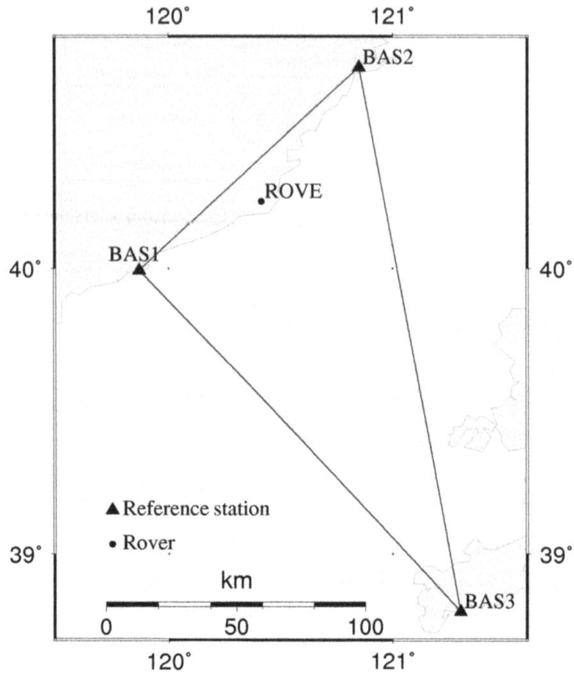

Fig. 3 The configuration of real-time CORS network used for this study

the real-time data stream simulator, the real-time data processing software for the reference network, as well as the real-time and rapid PPP software.

The real-time data stream which is composed of the observation data and the broadcast ephemeris are firstly simulated and sent out by internet using the TCP/IP protocol, each station has its own TCP/IP port as identity. Secondly, the real-time data processing software for the reference network receives and processes these data epoch-wise. The corrections generated according to the improved URTK method are sent to the rovers also by internet using the TCP/IP protocol. Lastly, the real-time and rapid PPP software is implemented as described in Sect. 3. The ambiguity resolution consisting of LAMBDA and ratio test is repeated for every epoch.

4.2 Accuracy Analysis of SPP

To ensure the robustness of the normal equation, the approximate coordinate of rover needs to be set at a suitable accuracy, and then, it is added into the normal equation as a simulated observation equation. The ambiguity resolution will be influenced if the coordinate accuracy doesn't meet the actual situation. Since the approximate coordinate of the rover is obtained by SPP, in this section we will focus on its positioning accuracy with the code correction and without it.

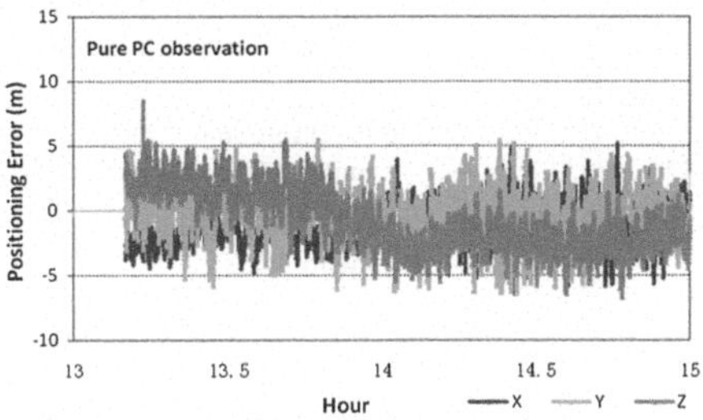

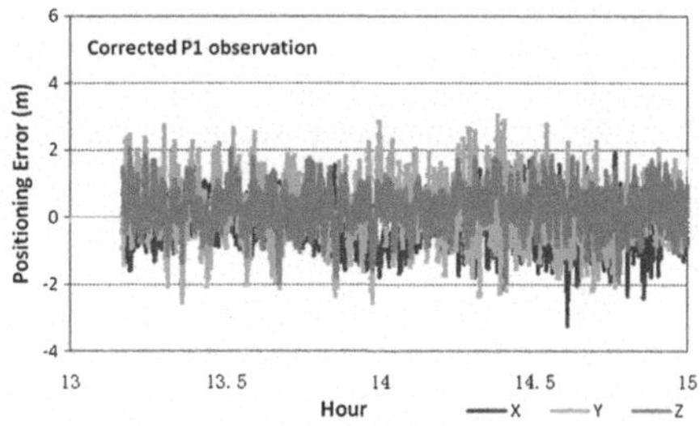

Fig. 4 The 3D positioning errors of the rover station in real-time SPP mode

Figure 4 shows the 3D positioning errors of the rover station in real-time SPP. The left chart is the result without code corrections. Because of the accuracy limit of code observation and broadcast ephemeris, the positioning accuracy of SPP is only 5 m. Compared with the original URTK method, rover's approximate coordinate obtained by SPP with the corrected P1/CA observation is much more accurate, and it is even better than 2 m in this test.

4.3 Positioning Results and Discussing

To demonstrate the efficiency of our new methods, all the data of the rover station are processed in real-time dynamic PPP mode with epoch-wise SD-ambiguity resolution as described in Sect. 3. According to the conclusions of Sect. 4.2, the coordinates of the rover station are estimated with a loose constraint of 5 and 2 m when using the original and improved URTK method, respectively.

Figure 5 presents the distribution of SD WL and L_1/LC ambiguity convergence time for the rover station according to the original and improved URTK method. The purple column is the results of SD WL ambiguity resolution, and the red line stands for the results of SD L_1/LC ambiguity resolution. According to the original URTK method, 96.94 and 90.08% of WL and L_1/LC ambiguities can be fixed within 5 s, and their average resolution time are 2.86 and 4.46 s, respectively. When the improved URTK method is used, the ambiguity convergence time for the rover station is speeded up significantly, 99.93 and 96.80% of WL and L_1/LC ambiguities can be fixed within 5 s, and their average resolution time are only 2.00 and 3.35 s.

Table 1 shows the positioning RMS with fixed SD WL and L_1/LC ambiguities between satellites according to the original and the improved URTK method. Using the 1 h and 50 min' observation data (1:10–3:00 pm local time), 1079 and 1373 times of tests were totally performed. When SD WL ambiguities are fixed, their positioning RMS in the north, east and up directions are (3.3, 2.2 and 9.2 cm), (2.3, 1.5 and 8.4 cm), respectively. Once the SD L_1/LC ambiguities are resolved, their RMS further improved to (1.4, 1.3 and 4.2 cm) and (1.3, 1.3 and 3.9 cm).

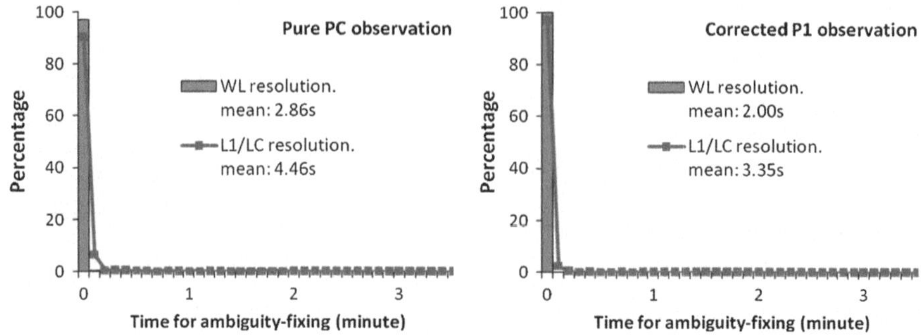

Fig. 5 Distribution of SD WL and L_1/LC ambiguity convergence time

Table 1 The positioning RMS with fixed SD WL and L_1/LC ambiguities between satellites, and the average ambiguity convergence time is also given

Modeling	Testing times		Ambiguity convergence time (s)		Positioning RMS (cm)					
					North		East		Up	
	Fixed	Unfixed	WL	L_1/LC	WL	L_1/LC	WL	L_1/LC	WL	L_1/LC
Original	1079	5	2.86	4.46	3.3	1.4	2.2	1.3	9.2	4.2
Improved	1373	4	2.00	3.35	2.3	1.3	1.5	1.3	8.4	3.9

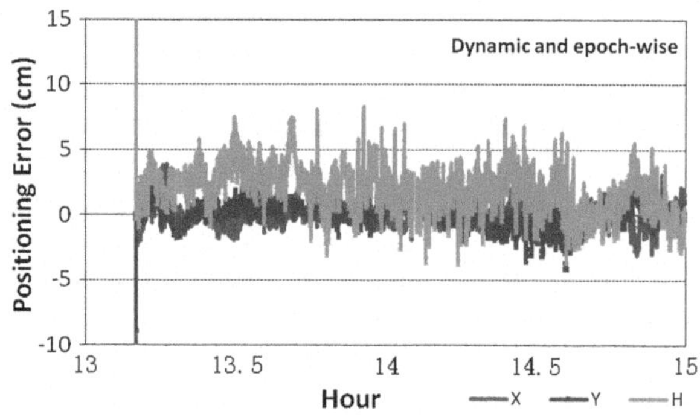

Fig. 6 The 3D positioning errors of the rover station in real-time PPP mode

The significantly large RMS of the WL solution is basically due to the enlarged observations noises through the combination.

From the performed data analysis and comparison, we can see that the accuracy of approximate coordinate is very important for ambiguity resolution both on the reliability and the convergence time. The improvement of URTK method and the proposed method for SD ambiguity resolution of the rover have been proved to be efficient.

Figure 6 shows the dynamic epoch-wise resolution of the station ROVE in real-time PPP mode. We can see that, the rover could obtain dynamic cm-level positioning in both horizontal and vertical directions after several seconds for ambiguity resolution. This further validates the availability of realizing rapid, cm-level and real-time PPP using CORS network.

5 Conclusions

In this paper, we improved the original URTK method by modeling the code UD observation residual, so as to obtain more accurate approximate coordinate and finally speed up ambiguity resolution.

Furthermore, a new method for rapidly fixing rover's ambiguities is proposed, which could solve the SD LC ambiguities between satellites in three steps: WL ambiguity resolution - L_1 ambiguity resolution - LC ambiguity fixed solution.

Based on the existing network RTK software and the real-time data stream simulator, the performances of our new methods have been proved to be efficient by the real-time test from a CORS network in China. Results demonstrate that, according to the improved URTK method, the rover could obtain rapid, cm-level and real-time PPP using CORS network that could be very flexible and useful for broader applications. In addition, our real-time test only used the broadcast ephemeris, it is expected to get better processing results once the real-time precise ephemeris and clock error products are available.

Acknowledgments This work is supported by the Open Foundation of State Key Laboratory of Information Engineering in Surveying, Mapping and Remote Sensing, Wuhan University (Grant No. 11P01), the National Natural Science Foundation of China (Grant No. 41004014) and the China Postdoctoral Science Foundation (Grant No. 20110491191).

References

1. Kouba, J., & Heroux, H. (2001). Precise point positioning using IGS orbit and clock products. *GPS Solutions, 5*, 12–28.
2. Gao, Y., & Chen, K. (2005). Performance analysis of precise point positioning using rea-time orbit and clockproducts. *Journal of Global Positioning Systems, 3*(1–2), 95–100.
3. Feng, Y. M., & Li, B. F. (2010). Wide area real time kinematic decimetre positioning with multiple carrier GNSS signal. *Science China Earth Sciences, 53*(5), 731–740.
4. Zumberge, J., Heflin, M., Jefferson, D., Watkins, M. M., & Webb, F. H. (1997). Precise point positioning for the efficient and robust analysis of GPS data from large networks. *Journal of Geophysical Research, 102*, 5005–5017.
5. Rizos, C., & Han, S. (2003). Reference station network based RTK systems—concepts and progress. *Wuhan University Journal of Natural Sciences, 8*(2), 566–574.
6. Ge, M. R., Zou, X., Dick, G., Jiang, W. P., Wickert, J., & Liu, J.N. (2010). An alternative network RTK approach based on undifferenced observation corrections. ION GNSS.
7. Teunissen P. J. G., Odijk, D., & Zhang, B. (2010), PPP-RTK: Results of CORS network-based PPP with integer ambiguity resolution, *Journal of Aeronautics, Astronautics and Aviation, Series A, 42*(4). 223–230.
8. Teunissen, P. J. G. (1995). The least-squares ambiguity decorrelation adjustment: A method for fast GPS integer ambiguity estimation. *Journal of Geodesy, 70*(1–2), 65–82.

9. Dai, L., Wang, J., Rizos, C., & Han, S. (2003). Predicting atmospheric biases for real-time ambiguity resolution in GPS/GLONASS reference station networks. *J Geodesy, 76*(11–12), 617–628.
10. Tang, W. M., Meng, X. L., Shi, S., & Liu, J. N. (2010). Assessment of a long range reference station ambiguity resolution algorithm for network RTK GPS positioning. *Survey Review, England: APR, 42*(316), 132–145.
11. Zou, X., Tang, W. M., Ge, M. R., & Liu, J. N. (2011), Equivalence of existing network RTK methods under the PPP mode, IUGG 2011.
12. Zou, X., Tang, W. M., Ge, M. R., & Liu, J. N. (2011 b). Method of network RTK based on undifferenced observation corrections and its functional realization in Cross-CORS service. *Acta Geodaetica et Cartographica Sinica, 40*. 1–5.
13. Sibson, R. (1978). Locally equiangular triangulations. *Computer Journal, 21*. 243–245.
14. Euler, H. J., & Schaffrin, B. (1990). On a measure of the discernibility between different ambiguity solutions in the static-kinematic GPS mode. In K. P. Schwarz & G. Lachapelle (Eds.), *Kinematic systems in geodesy, surveying and remote sensing, IAG Symposium 107* (pp. 285–295). Springer, New York.
15. Liu, J. N., & Ge, M. R. (2003). PANDA software and its preliminary result of positioning and orbit determination. *Wuhan University Journal of Natural Sciences, 8*(2), 603–609.
16. Shi, C., Zhao, Q. L., Geng, J. H., Lou, Y. D., Ge, M. R., & Liu, J. N. (2008), Recent development of PANDA software in GNSS data processing. Vol. 7285 72851S, in *Proceedings of SPIE*, ICEODPA.

Ionosphere-Free Combinations for Triple-Frequency GNSS with Application in Rapid Ambiguity Resolution Over Medium-Long Baselines

Jinlong Li, Yuanxi Yang, Junyi Xu, Haibo He and Hairong Guo

Abstract Aiming at rapid precise positioning by triple-frequency Global Navigation Satellite System (GNSS) over medium-long baselines, the application of triple-frequency ionosphere-free combinations in ambiguity resolution and precise positioning is analyzed. The analytical expression of linear coefficients of them is deduced, and the influences of various errors on ambiguity resolution and precise positioning are evaluated. Considering that the wavelength of optimal ionosphere-free combinations is too small, a method for decomposing the ambiguity of triple-frequency ionosphere-free combinations is presented, and the formula to calculate the virtual wavelength of them is given. In addition, an algorithm for narrow-lane ambiguity resolution is proposed, which is immune to systematic errors. It is proved that the resolutions of narrow-lane ambiguity by this proposed method are equal, with any wide-lane combinations that the sum of linear coefficients of them is equal to zero.

1 Introduction

Future Global Navigation Satellite Systems (GNSS), such as the US modernized Global Positioning System (GPS), the European GALILEO system and the Chinese Beidou or COMPASS system, all broadcast navigation signals in three or more

J. Li (✉) · J. Xu
Institute of Surveying and Mapping, Information Engineering University,
450052 Zhengzhou, China
e-mail: along0730@163.com

Y. Yang
Xi'an Research Institute of Surveying and Mapping, 710054 Xi'an, China

H. He · H. Guo
Beijing Global Information Center of Application and Exploitation, 100094 Beijing, China

J. Sun et al. (eds.), *China Satellite Navigation Conference (CSNC) 2012 Proceedings*, Lecture Notes in Electrical Engineering 160, DOI: 10.1007/978-3-642-29175-3_16,

frequencies, and triple-frequency carrier-phase measurements are available for GNSS precision positioning in the near future. It has been demonstrated that the computational burden of Ambiguity Resolution (AR) can be alleviated by using triple-frequency carrier-phase measurements, and the success rate and reliability of AR are also improved [1–7]. At present, various Three Carrier Ambiguity Resolution (TCAR) algorithms by geometry-free or geometry-based approach are effectual for AR over short baselines or wide-lane (WL) AR over long baselines [1, 2, 5, 8]. However, for long baselines, the geometry-free approach is affected by residual double-difference (DD) ionosphere error, and the geometry-based approach is affected by residual DD troposphere delay error, so both of them cannot resolve narrow-lane (NL) ambiguity practicably [6, 8–11]. For precision positioning over long baselines, the well-known ionosphere-free observation is often used to remove the ionosphere effect. It is often claimed that the crucial integer property of the DD carrier-phase ambiguity is lost when the ionosphere-free combination is formed, wherefore the related ambiguity usually expressed as the combination of WL and NL ambiguity for ambiguity fixing. The ionosphere-free ambiguity is fixed only when both its WL and NL ambiguity are fixed. Recently, Odijk presented another approach for ionosphere-free ambiguity resolution, which estimates the ionosphere-free ambiguity directly by only phase observations [12, 13]. Han and Rizos proposed a geometry-free AR algorithm without distance constraint in the condition of triple-frequency [14]. In addition, Zhang [15], Hatch [16] and Li [17, 18] derived a geometry-free and ionosphere-free technique to estimate NL ambiguity over long baselines respectively from two linearly independent ionosphere-free observations formed from three original carrier-phase measurements, the ionosphere-free observation formed from two ambiguity-corrected WL observations and the triple-frequency optimal ionosphere-free observation, two ambiguity-corrected WL observations and one NL observation. Because all systemic errors are removed, the NL ambiguity can be fixed by rounding the float ambiguity averaged in time domain. This paper is organized as follows. In the Sect. 2, the general definition of triple-frequency ionosphere-free combination is given and three kinds of ionosphere-free combination are defined, which are the dual-frequency ionosphere-free combination, the optimal triple-frequency ionosphere-free combination and the triple-frequency geometry-free and ionosphere-free combination. In the Sect. 3, the decomposition of ionosphere-free combination for narrow-lane ambiguity resolution is discussed. In the Sect. 4, the geometry-free and ionosphere-free approach for narrow-lane ambiguity resolution is derived from the decomposing of the geometry-free and ionosphere-free combination.

2 Ionosphere-Free Carrier-Phase Combinations for Triple-Frequency GNSS

If it is assumed that the $\nabla\Delta\phi_1, \nabla\Delta\phi_2$ and $\nabla\Delta\phi_3$ are the DD carrier-phase measurements in units of meter on f1, f2 and f3 respectively, and then the linear combination of carrier-phase measurements for the triple-frequency case can be defined as [3, 5, 19]:

$$\begin{aligned}\nabla\Delta\phi_{xyz} &= x\cdot\nabla\Delta\phi_1 + y\cdot\nabla\Delta\phi_2 + z\cdot\nabla\Delta\phi_3\\ &= (x+y+z)\cdot(\nabla\Delta\rho + \nabla\Delta T) - \mu_{xyz}\cdot\frac{\nabla\Delta K_1}{f_1^2} - \vartheta_{xyz}\cdot\frac{\nabla\Delta K_2}{2f_1^3}\\ &\quad + \lambda_{xyz}\cdot\nabla\Delta N_{xyz} + \nabla\Delta e_{xyz}\end{aligned} \tag{1}$$

where $\nabla\Delta\rho$ is the DD geometric range from receiver to satellite; $\nabla\Delta T$ is the DD troposphere delay error; $\nabla\Delta K_1$ and $\nabla\Delta K_2$ are the first- and second-order DD ionospheric parameter respectively. And the first- and second-order ionospheric scale factors, the ambiguity bias term in units of meter and the DD measurement noise of the linear combination can be respectively expressed as [5]:

$$\begin{cases}\mu_{xyz} = x + y\cdot\dfrac{f_1^2}{f_2^2} + z\cdot\dfrac{f_1^2}{f_3^2}\\ \vartheta_{xyz} = x + y\cdot\dfrac{f_1^3}{f_2^3} + z\cdot\dfrac{f_1^3}{f_3^3}\\ \lambda_{xyz}\cdot N_{xyz} = x\cdot\lambda_1 N_1 + y\cdot\lambda_2 N_2 + z\cdot\lambda_3 N_3\\ \nabla\Delta e_{xyz} = x\cdot\lambda_1\nabla\Delta\varepsilon_1 + y\cdot\lambda_2\nabla\Delta\varepsilon_2 + z\cdot\lambda_3\nabla\Delta\varepsilon_3\end{cases} \tag{2}$$

where λ_i, $\nabla\Delta N_i$ and $\nabla\Delta\varepsilon_i (i = 1, 2, 3)$ are the wavelength, the DD integer ambiguity and the DD measurement noise (including phase multipath errors) of the carrier-phase measurement on *fi*.

When it is assumed that the three original phase measurements are uncorrelated and have the same variance $\sigma_\varepsilon[\text{cy}]$ in units of cycles, the variance of the linear combination is computed as follows:

$$\sigma_{xyz}[\text{m}] = A_{xyz}\cdot 2\lambda_1\sigma_\varepsilon[\text{cy}] \tag{3}$$

where $A_{xyz} = \sqrt{x^2 + q_2\cdot y^2 + q_3\cdot z^2}$ is the noise amplification factor, and $q_2 = \lambda_2^2/\lambda_1^2$, $q_3 = \lambda_3^2/\lambda_1^2$.

In order to remove the first-order ionospheric effect, the coefficients x, y, z of the ionosphere-free carrier-phase combination must satisfy:

$$\begin{cases}x + y + z = s\\ \mu_{xyz} = 0\end{cases} \tag{4}$$

From Eq. (4), the coefficients of the ionosphere-free combination can be expressed as:

$$x_{\text{IF}} = x \quad y_{\text{IF}} = \frac{x\cdot f_2^2(f_3^2 - f_1^2) + s\cdot f_1^2 f_2^2}{f_1^2(f_2^2 - f_3^2)} \quad z_{\text{IF}} = \frac{x\cdot f_3^2(f_1^2 - f_2^2) - s\cdot f_1^2 f_3^2}{f_1^2(f_2^2 - f_3^2)} \tag{5}$$

From Eq. (5), we can know that there are infinite kinds of ionosphere-free combination for the triple-frequency case. However, the Optimal Ionosphere-Free (OIF) combination with not scaled geometric range and least noise variance is required for AR and precision positioning. Consequently, in addition to Eq. (4), the coefficients must further satisfy the conditions of $s = 1$ and $A_{xyz} = \min$. As a result, the coefficients of OIF can be expressed as:

$$\begin{aligned} x_{\mathrm{OIF}} &= f_1^2 \frac{f_2^2(f_1^2 - f_3^2) + f_3^2(f_1^2 - f_2^2)}{f_1^2(f_2^2 - f_3^2)^2 + f_2^2(f_3^2 - f_1^2)^2 + f_3^2(f_1^2 - f_2^2)^2} \\ y_{\mathrm{OIF}} &= f_2^2 \frac{f_1^2(f_2^2 - f_3^2) + f_3^2(f_2^2 - f_1^2)}{f_1^2(f_2^2 - f_3^2)^2 + f_2^2(f_3^2 - f_1^2)^2 + f_3^2(f_1^2 - f_2^2)^2} \\ z_{\mathrm{OIF}} &= f_3^2 \frac{f_1^2(f_3^2 - f_2^2) + f_2^2(f_3^2 - f_1^2)}{f_1^2(f_2^2 - f_3^2)^2 + f_2^2(f_3^2 - f_1^2)^2 + f_3^2(f_1^2 - f_2^2)^2} \end{aligned} \tag{6}$$

Furthermore, if s is given the value of zero in Eq. (5), the geometric range and troposphere delay error of the ionosphere-free observation can be further eliminated, forming a triple-frequency geometry-free and ionosphere-free (GIF) combination. According to Eq. (5), the coefficients of GIF can be expressed as:

$$x_{\mathrm{GIF}} = x \qquad y_{\mathrm{GIF}} = x \cdot \frac{f_2^2(f_3^2 - f_1^2)}{f_1^2(f_2^2 - f_3^2)} \qquad z_{\mathrm{GIF}} = x \cdot \frac{f_3^2(f_1^2 - f_2^2)}{f_1^2(f_2^2 - f_3^2)} \tag{7}$$

When x take some special value, Eq. (7) can give the same coefficients as literature [16] and [20]. Because x value has no influence on the analysis of the following text, $x = 1$ in the following text, i.e.

$$x_{\mathrm{GIF}} = \frac{f_2^2 - f_3^2}{f_2^2 - f_3^2} \frac{f_1^2}{f_1^2} \qquad y_{\mathrm{GIF}} = \frac{f_3^2 - f_1^2}{f_2^2 - f_3^2} \frac{f_2^2}{f_1^2} \qquad z_{\mathrm{GIF}} = \frac{f_1^2 - f_2^2}{f_2^2 - f_3^2} \frac{f_3^2}{f_1^2} \tag{8}$$

From the above, the carrier-phase observation with the coefficients expressed as Eqs. (6) and (8) can be expressed as

$$\nabla\Delta\phi_{\mathrm{OIF}} = \nabla\Delta\rho + \nabla\Delta T - \vartheta_{\mathrm{OIF}} \cdot \frac{\nabla\Delta K_2}{2f_1^3} + \lambda_{\mathrm{OIF}} \cdot N_{\mathrm{OIF}} + \nabla\Delta e_{\mathrm{OIF}} \tag{9}$$

$$\nabla\Delta\phi_{\mathrm{GIF}} = -\vartheta_{\mathrm{GIF}} \cdot \frac{\nabla\Delta K_2}{2f_1^3} + \lambda_{\mathrm{GIF}} \cdot N_{\mathrm{GIF}} + \nabla\Delta e_{\mathrm{GIF}} \tag{10}$$

In addition, the dual-frequency Ionosphere-Free (IFij) combination formed from the carrier-phase measurements on fi and fj can be expressed as [12]:

$$\begin{aligned} \nabla\Delta\phi_{\mathrm{IF}ij} &= \frac{f_i^2}{f_i^2 - f_j^2} \cdot \nabla\Delta\phi_i - \frac{f_j^2}{f_i^2 - f_j^2} \cdot \nabla\Delta\phi_j \\ &= \nabla\Delta\rho + \nabla\Delta T - \vartheta_{\mathrm{IF}ij} \cdot \frac{\nabla\Delta K_2}{2f_1^3} + \lambda_{\mathrm{IF}ij} \cdot N_{\mathrm{IF}ij} + \nabla\Delta e_{\mathrm{IF}ij} \end{aligned} \tag{11}$$

Table 1 The linear coefficients, error characteristics and wavelength of GNSS IF combinations

	$\nabla\Delta\phi_{IF}$	x	y	z	ϑ_{xyz}	A_{xyz}	σ_{xyz}[m]	λ_{IF}[m]	λ_E[m]
GPS	IF12	2.5457	−1.5457	0.0	−0.721	3.227	0.0123	0.0063	0.1070
	IF13	2.2606	0.0	−1.2606	−0.767	2.821	0.0107	0.0028	0.1089
	IF23	0.0	12.2553	−11.2553	−1.126	21.784	0.0829	0.1247	0.1247
	OIF	2.3522	−0.4965	−0.8557	−0.752	2.693	0.0102	5.30E-9	0.1083
	GIF	1.0	−5.4213	4.4213	0.159	9.190	0.0350	1.05E-5	−0.0070
COMPASS	IF12	2.9437	−1.9437	0.0	−0.679	3.793	0.0146	7.41E-4	0.1059
	IF13	2.4872	0.0	−1.4872	−0.729	3.144	0.0121	6.26E-4	0.1083
	IF23	0.0	10.5895	−9.5895	−1.004	17.990	0.0691	0.0404	0.1211
	OIF	2.6087	−0.5175	−1.0912	−0.716	3.034	0.0117	2.05E-10	0.1077
	GIF	1.0	−4.2577	3.2577	0.110	6.797	0.0261	2.29E-6	−0.0051

For the three kinds of IF observation in Eqs. (9–11), whose ambiguity can be fixed by the same method as literature [12]. For example, for the IF carrier-phase observation formed from the GPS L1 and L2 phase measurements [12]

$$\lambda_{IF12} \cdot N_{IF12} = \frac{f_1^2 \cdot \lambda_1 N_1}{f_1^2 - f_2^2} - \frac{f_2^2 \cdot \lambda_2 N_2}{f_1^2 - f_2^2} = \frac{77\lambda_1}{77^2 - 60^2} \cdot (77N_1 - 60N_2) \tag{12}$$

where $\lambda_{IF12} = \dfrac{77\lambda_1}{77^2 - 60^2}$ is the virtual wavelength of the IF12.

For modernized GPS, f1(L1) = 1545.72 MHz, f2(L2) = 1227.60 MHz and f3(L5) = 1176.42 MHz, and for COMPASS-II [21], f1(B1) = 1561.098 MHz, f2(B3) = 1268.52 MHz and f3(B2) = 1207.14 MHz, the linear coefficients, error characteristics and wavelength of the double-frequency IF combinations (IF12, IF13, IF23), triple-frequency optimal IF combination (OIF) and the triple-frequency geometry-free and ionosphere-free combination are listed in Table 1. In Table 1, the σ_{xyz} values are computed with $\sigma_\varepsilon = 0.01$ cy, λ_{IF} is the virtual wavelength of IF combination, λ_E is the virtual wavelength of the NL combination.

From Table 1, we know that:

(1) The second-order ionospheric scale factors of the ionosphere-free combinations IF12, IF13 and OIF are less than 1, which means that the second-order ionospheric effect of them is less than that of carrier-phase measurement on f1. So the effect of the second-order ionospheric error with cm order of magnitude on precision positioning can be ignored. However, the virtual wavelength of the ionosphere-free combination is so small that the effect of the second-order ionospheric error with cm order of magnitude on ambiguity resolution is greater than 1 cycle. Therefore, the effect of the second-order ionospheric error on ambiguity resolution should not be overlooked.

(2) The variance of OIF is slightly less than that of dual-frequency ionosphere-free combination (IF12 and IF13), but the virtual wavelength of OIF is greatly less than that of IF12 and IF13. Therefore, it is not sensible to resolve the OIF ambiguity directly.

(3) Except for IF23, the virtual wavelength of the else IF combination are less than 1 cm, and the virtual wavelength of COMPASS IF combination is much less than that of GPS IF combination. Consequently, it is almost impossible to fix the IF ambiguity by geometry-based model or the GIF ambiguity by geometry-free model.

3 The Decomposition of Ionosphere-Free Combination

Although the ionosphere-free observation is immune to the first-order ionospheric effect, its virtual wavelength is too small to solve the related ambiguity practically. In the triple-frequency case, two WL ambiguities can be fixed reliably by geometry-free approach in tolerant time, and this AR processes are not affected by the first-order ionospheric error, orbit error, troposphere delay error and satellite geometric configuration, which means that WL ambiguities resolution is independent of baseline length. However, the determination of the NL ambiguity over long baseline is still an open problem. Similar to the double-frequency case [8, 22–24], it is assumed that the triple-frequency ionosphere-free combination can be expressed as the linear combination of one NL ambiguity and two WL ambiguities

$$\lambda_{\mathrm{IF}} \cdot N_{\mathrm{IF}} = c_{ijk}^{\mathrm{IF}} \cdot N_{ijk}^{\mathrm{NL}} + c_{lmn}^{\mathrm{IF}} \cdot N_{lmn}^{\mathrm{WL}} + c_{uvw}^{\mathrm{IF}} \cdot N_{uvw}^{\mathrm{WL}} \tag{13}$$

where the subscript $i, j, k, l, m, n, u, v, w$ are the coefficients of the NL and WL combinations, and they are integers; c_{ijk}^{IF}, c_{lmn}^{IF}, c_{uvw}^{IF} are the linear coefficients of the NL and WL ambiguities. According to Eq. (13), the relationship of the linear coefficient vector of the ionosphere-free combination $\boldsymbol{W}_{\mathrm{IF}}$, the linear coefficient vector of the NL and WL ambiguities $\boldsymbol{C}$ and the linear coefficient matrix of the NL and WL combination $\boldsymbol{T}$ can be expressed as:

$$\boldsymbol{W}_{\mathrm{IF}} = \begin{bmatrix} x_{\mathrm{IF}}\lambda_1 & y_{\mathrm{IF}}\lambda_2 & z_{\mathrm{IF}}\lambda_3 \end{bmatrix} = \begin{bmatrix} c_{ijk}^{\mathrm{IF}} & c_{lmn}^{\mathrm{IF}} & c_{uvw}^{\mathrm{IF}} \end{bmatrix} \begin{bmatrix} i & j & k \\ l & m & n \\ u & v & w \end{bmatrix} = \boldsymbol{CT} \tag{14}$$

From Eq. (14), as long as the linear coefficients of NL and WL combinations is linear independent, the decomposition in Eq. (13) come into existence. So the linear coefficient vector $\boldsymbol{C}$ can derived from

$$\boldsymbol{C} = \boldsymbol{W}_{\mathrm{IF}}\boldsymbol{T}^{-1} = \frac{1}{\det \boldsymbol{T}} \boldsymbol{W}_{\mathrm{IF}}\boldsymbol{T}^{*} \tag{15}$$

where $\boldsymbol{T}^{-1}$, $\det \boldsymbol{T}$ and $\boldsymbol{T}^{*}$ are the inverse matrix, determinate and adjoint of $\boldsymbol{T}$. When two WL and one NL combination whose coefficients are linear independent are given, the linear coefficient vector $\boldsymbol{C}$ can obtained from Eq. (15). And the linear coefficient of the NL ambiguity c_{ijk}^{IF} is namely the virtual wavelength of the NL ambiguity $\lambda_{\mathrm{E}}^{\mathrm{IF}}$.

For GPS and COMPASS, the sum of the coefficients of WL combination which can be fixed quickly over long baselines commonly equals to zero [7, 25]. Moreover, in order to ensure linear independence of the coefficients of the NL and WL combinations, the sum of the coefficients of the NL combinations cannot equal to zero. Generally, it is assumed that the coefficients of the NL and WL combination satisfy:

$$\begin{cases} i+j+k = S_{\mathrm{NL}} \\ l+m+n=0 \\ u+v+w=0 \end{cases} \tag{16}$$

Thereby,

$$\det \boldsymbol{T} = \det \begin{bmatrix} i & j & k \\ l & m & n \\ u & v & w \end{bmatrix} = \det \begin{bmatrix} S_{\mathrm{NL}} & j & k \\ 0 & m & n \\ 0 & v & w \end{bmatrix} = S_{\mathrm{NL}}(mw - nv) \tag{17}$$

According to Eqs. (15) and (16), we can know:

$$\lambda_{\mathrm{E}}^{\mathrm{IF}} = c_{ijk}^{\mathrm{IF}} = \frac{1}{\det \boldsymbol{T}} \det \begin{bmatrix} x_{\mathrm{IF}}\lambda_1 & y_{\mathrm{IF}}\lambda_2 & z_{\mathrm{IF}}\lambda_3 \\ l & m & n \\ u & v & w \end{bmatrix} = \frac{(x_{\mathrm{IF}}\lambda_1 + y_{\mathrm{IF}}\lambda_2 + z_{\mathrm{IF}}\lambda_3)(mw - nv)}{\det \boldsymbol{T}} \tag{18}$$

Substituting Eq. (17) to Eq. (18),

$$\lambda_{\mathrm{E}}^{\mathrm{IF}} = \frac{x_{\mathrm{IF}}\lambda_1 + y_{\mathrm{IF}}\lambda_2 + z_{\mathrm{IF}}\lambda_3}{S_{\mathrm{NL}}} \tag{19}$$

From Eq. (19), when two WL ambiguities have be fixed, the virtual wavelength of the IF combination is in inverse proportion to the sum of the coefficients of the NL combinations. And the smaller the sum is, the longer the virtual wavelength is. So the sum is normally set as 1. According to Eq. (19), when the NL ambiguity with $S_{\mathrm{NL}} = 1$ is estimated by the decomposition of the IF combination, the virtual wavelength of the IF combination is listed in the last column of Table 1.

From Table 1, estimation of IF ambiguity is replaced by that of NL ambiguity by decomposition of IF combination, which can increase the virtual wavelength of IF combination. Especially, the virtual wavelength of the COMPASS IF combination is very small, but that of its NL combination is comparable with that of the GPS NL combination. However, the virtual wavelength of triple-frequency NL combinations is slightly greater than that of dual-frequency NL combinations. Hence, there are the same problems as dual-frequency NL ambiguity resolution in the triple-frequency case, which is that the ambiguity resolution is affected by orbit error, the DD troposphere delay error and the satellite geometric configuration. So the NL ambiguity over medium-long baselines can not be rapidly fixed even in the triple-frequency case.

Table 2 The error characteristics of the geometry-free and ionosphere-free combination

	Standard deviation [cy]	$\vartheta_{GIF}/\lambda_E^{GIF}$	Standard deviation [m]	$(\lambda_E^{GIF}/\lambda_E^{GIF})\cdot\vartheta_{GIF}$
GPS	5.016	−22.812	0.543	−2.471
COMPASS	5.073	−21.430	0.546	−2.307

4 Geometry-Free and Ionosphere-Free Approach for Narrow-Lane Ambiguity Resolution

The benefits of the triple-frequency ionosphere-free combination ambiguity decomposition to narrow-lane ambiguity resolution by geometry-based model over medium-long baseline are limited. However, the triple-frequency ionosphere-free combination are superior than that of dual-frequency in one way that it can eliminate the geometric error further, and form the geometry-free and ionosphere-free combination. By the decomposing of the ambiguity of geometry-free and ionosphere-free combinations, the narrow-lane ambiguity can be estimated by a geometry-free approach. From Eqs. (13) and (10), we can get:

$$\hat{N}_{ijk}^{NL} = \frac{\nabla\Delta\phi_{GIF} - c_{lmn}^{GIF}\cdot N_{lmn}^{WL} - c_{uvw}^{GIF}\cdot N_{uvw}^{WL}}{\lambda_E^{GIF}} = N_{ijk}^{NL} + \frac{\nabla\Delta e_{GIF}}{\lambda_E^{GIF}} - \frac{\vartheta_{GIF}}{\lambda_E^{GIF}}\cdot\frac{\nabla\Delta K_2}{2f_1^3} \tag{20}$$

Obviously, when two wide-lane ambiguities are fixed, the resolution of narrow-lane ambiguity by using triple-frequency geometry-free and ionosphere-free combination are not affected by orbit error, troposphere delay error, the first-order ionosphere error, satellite geometric configuration and so on, that is to say, it only influenced by the phase noise, multipath error and the second-order ionosphere error.

Assume that the second-order ionosphere error can be neglected, and the standard deviation of the narrow-lane ambiguity in Eq. (20) can be expressed as

$$\sigma_{ijk} = S_{NL}\frac{\sqrt{(x_{GIF}\lambda_1)^2+(y_{GIF}\lambda_2)^2+(z_{GIF}\lambda_3)^2}}{x_{GIF}\lambda_1 + y_{GIF}\lambda_2 + z_{GIF}\lambda_3}\cdot 2\sigma_\varepsilon \tag{21}$$

From Eq. (21) we know that when two wide-lane ambiguity are fixed, the standard deviation of $\hat{N}_{ijk}^{NL}$are only related to the phase noise and the sum of the coefficients of the narrow-lane combination. The smaller the sum of the coefficients of the narrow-lane combination is, the smaller standard deviation is. The standard deviation of narrow-lane ambiguity and the influence coefficients calculated by Eqs. (20) and (21) are listed in Table 2 (column 2 and 3).

It can be seen from Table 2 that the standard deviation of narrow-lane ambiguity will be larger than 5 cycles when the standard deviation of phase measurement is 0.01 cycles. However, all the systematic errors are eliminated. So the narrow-lane ambiguity can be determined by directly rounding of the average of multi epoch estimators.

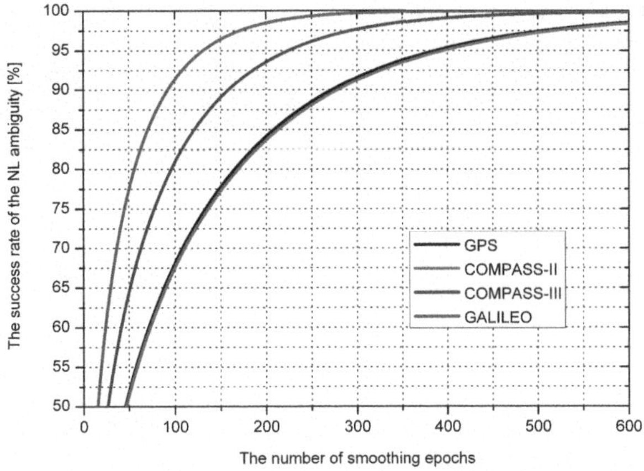

Fig. 1 The success rate of narrow ambiguity for different GNSS (σ_{ε}=0.01 cy)

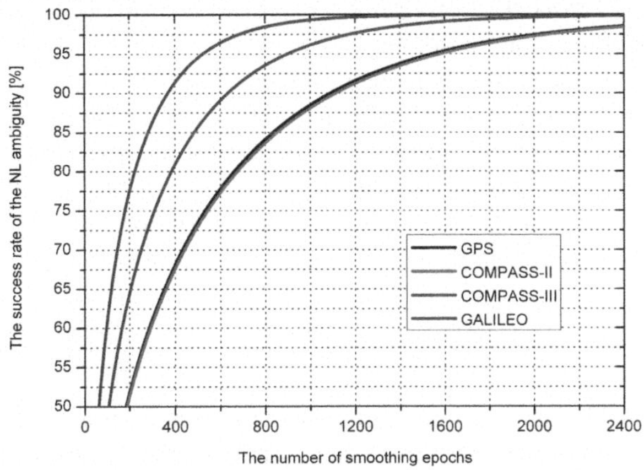

Fig. 2 The success rate of narrow ambiguity for different GNSS (σ_{ε}=0.02 cy)

When the standard deviation of phase measurement are 0.01 and 0.02 cy, the relations between the success rates of narrow-lane ambiguity and smoothing epochs are plotted in Figs. 1 and 2 for GPS, COMPASS-II, COMPASS-III(f1(B1) = 1575.42 MHz, f2(B3) = 1268.52 MHz, f3(B2) = 1191.795MHz) [21] and GALILEO (f1(E1) = 1575.42 MHz, f2(E6) = 1278.75 MHz, f3(E5a) = 1176.45 MHz).

Figures 1 and 2 show that the time needed for GPS and COMPASS narrow-lane ambiguity resolution are about the same. Supposing the sample rate is 1 s, and the correlation between different epochs are not considered, when $\sigma_{\varepsilon} = 0.01$ cy, theoretically the success rates will be above 98% when the time span is about 10 min. For COMPASS-III and GALILEO, only 6 and 4 min will be needed respectively. When $\sigma_{\varepsilon} = 0.02$ cy, the smoothing time span needed will be 40, 22 and 13 min for these systems. Hence, the narrow-lane ambiguity resolution success rates are very sensitive to the phase noise, and the frequency distribution of COMPASS-III

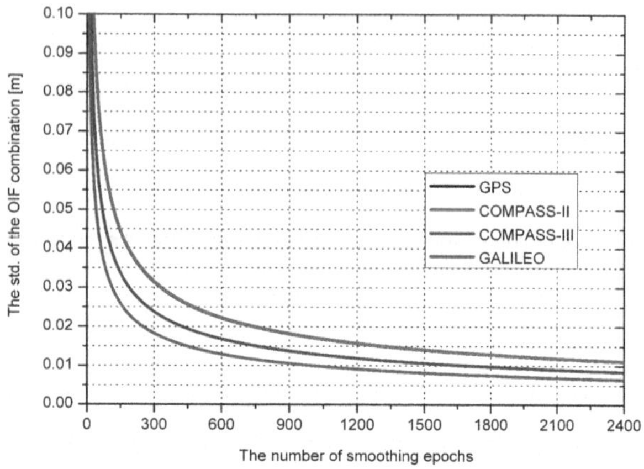

Fig. 3 The std. of optimal ionosphere-free combination ($\sigma_\varepsilon = 0.01$ cy)

and GALILEO(E1,E6,E5a) are more suitable for ambiguity resolution over medium-long baseline.

Besides, the influence of 1 cm DD ionosphere error to the narrow-lane ambiguity resolution is larger than 0.02 cy. Considering the effects of multipath error, the smoothing time for narrow-lane ambiguity resolution may be even longer. In that case, narrow-lane ambiguity does not be fixed necessarily.

Assuming the virtual wavelength of the optimal triple-frequency ionosphere-free combination is $\lambda_{\mathrm{E}}^{\mathrm{OIF}}$, then from Eq. (20) we get

$$\lambda_{\mathrm{E}}^{\mathrm{OIF}} \hat{N}_{ijk}^{\mathrm{NL}} = \lambda_{\mathrm{E}}^{\mathrm{OIF}} N_{ijk}^{\mathrm{NL}} + \lambda_{\mathrm{E}}^{\mathrm{OIF}} \frac{\nabla\Delta e_{\mathrm{GIF}}}{\lambda_{\mathrm{E}}^{\mathrm{GIF}}} - \lambda_{\mathrm{E}}^{\mathrm{OIF}} \frac{\vartheta_{\mathrm{GIF}}}{\lambda_{\mathrm{E}}^{\mathrm{GIF}}} \cdot \frac{\nabla\Delta K_2}{2 f_1^3} \tag{22}$$

The standard deviation of the optimal triple-frequency ionosphere-free observation when the narrow-lane ambiguity is not fixed and the influence coefficients of the second-order ionosphere delay calculated by Eq. (22) are shown in Table 2 (See column 4 and 5). When the narrow-lane ambiguity is not fixed, the effective noise of optimal ionosphere combination is about 0.54 m, the observation noise level of optimal triple-frequency can be acquired only after a certain smoothing time span. When the standard deviation of phase observation are 0.01 and 0.02 cycles, the relations between the standard deviation of the noise of optimal ionosphere combination when the narrow-lane ambiguity is not fixed and smoothing epochs are plotted in Figs. 3 and 4.

It depicts in Figs. 3 and 4 that, when $\sigma_\varepsilon = 0.01$ and 0.02 cycles, only about 5 and 10 min are needed respectively to acquire ionosphere observations with accuracy superior than 5 cm, and the dm-cm level fast positioning can be achieved in a short time.

The main steps for the medium-long baseline ambiguity resolution based on geometry-free and ionosphere-free can be summarized as follows: Firstly, two WL ambiguities are fixed by a geometry-free and ionosphere-free observation. And then the narrow-lane ambiguity based on the decomposition of triple-frequency

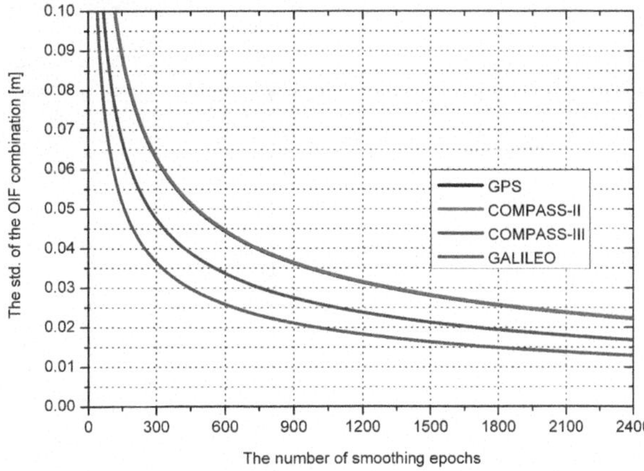

Fig. 4 The std. of optimal ionosphere-free combination ($\sigma_\varepsilon = 0.02$ cy)

geometry-free and ionosphere-free combination ambiguity are estimated. The geometry-free model are used in the whole process, so it is not affected by the state of receiver, the first-order ionosphere error, orbit error, troposphere delay, satellite geometry and the visible satellite number, only four satellites are needed for medium-long baseline precise positioning. It should be noted that the method is affected by cycle slip and the correlative noise between different epochs because of the smoothing of multi epochs.

5 Computation and Analysis

The real GNSS triple-frequency measurements are unavailable now, so only the GPS dual-frequency data are used to verify the success rate of single epoch and multi epoch smoothed wide-lane ambiguity resolution of the new method proposed in this paper. The influence of sample rate to multi epoch smoothed wide-lane ambiguity is analyzed. The GPS dual-frequency datas of CORS site located in America are used, the time span is 24 h. The baseline length, sample interval, receiver and antenna information are shown in Table 3.

Four schemes are employed for the computation:

Scheme 1: the success rate of single epoch ambiguity estimation;
Scheme 2: the success rate of multi epoch averaged ambiguity estimation for 2 min time span;
Scheme 3: the success rate of multi epoch averaged ambiguity estimation for 5 min time span;
Scheme 4: the success rate of multi epoch averaged ambiguity estimation for 10 min time span

The results are shown in Table 4, the STD refers to the standard deviation of ambiguity estimation calculated according to the wide-lane ambiguity estimation

Table 3 The types of receiver and antenna

Group	Baseline length	Sample rate	Receiver 1	Antenna 1	Receiver 2	Antenna 2
(b)	102 km	1 s	LEICA RX1200PRO	LEIAT504 LEIS	LEICA RX1200PRO	LEIAT504 LEIS
(c)	1471 km	1 s	TRIMBLE NETRS	TRM41249.00	TRIMBLE NETRS	TRM41249.00
(a)	982 km	5 s	TRIMBLE NETRS	TRM41249USCG	**ASHTECH Z-XII3**	ASH700718B
(c)	657 km	5 s	TRIMBLE NETRS	TRM41249USCG	TRIMBLE NETRS	TRM41249USCG
(c)	716 km	5 s	TRIMBLE NETRS	TRM41249USCG	TRIMBLE NETRS	TRM41249USCG
(b)	941 km	10 s	LEICA GRX1200GGPRO	LEIAX1202GG	LEICA GRX1200GGPRO	LEIAT504 LEIS
(c)	502 km	15 s	TRIMBLE NETRS	TRM29659.00	TRIMBLE NETRS	TRM29659.00
(d)	974 km	30 s	**TRIMBLE 4000SSI**	TRM33429.00+GP	**ROGUE SNR-8100**	AOAD/M_T

Table 4 The success rate of wide-lane ambiguity rounding

Group	Sample rate	Baseline length	STD	PC	Scheme1	Scheme 2	Scheme 3	Scheme 4
(b)	1 s	102 km	0.3745	81.84	82.48	85.28	91.11	96.03
(c)	1 s	1471 km	0.5821	61.36	64.90	91.47	96.21	98.04
(a)	5 s	982 km	0.9648	39.90	43.35	63.49	77.21	84.55
(c)	5 s	657 km	0.5240	66.11	69.29	94.15	97.17	98.30
(c)	5 s	716 km	0.4938	68.97	71.32	96.75	98.51	99.63
(b)	10 s	941 km	0.3764	81.72	84.88	87.91	93.82	97.36
(c)	15 s	502 km	0.4873	69.59	78.14	96.35	98.58	99.53
(d)	30 s	974 km	0.5804	62.10	69.67	89.99	96.57	98.19
Average (Group(a) is excluded)			0.4884	70.24	74.38	91.70	96.00	98.15

series. PC is the success rates of single epoch calculated by STD. The baseline is categorized into 4 groups according to the receiver type and the standard deviation of ambiguity estimation. From Table 4 we can see that

(1) For baselines in group (b), the standard deviation of wide-lane ambiguity estimation is approximately 0.37 cycles, theoretically the standard deviation of the averaged float ambiguity of 4 epochs will be less than 0.2 cycles, and the success rate will be above 98%. However, the success rate of multi epoch averaged ambiguity estimation for 2 min time span is about the same as that of single epoch, even for 10 min time span, the success rate is only 96%. So the observations of different epochs are correlated in group (b), and this is adverse for enhancing the success rates by the average of multi epochs. Table 3 shows that all the receivers used in group (b) are LEICA receivers. May phase smoothed pseudorange algorithm is used before hand in the receiver which results in the strong correlation between epochs.
(2) The success rate of multi epoch averaged ambiguity estimation for 2, 5 and 10 min time span are 91.7, 96, 98.15% respectively except for group (a), so 10 min are needed for the fast estimation of wide-lane ambiguity with 98% success rates.

6 Conclusion

The general form of the coefficients of the triple-frequency ionosphere-free linear combinations is constructed in this paper, and then three kinds of ionosphere-free combination are defined. Besides, their application in narrow-lane ambiguity resolution over medium-long baseline is analyzed. The following conclusion can be derived from the computation and analysis:

(1) The optimal triple-frequency ionosphere-free combination has no advantage in precise positioning and fast ambiguity resolution with geometry-based model over medium-long baseline when compared with the dual-frequency ionosphere-free combination.
(2) In triple-frequency case, the geometry-free and ionosphere-free combination can be formed, and the corresponding geometry-free and ionosphere-free model for ambiguity resolution is constructed. The ambiguity resolution based on the geometry-free and ionosphere-free model is not affected by the first-order ionosphere error, the troposphere delay error and the orbit error and so on, so it is suitable for ambiguity resolution over medium-long baselines.
(3) By using the presented method in this paper, the standard deviation of narrow-lane ambiguity is only related to the sum of the coefficients of the narrow-lane combination. The smaller the sum of the coefficients of the narrow-lane combination is, the smaller standard deviation of narrow-lane ambiguity will be.
(4) For different GNSS system, the characteristic of their frequency distribution is different, which will significantly affect the performance of ambiguity resolution. It can be known from the analysis that the GALILEO (E1, E6, E5a) has the best performance for narrow-lane ambiguity resolution over long baselines, then COMPASS-III. The results for COMPASS-II and GPS are similar.

Acknowledgments This project is sponsored by The Natural Science Foundation of China (41020144004, 41104022).

References

1. Forssell, B., Martin-Neira, M., & Harris, R. (1997). Carrier phase ambiguity resolution in GNSS-2. *Proceedings of ION GPS-97, 16–19 September 1997, Kansas City* (pp. 1727–1736).
2. Vollath, U., Birnbach, S., & Landau, H. (1998). Analysis of three carrier ambiguity resolution (TCAR) technique for precise relative positioning in GNSS-2. *Proceedings of ION GPS 1998* (pp. 417–426).
3. Feng, Y. (2008). GNSS three carrier ambiguity resolution using ionosphere-reduced virtual signals. *Journal of Geodesy, 82*, 847–862.
4. Todd, R., & El-Sheimy, N. (2007). Optimal linear combinations of triple frequency carrier phase data from future global navigation satellite systems. GPS Solutions, *11*, 11–19.
5. Wu, Y. (2005). *Theory and application on multi-frequency data processing of GNSS 2*. Wuhan: Wuhan University.
6. Li, B. (2010). *Theory and method for parameter estimation for mixed integer GNSS function and stochastic models*. Shanghai: Tongji University.
7. Li, J. (2011). *Research on the algorithms of GNSS triple frequency precise positioning*. Zhengzhou: Information Engineering University.
8. Hatch, R., Jung, J., Enge, P., & Pervan, B. (2000). Civilian GPS: the benefits of three frequencies. *GPS Solutions, 3*(4), 1–9.
9. Teunissen, P. J. G., Joosten, P., & Jonkman, N. F. (2000). Evaluation of long range GNSS ambiguity resolution. *Proceedings of the 8th European CGSIC IISC Meeting, 2–3 December, 1999, Prague, Czech Republic* (pp. 171–177).

10. Jonkman, N. F., Teunissen, P. J. G., Joosten, P., & Odijk, D. (1999). GNSS long baseline ambiguity resolution: impact of a third navigation frequency. *Accepted for publication in IAG Proceedings of IUGG 22nd General Assembly, 19–30 July, 1999, Birmingham, UK.*
11. Teunissen, P. J. G., & Jonkman, N. F. (2001). Will geometry-free full ambiguity resolution be possible at all for long baselines. *Proceedings of ION-NTM 2001, 22–24 January, 2001, Long Beach, CA* (pp. 271–280).
12. Odijk, D., Teunissen, P. J. G., & Tiberius, C. C. J. M. (2002). Triple-frequency ionosphere-free phase combinations for ambiguity resolution. *Proceedings of the European Navigation Conference ENC-GNSS 2002, Copenhagen, Denmark, 27–30 May 2002* CD ROM.
13. Dennis, O. (2003). Ionosphere-free phase combinations for modernized GPS. *Journal of Surveying Engineering, 129*(4), 165–173.
14. Han, S., & Rizos, C. (1999). The impact of two additional civilian GPS frequencies on ambiguity resolution strategies. *Proceedings of ION Annual Technical Meeting, Cambridge, MA, 28–30 June, 1999* (pp. 315–321).
15. Zhang, W. (2005). Triple frequency cascading ambiguity resolution for modernized GPS and Galileo. UCGE Reports, Number 20228, July, 2005, URL: http://www.geomatics.ucalgary.ca/links/GradTheses.html, pp. 81–82.
16. Hatch, R. (2006). A new three-frequency, geometry-free technique for ambiguity resolution. *Proceedings of ION GNSS 2006, 26–29 Sept, Fort Worth, TX* (pp. 309–316).
17. Li, B., Shen, Y., & Zhou, Z. (2009). A new method for medium and long range three frequency GNSS rapid ambiguity resolution. *Acta Geodaetica et Cartographica Sinica, 38*(4), 296–301.
18. Li, B., Feng, Y., & Shen, Y. (2010). Three carrier ambiguity resolution: distance-independent performance demonstrated using semi-generated triple frequency GPS signals. *GPS Solutions, 14*, 177–184.
19. Wang, Z., & Liu, J. (2003). Model of inter-frequency combinations of Galileo GNSS. *Geomatics and Information Science of Wuhan University, 28*(6), 723–727.
20. Simsky, A. (2006) Three's the charm: Triple-frequency combinations in future GNSS. Inside GNSS, July/August, 38–41.
21. China Satellite Navigation Office. (2010). BeiDou (COMPASS) Navigation satellite system development. Munich Satellite Navigation Summit 2010, 9–11 March, 2010, Munich.
22. Blewitt, G. (1989). Carrier phase ambiguity resolution for the global positioning system applied to geodetic baselines up to 2000 km. Journal of Geophysical Research, *94*(B8), 10187–10203.
23. Ge, M., Gendt, G., Rothacher, M., Shi, C., & Liu, J. (2008). Resolution of GPS carrier-phase ambiguities in precise point positioning (PPP) with daily observations. *Journal of Geodesy, 82*, 389–399.
24. He, H. (2002). *Precise kinematic GPS surveying and quality control.* Zhengzhou: Information Engineering University.
25. Cocard, M., Bourgon, S., Kamali, O., & Collins, P. (2008). A systematic investigation of optimal carrier-phase combinations for modernized triple-frequency GPS. *Journal of Geodesy, 82*, 555–564.

Preliminary Analysis of Real-Time Orbit and Clock Error Based on BNC

Guangxing Wang and Qile Zhao

Abstract Data stream broadcast with BNC, correction algorithm of broadcast ephemeris and the conception of SISRE was introduced in this paper. With BNC, GPS broadcast ephemeris and orbit/clock corrections was retrieved real time, and real-time orbit and clock product with high precision was obtained. Preliminary analysis was conducted among results with different broadcasters, which prepare for real-time PPP.

1 Introduction

As global navigation satellite system (GNSS) was continuously improved, real-time positioning with high precision, which could provide technological support for hazard monitoring and precision agriculture, became the hot issue of satellite navigation and positioning realm [1].

In the past, precise product of orbit and clock error provided by IGS was applied to post-time PPP, and intensive study had been conducted by scholars at home.

In recent years, GNSS real-time application services with high precision was studied by relevant institutes both at home and abroad. IGS set real-time group to conduct IGS Real-Time Pilot Project [2], which was aiming at establishing real-time product services and supporting the development of real-time GNSS application.

G. Wang (✉) · Q. Zhao
GNSS Research Center, Wuhan University, 430079 Wuhan, China
e-mail: gxwang0113@126.com

J. Sun et al. (eds.), *China Satellite Navigation Conference (CSNC) 2012 Proceedings*, Lecture Notes in Electrical Engineering 160, DOI: 10.1007/978-3-642-29175-3_17,

BKG Ntrip Client (BNC) is a program for simultaneously retrieving real-time GNSS data streams from Networdked transport of RTCM via internet protocol (NTRIP) broadcasters. Data streams, after being decoded and converted, are able to be used for real-time positioning. BNC has been developed for the federal agency for cartography and geodesy (BKG).

In this paper, real-time orbit and clock error for GPS satellites was computed with BNC data streams and preliminary analysis was conducted.

2 Real-Time Orbit and Clock Error

2.1 BNC Data Streams

BNC streams carry observations, orbit/clock correctors, raw data and RINEX observation files, which are all encoded in RTCM format, and the so called 'state space representation' (SSR) messages have been implemented in BNC.

RTCM Version 3.x could be applied to broadcast correction information for satellite orbit and clock, and message type 1,060 contains combined orbit and clock corrections to GPS broadcast ephemeris. Specifically, the first five parameters in each broadcast corrections record are message type number, SSR message updated interval indicator, GPS week, second in GPS week, GNSS indicator and PRN, and they are followed by issue of data (IOD), three polynomial coefficients (constant, first order, and second order term) for clock correction, three components of orbit correction (radial, along-track, and cross-track), and three velocity components of orbit correction (radial, along-track, and cross-track), all referring to broadcast ephemeris. Orbit corrections are referred to a spacecraft-fixed right-handed orthogonal coordinate system, directions vectors of whose three axes are defined according to satellite position and velocity vectors in earth-centered earth-fixed (ECEF) reference system.

The orbit and clock corrections do not include local effects (like ocean loading or solid earth tides) or atmosphere effects (ionosphere and/or troposphere).

2.2 Computation of Real-Time Satellite Position

Computation of real-time satellite positioning took place in two steps: obtaining satellite positioning with broadcast ephemeris and adding orbit corrections. The first step was discussed in detail by Li and Huang [3], and the only difference was that broadcast ephemeris here came from BNC data streams instead of ephemeris files.

Since broadcast ephemeris is updated every 2 h, some satellites might have more than one set of navigation data referring to different time, with different IOD value. Before adding orbit corrections, the IOD value of orbit corrections should be compared with that of broadcast ephemeris. The correction could be conducted only when the two IOD values were consistent. As was explained by Li and Huang [3], satellite positions obtained from broadcast ephemeris referred to ECEF, while correctors released by BNC referred to spacecraft-fixed system, which, as above mention, was a right-handed orthogonal coordinate system formed by radial, along-track, and cross-track axes. Therefore, corrections should be transformed to ECEF first.

If $[X_0, Y_0, Z_0]^T$ was satellite position obtained from broadcast ephemeris, while $[x_0, y_0, z_0]^T$ and $[v_x, v_y, v_z]^T$ denoted orbit correctors and their velocities provided by BNC, satellite positions after correcting could be expressed as

$$\begin{bmatrix} X_P \\ Y_P \\ Z_P \end{bmatrix} = \begin{bmatrix} X_0 \\ Y_0 \\ Z_0 \end{bmatrix} - R \begin{bmatrix} x_S \\ y_S \\ z_S \end{bmatrix} \quad (1)$$

where R is the rotation matrix from spacecraft-fixed to ECEF system, and $[x_S, y_S, z_S]^T$ denoted orbit correctors referring to the former one. $[x_S, y_S, z_S]^T$ could be expressed as

$$\begin{bmatrix} x_S \\ y_S \\ z_S \end{bmatrix} = \begin{bmatrix} x_0 \\ y_0 \\ z_0 \end{bmatrix} + \begin{bmatrix} v_s \\ v_y \\ v_z \end{bmatrix} \Delta t \quad (2)$$

where Δt is the time difference between calculating time and correctors' time. This difference occurred because the corrections of satellite orbit/clock were broadcast every 5 or 10 s, depending on the selected data streams.

2.3 Computation of Real-Time Satellite Clock Error

Computation of real-time satellite clock error was similar with that of real-time satellite position. IOD match was unnecessary, since it had been done when computing satellite position in the above. The first step was to obtain satellite clock error t_0 from broadcast ephemeris as

$$t_0 = a_0 + a_1 \Delta t + a_2 \Delta t^2 \quad (3)$$

where a_0, a_1 and a_2 were three polynomial coefficients decoded from broadcast ephemeris data stream, and Δt is the time difference between calculating time and epoch time.

Then, satellite clock error correction t_C was calculated with c_0, c_1 and c_2 decoded from RTCM message type 1,060 of BNC data streams. t_C could be expressed as

$$t_C = c_0 + c_1 \delta t + c_2 \delta t^2 \quad (4)$$

where δt had similar meaning with Δt in Eq. (2). Therefore, the final value of satellite clock error t_P could be expressed as

$$t_P = t_0 + t_C \tag{5}$$

Through the above methods, real-time satellite orbit and clock error with higher precision could be obtained with the help of data streams of BNC.

The temporal change of satellite clock error was somewhat complicated. What's more, the extrapolation precision of polynomial was limited, and clock error correctors were released with latency. Therefore, the precision of clock error might be affected, if the extrapolation time were too long.

2.4 Precision Assessment

Since significant relevance existed between real-time orbits and clock error, only the comparison between real-time and precise orbit/clock could not give full evidence of the real-time products' capacity in providing precision positioning service. Users paid more attention to the precision level of positioning that real-time products could provide.

Lou [4] divided the errors affecting GPS positioning into two parts: signal-in-space range error (SISRE) and user equivalent range error (UERE). The former was caused by inaccuracy or broadcast ephemeris, while the latter was the estimated error in ranging, including noise of receiver, troposphere, ionosphere, multi-path effects, and so on. Since this paper was concern with the impact of real-time orbit and clock error on positioning precision, only SISRE was discussed.

SISRE was given by [4]

$$SISRE = \sqrt{(R - CLK)^2 + \frac{1}{49}(A^2 + C^2)} \tag{6}$$

where R, A and C denoted radial, along-track and cross track error of satellite orbit, respectively, and CLK was satellite clock error.

The above equation showed that SISRE was mainly relied on the magnitude of radial component of orbit error and satellite clock error, as well as their coupling. SISRE could be employed to measure the precision potential that real-time products were able to provide.

3 Results and Analysis

In order to investigate the quality of corrected orbit/clock products, data from 0:00 to 16:00 UTC, Nov. 12th, 2011 was selected as the sample. Both broadcast ephemeris and orbit/clock corrections were retrieved during the time. For post-time comparison, broadcast ephemeris in format of RINEX files and orbit/clock corrections files were saved synchronously, and a self-defined index file was generated for post-time simulation.

Using the above mentioned method, real-time satellite orbit and clock error products were obtained, and the sampling interval was 1 s. Comparison was conducted between the above products and final products released by IGS. The real-time products generated from data streams CLK15, CLK50 and CLK81 were selected for statistics, since data from these three streams were relatively complete.

3.1 Precision of Satellite Orbits

Precisions of satellite orbits were measured by the root mean square (RMS) of discrepancy between computed real-time orbits and IGS final orbit products. The sampling interval of the former was 1 s, while that of the latter was 30 s. Therefore, interpolation was needed.

RMS of discrepancy between the two set of orbits could be calculated as

$$RMS = \sqrt{\frac{\sum_{i=1}^{n} \Delta_i \Delta_i}{n}} \tag{7}$$

where Δ_i was the discrepancy at the ith epoch, and n was the total of epochs. The precision of real-time orbits obtained with CLK15, CLK50 and CLK81 were illustrated as Fig. 1.

As was shown in Fig. 1, radial precisions of satellite were relatively high as a whole, and RMS value of most satellites were <3 cm. Cross-track precision took the second place, with RMS value of most satellites <5 cm, and was followed by along-track precision. Generally speaking, precision of orbits with CLK15 was superior to CLK50 and CLK81, except satellites G05 and G20. Several satellites were not presented in the figure because of data deficiency.

3.2 Precision of Satellite Clock Error

"Second order difference" method was employed to assessing the precision of satellite clock error [5]. Similarly, interpolation was also needed since the interval of final clock error products by IGS was 30 s. Specifically, the first step was to select an arbitrary satellite, G02 for example, as the reference satellite, whose clock would be referred as the reference clock. Then, calculate the first order difference between each other satellite and G02, with both the above real-time clock error and IGS products, respectively. Thus, the effect of the reference clock was eliminated. The second order difference existed between the real-time clock and IGS products, satellite by satellite, at corresponding epoch. "Second order

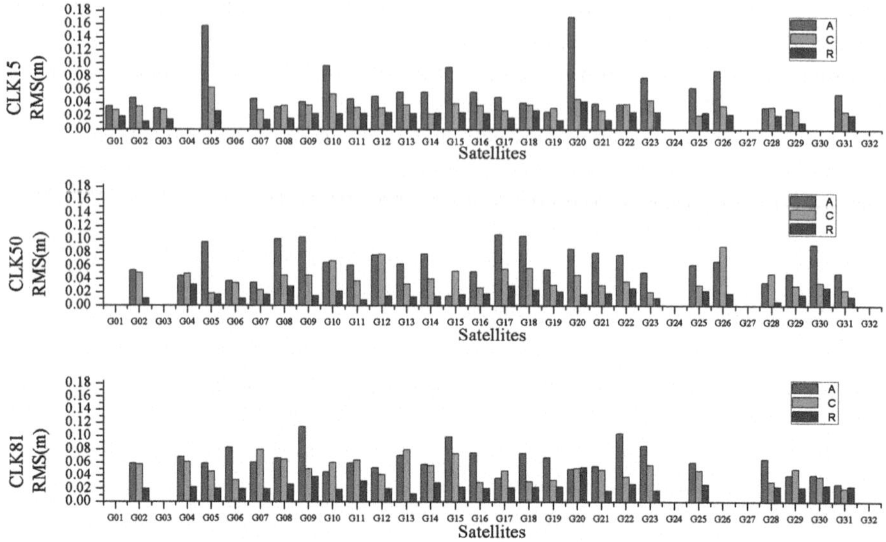

Fig. 1 The results of real-time orbit (*A*: along-track, *C*: cross-track, *R*: radial)

difference" method was effective in indicating the consistency between the above computed real-time clock error and IGS final products.

Precisions of satellite clock error were measured by the standard deviation (SD), which was different from the case of orbits. Comparing with RMS, SD was able to indicate the statistical precision of clock errors with systematic bias, and it reflected the noise level of satellite clock errors. Since each satellite clock error was matched by an initial ambiguity in non-difference positioning, systematic bias could be absorbed into ambiguity. Thus it would not affect the precision of positioning. Therefore, SD was of more practical significance for PPP users [5]. SD could be expressed as

$$SD = \sqrt{\frac{\sum_{i=1}^{n}(\Delta_i - \bar{\Delta})(\Delta_i - \bar{\Delta})}{n-1}} \tag{8}$$

where Δ_i was the discrepancy between real-time clock error and IGS products at the ith epoch, and n was the total of epochs. The precisions of satellite clocks were illustrated as Fig. 2.

As was shown in Fig. 2, precisions of satellite clock errors with CLK50 and CLK81 were higher than that with CLK15. SD of most satellite clock errors with CLK15 fell between 0.2 and 0.4 ns, while those with CLK50 and CLK81 fell around 0.2 ns.

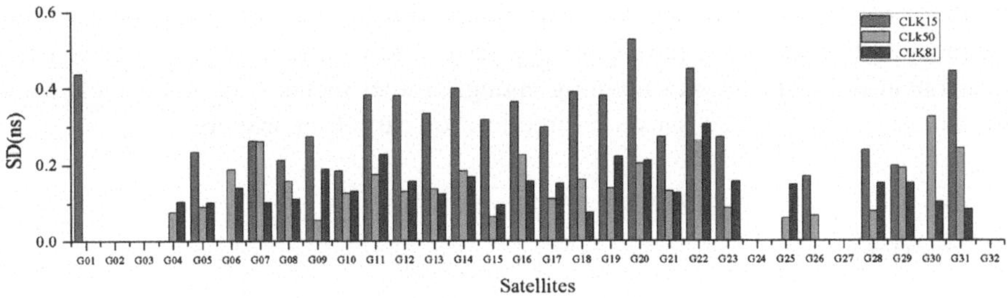

Fig. 2 The results of real-time clock error

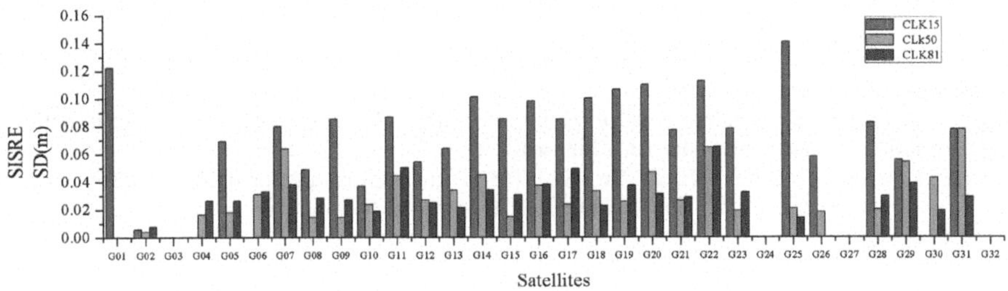

Fig. 3 The results of SISRE

3.3 SISRE Results

To investigate the positioning precision potential that the real-time product could provide, SISRE results were analyzed. According to Eq. (6), SISRE of each satellite at each epoch was calculated, and the SD values were illustrated in Fig. 3.

As is shown in Fig. 3, SD of SISRE with CLK15 fell around 7 cm, while those with CLK50 and CLK81 fell around 3 cm. Combined with Fig. 2, it was evident that SISRE was more sensitive to clock error.

4 Conclusions

Satellite orbits and clocks directly obtained from broadcast ephemeris were obviously improved with the real-time orbit/clock correction retrieved from BNC. If proper data steam was chosen, the discrepancy with IGS final products can fall at 2–7 cm and 0.2 ns, which might be employed in real-time PPP. The comparisons showed that the error of satellite clock had more significant impact on SISRE. In addition, precisions of satellite clocks were statistically different when selecting different data streams of corrections. Therefore, the quality of corresponding data streams need improving.

Statistical results provided the preliminary comparisons of different real-time products. However, the results were not perfect, since the time span selected for experiment in this paper was not long enough. Experiments with longer time span are still necessary, which shows a direction for current endeavors.

References

1. Zhang, X., Li, X., Guo, F., Li, P., & Wang, L. (2010). Server-based real-time precise point positioning and its application. *Chinese Journal of Geophysics, 53*(6), 1308–1314 (in Chinese).
2. Shi, C., Lou, Y., Song, W., & Cai, H. (2009). A wide area real-time differential GPS prototype system and the initial results. *Geomatics and Information Science of Wuhan University, 34*(7), 778–781 (in Chinese).
3. Li, Z., & Huang, J. (2005). *GPS surveying and data processing*. Wuhan: Wuhan University Press (in Chinese).
4. Lou, Y. (2008). Research on real-time precise GPS orbit and clock offset determination. Dissertation Wuhan University (in Chinese).
5. Lou, Y., Shi, C., Zhou, X., & Ye, S. (2009). Realization and analysis of GPS precise clock products. *Geomatics and Information Science of Wuhan University, 34*(1), 288–291 (in Chinese).

A Method on Constellation On-Orbit Backup of Regional Navigation Satellite System

Laiping Feng, Wenhai Jiao, Xiaolin Jia, Xianbing Wu and Kai Ren

Abstract The constellation of regional navigation satellite system are based on the mixed configuration, and the initial design does not take the redundancy into account, so the system reliability is difficult to be guaranteed due to various satellite faulty factors. To ensure the constellation performance under the satellite failure conditions, the method on IGSO backup is studied and investigated in this paper. On the design constraints of only two surplus satellites, the IGSO orbit position and backup options are given through comparison and analysis. At the same time, from the view of meeting the global positioning satellite test and playing a backup role for IGSO, the orbital slots of four MEO of the basic constellation are adjusted and the on-orbit backup solution of 5GEO/5IGSO/4MEO is designed, which can effectively ensure the continuity and availability of system, and provide the reference for the optimization of system backup program.

L. Feng (✉) · K. Ren
Institute of Surveying and Mapping, Information Engineering University,
450052 Zhengzhou, China
e-mail: fenglaiping@163.com

L. Feng · X. Jia · X. Wu
Xi'an Research Institute of Surveying & Mapping,
710054 Xi'an, China

W. Jiao
Beijing Institute of Tracking and Communication Technology,
100097 Beijing, China

K. Ren
Xi'an Technical Division of Surveying & Mapping,
710054 Xi'an, China

J. Sun et al. (eds.), *China Satellite Navigation Conference (CSNC) 2012 Proceedings*,
Lecture Notes in Electrical Engineering 160, DOI: 10.1007/978-3-642-29175-3_18,

1 Introduction

Based on the construction experience and development of the GPS, GLONASS and other satellite navigation systems, the redundancy design of the constellation is great significant to improve the system reliability, and the constellation on-orbit backup is necessary to ensure that the system operation is continuous and stable [1]. The constellation of BeiDou regional navigation system is mixed configuration of GEO/IGSO/MEO (referred to as the baseline constellation), the track of subsatellite point and the satellite ID are shown in Fig. 1, and the initial design does not take the redundancy into account, so the system reliability is difficult to be guaranteed and the navigation service disruption is inevitable due to various satellite faulty factors such as orbital maneuver, space environment and the ephemeris upload failure [2], therefore, the constellation backup is necessary to improve the system reliability.

Because the contribution to the system by various type satellites of regional constellation is different, the backup problem of the regional constellation is much more complex than the global system. GEO and IGSO are the key to realize the navigation performance and the focal point of the constellation on-orbit backup. Because the International Telecommunication Union (ITU) strictly control the GEO satellite slot, this paper focus on resolving the IGSO satellite backup. In addition, the MEO satellites will be a certain extent on the backup for GEO and IGSO satellites, which will improve the system performance but not significant. That four independent MEO satellites are used to positioning is of great significance for the global system transition test. Therefore, in this paper, the slots of these MEO satellites are adjusted on the basis of the two backup IGSO satellites, and a new solution is proposed in which the four MEO satellites can position independently under the normal state and play a backup role by the orbit control if the IGSO is failure.

2 Constellation Performance Analysis Method

2.1 IGSO Backup Options

Constellation backup is divided into on-orbit backup and ground backup, and both of them have advantages and disadvantages. The on-orbit backup can effectively improve the system continuous service ability, but BeiDou regional navigation satellite system has complex constellation, one on-orbit backup satellite will only make a backup of a specific satellite, which leads to the less flexibility. The ground backup can mobile launch and flexible configuration, but more stringent requirements for emergency launch capability, the replacement of failure satellites will require a long period of time, during which the system performance will decline.

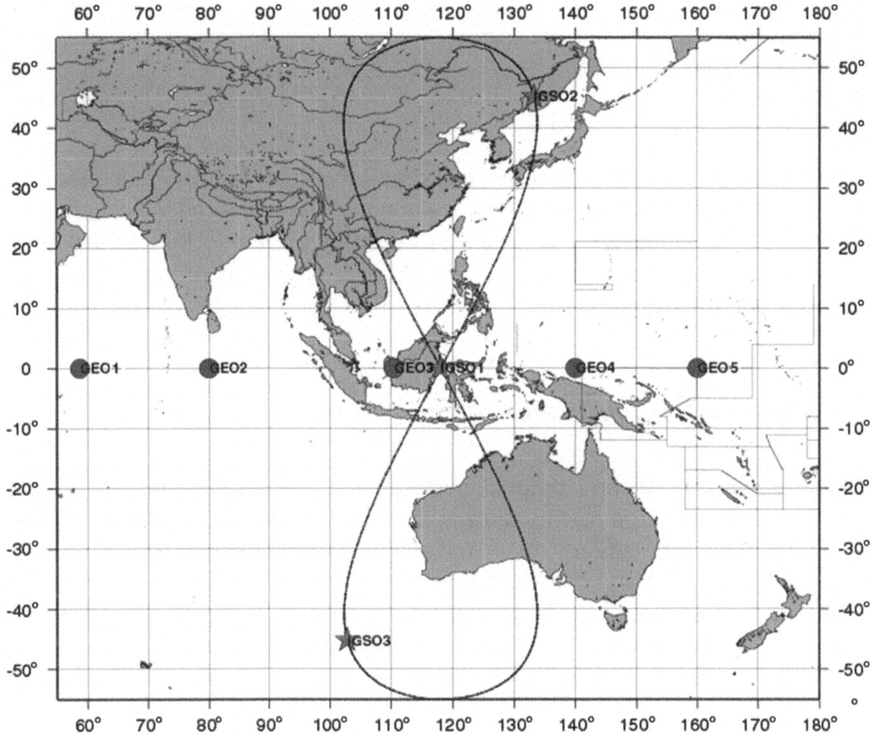

Fig. 1 The track of subsatellite point of baseline constellation

Therefore, a certain number of on-orbit backup satellites are necessary to ensure the continuous reliability of the system operation.

The slots selection of the two IGSO satellites used for backup are in two ways as follows: (1) The slots locate in the new orbital plane differ from the IGSO of the baseline constellation, and the geographical longitude of ascending node is coincident with them, which is 118° [3, 4], (2) The backup IGSO satellites and two IGSO satellites of initial design are in the same orbit planes respectively with difference argument of latitude (phase). This difference caused the geographical longitude of ascending node drift east or west, and the ground track forms another "8" shape. Because the IGSO satellites are the inclined geosynchronous orbit, so the offset of the geographical longitude of ascending node (GLAN) is the same as the difference of argument of latitude.

For above pattern (1), the condition of the two IGSO satellites repeating the same ground track is [1]:

$$\begin{cases} i_1 = i_2 \\ \Omega_1 + u_{01} = \Omega_2 + u_{02} \end{cases} \quad (1)$$

For above pattern (2), the condition of the two IGSO satellites locating in the same orbit plane is:

$$\begin{cases} i_1 = i_2 \\ \lambda_1 - u_{01} = \lambda_2 - u_{02} \end{cases} \tag{2}$$

The above two equations essentially express the relationships among the right ascension of the ascending node Ω, the argument of latitude u and the geographical longitude of ascending node λ, which is the basis to choose the optimal IGSO backup slots. The following will study how select optimal IGSO backup slots by comparing these two methods.

2.2 Constellation Performance Indicators

To realize the constellation optimal performance, the three levels are used to describe the constellation performance in this paper, maximum of PDOP, regional availability and constellation value (Constellation Value, referred to as CV) [5–7]. The maximum of PDOP value is defined as all the maximum of PDOP values through the observation period for a specific station; The regional availability is defined as the average of all the instantaneous availabilities through the interval $(t_0, t_0 + \Delta T)$ for the specific location l, which is used to evaluate the extent that the constellation can meet the accuracy requirements in the region in the form of time percentage, as the formula (3) show.

$$\bar{\beta}(l) = \frac{1}{\Delta T} \sum_{t=t_0}^{t_0+\Delta T} \beta(l, t) \tag{3}$$

And $\beta(l, t)$ is expressed as follows.

$$\beta(l, t) \equiv bool\{\alpha(l, t) \geq \alpha_{\min}\} \tag{4}$$

Where: $bool(X)$ is Boolean function, which equals to 1 if X is true else equals to 0.

The CV is defined as the average of the area percentages of the region accounting for in which the DOP is less than the certain threshold through the observation period, and the CV is expressed as formula (5), which reflects the the geometric characteristics, continuous visibility and performance of the constellation. This paper focuses on the CV for constellation optimization.

$$CV = \frac{\sum_{t=t_0}^{t_0+\Delta T} \sum_{i=1}^{L} bool(DOP_{t,i} \leq DOP_{\max}) \times area_i}{\Delta T \times Area} \times 100\% \tag{5}$$

where: ΔT is the total simulation time, L is the number of grid, $Area$ is the total area of the service region, $area_i$ is the area of the ith grid, and $Area$ can be expressed as the number of grid points when the grid is divided thickly enough

Table 1 CV of 5G/3I/4M with different failure satellite

Failure satellite	Key	General
–	100.0	91.4
G1	93.8	88.1
G2	96.1	87.9
G3	95.0	89.4
G4	93.3	88.6
G5	98.1	89.1
I1	93.2	85.0
I2	95.6	86.3
I3	93.7	85.8

Note 1: "G" represents GEO, and "I" denote IGSO
Note 2: "–" denote baseline constellation

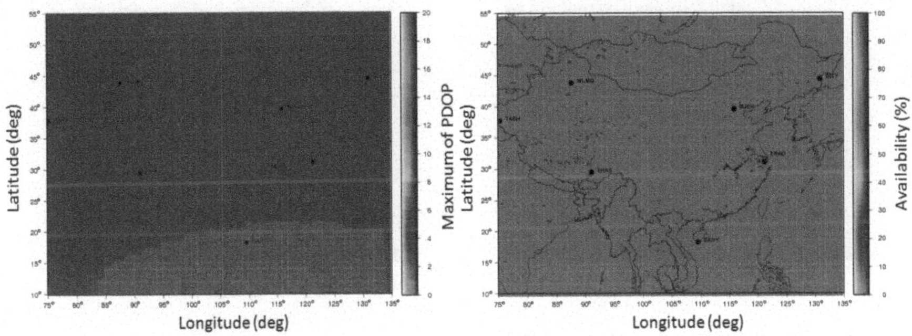

Fig. 2 Maximal PDOP and availability of healthy constellation

[8–10]. Formula (5) indicates the CV is related to the service area, the threshold value, the simulation revolution and the observation sampling rate. The service area is divided into the key services area and the general services area. The key services area is from 75° to 135° east longitude and 10°–55° north latitude, while the general services area is from 55° to 80° east longitude and 55° south latitude –55° north latitude. The PDOP threshold is set to 5.0 for the key services area while 7.0 for the general services area. According to the MEO revolution the simulation revolution is set to 7 days and the sampling rate is 10 min.

Based on the above indicators, Table 1 shows the CV of the key services area and the general services area when each satellite of the basic constellation is absence. The results show that the GEO satellites and IGSO satellites play an important role in normal operation of the constellation and their failure cause the loss of CV about 5% in the key services area while 10% in the general services area. In order to express the impact of the positioning performance more intuitively, Figs. 2–4 give the maximum of PDOP value and the regional availability of the key services area when the constellation is complete or IGSO1 and GEO4 are absence. Figure 3 shows that IGSO1 failure leads to the PDOP values more than 10.0 in most areas of China and the emergence of the positioning accuracy discontinuous, at the

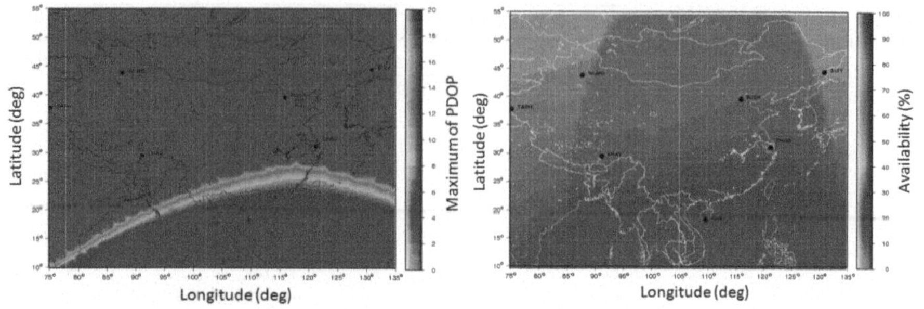

Fig. 3 Maximal PDOP and availability with IGSO1 Failure

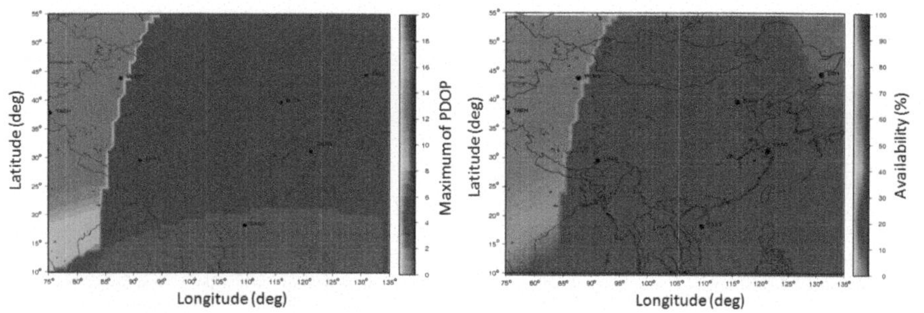

Fig. 4 Maximal PDOP and availability with GEO4 failure

same time, the availability has decreased by 10%. Figure 4 shows that GEO4 failure leads to a serious decline of the availability in the coverage area.

Thus, the GEO satellites and IGSO satellites play a prominent role in the regional navigation system, however, the slots of GEO satellites are strictly limited by ITU and the backup can not be configured freely, the slot selection of backup IGSO satellites are discussed emphatically in this paper.

3 The Slot Selection of Backup IGSO

3.1 The Same Orbital Plane

As the backup IGSO and the IGSO of baseline constellation are in the same orbital plane, the relationship of the geographical longitude of ascending node between them is determined by the difference of the argument of latitude which further affect the constellation performance, so different constellation performance can be obtained from different of the geographical longitude of ascending node. Assume that one backup IGSO is set in each orbital plane of baseline constellation, in order

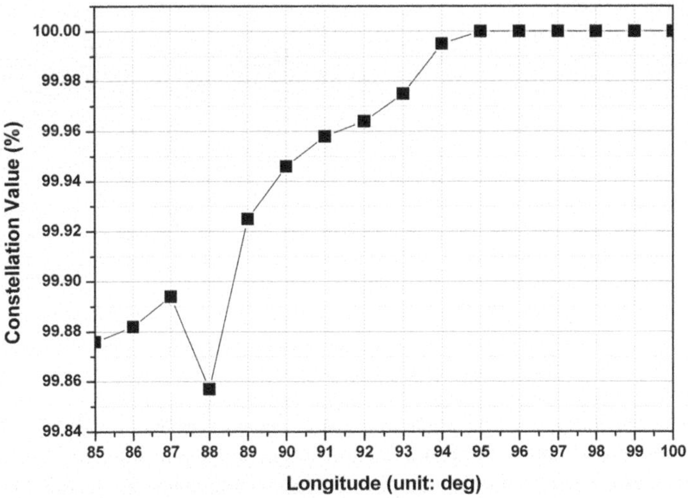

Fig. 5 CV of key area of 5G/6I constellation with single IGSO failure

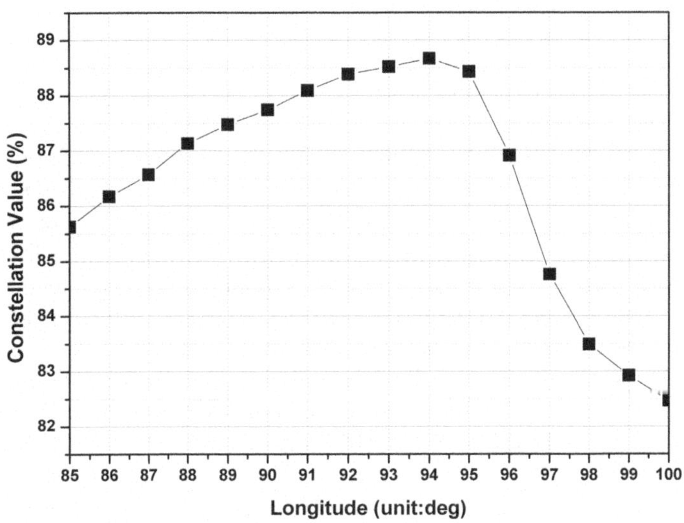

Fig. 6 CV of general area of 5G/6I constellation with single IGSO failure

to get the best geographical longitude of ascending node, the CV of the geographical longitude of ascending node are calculated range from 85 to 100°. To achieve the purpose of baseline constellation backup, the corresponding CV is analyzed when each IGSO is failure. Figures 5 and 6 show the results of the key areas and the general areas.

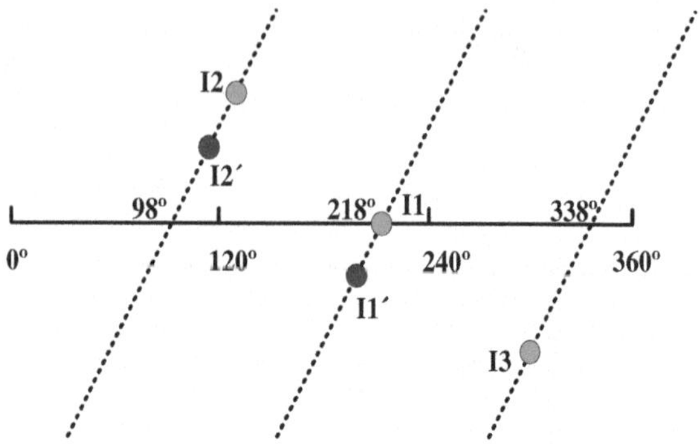

Fig. 7 Relation of backup IGSO and baseline IGSO slot

Synthetically analyzing the performance of the key areas and the general areas, the maximum of CV can be attained when the geographical longitude of ascending node is 95° east longitude, at this time, the relationship among five IGSO is showed in Fig. 7 under J2000.0. $I1'$ and $I2'$ are two backup IGSO which are in the same orbital plane with $I1$ and $I2$ of baseline constellation respectively. The difference of the argument of latitude is 23° and the difference of GLAN are also 23°.

3.2 The Different Orbital Plane

For the same orbital plane, the optimal positioning performance can be obtained when the geographical longitude of ascending node of the backup IGSO satellites is 95°east longitude, while different orbital plane can also get backup at 95° east longitude. In order to compare the positioning performance in both cases, the backup constellation performance of different orbital plane is calculated as follows. Traversing method is used for the performance analysis, in the condition of the phase difference of 120° the mean anomaly traverse from 0 to 120° and take the failure of the IGSO of the baseline constellation into account. Figure 8 shows the changes of the CV according to the mean anomaly when each IGSO in the baseline constellation is failure, Fig. 9 shows the changes of average CV integrated three IGSO satellites, Figs. 10 and 11 show the CV of the general services area.

Through the above analysis, the conclusion can be draw as follows: When the ascending nodes of two backup IGSO in different orbital planes are 105° and 345° (J2000.0) respectively, the performance is best, and the corresponding mean anomaly of the backup satellites are 90° and 210° respectively. The navigation performance of this backup method is consistent with another backup method of two IGSO of the same orbital planes with the phase difference of 23°, the difference lies in the corresponding relationship between the two above methods and the baseline orbit.

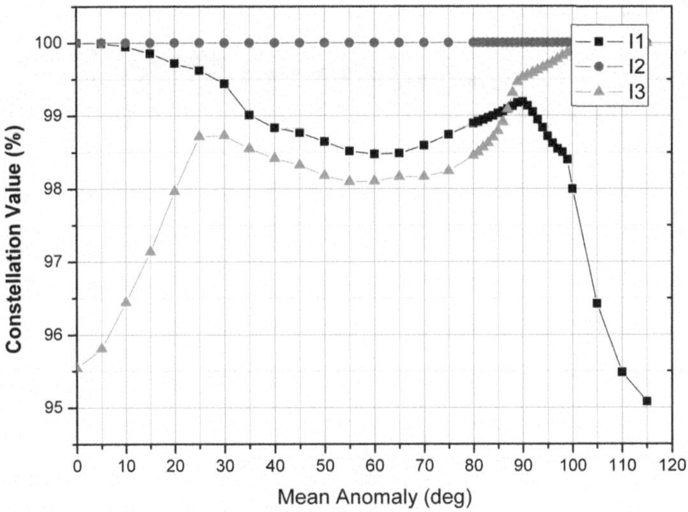

Fig. 8 The variety of CV with mean anomaly in key area

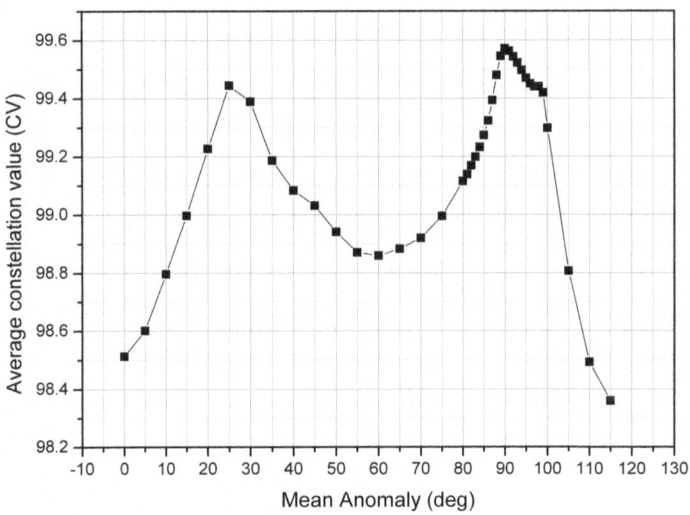

Fig. 9 The variety of average CV with mean anomaly in key area

3.3 Results Comparison

With the comprehensive comparison of the 5G/5I/4M/calculated results of the same orbital plane and the different orbital shown in Table 2, Figs. 12 and 13, the performance difference between the same orbital plane backup and the different

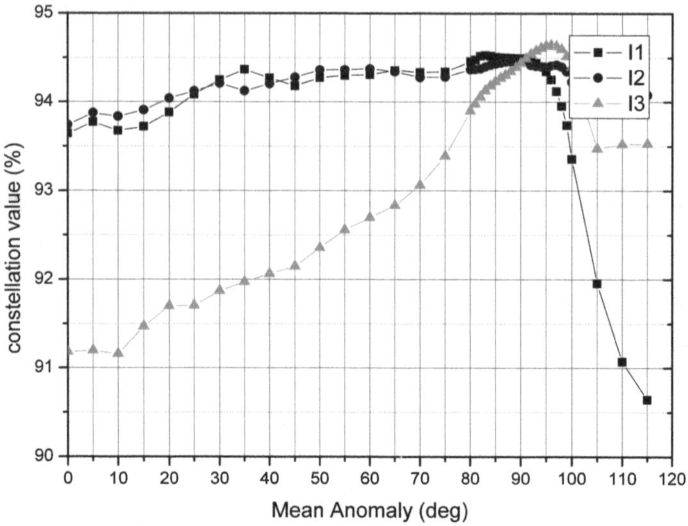

Fig. 10 The variety of CV with mean anomaly in general area

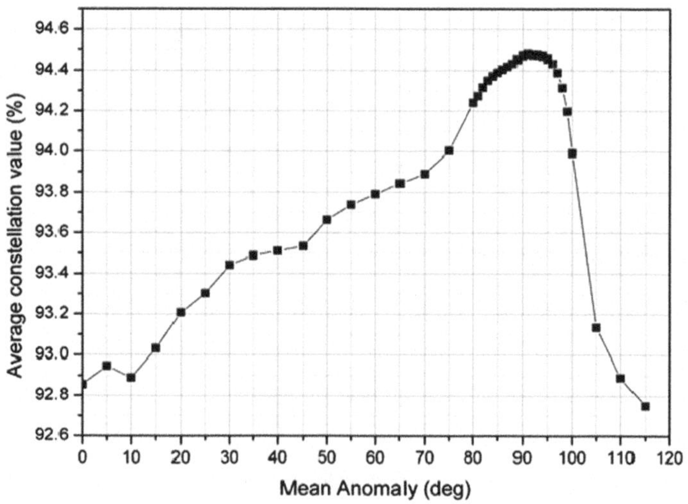

Fig. 11 The variety of average CV with mean anomaly in general area

orbital plane backup is small. If the two IGSO satellites simultaneously failed, the different orbital plane backup is slightly better than the same orbital plane backup. In addition, the different orbit backup can not transfer to baseline orbit, so it has poor flexibility compared with the same orbital plane.

Table 2 CV comparison between same orbital plane and different orbital plane

Failure satellite	Same		Different	
	Key	General	Key	General
–	100.0	97.7	100.0	97.5
G1	95.0	95.2	95.3	95.0
G2	98.3	96.6	98.3	96.4
G3	98.2	96.1	98.4	96.1
G4	96.3	95.5	96.4	96.3
G5	98.1	92.7	98.2	93.1
I1	100.0	94.5	99.2	94.5
I2	99.8	94.9	100.0	94.5
I3	99.7	95.0	99.5	94.4
I1/I2	99.8	90.8	98.0	90.9
I1/I3	97.5	91.1	95.5	89.4
I2/I3	96.6	90.6	99.5	90.6

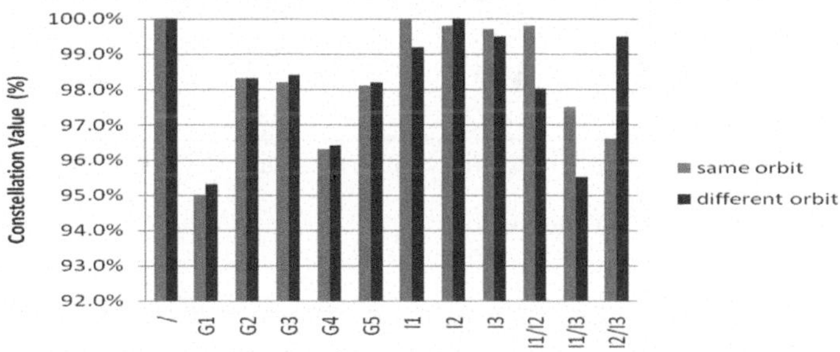

Fig. 12 Comparision of CV between same orbital plane and different plane in key area

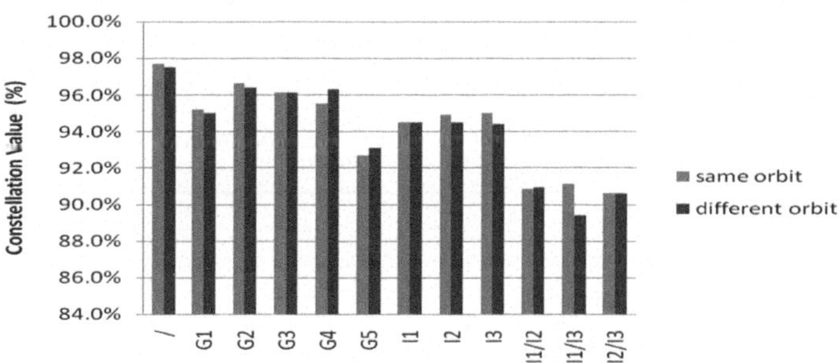

Fig. 13 Comparision of CV between same orbital plane and different plane in general area

Thus, the IGSO backup scheme is reached, and the relationship of the substellar point between the backup IGSO and the baseline constellation is showed in Fig. 14, the geographical longitude of ascending node is east longitude 95°.

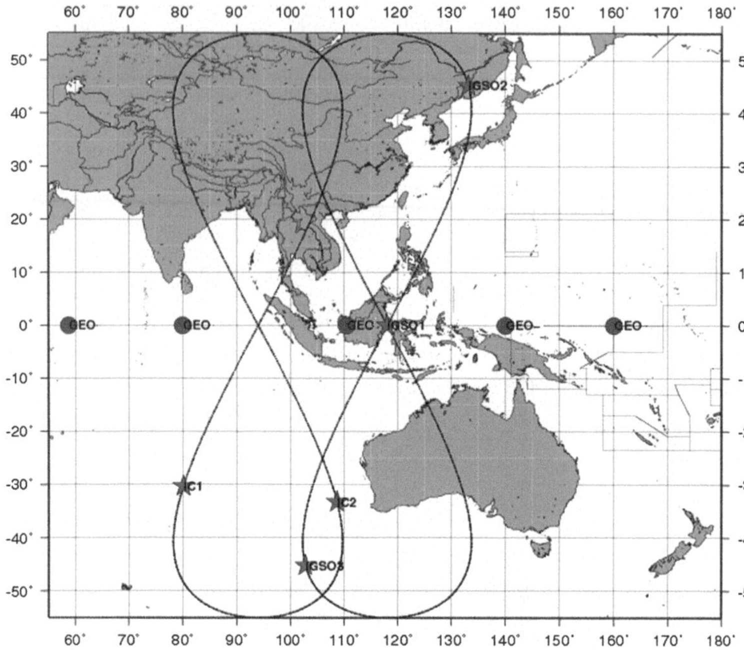

Fig. 14 subastral point of IGSO after backup

Table 3 CV of each failure satellite in 5G/3I (%)

Failure satellite	5G/3I
–	100
G1	84.4
G2	89.9
G3	85.8
G4	83.6
G5	95.0
I1	85.4
I2	85.4
I3	85.4

4 Slot Adjustment of MEO

Two of Four MEO satellites were placed in the 7 and the 8th slot of the first orbital plane, and the other two were placed in the 3rd and the 4th slot of the second orbital plane in baseline constellation. Not considering MEO, the performance of 5G/3I in the key regional with single satellite failure is shown in Table 3. Through the comparison with Table 1, when all satellite are healthy, MEO do not play a role. As certain satellite failure, to a certain extent MEO played indirect backup

Table 4 Positioning results satisfying the demands of certain threshold

Station name	Before					After				
	Arc	min (s)	max (s)	Total (s)	Percentage	Arc	min (s)	max (s)	Total (s)	Percentage
KaShi	2	971	1,277	2,248	0.37	8	538	3,980	20,633	3.41
SanYa	10	439	5,127	37,556	6.21	11	454	8,673	52,104	8.62
BeiJing	4	40	1,388	2,734	0.45	8	592	4,093	22,012	3.64
SuiYang	1	79	79	79	0.01	6	356	2,099	8,218	1.36
ChengDou	7	516	2,397	9,157	1.51	8	304	4,610	24,440	4.04

Table 5 CV of 5G/I5/4M after adjustment of MEO in key area (%)

Failure satellite	Before	After
–	100.0	100.0
G1	94.3	95.5
G2	97.9	98.4
G3	97.7	98.3
G4	95.5	96.3
G5	97.9	98.1
I1	100.0	100.0
I2	99.8	99.9
I3	98.6	99.8
I1/I2	99.8	99.9
I1/I3	97.4	99.6
I2/I3	95.8	97.9

role. For example, for IGSO failure, CV is improved from 85.4 to 93.2% in the key region because four MEO role.

Based on the results of the second and third parts in this paper,the slot of IGSO1 and IGSO2 have got backup. The other IGSO may be reinforced with four MEO satellites indirectly, due to minor influence to improve performance of normal constellation, more meaningful method are puts forward, that four MEO were use to position for global positioning test with normal constellation and backup GEO or IGSO with their failure indirectly. The result indicates for global position test, the slot of four MEO may select the 2nd and 3rd slot of the first orbital plane, and the 4 and 5th slot of the first orbital plane, while for backup opinion, four MEO may select the 2nd and 4th slot of the first orbital plane, and the 6 and 8th slot of the third orbital plane.

Table 4 shows the positioning time comparison of five stations in the region before and after the MEO adjustment in the condition that the PDOP is <7.0. Positioning performance is improved by adjusting the slot of four MEO satellite. For example, before adjustment, a short time positioning can be done in SanYa which is located in low latitudes, and the percentages of time meeting the threshold have been significantly improved after adjusting.

The improvement of backup is analyzed by adjustment of MEO respectively to the 2nd and 3rd slot of the first orbital plane, and the 4 and 5th slot of the first orbital plane. CV of key region are given in Table 5, and result show CV value increased by 1.2% with IGSO3 failure, and by more 2% with two IGSO failure.

Moreover, it can be drawn from the calculation results that only a simple phase adjustment of the same orbital plane is needed from the optimal positioning performance under normal conditions to the optimal backup performance, which has a strong engineering feasibility.

5 Conclusion

In this paper, the on-orbit backup method of the regional navigation satellite system is studied under the constraint of two surplus IGSO backup satellites, through calculation and analysis, the conclusions are drawn as follows:

(1) The optimal slots of the backup IGSO satellites is 95°. The backup IGSO satellites and the baseline IGSO satellites locate in the same orbital plane with 23°difference of the argument of latitude.

(2) When the right ascension of the ascending node of two backup IGSO in different orbital planes are 105 and 345° (J2000.0) respectively, the performance is best, and the corresponding flat angles of the backup satellites are 90 and 210° respectively. The navigation performance of this backup method is consistent with another backup method of two IGSO of the same orbital planes with the phase difference of 23°, the difference lies in the corresponding relationship between the two above methods and the baseline orbit.

(3) Adjusting the slots of four MEO satellites can meet the need of independent positioning and maximize the backup role on GEO and IGSO through the phase control in case that the GEO or IGSO is failure.

In the sight of the positioning performance for users, the method of IGSO satellites backup for regional constellation and the MEO satellites adjustment is studied in this paper, and the on-orbit backup solution of "5GEO/5IGSO/4MEO" is proposed, which have reference value for the regional system construction.

References

1. Zhang, Y., Fan, L., & Zhang, Y. (2008). *Theory and design of constellations*. Beijing: Science Press.
2. Li, G., & Li, J. (2010). Analysis of availability in China's territory volume based on the satellite service interruption. *The 1st China Satellite Navigation Conference*.
3. International Telecommunication Union Radio Communication Bureau (2005). *Third resolution 609 (WRC03) Consultation Meeting* (pp. 14–16). Munich, Germany.
4. Grelier, T., Ghion, A., & Dantepal, J. (2007). *Compass signal structure and first measurements* (p. 3015). Fort Worth, TX: Pro.ION GNSS.
5. Wang, L. (1999). Design and comparison of constellation on regional navigation satellite system. *Beijing: Journal of Spacecraft TT & C Technology*, 18(4), 1–8.
6. Jia, X. (2001). *Study on the navigation satellite the system's constellation design and orbit determination*. ZhengZhou: Institute of Surveying Mapping, Information Engineering University.
7. Liu, L., & Hu, S. J. (2006). *An introduction of astrodynamics*. Nanjing: Nanjing University Press
8. Department of Defense of United States of America (2008). *Global positioning system standard positioning service performance standard* (4th edn). September (S/OL). http:// pnt.gov/ public/docs.
9. Hen, G. W. (2007). Envisioning a future GNSS system of systems. Inside GNSS.
10. Meng, L., & Yu, S. H. *The impact of compass on future GNSS: A perspective from Asia*. Taiwan: Department of Geomatics, National Cheng Kung University.

SHA: The GNSS Analysis Center at SHAO

Junping Chen, Bin Wu, Xiaogong Hu and Haojun Li

Abstract Today, most precise GNSS products, including orbits and clocks, are provided by the International GNSS Service (IGS) and its Analysis Centers (ACs). Each AC provides its products to the AC Coordinator (ACC) for combination. ACs develop their own software packages by implementing different strategies, which as a result improving the robustness of the combined products. Following the IGS AC strategy and to fulfill the requests of satellite missions in China, we set up the GNSS Analysis Center at Shanghai Astronomical Observatory (SHAO). Currently our GNSS routine analysis includes: Global GPS+GLONASS data processing, GLOBAL+CMONOC GPS data processing. In the first routine, we use ~110 global stations, of which ~50 have GLONASS observations, to derive the integrated and consistent GNSS products. In the second routine, we combine the IGS network used the first routine and the Crustal Movement Observation Network of China (CMONOC) network, GPS only solution is performed using ~300 stations. This paper introduces the details of the Analysis Center and presents the latest results.

1 Introduction

With the improvement of accuracy and precision, GNSS has contributed to the mm-level applications as: earth dynamics monitoring, global reference frame definition, natural disaster monitoring, weather forecasting etc. In all these

J. Chen (✉) · B. Wu · X. Hu · H. Li
Shanghai Astronomical Observatory (SHAO), Chinese Academy of Sciences (CAS), Shanghai, China
e-mail: junping.chen@shao.ac.cn

J. Sun et al. (eds.), *China Satellite Navigation Conference (CSNC) 2012 Proceedings*, Lecture Notes in Electrical Engineering 160, DOI: 10.1007/978-3-642-29175-3_19,

applications, precise GNSS (Global Navigation Satellite System) products including orbits and clocks play a fundamental role. Today, most precise GNSS products are provided by the IGS [1] and its Analysis Centers. Each AC provides its products to the AC Coordinator for combination. ACs develop their own software packages by implementing different strategies, which as a result improving the robustness of the combined products.

IGS products and its associating service becomes ever important in GNSS research. However, IGS does not guarantee such service. Since November 2011, the web page and FTP of the IGS are not open to Chinese IPs. GNSS community in China is thus facing a situation that other IGS services may be interrupted due to policy change of the NASA. To improve the availability of precise GNSS products and to shorten the time of products releasing, we set up the GNSS Analysis Center at Shanghai Astronomical Observatory (SHAO) [2]. The analysis center (abbreviated as SHA) follows the IGS AC strategies and aims to fulfill the requests of satellite missions in China. Currently, our GNSS routine analysis includes: Global GPS+GLONASS data processing, GLOBAL+CMONOC GPS data processing. In the first routine, we use ~110 global stations, of which ~50 have GLONASS observations, to derive the integrated and consistent GNSS products. In the second routine, we combine the IGS network used the first routine and the CMONOC network, GPS only solution is performed using ~300 stations. We presents the latest results of SHA and presents its applications.

2 Status of IGS

The applications of GNSS explores from the original navigation to other areas like meteorology, precise positioning, etc. It has been applied to the ITRF definition [3], Earth Geo-dynamics monitoring [4], natural disaster monitoring [5], precise orbiting for LEOs [6], Earth rotation parameter estimation [7] and atmosphere monitoring [8], etc.

Providing the most precise GNSS products, IGS and its participating agencies operates more than 400 GNSS sites. Data centers are responsible for the disseminating of observations and products. The precise IGS products includes: precise GNSS orbits and clocks, station coordinates and velocities, earth rotation parameters, atmosphere products (ZTD and TEC maps), etc. The IGS AC distributes in Europe (5), Canada (1) and USA (5). Each AC differs in data sets, software packages, and data analysis strategies, of which the results are compared and combined.

Current precision of IGS core products are listed in Table 1 [9].

Table 1 Precision of IGS core products (IGS stands for Final products; IGR is Rapid products; IGUA is the estimating part of Ultra rapid products; IGUB is the predicting part of Ultra rapid products, i.e. real-time products)

	GPS orbits (cm)	GPS clocks (ns)	Pole (mas)	Lod (μs)	Coordinates (mm)/ velocities (mm/year)
IGS	2.5	0.02	0.03	10	Horizontal: 3.0/2.0
IGR	2.5	0.03	0.04	10	Height: 6.0/3.0
IGUA	3.0	0.05	0.05	10	
IGUB	5.0	1.5	0.20	50	

3 Challenges of GNSS Data Analysis

With the development of the GNSS technology, especially with the coming new signals and new constellations, GNSS data analysis is facing new challenges:

(1) Multi-system data analysis and handling of new frequencies/data types

Most of IGS products are generated based on GPS ionosphere-free combination observations. Recently, five ACs start to provide GLONASS products with final orbits precision of 5 cm and real-time orbits of 10 cm. The inclusion of more satellite systems like GLONASS and Compass/Beidou improves the coverage of satellite constellations and can improve the precision of the common parameters (e.g. coordinates) at the stations. The general strategy for integrated GPS+GLONASS data processing follows the convention that GPS system is selected as the reference system and system biases are estimated for GLONASS satellites. With the coming Galileo system, more bias parameters have to be defined and the data processing will become ever complicated.

At the time of this writing, GPS constellation contains seven satellites of Block IIR-M type and two Block IIF satellites, where Block IIR-M satellites have new L2C observations and Block IIF satellites provide observations at L5 frequency. Galileo system has even more frequencies and observation types. New types of observation may lead to a revolution of the current IGS AC strategies, which are always based on ionosphere-free L3 observations.

(2) Data analysis of huge networks

IGS provides data of more than 400 sites, each AC processes only sub-network (<250) of these stations due to limitation of computation capability. However, many regions operate much denser networks: e.g., the Southern California Integrated GPS Network (SCIGN) [10] with ~250 stations, the European EUREF [11] network with ~250 stations, the GEONET [12] network in Japan with more than 1200 stations and the CMONOC network in China with ~260 stations. The integrated data processing of these huge networks requires new strategies.

(3) Real-time GNSS

Since 2007, the IGS operates the IGS-RTPP [13]. IGS-RTPP aims to gather and distribute real-time data and products associated with GNSS satellite constellations.

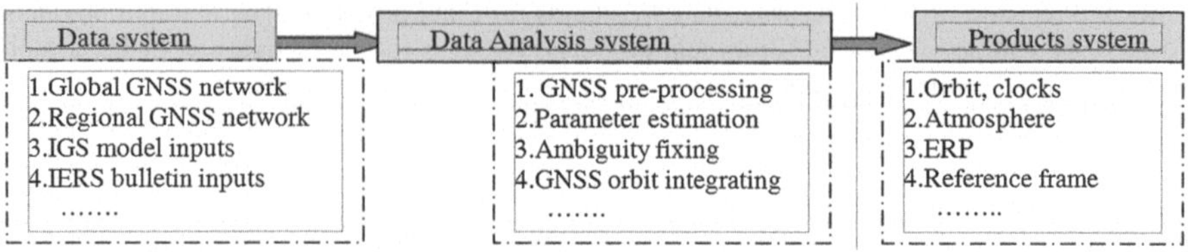

Fig. 1 Flowchart of the routine GNSS data analysis at SHAO

The primary products envisioned for the project are multi-frequency observation data and precise satellite clocks made available in real-time.

Under IGS-RTPP collaborations, there are currently more than 100 stations providing real-time streams. The RTPP AC retrieves real-time streams through the open internet protocol and estimates real-time satellite orbits and clocks. The IGS-RTPP is still at piloting stage with satellite clock sampling of 5 s and a latency of 10–15 s (including latency of stream, time of data analysis and internet communication). Due to the limitation of computation capabilities, most ACs use ~50 stations, the availability and robustness of their products are major problems.

4 GNSS Data Analysis Center at SHAO (SHA)

SHAO supports the Group of GNSS Data Analysis and Applications (GGDAA) in recent years. The GGDAA works with the above-mentioned challenges and starts routine GNSS data analysis since June 2011. Figure 1 shows the flowchart of the data analysis system.

In Fig. 1, data system automatically downloads GNSS observations and input tables for data analysis; the analysis system performs routine data analysis and provides parameter estimations; the products system generate all GNSS products in given internal and external formats.

Using the above platform, routine data analysis is performed using the IGS network with ~110 stations and CMONOC network with ~260 stations. Figures 2 and 3 illustrate the two networks.

5 Products of SHA

5.1 GNSS Clocks

SHA starts to provide precise GNSS products since Doy 165, 2011. To validate the clock precision, we compare our GPS clocks to the IGS final clocks, our GLONASS clocks are compared to the GFZ final clocks as there are no combined

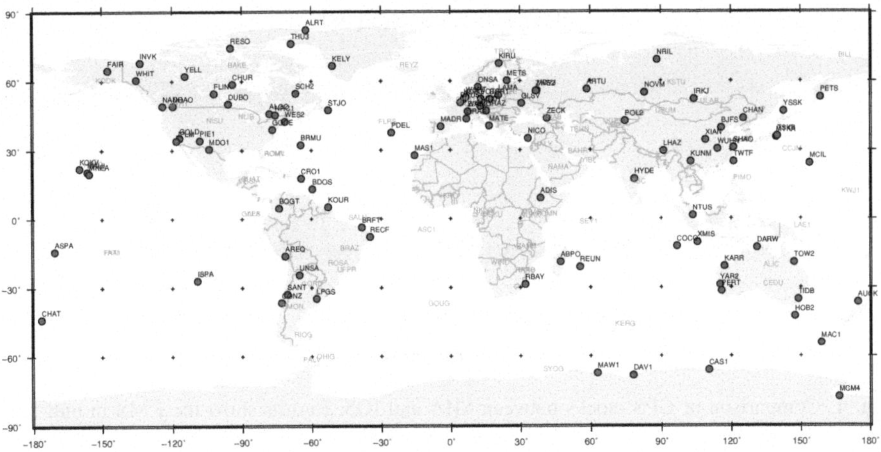

Fig. 2 IGS network processed in the GNSS routine of SHA

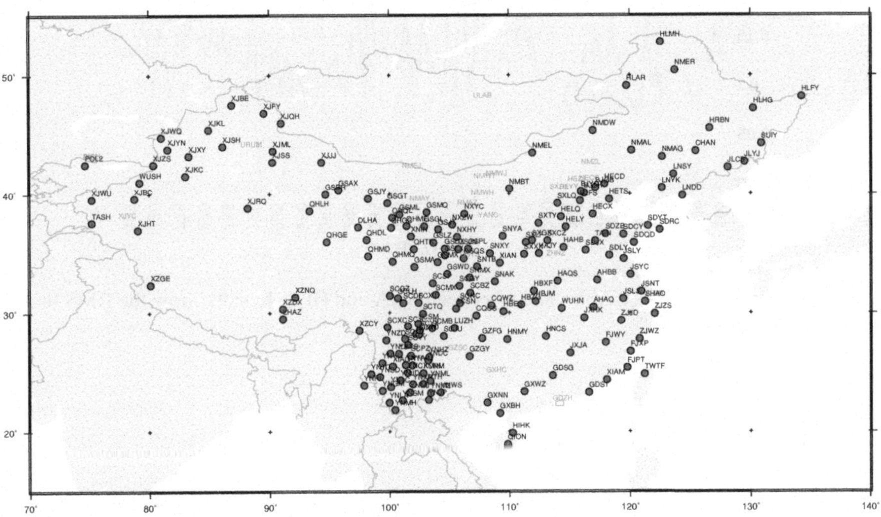

Fig. 3 CMONOC network processed in the GNSS routine of SHA

IGS GLONASS clocks. Figures 4 and 5 show the RMS of the comparisons, where we see the precision of GPS clocks of SHA is at about 0.05 ns and the precision of GLONASS clocks at about 0.15 ns.

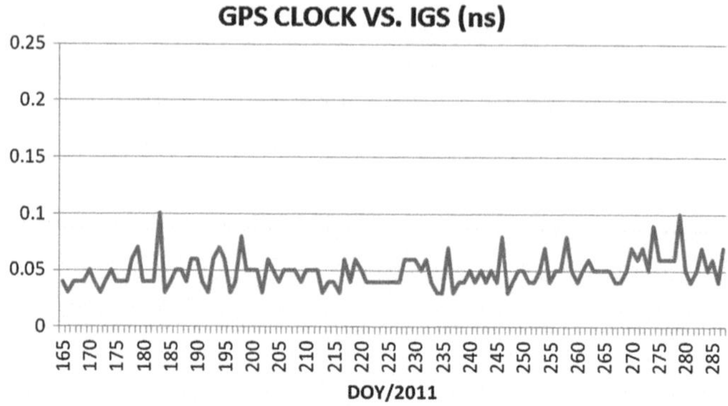

Fig. 4 Comparison of GPS clocks between SHA and IGS. Results show the RMS in mm

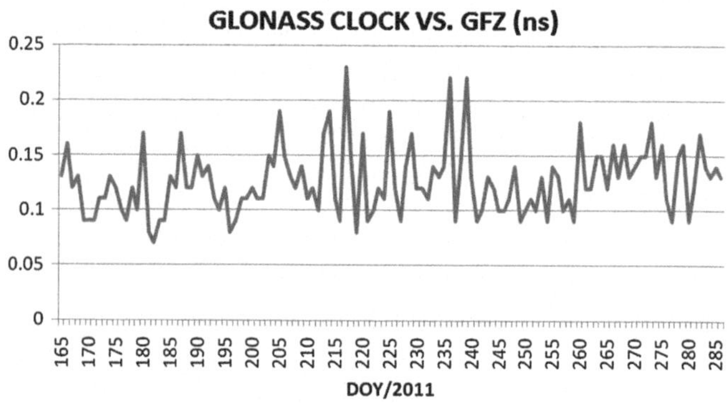

Fig. 5 Comparison of GLONASS clocks between SHA and GFZ. Results show the RMS in mm

5.2 GNSS Orbits

To validate the GNSS orbits precision, we compare our GPS orbits to the IGS final orbits, our GLONASS orbits are compared to the GFZ final orbits. Figures 6 and 7 show the RMS of the comparisons, where we see the precision of GPS orbits of SHA is at 2 cm and the precision of GLONASS orbits at 4 cm.

5.3 GPS/GLONASS Time Offset

The Time Offset (TO) between GPSt and GLONASST can be derived from the following equations:

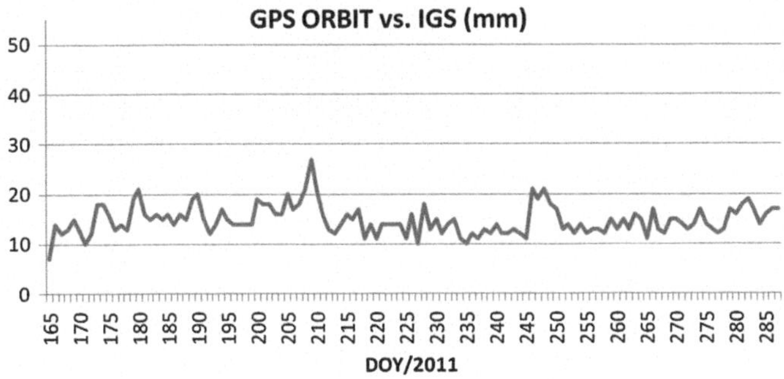

Fig. 6 Comparison of GPS orbits between SHA and IGS. Results show the RMS in mm

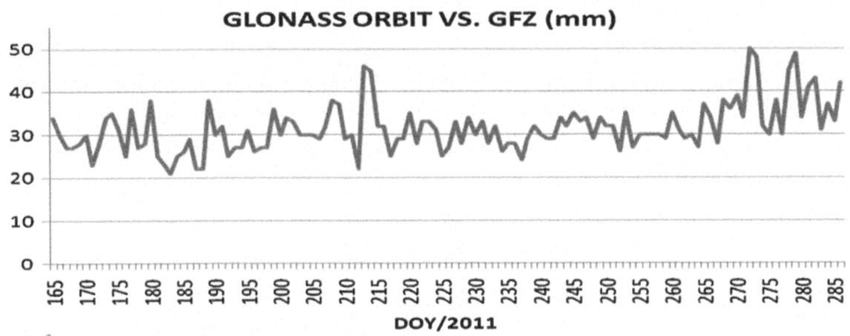

Fig. 7 Comparison of GLONASS orbits between SHA and GFZ. Results show the RMS in mm

$$\begin{aligned} \mathrm{TO} &= \mathrm{GLONASST} - \mathrm{GPSt} \\ &= [\mathrm{GLONASST} - \mathrm{UTC(SU)}] \\ &\quad - [\mathrm{GPST\text{-}UTC(USNO)}] \\ &\quad + [\mathrm{UTC(SU)\text{-}UTC(USNO)}] \end{aligned} \tag{1}$$

In (1), the first two terms GLONASST-UTC(SU) and GPSt-UTC(USNO) are monitored at monitoring stations of each satellite system and they are encoded in the navigation files using predicted values. The latter part UTC(SU)-UTC(USNO) is in the order of few ns and can be retrieved in the BIPM bulletin with a latency of ~1 month.

SHA performs the integrated processing of GPS and GLONASS observations, through which the TO between GPSt and GLONASST could be monitored using the following equations:

$$\begin{cases} \mathrm{TO(i)} = \delta^{r} - \delta^{\mathrm{brdc}} \\ \mathrm{TO1} = \dfrac{1}{\mathrm{n}} \sum\limits_{\mathrm{i-1}}^{\mathrm{n}} \mathrm{TO(i)} \\ \mathrm{TO2} = \mathrm{Median}[\mathrm{TO(i)}] \end{cases} \tag{2}$$

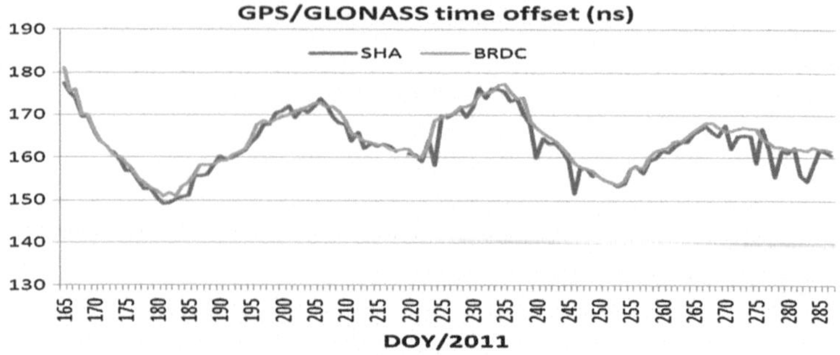

Fig. 8 GPS/GLONASS time offset in ns, where SHA illustrates TO derived from Eq. (2) using the SHA integrated GNSS solutions; BRDC presents TO calculated from Eq. (1) with the first two terms from daily navigation files and the term UTC(SU)-UTC(USNO) is being ignored

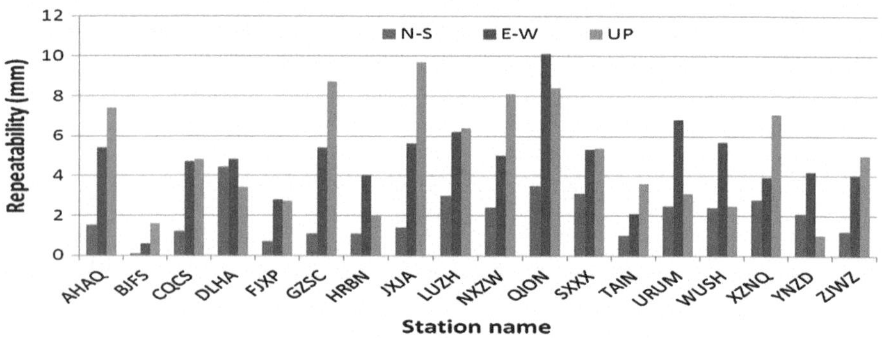

Fig. 9 PPP coordinates repeatability using precise orbits and clocks of SHA

In (2), the term δ^r is the GLONASS clock from SHA estimation, which is under the time frame of GPSt; δ^{brdc} is the GLONASS clock from the broadcast, which is under the time frame of GLONASST. The following figure shows the TO between GPSt and GLONASST using (1) and (2). In Fig. 8, the term UTC(SU)-UTC(USNO) is ignored due to its long latency, which accounts for additional few ns errors. The agreement between the two approaches is better than 10 ns.

5.4 Station Coordinates

Selecting 18 stations from the CMONOC network, Fig. 9 shows the PPP coordinates repeatability from DOY 165 to 167, 2011.

6 Conclusion

We present the current challenges of GNSS data analysis. The GNSS data analysis center at SHAO is introduced and the routine results of SHA are presented. The GNSS routine processing of SHA provides integrated solutions, where all products are based on common references. The products of SHA are at the same precision level of IGS products and have been applied to satellite missions of China. Related researches are being carried out within the GGDAA group based on the routine results.

Acknowledgments This research is supported by 100 Talents Programme of the Chinese Academy of Sciences. IGS community is acknowledged for providing Rinex data.

References

1. Dow, J. M., Neilan, R. E., & Rizos, C. (2009). The international GNSS service in a changing landscape of global navigation satellite systems. *Journal of Geodesy, 83*, 191–198. doi:10.1007/s00190-008-0300-3.
2. Shanghai Observatory GNSS Analysis Center. Retrieved from www.shao.ac.cn/shao_gnss_ac
3. Zuheir, A., Xavier, C., & Laurent, M. (2011). ITRF2008: An improved solution of the international terrestrial reference frame. *Journal of Geodesy*. doi:10.1007/s00190-011-0444-4.
4. Moreno, M., et al. (2011). Heterogeneous plate locking in the South-Central Chile subduction zone: Building up the next great earthquake. *Earth and Planetary Science Letters, 305*(3–4), 413–424.
5. Chen, J., Bender, M., Beyerle, G., Dick, G., Falck, C., Ge, M., et al. (2010). GNSS activities for natural disaster monitoring and climate change detection at GFZ—an overview. In E. Chuvieco, J. Li, & X. Yang (Eds.), *Advances in Earth Observation of Global Change*, Springer (pp. 159–172).
6. Bock, H. (2003). Efficient methods for determining precise orbits of low earth orbiters using the global positioning system. Ph.D. thesis, Astronomical Institute University of Berne, Switzerland.
7. Mireault, Y., Kouba, J., & Ray, J. (1999). IGS earth rotation parameters. *GPS Solutions, 3*(1), 59–72. doi:10.1007/PL00012781.
8. Gendt, G., Reigber, C., & Dick, G. (2001). Near real-time water vapor estimation in a German GPS network-first results from the ground program of the HGF GASP project. *Physics and Chemistry of the Earth, Part A: Solid Earth and Geodesy, 26*(6–8), 413–416.
9. IGS. (2011). http://igscb.jpl.nasa.gov/
10. SCIGN. Retrieved from http://www.scign.org/
11. EUREF. Retrieved from http://epncb.oma.be/
12. GEONET. Retrieved from https://www.geonetjapan.com/
13. Caissy, M. (2006). The IGS real-time pilot project—perspective on data and product generation. *Report at Streaming GNSS Data via Internet Symposium*, Frankfurt, February 6–7, 2006.

MEO and HEO Satellites Orbit Determination Based on GNSS Onboard Receiver

Tao Geng, Xing Su and Qile Zhao

Abstract The method of Middle Earth Orbit (MEO) and High Earth Orbit (HEO) satellites orbit determination based on Global Navigation Satellite System (GNSS) onboard receiver has several disadvantages such as less navigation satellites signal, poor geometrical coverage, and weaker navigation signal. Using the weak signal receiver to get the side-lobe signal is a good way. But as a result of weak receiving power and poor C/N0 (Carrier-to-Noise ratio) of the side-lobe signal, the observation noise was increased considerably, and the stochastic models of the main beam and side-lobe signal are different. Based on the Chinese COMPASS navigation system, this thesis analyses the restrictions of the aspects of GNSS signal geometries visible conditions. The simulation noise sigma model is based on the C/N0 of the observation as an exponential function. The orbits were determined by combining the main beam and side-lobe signal. The beam angle of main beam and side-lobe is 0°–21.3° and 21.3°–45°, respectively, navigation signal transit power is 26.8 W (Watt), receiving C/N0 threshold is 25 dB Hz and the system error of inter-satellite link is 20 cm. The precision of orbit determination of MEO is better than 0.3 m, and the GEO is 1.0 m. Lastly, navigation satellites integrated orbit determination was carried out using ground observations from six regional stations in China and inter-satellite link observations which is simulated by installing onboard receivers on each of the navigation satellites.

T. Geng (✉) · X. Su · Q. Zhao
Research Center of GNSS, Wuhan University, Wuhan, China
e-mail: gt_gengtao@whu.edu.cn

X. Su
e-mail: longtian00@163.com

Q. Zhao
e-mail: zhaoql@whu.edu.cn

J. Sun et al. (eds.), *China Satellite Navigation Conference (CSNC) 2012 Proceedings*, Lecture Notes in Electrical Engineering 160, DOI: 10.1007/978-3-642-29175-3_20,

1 Introduction

Satellites tracking system contains ground-based and space-based tracking nets. Up to now, the orbit determination for most of the MEO and HEO has depended on ground-based tracking nets. The distribution of monitoring stations has an important influence on the satellites tracking time, tracking extent and the precision of orbit determination. And the construction and running fee of the stations are also huge. Using space-based tracking nets is becoming one of the most important ways of satellites orbit determination.

At present space-based tracking nets have mainly two mode, namely space-borne receiver mode and inter-satellite two-way ranging mode. Inter-satellite two-way ranging mode uses Ultrahigh Frequency (UHF) or Ka band and data receiving and sending units to measure distance and communicate with other satellites [1–3]. Space-borne receiver receives the signals transmitted from the navigation satellite for building inter-satellite links.

The technology of using the space-borne receiver on the low-orbit satellites is very perfect, which has advantages such as the enough visible satellites, the good effect of geometry coverage and the stronger signal power [4, 5]. However, the research of space-borne receiver on the Geostationary Earth Orbit (GEO), Inclined Geosynchronous Orbit (IGSO), Geostationary Transfer Orbit (GTO), Medium Earth Orbit (MEO) satellites whose orbits are higher than 20000 km, started very late. In 2002, Moreau analyzed the characters of signals received by the GPS receiver of the AMSAT OSCAR-40 satellites, and proved the feasibility of using GPS receiver to determine the high-orbit spacecraft orbit [6]. In 2005, using the simulated data, Jennifer and Charles verified the precision of GEO determination orbit based on the GPS satellites navigation system [7]. In 2010, Peter and Sten put forward to equipping a small GEO satellite with two GPS receivers for the demand of GEO satellites in the transfer orbit [8].

Compared to the low-orbit satellites, the MEO and HEO mainly receive the navigation signals from satellites on the opposite side of the earth, which has some shortages such as the weaker signal power, the fewer number of visible satellites and the poorer geometry coverage. Meanwhile, because of the limitation of the transmit angle and power of navigation satellites, the range of received main beam signals for MEO and HEO is narrow. In order to increasing the number of visible satellites, the weak signal receiver is used to receive side lobe signals. But this method will lead to the different observation noises of main beam and side lobe signals, and then the integrated orbit determination based on main beam and side lobe signal observations will face the problem of the different stochastic models of them.

From 2001, our country has launched more than 20 MEO and HEO including Chang E lunar probe, Sino Sat, Asia-Pacific Sat, Broadcast TV Sat and so on. Along with the construction of China COMPASS navigation system, the study on the orbit determination of MEO and HEO based on GNSS space-borne receiver is very imperative.

This chapter introduces the fundamental principle of the MEO and HEO orbit determination based on space-borne receivers in second chapter; the third chapter provides the simulation conditions including navigation satellites constellation, MEO and HEO experiment satellites and ground tracking stations; the fourth chapter analyzes the geometry visibility and the orbit determination PDOP value of MEO and HEO, and then determines the MEO and HEO orbit based on the COMPASS navigation system; the fifth chapter builds inter-satellite links based on the space-borne receiver of the navigation satellites, and then carries out experiments of integrated orbit determination of COMPASS navigation system; the sixth chapter conclusions are drawn.

2 Basic Theories

The main beam of the transmitted GPS signals is approximately 21.3° wide and directed to the center of the Earth. Because the Earth only subtends a half angle of 13.9°, there is approximately 8° of signal spillover.

Figure 1 [9] illustrates the geometry for tracking these limb crossing GPS signals by a receiver high above the altitude of the GPS constellation. The only GPS signals reaching the receiver at these high altitudes originate from satellites on the opposite side of the Earth. The limit of the main beam corresponds to a limb-crossing altitude of approximately 3000 km, above which the GPS signal visibility and receive power begins to drop off rapidly.

A chapter written by Czopek [10] has become the primary reference regarding the Block II A GPS satellite antenna patterns (Fig. 2). The side lobe signals are generally about 20 dB weaker than signals transmitted from the main beam; however, it is possible that a receiver could actually track a GPS satellite from these side lobes when antenna orientations and ranges are favorable.

Relative to the main beam signal, the side lobe signal has a weaker power and poorer C/N0. The accuracies of the observations, such as pseudo-range and carrier phase, are closely related to the C/N0 (Fig. 3) [11]. So the observation error of the side lobe signal is much bigger than that of the main beam, and then the integrated orbit determination based on main beam and side lobe signal observations will face the problem of the different stochastic models of them.

3 Data Process

The Position and Navigation Data Analyst (PANDA) software developed independently by the GNSS Research Centre of Wuhan University in China is adopted for the orbits and observations simulation and processing in this study [12, 13].

The COMPASS constellations in 2020 including 3 GEOs + 3 IGSOs + 24 MEOs were used [14]. The number 01, 02, and 03 satellites are GEO,

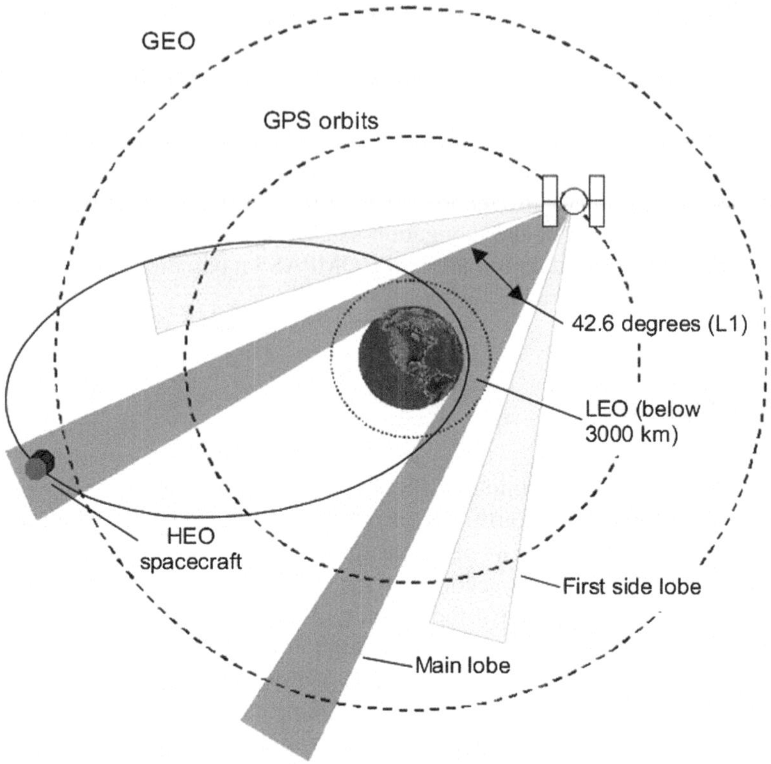

Fig. 1 LEO and HEO receive navigation signal map

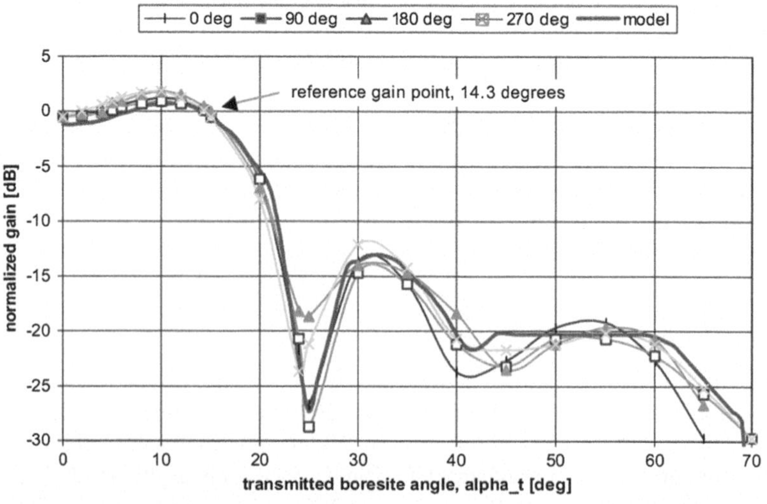

Fig. 2 GPS transit antenna gain pattern model

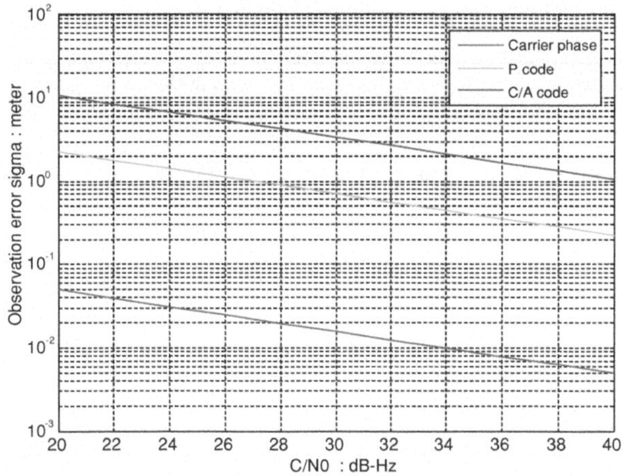

Fig. 3 Pseudo-range and carrier RMS PR error versus C/N0

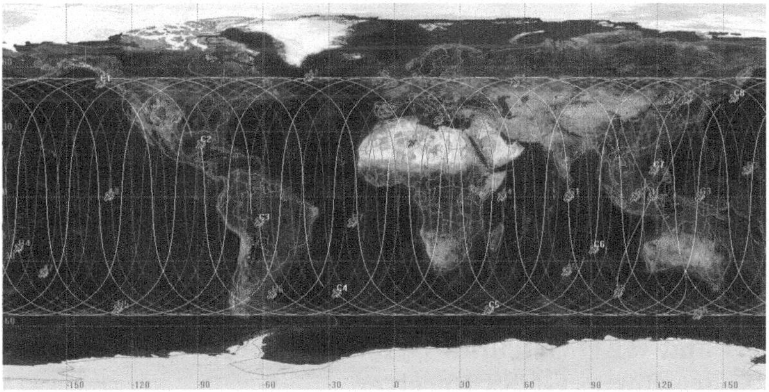

Fig. 4 COMPASS navigation satellite sub-point track

and are launched above the East longitude 80°, 110.5°, and 140°, respectively. The number 04, 05, and 06 satellites are IGSO, which are distributed on three different orbit planes but the same sub-point track. The number 07–30 are MEO, each of them has a 27878 km semi-major axis and 55° inclination orbit and is well-distributed on three different orbit planes. The sub-point tracks of the constellation are presented in Fig. 4.

The experiment satellites were one MEO and three GEO, their orbit parameters are presented in Table 1.

The six ground stations used for COMPASS orbit determination are distributed in China region, as presented in Fig. 5.

Table 1 MEO and GEO experiment satellites orbital element

Orbit parameter	MEO	GEO
Semi-major axis (km)	22000	42164
Eccentricity	0.0015	0
Inclination (°)	51.33	0
Right ascension of the ascending node (°)	75.71	0E, 120E, 240E
Argument of perigee (°)	93.86	0
Mean anomaly (°)	227.2	0

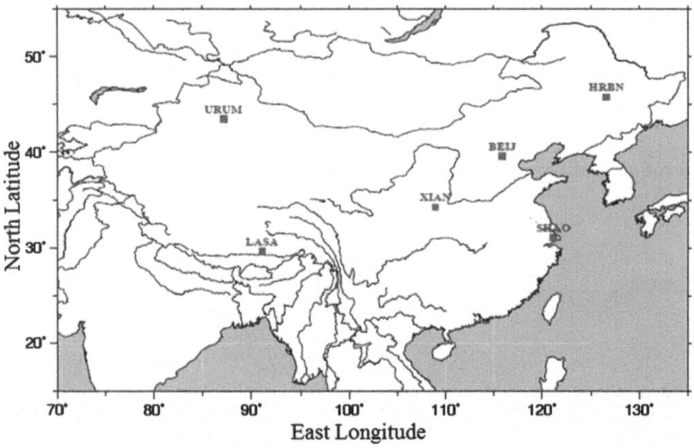

Fig. 5 Distribution of ground tracking stations

4 MEO and HEO Satellites Orbit Determination Based on Space-Borne GNSS Receiver

4.1 GNSS Signal Visibility

Several different geometric and signal strength flags are evaluated at each time; VIS_{Earth} is the visibility subject to obstruction by the Earth; VIS_{Atm} is the visibility subject to the atmosphere mask altitude; VIS_{Ant} is the visibility subject to the transmitting and receiving antenna masks; and $VIS_{C/N0}$ is the visibility based on the defined tracking threshold for the receiver. The signal is visible if each of the geometric and signal strength flags is valid, as determined by VIS:

$$VIS = VIS_{Earth} \& VIS_{Atm} \& VIS_{Ant} \& VIS_{C/N0} \tag{1}$$

The C/N_0 in dB Hz is computed by subtracting the noise spectral density from the received carrier power by the following Eq. (2) [8]:

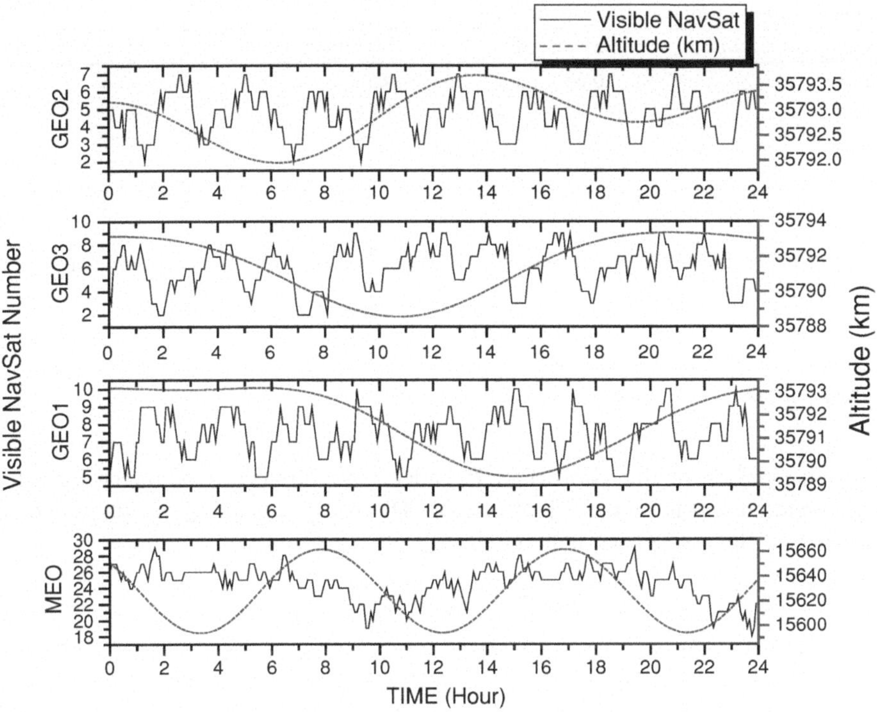

Fig. 6 Visible navigation satellites of the four experiment satellites

$$\begin{aligned} RP &= EIRP - L_t - L_D - L_e + G_r - L_r \\ C/N0\ [\text{dB Hz}] &= RP - 10\log T_{sys} + 228.6 \end{aligned} \quad (2)$$

RP is the received power. EIRP is the effective isotropic radiated power of the navigation satellites. L_t is the loss due to the attenuation of the navigation satellite antenna with respect to the reference gain. L_D is the attenuation due to free space propagation losses. L_e is the atmospheric path loss. G_r is reference gain of the receiving antenna. L_r is the loss due to the attenuation of the receiving antenna referenced to peak the peak gain. T_{sys} is the equivalent system noise temperature.

The main transmitting beam of the navigation satellite antennas is typically assumed to measure a half-angle of 21.3°, and the side lobe was 21.3°–45°. The space-borne receiving antennas were modeled as a single, omni-directional antenna with zero dB gain in all directions. The receiving C/N0 threshold is 25 dB Hz. The signal visibility analysis results of the four experiment satellites based on the COMPASS constellation are shown in Fig. 6.

The number of visible satellites of the MEO is among 18–28, more than the GEO because of the different orbit altitudes. The number of visible satellites of GEO2 is about 2–7, almost less 2 or 3 satellites than that of GEO1 and GEO3.

Table 2 PDOP of four experiment satellites

Experiment satellite	Mean value	Minimum value	Maximum value
MEO	0.62	0.56	0.72
GEO1	2.34	1.41	8.94
GEO2	3.57	1.84	15.11
GEO3	3.03	1.51	59.23

The reason is that GEO2 is above the China region and cannot receive the signal of GEO navigation satellites.

4.2 PDOP

Position Dilution of Precision (PDOP) reflects the observing geometry for orbit determination and is expressed as Eq. (3). The p_{11}, p_{22}, and p_{33} are the diagonal elements of the covariance matrix related to the satellite state.

$$PDOP = \sqrt{p_{11} + p_{22} + p_{33}} \tag{3}$$

The PDOP values of the four experiment satellites based on the COMPASS system were summarizing in the Table 2. The PDOP value of the MEO is better than that of GEO because of its lower altitude. GEO2 is the worst one as it could not receive the GEO navigation signal, but its mean value is not too bad, about 3.5.

4.3 Orbit Determination Precision

Firstly, we use the 3 days observation data of the six ground tracking stations in China to determine the navigation satellites orbits. The pseudo-range error was set as 1 m, the carrier phase error was 0.01 cycles, and the system error was period term with 3 cm amplitude.

Secondly, the experiment satellites observation data based on GNSS onboard receiver were simulated using the constraints of the satellite visible conditions and received C/N0 threshold described as Sect. 4.1, and the system error was set as 20 cm. The pseudo-range and carrier phase observation stochastic noises are closely related to the received C/N0. The pseudo-range stochastic noise is computed as follows [11]:

$$\sigma_n^{(j)}(t_i) = \begin{cases} 0.2 \times random(iseed), C/N0 \geq 40 \\ \sqrt{-0.642 + 505 \times 10^{-\frac{C/N0}{10}}} \times random(iseed), C/N0 < 40 \end{cases} (m) \tag{4}$$

where *random(iseed)* is a random number obeying standard Gaussian distribution. The carrier phase stochastic noise is computed as follows:

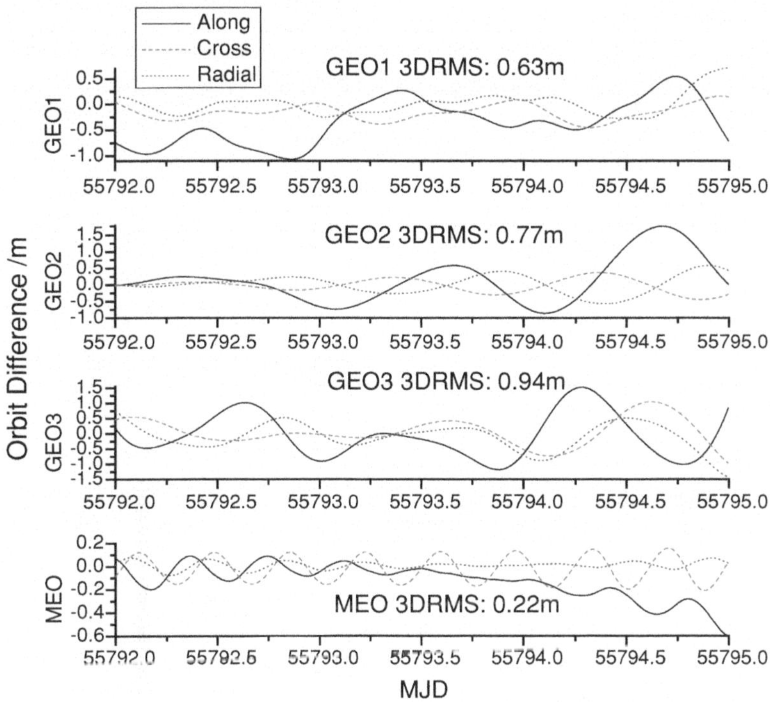

Fig. 7 Orbit determination precision of 4 experiment satellites

$$\sigma_{n,c}^{(j)}(t_i) = \begin{cases} 0.01 \times random(iseed), C/N0 \geq 34 \\ \sqrt{0.244 \times 10^{-\frac{C/N0}{10}}} \times random(iseed), C/N0 < 34 \end{cases} (cycle) \quad (5)$$

Finally, four experiment satellites orbit determination were carried out using six ground stations and inter-satellite observations based on GNSS onboard receiver. The orbit differences on the three directions (radical, along and cross) are presented in Fig. 7. They were almost within 1 m. Among the three directions, RMS on along direction was the biggest one, and the value was 0.19 and 0.65 m for the MEO and GEO, respectively. On the radical and cross directions, RMS of MEO were 0.03 and 0.11 m, and GEO averagely 0.29 and 0.31 m.

5 Navigation Constellation Orbit Determination Based on Space-Borne GNSS Receiver

In Sect. 4, the GNSS onboard receivers were installed on the MEO and HEO satellites to receive the navigation signals for the satellites orbit determination. If the receivers were installed on the navigation satellites to build the inter-satellite

Table 3 Simulated prior information of observations

Type	Satellite-ground observation	Inter-satellite observation (GNSS onboard receivers)
Pseudo-range	1 m	Eq. (4)
Carrier phase	0.01 cycle	Eq. (5)
Satellite clock	0.1 ns	0.1 ns
Receiver clock	0.1 ns	0.1 ns
Systematic error	Period term with 3 cm amplitude	Period term with 20 cm amplitude
GNSS signal visibility	Elevation cut-off 5°	Constraints of the satellite visible conditions and received C/N0 threshold described as Sect. 4.1
Beam angle	–	Main beam: 0° to 21.3°; side lobe: 21.3°–45°

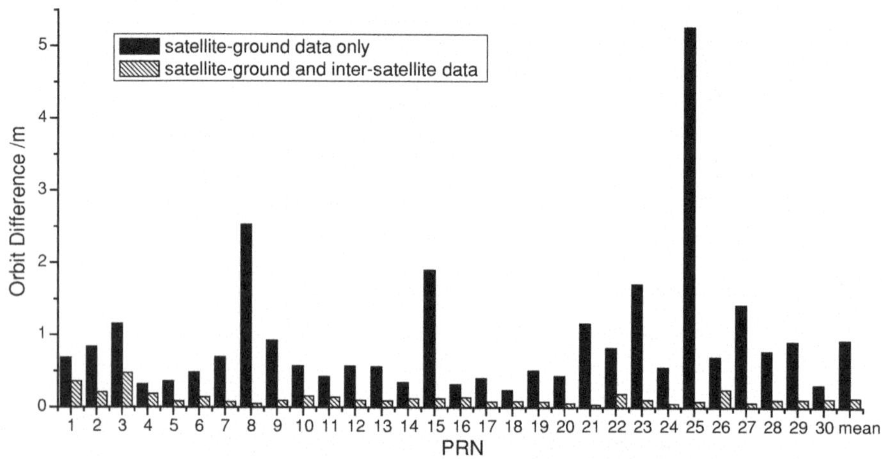

Fig. 8 Orbit determination precision of navigation constellation based on onboard receiver

link and the frequencies interference problem of received and transmitted signals was resolved, the navigation satellites orbits could be autonomously determined by the inter-satellite link data.

The ground tracking network composed of six regional stations (Fig. 5) in China. The satellite-ground and inter-satellite observation data (GNSS onboard receivers) were simulated based on the priori constraints in Table 3. Then navigation satellites integrated orbit determination was carried out using satellite-ground and inter-satellite observations.

The statistical results of navigation satellite orbit determination are shown in Fig. 8. Comparing to using the satellite-ground data only, the accuracies of the GEO, IGSO and MEO have a significant improvement indicated by the reduction of RMS from 0.93 to 0.13 m, which represents an improvement of about 86%. The improvement for MEOs is the most, which is about 90%.

6 Summaries

In this chapter, the theories and methods of MEO and HEO orbit determination based on GNSS onboard receiver were proposed. According to COMPASS navigation system, the MEO and HEO satellites visibility, PDOP, orbit determination precision were analyzed. Finally, navigation satellites integrated orbit determination based on GNSS onboard receiver was discussed.

1. Compared with the LEO, the MEO, and HEO mainly received the navigation signals on the opposite of the Earth. So some new problems were appear, such as the weaker power of received signal, less number of visible navigation satellites, poorer geometric structure, different observation errors between the mean beam and side lobe signal.
2. Under the conditions that the beam angle of main beam is $0°-21.3°$ and that of side-lobe $21.3°-45°$, navigation signal transmit power is 26.8 W, and the system error of inter-satellite link is 20 cm, the MEO could receive 18–28 navigation satellites signals, and its orbit determination precision was 0.22 m. But to the GEO, the received number is 2–10, and the precision lies in 0.6–0.9 m.
3. The accuracy of navigation satellite integrated orbit determination increased a lot after adding the inter-satellite observation data based on GNSS onboard receiver, compared to using ground tracking data only. The orbit determination precision was improved from 0.93 to 0.13 m.

Acknowledgments This study is sponsored by the Fundamental Research Fund for the Central Universities, Specialized Research Fund for the Doctoral Program of Higher Education (No. 20110141120076), and Open Research Fund of the Key Laboratory of Geospace Environment and Geodesy, China (No. 10-02-01).

References

1. Bernstein, H., Bowden, A. F., & Gartside, J. H. (1993). GPS user position accuracy with block IIR autonomous navigation. *Proceeding of the ION GPS-93* (pp. 1389–1399). Alexandria: Institute of Navigation.
2. Rajan, J. A., Brodie, P., & Rawicz, H. (2003). Modernizing GPS autonomous navigation with anchor capability. *ION GPS/GNSS 2003, Portland*: [*s. n.*], pp. 1534–1543.
3. Shuai, P. (2009). *X-ray pulsar-based navigation system theory and methods*. Beijing: China Space Press.
4. Konig, R., Reigber, C., & Neumayer, K. H. (2003). Satellite dynamics of the CHAMP and GRACE LEOs as revealed from space- and ground-based tracking. *Advances in Space Research, 31*(8), 1869–1874.
5. Zhao, Q. (2004). *Research on precision orbit determination theory and software of both GPS navigation constellation and LEO satellites*. Wuhan: Wuhan University.
6. Moreau, M. C. (2002). Results from the GPS flight experiment on the high earth orbit AMSAT OSCAR-40 spacecraft. *ION GPS 2002 Conference, September 2002*, Portland, OR.

7. Ruiz, J. L., & Frey, C. H. (2005). Geosynchronous satellite use of GPS. *ION GNSS* 18th, Sept 2005, Long Beach, CA.
8. Zentgraf, P., & Berge, S. (2010). Preparing the GPS-experiment for the small GEO mission. *33rd Annual AAS Guidance and Control Conference*, February 2010, Breckenridge, Colorado.
9. Moreau, M. C. (2001). GPS receiver architecture for autonomous navigation in high Earth orbits. Ph.D. Dissertation, University of Colorado, Boulder, CO.
10. Czopek, F. M. (1993) Description and performance of the GPS Block I and II L-Band antenna and link budget. *ION GPS 1993*.
11. Hartinger, H., & Brunner, F. K. (1999). Variances of GPS phase observations: The SIGMA-ε model. *GPS Solutions, 2*(4), 35–43.
12. Liu, J., & Ge, M. (2003). PANDA software and its preliminary result of positioning and orbit determination. *Wuhan University Journal of Natural Science, 8*(2), 603–609.
13. Geng, T. (2009). *Real-time precise orbit determination theory for navigation satellite and its experimental application based on regional reference stations*. Wuhan: Wuhan University.
14. Ran, C. (2010). Plan of compass development. Beijing: Chinese Compass Satellite Navigation Systems Management Office.

The Algorithm Research of Precise Point Positioning Based on Undifferenced Corrections of Reference Network

Shuhong Jia, Shirong Ye, Yanyan Liu and Chao Xiong

Abstract This paper use PANDA software to compute the undifferenced ambiguity and receiver clock bias of reference stations. And then the undifferenced corrections of reference stations can be obtained. After modeling for each satellite and passing the corrections to the rover station, the solution of single difference between satellites is computed. Now the errors about satellite and receiver have been cancelled, the user station's instantaneous ambiguity could be obtained very fast. The proposed method is validated with the data of Hong Kong SatRef network and the west coast of American's IGS network by coding with C++. The results show that the time of convergence is greatly reduced and the accuracy of positioning can reach to cm-level or sub dm-level.

1 Introduction

Precise Point Positioning of Global Positioning System (GPS) can just use only one single receiver to get high precision coordinates of the station, which is research focus on international because of its cost-effciency, global coverage and its accuracy meeting the requirements of most applications. However, the sparse reference network employed cannot provide accurate atmosphere delays for positioning with higher precision and for reliable ambiguity-fixing. Hence, ionospheric delays have to be eliminated through ionosphere-free combination and zenith tropospheric delays must be estimated together with other parameters. As a consequence current PPP needs at least 30 min to obtain a converged solution [1].

S. Jia (✉) · S. Ye · Y. Liu · C. Xiong
GNSS Centre of Wuhan University, 129 Luo Yu Road, 430079 Wuhan, China
e-mail: jiashuhong2008@sohu.com

J. Sun et al. (eds.), *China Satellite Navigation Conference (CSNC) 2012 Proceedings*, Lecture Notes in Electrical Engineering 160, DOI: 10.1007/978-3-642-29175-3_21,

Conversely, Network Real-Time Kinematic (NRTK) service has not only short convergence time, but also high precision that PPP cannot match, but a dedicated dense reference network is required which restricts its unique and homogeneous service for large regions, meanwhile the network solution is time-consuming.

Many scholars have bring forward the idea of aligning PPP and RTK in which the combination can complement each other in the positioning. A new real-time precise positioning service system enhanced the global PPP with regional augmentation was proposed by Ge in 2009. This method generates zero-difference measurements corrections of a local network, and broadcast the corrections to the user's station to fix the instantaneous ambiguity quickly. This article is based on this approach to program development to verify the accuracy of this thinking.

2 PPP Based on Zero-Difference Measurements Corrections of a Local Network

2.1 The Computation of Zero-Difference Correction

The carrier phase observation equation is given by the following:

$$L_r^s = \rho_r^s + c\delta t_r - c\delta t^s + b_{L,r} - b_L^s + \delta_{orb,r}^s + \delta_{trop,r}^s + \delta_{rel,r}^s - \delta_{ion,r}^s + \lambda N_r^s + \varepsilon \tag{2.1}$$

In Eq. (2.1), L_r^s is the carrier phase observation, ρ_r^s is the distance from satellite to receiver, δt_r and δt^s are the receiver clock and satellite clock, respectively, $\delta_{orb,r}^s$, $\delta_{trop,r}^s$, $\delta_{ion,r}^s$, $\delta_{rel,r}^s$ are the orbit error, tropospheric delay, ionospheric delay and relativity effect, $b_{L,r}$, b_L^s are the uncalibrated phase delay (UPD) of receiver and satellite, N_r^s is the integer ambiguity, ε is the observation error. The superscript s represents satellites while the subscript r represents receivers.

According to the site coordinates of reference stations to generate a delaunay triangle, the model is built for each satellite in each sub-network (contain 3 reference stations), constructing a pyramid illustrated in Fig. 1 [4]. In this little area, the satellite ephemeris error, satellite clock error, satellite UPD, satellite antenna phase centre error, relativity error are considered having the same influence to reference station and rover station. The tropospheric delay and the ionospheric delay are modeled too effectively. Multipath error is decided by the circumstance around the station, performance of receiver, and the length of the observation time. So the bias of ionospheric and tropospheric and these errors about satellite can be viewed as a large error correction which is expressed as follows.

$$\delta_m^s = -c\delta t^s - b_L^s + \delta_{rel,r}^s + \delta_{orb,r}^s + \delta_{trop,r}^s - \delta_{ion,r}^s \tag{2.2}$$

The zero-difference ambiguity must be computed if we want to get the correction from Eq. (2.1). To get the zero-difference ambiguity, there are two

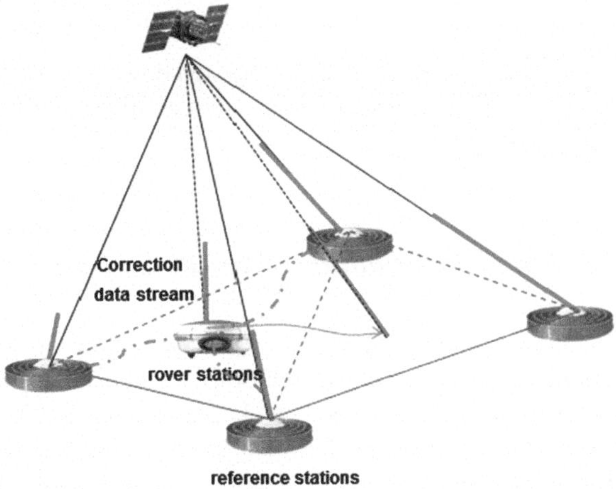

Fig. 1 Model for a satellite based on reference stations

methods. One is use data processing software like bernese to obtain the double-difference ambiguity and projecting it to zero-difference ambiguity which is more extensive because most data software can calculate double-difference ambiguity, meanwhile it is also suitable for single-frequency. Another one is compute the zero-difference ambiguity and use it directly. The latter approach is used in this article with the software PANDA which is researched and developed independently by GNSS Research Centre, Wuhan University. The zero-difference of ionosphere-free ambiguity at each epoch could be gained by using single epoch processing model [2].

Unable to verify the correctness of the ambiguity solved out by PANDA, To view the initial data is reliable, The solutions of single epoch coordinates are compared between PANDA and BERNESE, concluded that the solution solved by PANDA have relatively large fluctuations with the RMS of three directions (XYZ) up to 6–7 cm. This will introduce large bias into the post positioning, so just select the more smooth and stable periods to involved in computing.

2.2 Correction Error Model

Building correction model in accordance with 3 parameters plane fitting model includes an elevation factor. Both estimated integrated error with the linear variation of plane location and considered the troposphere error with the linear variation of elevation.

The troposphere accuracy of this model has been proved that is better than 1 cm in 60 km × 60 km range [3]. In addition to the three vertex of triangle that the rover station located in, another reference station closer to the rover station is needed to obtained the following formula parameters.

$$\delta_m^s = ax + by + ch + d \tag{2.3}$$

In the Eq. (2.3), δ_m^s is the integrated error, x, y are the plane coordinates of rover station, h is the vertical coordinates of rover station, a, b, c, d are the constant coefficient.

2.3 Rapid Positioning of Rover Station

The corresponding error is interpolated within the triangle according to the position of user station. Introducing this correction into the user observation equation errors about satellite like clock biases and UPDs are fully cancelled for the correlation of the error between reference station and rover station. Then make the single difference between satellite to eliminate the user's receiver clock error and receiver hardware delay. After that the remaining items are just ambiguity and receiver position. Now the observation of user corrected by the model can perform PPP with rapidly ambiguity-fixing to get the precise position.

2.4 The Design and Implementation of Software

Based on the approaches and algorithms for generating zero-difference corrections from a regional network, a software in C++ was developed. Read into the files of the observations, ephemeris and PANDA solvering. First make preprocessing of reference observation, detect and repair cycle slips, exclude unusual observations, establish the observation equation, eliminate ambiguity and the receiver clock errors then generate the zero-difference correction and broadcast them to the rover station and estimate the parameters as the three-dimensional coordinates of the station and ambiguity parameters, the error which do not modeled accurately absorbed by other parameters through estimation. Finally calculate the results of the rover's location (Fig. 2).

3 Examples and Results

3.1 Example 1

Select HKFN, HKSL, HKMW, HKSC, HKWS of HongKong Satellite Positioning Reference Network as reference station in 2010, day of year 174, 175, the rest station of the network are regarded as the rover station except HKNP and HKOH for not in the triangles showed in the Fig. 3. The average distance between the

Fig. 2 The algorithm flow chart of PPP based on undifferenced corrections of reference network

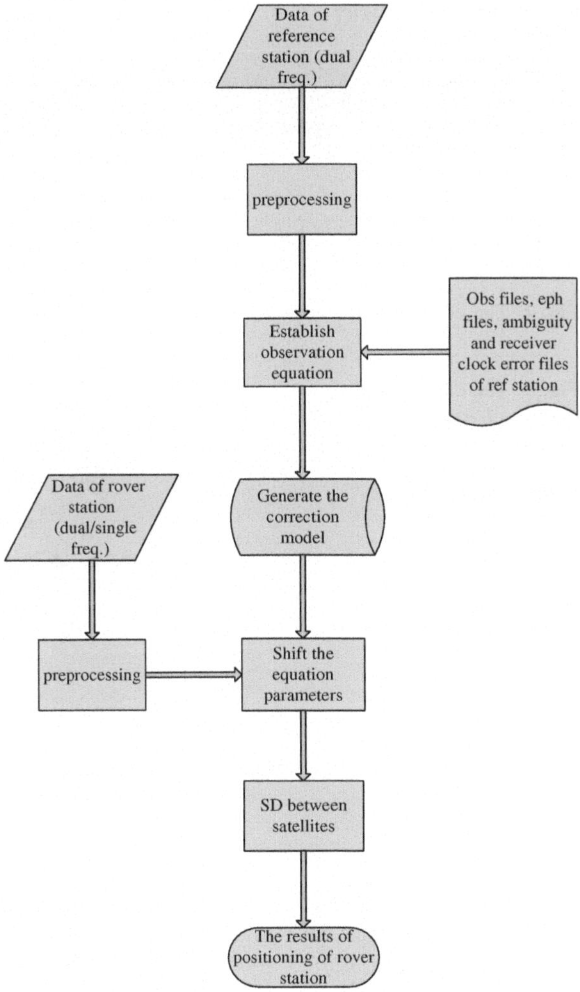

Fig. 3 The distribution of stations of Hong Kong district

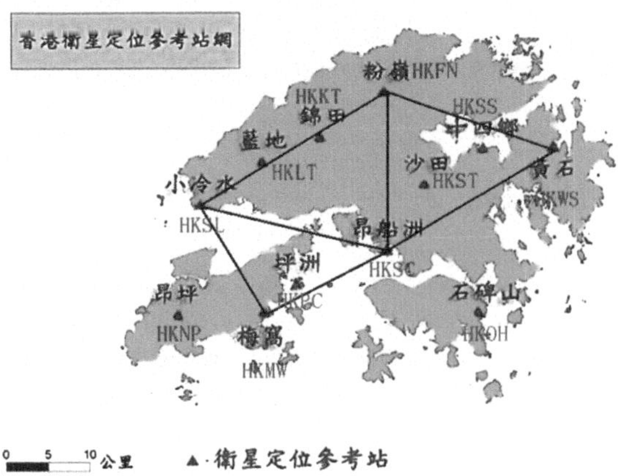

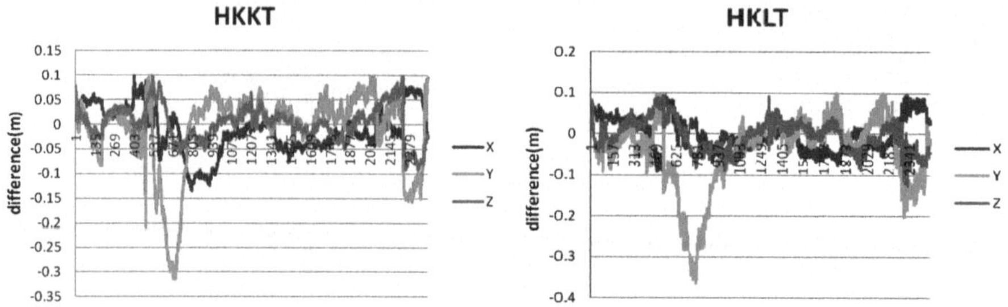

Fig. 4 The coordinate deviations' time series of HKKT and HKLT

Fig. 5 The distribution of stations of American district

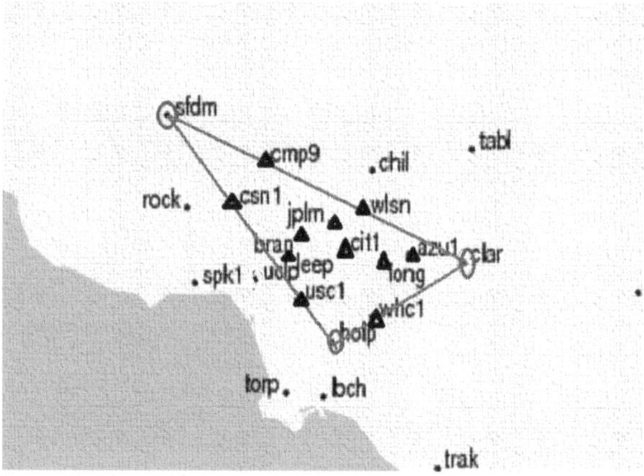

reference station is about 20 km, the sampling rate is 5 s. The results are compared with BERNESE result, the time series errors of HKKT and HKLT in 175 day are provided in the Fig. 4.

For the appearance of spine in the y direction of multiple rover station, it is probably due to the relatively large perturbation of the ionospheric second-order term at noon while the integrated errors cannot eliminate or weaken it effectively.

3.2 Example 2

Select three IGS stations: holp, sfdm and clar located in U.S. west coast as reference station in 2010, day of year 29 showed in Fig. 5. The average distance between the reference station is about 30 km, the sampling rate is 15 s, the other 9 stations (azu1, bran, cmp9, cit1, csn1, jplm, long, leep, wlsn) are selected as the rover stations. The time series errors of cit1 (in the centre of triangle) and csn1 (on the edge of triangle) are provided in the Fig. 6.

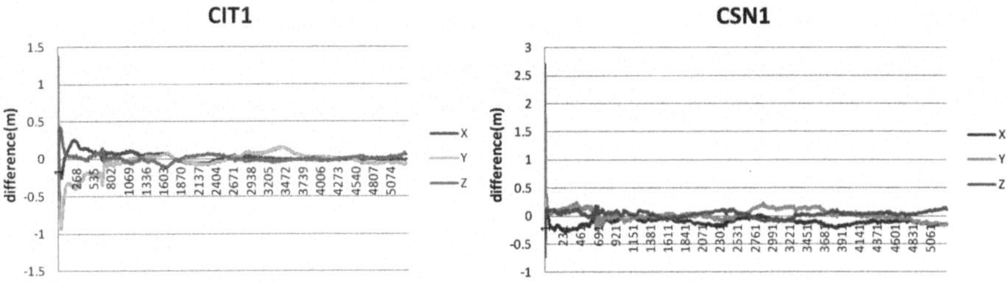

Fig. 6 The coordinate deviations' time series of CIT1 and CSN1

4 Conclusions

It can be illustrated that convergence time is become shorter within 20 epochs. However, the accuracy of positioning results is not very high because of the single epoch ambiguity resolution solved by PANDA is not accurate which make the initial error of the system up to 7–8 cm and the UPDs have not been corrected leaving the bias absorbed by the ambiguity and coordinates of stations. So accuracy is only cm level or sub-dm level.

Acknowledgments This work was supported by the National Natural Science Foundation of China (41074008). The authors of this paper also acknowledge Dr.Maorong Ge from Helmholtz-Centre Potsdam -German Research Centre for Geosciences for providing the article of References [1].

References

1. Ge, M., Dousa, J., Li, X., Ramatschi, M., Nischan, T., & Wicker, J (2011). A novel real-time precise point positioning service system: Global precise point positioning with regional augmentation. *Third International Galileo Colloquium*, August 31–September 2, Copenhagen, Denmark.
2. Tao, G., Qile, Z., Liu, J., & Du, R. (2007). Real-time precise point positioning based on PANDA software. *Geomatics and Information Science of Wuhan University, 4*, 32–34.
3. Xiong, Y., Huang, D., Ding X., & Yin, H. (2006). Research on the modeling of tropospheric delay in virtual reference station. *Acta Geodaetica et Cartographica Sinica, 35*(2), 118–121
4. Zou, X. (2010). *Study on GNSS precise point positioning with single-frequency receiver*. Ph.D. dissertation, Wuhan University.

Secular Changes in Differential Code Bias of COMPASS System

Nan Xing, Xiaoli Wu, Xiaogong Hu and Ranran Su

Abstract Differential Code Bias (DCB), named Timing Group Delay (TGD) for satellite and Inter-Frequency Bias (IFB) for receiver, is a crucial parameter in COMPASS system, which is needed in satellite clock correction and ionospheric delay modeling. Though DCB parameters were measured in laboratory initially, it may change with working environment as well as with time elapsed, necessitating DCB estimation for the deployed satellites and receivers. DCB estimates can be obtained by combining COMPASS data in three frequencies and accurate ionospheric Total Electron Content (TEC) models (e.g. Global Ionosphere Model, GIM from International GNSS Service). However, TEC model errors as revealed by discrepancies with GPS observations would translate into and contaminate DCB estimates. This study concludes that the error GIM brought into TGD is under 1 ns that is equivalent to a positioning error of 0.18 m, an ionospheric time delay correction error of 0.4–0.7 m and TEC models error of 3TECU. Analysis of COMPASS data for 290 days in 2011 shows a secular change in TGDs with amplitude of about 1.5 ns for GEO satellites and 4 ns for IGSO satellites respectively and in IFBs with amplitude of 3 ns. The DCB time series could be well fitted by a second-degree polynomial with residuals of 0.4–0.6 ns for TGDs and 0.6–0.9 ns for IFBs, based on which the TGD could be predicted with accuracy of 0.5 ns for GEO satellites and 0.5–1 ns for IGSO satellites respectively.

N. Xing (✉) · X. Hu
Shanghai Astronomical Observatory, Chinese Academy of Sciences, 200030 Shanghai, China
e-mail: nanx@shao.ac.cn

N. Xing
Graduate University of Chinese Academy of Sciences, 100049 Beijing, China

X. Wu · R. Su
Beijing Global Information Center of Application and Exploitation, 100094 Beijing, China

J. Sun et al. (eds.), *China Satellite Navigation Conference (CSNC) 2012 Proceedings*, Lecture Notes in Electrical Engineering 160, DOI: 10.1007/978-3-642-29175-3_22,

1 Introduction

High precise positioning, navigation, timing services of COMPASS system depend on accurate signal error correction. According to the preliminary User Interface Control Document, navigation message includes not only satellites ephemeris, but also system bias information, based on which signal errors can be partially corrected, such as, satellite clock, parameters in Klobuchar model for ionospheric delay calculation and TGD. TGD is the bias of transmission time delay from the satellite clock to the satellite antenna phase center between different frequencies [1, 2]. COMPASS system broadcasts GNSS signals on three frequencies (named as B1, B2, B3 hereafter), namely each satellite has two independent TGDs. The broadcasted satellite clock is contained by transmission time delay on the third frequency B3, which means users on B3 can directly correct satellite clock with this information while users on the other frequencies also need TGD parameters as to calculate their own specific proper clocks. When positioning with single frequency, users on B1 or B2 would care about TGDs [3] whose uncertainty are about several or tens of nanoseconds.

As TGDs to satellite, there are also some biases to receivers of each station, e.g., IFBs, which has barely difference with TGDs in order of magnitude. Since TGD and IFB share similar properties and must be computed at the same time, people often refer them collectively as DCB [4]. Generally DCBs and ionospheric delay information are derived from dual frequency geometry-free combination and they are coupled with each other, indicating if DCB was neglected when calculating ionospheric parameter Total Electron Content (TEC), tens of TEC Unit (TECU) error could be introduced depending on different frequencies.

Moreover, COMPASS system provides Klobuchar model parameters for general users, and grid ionospheric delay for wide area differential services to correct ionospheric delay, both of which need DCBs in preprocess course. Thus, DCBs are indispensable parameters both in the study of ionosphere model and its application in relative projects.

The value of TGD for a satellite or IFB for a receiver is measured in laboratory initially yet varies with its environment in work. DCB need to be fixed continuously to decrease the deviation. Similar problem also happened in GPS system. Thus Jet Pulsation laboratory (JPL) keeps updating TGDs since 1999, and IGS organized five analyses centers to provide DCBs everyday as criteria [5].

Recently, two kinds of methods are proposed to separate DCB from ionospheric delay: one way is using known TEC value derived from existed model such as, Global Ionosphere Map (GIM), International Reference Ionosphere (IRI) and etc., or interpolated by observations as mask to obtain DCBs; on the other hand, one can consider DCBs as constants and TEC as spatial variances modeled with regional polynomial model, regional trigonometric model, global spherical harmonics model or global assimilative ionospheric model to get the solution of DCBs and TEC simultaneously. The former kind depends on external data that would make DCBs more stable but bring a latency of several days to DCBs process. The latter one can be used to monitor real-time TEC and DCBs, however, under some special situations such as absence of station or satellites and almost unchanged configuration of stations and satellites like Geostationary Earth orbiters (GEOs), the relevance of TEC and DCBs are so strong that their uncertainty of estimation will increase.

In this paper, we mainly focus on accuracy and secular changes of DCBs of COMPASS system, especially on TGDs. The algorithm to solve DCBs is presented in Sect. 2. In Sect. 3, in order to assess the error of DCBs derived from the method that use GIM model to calculate TEC, we introduce another method that interpolate GPS-TEC observation instead to compare the DCBs of each method. And we estimate the error brought by DCBs to positioning and grid ionospheric delay. Finally, brief analysis of long-term change of DCBs during January and October in 2011 is given in Sect. 4.

2 DCB Solution Algorithm

Pseudo-range at the ith frequency of Compass system can be described as Eq. 1:

$$\begin{aligned} P_i = \rho_j^k + \delta t_j \cdot c - \delta t^k \cdot c + d_{ion,i} \\ + d_{trop} + \left(\tau_{j,i} + \tau_i^k\right) \cdot c + \varepsilon_i \end{aligned} \quad (1)$$

where ρ_j^k is geometry distance between the jth station and kth satellite, δt_j is receiver clock of jth station, δt^k is kth satellite clock, $d_{ion,i}$ is line of sight ionosphere delay, d_{trop} is tropospheric delay, $\tau_{i,j}$ and τ_i^k are transmission time delay of jth station and kth satellite respectively at ith frequency, c is speed of light and ε_i is observational error that includes multipath error and etc.

Line of sight ionosphere delay can be written in the form

$$d_{ion,i} = \frac{40.28}{f_i^2} \cdot TEC \quad (2)$$

where f_i is signal frequency. Consequently, geometry-free combination differentiated by pseudo-range at pth and qth frequencies can be described as following:

$$\begin{aligned} P_{p,q} = 40.28\left(\frac{1}{f_p^2} - \frac{1}{f_q^2}\right) \cdot TEC \\ + TGD \cdot c + IFB \cdot c \end{aligned} \quad (3)$$

where $TGD = \tau_{j,p} - \tau_{j,q}$ and $IFB = \tau_p^k - \tau_q^k$.

For simplification, we assume ionosphere is a spherical shell with altitude of 375 km. Thus, line of sight TEC can be projected to vertical direction on the ionospheric pierce point (IPP) where the line intersects the shell, and this projected value is often called Vertical TEC (VTEC) [6]. In fact, TEC obtained from model or interpolated by observation is usually VTEC.

After getting rid of ionospheric delay in Eq. 3, the combination of a pair of TGD of *j*th satellite and IFB of *k*th receiver could be easily obtained, denoting as y in the following. We can then use a set of such combinations from *n* satellites and *m* receivers to form Eq. 4:

$$y_{l,1} = B_{l,m+n} dcb_{m+n,1} + \varepsilon_{l,1}, dcb = \begin{pmatrix} tgd_1 \\ \cdots \\ tgd_n \\ ifb_1 \\ \cdots \\ ifb_m \end{pmatrix} \tag{4}$$

where y has l elements determined by quantity of observations, $\boldsymbol{B}$ is the coefficient matrix, *dcb* is a vector constituted by n TGDs and m IFBs and ε is error of the combination. Equation 4 is rank deficient because the variables are relevant with each other. One additional independent condition is required to make it rank full. In GPS system, sum of TGDs of 32 satellites are constrained to zero. With the augment of COMPASS system, GPS strategy will cause leaps of TGDs every time a new satellite joins to the constellation. Therefore, we fix the TGD of Sat1 with its initial value as the constraint.

To estimate the accuracy of TGDs and IFBs, error propagation theory [7] is applied here. The covariance matrix $\boldsymbol{D}$ of dcb can be calculated according to Eq. 5:

$$D = \frac{\varepsilon^T P \varepsilon}{l - m - n} (B^T P B)^{-1} \tag{5}$$

where $\boldsymbol{P}$ is a matrix of weight we used. Thus the error σ_i of the *i*th element of dcb, i.e. TGD or IFB, can be computed as following.

$$\sigma_i = \sqrt{D(i,i)} \tag{6}$$

3 Accuracy of DCB and its Influence to PNT

Right now broadcasted TGD and IFB parameters of COMPASS system are estimated based on GIM model of IGS [8] (called method A hereafter). According to Eqs. 5 and 6, theoretical accuracy of TGD for method A is about 0.2 ns, which actually underestimates the real error. The difference is mainly caused by the error related to GIM model. Due to shortage of stations in China accounted by IGS, error of GIM model is about 2–3 TECU. As approaching to solar maximum, this error might climb to 10 TECU or more when abnormal activity happens in ionosphere. To quantify this error, GPS data from 200 stations of a network in China are adopted to interpolate TEC at IPP of COMPASS system so that we can get high precise DCB in another way (called method B hereafter).

Table 1 Comparison of DCB residuals derived from GIM method and interpolation method

	10.12 (ns)	10.13 (ns)	10.14 (ns)
Method A	1.97	1.71	1.80
Method B	0.26	0.28	0.26

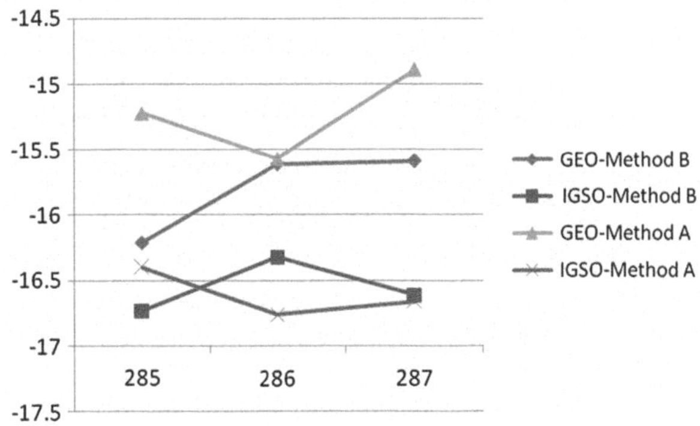

Fig. 1 DCB for a GEO and an IGSO satellite estimated by GIM method and interpolation method

DCB root mean square (RMS) value of two methods during October 12 and 14th in 2011 is shown in Table 1. DCB RMS for method B is less than 0.3 ns, for method A is about 1.8 ns. This discrepancy is 1.5 ns on average of all stations and 4 ns at maximum for some station near 25°N.

The differences of DCB between two methods are less than 1 ns for all satellites implying that both methods are reliable. The values of a GEO and an Inclined Geosynchronous Orbit (IGSO) are shown in Fig. 1 as an example.

Due to the daily changes of TEC, there should be a slight discrepancy between DCB results from 24-hour data and from nighttime data only. It implies that nighttime data may prevent DCB from being contaminated by maximum error of TEC in noon.

Although accuracy of DCB derived from method B is a little higher than method A, the constraint of acquiring the data makes method B inconvenient to provide routine DCB.

Error of DCBs will affect the following services for users: (1) positioning accuracy for single frequency users at B1 and B2; (2) accuracy of Klobuchar parameters and grid ionospheric delay.

As mentioned above, TGD parameters are needed to correct satellite clock at B1 or B2 frequency. With the knowledge of other source of User Equivalent Range Error (UERE), 1 ns error in TGD will bring 0.03 m into UERE. Under the constellation of 3 GEOs and 4 IGSOs, Position Dilution of Precision (PDOP) is about 6 so that positioning error caused by TGD is about 0.18 m.

Based on Eq. 1, error of DCB will be magnified by 2.7–3.5 times. Hence the error introduced to ionospheric Klobuchar model is around 3 TECU and to grid ionospheric delay is about 0.4–0.7 m.

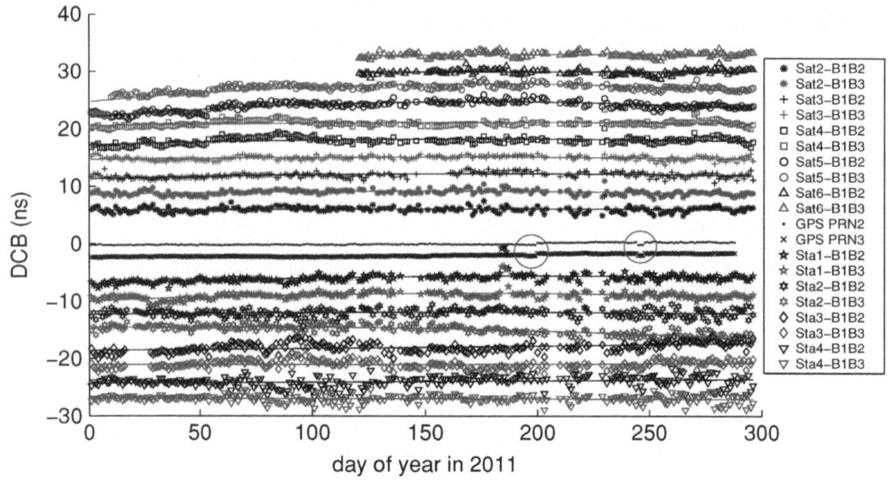

Fig. 2 DCB secular changes

4 DCB Secular Change and Prediction Accuracy

DCB, as basically the bias caused by hardware in satellites and receivers, should be barely variable. However, the change of working environment may cause a secular changes to DCB. Figure 2 shows DCB secular changes (B1, B2 group and B1, B3 group) for five satellites and four receivers of COMPASS system and two satellites of GPS system for the first 290 days in 2011. Sat2 and Sat3 are GEO satellites, the rest three are IGSO satellites. For this figure, our aim is to present the variability of DCBs for each satellites or receivers, thus proper adjustment on y-axis was made to put all time series together. Note that we abandoned two satellites of COMPASS system in this analysis, because we only have 2 months data for one of them and TGD was calibrated as constant for the other.

We find that TGDs of GEO satellites have linear variation trend and smaller variation scope, i.e., only 1.5 ns within 280 days, which is obviously more stable than IGSO whose TGDs tends to change with quadratic function and the variation scope could extend to 4 ns. Our results imply that broadcasted TGD need to be updated within 2 months. Otherwise the accuracy of TGD parameters for IGSO would decrease about 1 ns.

To analyze the stability of our results, we fit all the TGDs with first-degree or second-degree polynomial (best-fit lines are shown in Fig. 2). We present the best-fit coefficients in Table 2. The notations in the second column of Table 2 are for different frequency group, i.e., A denotes B1, B2 group and B denotes B1, B3 group. The third to fifth columns are best-fit coefficients for second-degree polynomial. The sixth and seventh columns give the RMS when we apply first and second degree polynomial respectively. It is found that RMS of TGD is 0.4 ns, which is slightly lower than IGSO whose RMS is about 0.6 ns. We also find that

Table 2 Predicted DCB accuracy

SN	FF	Coefficient fitted by Second degree polynomial			RMS (ns) for different degree polynomial		Coefficient for DCB prediction			Prediction precise (ns) for different days				
		0th (ns)	1st (ps)	2nd (fs)	First	Second	0th (ns)	1st (ps)	2nd (fs)	2	7	14	30	60
2	A	15.4	3.41	−9.00	0.47	0.47	17.2	−1.53	3.70	0.89	0.66	0.54	0.49	0.45
	B	3.9	9.09	−2.90	0.45	0.4	2.7	2.39	−7.00	0.74	0.51	0.44	0.47	0.49
3	A	14.9	1.06	−2.90	0.44	0.39	11.0	5.31	−1.35	0.23	0.5	0.42	0.48	0.54
	B	4.3	5.83	−1.70	0.37	0.35	2.7	2.39	−6.40	0.29	0.69	0.53	0.48	0.47
4	A	11.5	1.21	−3.30	0.56	0.52	12.1	1.83	−1.00	0.69	0.49	0.36	0.44	0.49
	B	11.8	7.00	−1.80	0.47	0.45	12.2	−2.85	1.60	0.48	0.4	0.31	0.46	0.55
5	A	12.8	3.08	−8.70	0.73	0.44	8.2	8.16	−2.23	0.56	0.45	0.44	0.79	1.11
	B	15.6	3.56	−1.00	0.93	0.66	13.0	5.93	−1.57	0.46	0.42	0.37	0.59	0.64
6	A	17.0	−2.74	9.00	0.45	0.45	12.0	5.40	−1.45	0.47	0.53	0.49	0.91	1.26
	B	16.0	9.98	−2.10	0.41	0.4	13.0	4.42	−1.14	0.35	0.5	0.46	0.64	0.82
I	A	−17.49	15.73	−39.23	0.71	0.66	–	–	–	–	–	–	–	–
	B	−18.46	15.99	−46.01	0.73	0.66	–	–	–	–	–	–	–	–
II	A	−12.7	5.15	−15.44	0.65	0.64	–	–	–	–	–	–	–	–
	B	−11.1	5.25	−39.14	0.59	0.53	–	–	–	–	–	–	–	–
III	A	−16.29	2.62	6.46	0.78	0.78	–	–	–	–	–	–	–	–
	B	−16.66	−0.49	4	0.67	0.67	–	–	–	–	–	–	–	–
IV	A	−21.11	2.19	0.83	0.91	0.91	–	–	–	–	–	–	–	–
	B	−21.1	−0.43	−0.47	0.67	0.67	–	–	–	–	–	–	–	–

SN Satellite or station number, arabic number for TGDs and roman number for IFBs
FF Frequency flag

introduce of second-degree polynomial does not change RMS of TGD too much except for Sat5, which means linear variation trend could still be a good approximation for most of the satellites.

The two satellites of GPS system (PRN2 and PRN3) are presented in the middle of Fig. 2, whose TGD results are from IGS. There are two discontinuous TGD change for those two satellites, circled in Fig. 2, at around 200 days and 250 days and both are due to the sudden change of PRN1 TGD. To be specific, sudden change at 200 days is because of the upgrading of GPS system and the one at 250 days is due to lack of PRN1 data. Dividing TGD data into three segments by the two discontinuous points, we separately fit each segment and find that RMS of TGD is about 0.2 ns, lower than COMPASS system. We propose that the main reason for this should be related to the data quality and the number of stations that in charge of TGD estimation. Since COMPASS system currently is regional satellite navigation system, the number of stations is relative small, and data quality from several stations that are still under construction is not good enough. When calculating TGD, we will assign relatively lower weight to bad data or even remove them; this may increase the daily variation of TGD results. Moreover, Fig. 2 shows TGD results for COMPASS satellites slightly increase at 150 and 220 days, this should relate to the newly adding of satellites at those times.

Similar analysis was done to IFBs, four of which is shown at the bottom of Fig. 2. The four receivers are located in the northeast, southern, western and center of China respectively. The variation scope of IFBs is comparable with TGDs, namely 3 ns around. But RMS of IFBs fitted by second-degree polynomial are 0.6–0.9 ns that are greater than TGDs in average.

The reason for TGD secular change is still not clear, however, based on our best-fit results we can do predictions for it. Columns 11–15 in Table 2 present the accuracy of our prediction for 2, 7, 30 and 60 days respectively. To analyze all of the five COMPASS satellites, we only adopt data from 120 days to 250 days. The results show that, for GEO satellites both short-term and long-term prediction has accuracy around 0.5 ns; while for IGSO satellites, especially for Sat5 and Sat6, accuracy of short-term and long-term prediction is around 0.5 and 1 ns respectively, implying the accuracy decrease with the extending of prediction period.

5 Summary

In this paper, we mainly analyze the accuracy and secular changes of DCBs, especially TGDs, of COMPASS system. Theoretical accuracy of TGD based on GIM model of IGS is about 0.2 ns. However, TEC model errors as revealed by discrepancies with GPS observations are around 1.5 ns and bring 1 ns of error into TGD.

Error of DCBs will affect positioning accuracy for single frequency users at B1 and B2 as well as accuracy of Klobuchar parameters and grid ionospheric delay. We find 1 ns error in TGD could introduce a positioning error of 0.18 m, an ionospheric time delay correction error of 0.4–0.7 m and TEC models error of 3TECU.

Finally, we analyzed COMPASS data for 290 days in 2011 and it shows a secular change in DCBs with amplitude of about 1.5 ns for GEO satellites and 4 ns for IGSO satellites respectively and 3 ns for receivers in average. The TGD time series could be well fitted by a second-degree polynomial with residuals of 0.4–0.6 ns while IFBs of 0.6–0.9 ns, based on which, the TGD could be predicted. We find that for GEO satellites both short-term and long-term prediction has accuracy around 0.5 ns while for most of IGSO satellites accuracy of short-term and long-term prediction is around 0.5 and 1 ns respectively.

References

1. Matsakis, D. (2007). The Timing Group Delay Correction (TGD) and GPS timing biases. *Proceedings of the 63rd Annual ION National Technical Meeting*, April 23–25, 2007. Cambridge, Massachusetts, USA (Institute of Navigation, Alexandria, Virginia).
2. NAVSTAR GPS Space Segment/Navigation User Interface. (2003). ICD-GPS-200 Rev. C-005R1, ARINC Research Corporation, El Segundo, California, USA.

3. Dai, W., Jiao, W., Jia, X. (2009). Application research for compass navigation satellite interfrequency bias correction terms. *Journal of Zhengzhou Institute of Surveying and Mapping, 26*(5).
4. Huang, S., Wang, X., Liu, L., Ping, J., & Dong, E. (2010). Separation solution methods of TGD and IFB for compass system. *Geomatic Science and Engineering, 3*, 28–32.
5. Feltens, J. (2003). The international GPS service (IGS) ionosphere working group. *Advances in Space Research, 31*(3), 635–644.
6. Xing, N., Wang, X., & Hu, X. (2011). Study of ionosphere using GPS data and its application to analysis about earthquake. *Chinese Journal of Space Science, 31*(2), 236–245.
7. Measurement and Adjustment Research Group of School of Geodesy and Geomatics of Wuhan University (2007). *The theory of errors and least squares adjustment*. Wuhan: Wuhan University Press.
8. Wu, X., Ping, J., & Liu, L. (2011). Estimation of hardware delay solution of regional satellite navigation system. *Geomatics and Information Science of Wuhan University, 10*, 1218–1221.

Seasonal Variations Analysis of the Origin and Scale of International Terrestrial Reference Frame

Yan-yan Li, Shu-li Song, Wen-yao Zhu and Juan Zhao

Abstract Based on the time series of station coordinates provided by international SLR service (ILRS) over 1993–2011 using a set of global 92 ILRSA fiducial stations and by international VLBI service (IVS) over 1980–2009 using a set of global 113 VLBI fiducial stations, we generated a time series of the geocenter motion (GCM) and scale factor variations in terms of the network shift approach, which has been the longest and highest accurate one until now. Fourier analysis and wavelet analysis were employed to analyze the nonlinear variations of GCM and scale factor time series. The results indicate that the variations of GCM and scale are dominated by annual and semiannual motions, and show certain long periodic variations. The amplitude and phase of the annual and semiannual motions has certain changes in different years.

1 Introduction

The Terrestrial Reference System (TRS) is a fundamental basis for the geosciences [1]. Monitoring and researching the various movements and changes of the earth and its environments needs to be described in a Terrestrial Reference Frame (TRF) fixed on the earth [2, 3]. The primary standards of a TRS (at the theoretical level)

Y. Li (✉) · J. Zhao
Department of Astronomy, Beijing Normal University,
No. 19, XinJieKouWai St., HaiDian District, 100875 Beijing, China

S. Song · W. Zhu
Shanghai Astronomical Observatory, Chinese Academy of Sciences,
80 Nandan Road, 200030 Shanghai, China
e-mail: yyli@shao.ac.cn

J. Sun et al. (eds.), *China Satellite Navigation Conference (CSNC) 2012 Proceedings*,
Lecture Notes in Electrical Engineering 160, DOI: 10.1007/978-3-642-29175-3_23,

or a TRF (at the realization level) are the origin, the scale, the orientation and their time evolution [4].

The origin of the International Terrestrial Reference System (ITRS) is defined as the center of mass of the whole earth, including atmosphere, oceans and surface groundwater (CM). The International Terrestrial Reference Frame (ITRF) as the realization of the ITRS is formed by a set of special earth reference points (fiducial stations), whose positions and velocities are known. According to IERS2003, the origin of the ITRF is the center of figure of the outer surface of the solid Earth (CF). The geocenter motion (GCM) is defined as CF movement relative to CM. Geocenter variations are produced by mass redistribution within the Earth system, especially surface mass load changes associated with the atmosphere, oceans, and the continental hydrological cycle [5–7]. Zhu et al. [2] point out that the origin of the latest ITRF2008, is neither CM nor CF. The origin includes the linear variations of GCM without considering the main terms of GCM, namely, seasonal variations.

Over the past 20 years, many researchers have focused on GCM. The data processing method of space geodesy techniques, such as satellite ranging (SLR), Global Positioning Satellites (GPS), and Doppler Orbitography Radiopositioning Integrated by Satellites (DORIS), is usually a dynamic geodetic method. The SLR solution origin is obviously CM. Therefore, the time series of the global station coordinates, computed by SLR, GPS and DORIS solutions, have become the primary means of monitoring GCM. GCM has been studied using the space geodetic data and geophysical data [4–15]. These studies and monitored results indicate that GCM mainly shows seasonal variations which are dominated by annual components. The amplitude of the annual components is smaller than 1 cm. There are great differences between various monitored results including geophysical prediction for the amplitude and phase of the annual GCM. Furthermore, the results of Altamimi et al. [4, 8] are significantly different from those of Angermann et al. [9] (see Table 1). The reasons can be explained as follows. (1) The complexity of GPS and DORIS observation errors, such as the uncertainty of the satellite antenna phase center, makes the reliability of the GCM solution low; (2) Although it is commonly accepted that SLR is currently the best space geodetic technique to measure the geocenter [16–18], the limited number of coordinates time series with good quality, caused by the restriction of the SLR observation conditions, seriously influence the accuracy of the network shift approach derived time series of the GCM; (3) The accuracy of geophysical forecast is seriously influenced by the modeling precision and annual changes of the geophysical factors, such as of atmosphere, snow, soil moisture, etc. [3].

Another problem for the establishment and maintenance of the ITRF is the scale. According to IERS conventions (2003), the scale of Conventional Terrestrial Reference System (CTRS) is defined as the scale of the local Earth frame in the sense of relativistic theory of gravitation. The coordinate time of the CTRS is the geocentric coordinate time (TCG), and the length unit of the CTRS is the SI m. VLBI data processing using a geometric method is basically unrelated to the earth's gravitational field. The scale factor mainly depends on the velocity of light, which makes the long-term stability of the VLBI scale factor better than that of SLR and other space techniques in the establishment and maintenance of a terrestrial reference frame. The scale of ITRF2005 depends on IVS, but the ITRF2008 scale depends on the average value of IVS and ILRS. As can be shown by many researches, there are obvious annual changes among the ITRF2005 scale, the ITRF2008 scale and the IVS, ILRS scales [2, 19]. Few international researches in this aspect have been studied. In recent years, Altamimi et al. [4, 8] obtained the amplitude and phase of the annual variation of the scale factor based on the time series of the scale factor derived from the IVS and ILRS coordinate transformation.

With the improvement of the monitored precision of GPS, VLBI, SLR and other space technologies, and the development of the crust movement and earth environment monitoring systems from the cm-level to the mm-level, the construction of mm-level terrestrial reference frame is imperative. Based on the time series of station coordinates provided by ILRS over 1993–2011 using a set of global 92 ILRSA fiducial stations and by IVS over 1980–2009 using a set of global 113 VLBI fiducial stations, we generated a time series of the GCM and scale factor variations in terms of the network shift approach, which has been the longest and highest accurate one until now. In this paper, Fourier analysis and wavelet analysis are employed to analyze the nonlinear variations of the GCM and scale factor time series. In this paper, we also discuss the effect of the asymmetric distribution of the ILRS and IVS fiducial stations in the northern and southern hemisphere on monitoring of the annual variations of the ITRF GCM and scale factor.

2 SLR and VLBI Station Network

The SLR and VLBI station network around the world has a poorly spatial distribution. The SLR station network with the number of observed normal points from 1993 until 2011 is shown in Fig. 1. The spatial distribution of SLR stations and the amount of SLR data observed over the past 19 years undergoes a lack of stations in the southern hemisphere. The number of SLR observation stations is in most weeks in the order of 20–25. There is a slight variation over the 19 years time span, which is more obviously visible for the number of observed normal points per week (see Fig. 2).

The VLBI station network with the number of observed normal points from 1980 until 2009 is shown in Fig. 3. The spatial distribution of VLBI stations and the amount of SLR data observed over the past 30 years also suffers from a lack of stations in the southern hemisphere. As can be seen from Fig. 4, the number of VLBI observation stations from 1980 to 1993 is in most weeks in the order of 3–8, from 1994–2009 in order of 9–18.

Fig. 1 SLR network stations

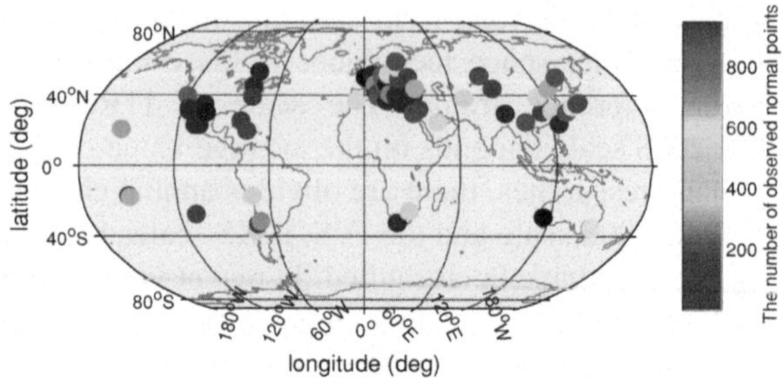

Fig. 2 SLR observation stations per week

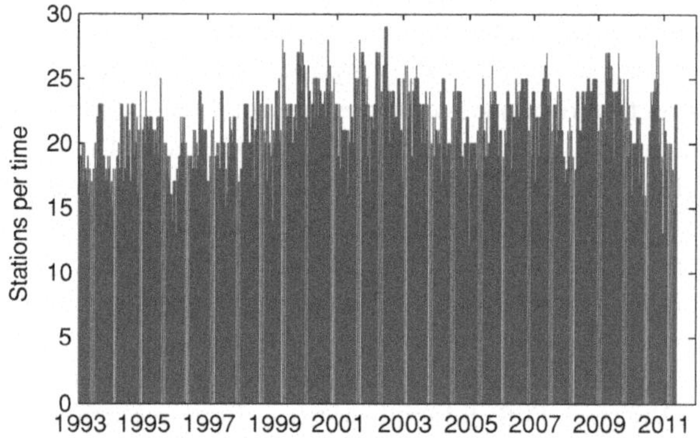

Fig. 3 VLBI network stations

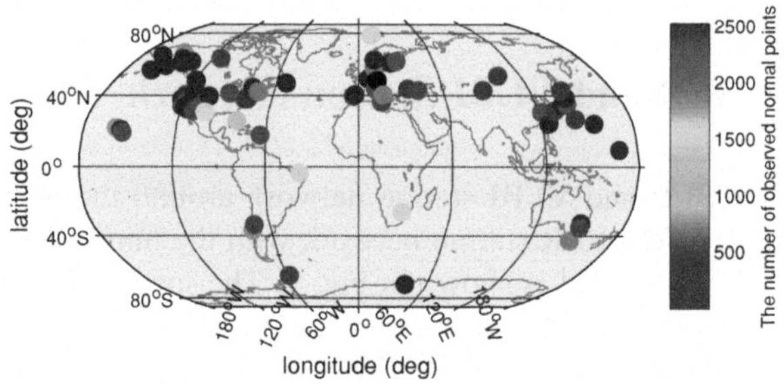

Fig. 4 VLBI observation stations per week

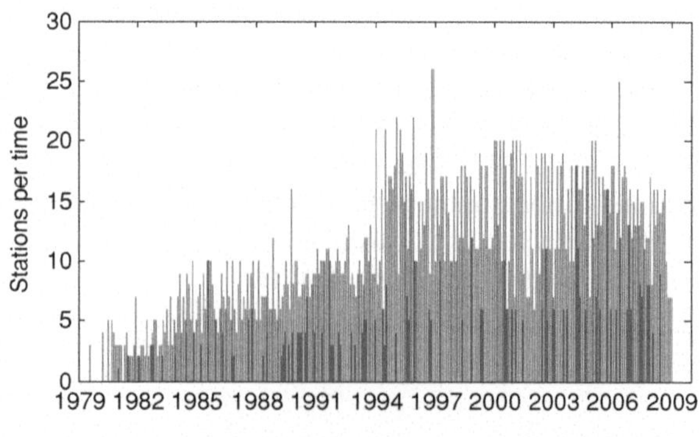

3 The Calculation Method of Time Series of GCM and Scale Factor Variations

The network shift approach, used in the Jet Propulsion Laboratory (JPL) operational analysis, obtains free-network solutions in a CM frame, and performs a seven-parameter transformation to align the solutions to ITRF. The estimated three translation parameters and one scale factor represent the GCM and scale movement, respectively [17, 20].

3.1 Mathematics Model

The transformation parameters of the Helmert seven parameters between two TRFs include three translations, one scale factor, and three rotations, termed T_x, T_y, T_z, D, R_x, R_y, R_z, respectively. The observation equation of the ITRF combined solution can be expressed as,

$$\begin{bmatrix} X^i(t) \\ Y^i(t) \\ Z^i(t) \end{bmatrix}_{ITRF} = \begin{bmatrix} X^i(t) \\ Y^i(t) \\ Z^i(t) \end{bmatrix} + \begin{bmatrix} T_x(t) \\ T_y(t) \\ T_z(t) \end{bmatrix} + \begin{bmatrix} D(t) & R_z(t) & R_x(t) \\ R_z(t) & D(t) & -R_y(t) \\ -R_x(t) & R_y(t) & D(t) \end{bmatrix} \begin{bmatrix} X^i(t) \\ Y^i(t) \\ Z^i(t) \end{bmatrix} \tag{3.1}$$

where $\begin{bmatrix} X^i(t) & Y^i(t) & Z^i(t) \end{bmatrix}^T_{ITRF}$ are the station coordinate vectors of the ITRF combined solution at epoch t, $\begin{bmatrix} X^i(t) & Y^i(t) & Z^i(t) \end{bmatrix}^T$ are the station coordinate time series from individual analysis centers (or individual international service centers for space technologies, such as IVS, ILRS, IGS, IDS, etc.), which act as the input values of the observation Eq. (3.1). Least square adjustment is common used to estimate the seven transformation parameters between two TRFs. For this purpose, Eq. (3.2) is given by

$$\begin{bmatrix} X \\ Y \\ Z \end{bmatrix}_{ITRF} = \begin{bmatrix} X \\ Y \\ Z \end{bmatrix} + A\theta \tag{3.2}$$

where θ is the vector of the seven transformation parameters. A is the design matrix of partial derivatives constructed upon approximate station positions $(\ldots, x_0^i, \ldots)$, where $1 < i < n$ and n is the number of stations:

$$A = \begin{bmatrix} \cdot & \cdot & \cdot & \cdot & \cdot & \cdot & \cdot \\ 1 & 0 & 0 & x^i & 0 & z^i & -y^i \\ 0 & 1 & 0 & y^i & -z^i & 0 & x^i \\ 0 & 0 & 1 & z^i & y^i & x^i & 0 \\ \cdot & \cdot & \cdot & \cdot & \cdot & \cdot & \cdot \end{bmatrix}$$

Least squares adjustment yields solutions for θ of Eq. (3.2) as follows:

$$\theta = (A^T P_x A)^{-1} A^T P_x (X_{ITRF} - X) \tag{3.3}$$

There are mainly three choices for the weight matrix: unit weights, $P_x = I$, where I is the identity matrix; cumulated diagonal terms of the two inverses of the variance matrices with solution X_{ITRF} and X; and cumulated inverses of the two full variance matrices.

3.2 Time Series of the GCM and Scale Factor

The outliers were removed from the coordinate time series prior to the similarity transformation. The seven transformations were done iteratively by rejecting those stations in a particular week, which have only few observations and/or large position residuals. Due to the lack of the SLR and VLBI observation stations, the parameter translations chose more than three participating stations for each global session. Under the above model, based on the time series of station coordinates provided by ILRS over 1993–2011 using a set of global 92 ILRSA fiducial stations and by IVS over 1980–2009 using a set of global 113 VLBI fiducial stations, we generated a time series of the GCM and scale factor variations in terms of the network shift approach, which has been the longest and highest accurate one until now.

4 Seasonal Variation Analysis of the GCM and Scale Factor

4.1 Data Preprocessing

Considering that the analysis results can be seriously influenced by the noise sampling signal, data preprocess must be conducted based on the wavelet analysis. So the time series of origin and scale will be processed based on wavelet. The procedure of the wavelet analysis is as follows.

(1) Wavelet coefficient decomposition
The signal is decomposed into N layers in terms of db3 wavelet based multi-scale analysis. Generally, the resolution decreases with the increasing of layer

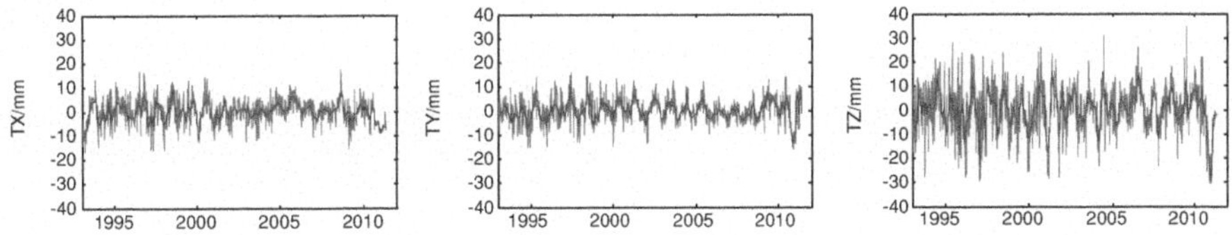

Fig. 5 GCM time series in X, Y, Z directions from 1993–2011 (*grey* original time series, *red* noise-reduced time series)

number. In order to keep the nonlinear feature and remove most of the noise of the decomposed time series, three layers are employed to obtain the coefficients of scale and wavelet in this paper.

(2) Tolerance determination
The tolerance used for noise remove in each layer is determined based on the ratio of signal to noise, namely, the standard deviation of decomposed wavelet coefficient. In this paper, layer tolerance is first determined, and then the soft tolerance is employed to decrease the noise.

(3) Signal reconstruction
The signal reconstruction is similar to signal decomposition, yet the direction is reverse.

In terms of the above procedure, the results of GCM and scale time series are shown in Figs. 5 and 6. Results indicate that the noise-reduced series is smoother than the original time series.

4.2 Fourier Analysis of the GCM and Scale Factor Variation

The power spectrum analysis is based on Fourier transform-based frequency domain analysis, which decomposes the time series of total energy into different frequency components. According to the variance contribution of the different frequency signals, we judge the sequence implied significant period. Power spectral analysis is based on the stationary random process, and it has some limitations on the non-stationary random process. Therefore, we assume that the GCM and scale changes in time series are stationary signals. The Fourier transformation is first used for the noise-reduced time series of the geocenter and scale factor. Then, the power spectral density of each frequency is calculated to determine the GCM and the change of the scale factor which can be used for finding the trend and the periodic terms, and estimating the contribution of periodic terms. The results are shown in Figs. 7 and 8.

For the center of the earth movement, all of the strongest components in the X, Y and Z directions are 358 days of the anniversary of the cycle and 170 days of semiannual component. The components in X and Y directions are

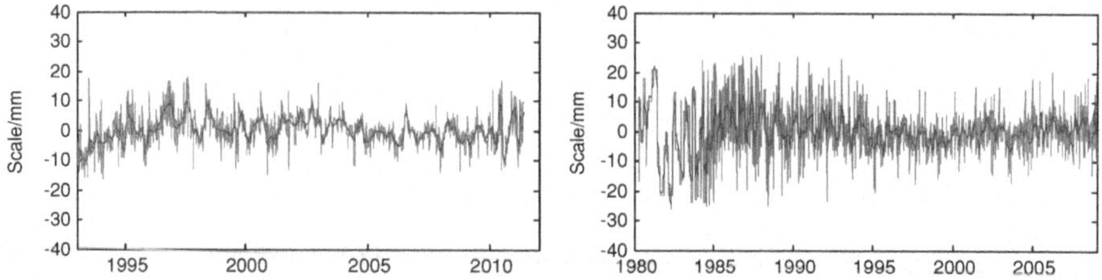

Fig. 6 ILRSA scale time series from 1993–2011and scale time series from 1980–2009 (*grey* original time series, *red* noise-reduced time series)

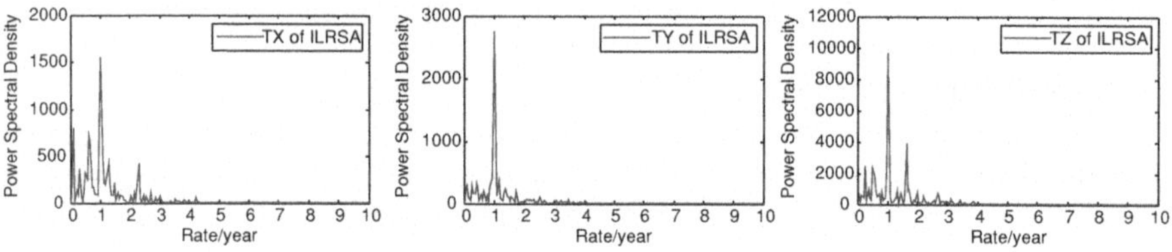

Fig. 7 Power spectrum density of ILRSA GCM in X, Y, Z directions with Fourier transformation

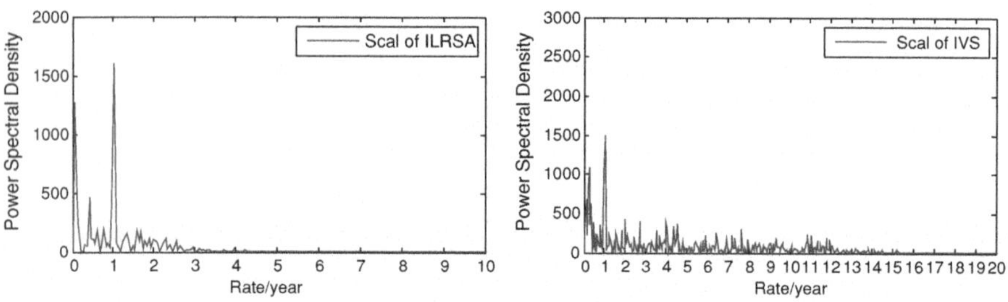

Fig. 8 Power spectrum density of ILRSA and IVS scales with Fourier transformation

about 3,584 days long period, and the component in the Z direction show a corresponding 7,168 days long period. The geophysical mechanism for these long cycles may come from crustal rebound after the ice age, ice and snow melt, or from core-mantle coupling in the deep earth. For the scale change, the strongest component of the IVS-scale change is 356 days annual variation and 182 days semiannual component. In the scale changes of ILRSA, in addition to the strong 358-day anniversary component, the strongest component is 7,168 days. The geocenter and scale change is mainly characterized by the long cycle and seasonal changes. However, some relatively weak cycle is obtained based on Fourier transform, which may be caused by noise and sampling frequency [15].

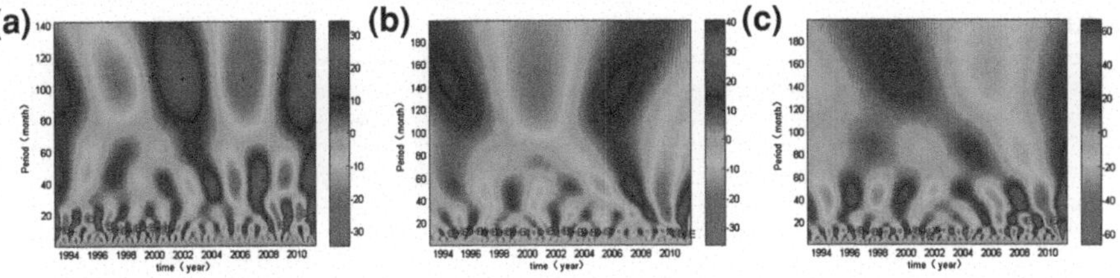

Fig. 9 The CWT results of ILRSA geocenter motion in X, Y, Z direction. **a** X-direction. **b** Y-direction. **c** Z-direction

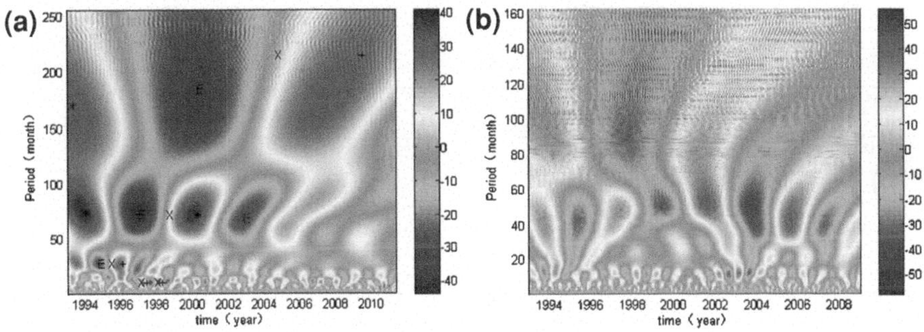

Fig. 10 The CWT results of ILRSA scale factor (**a**) and IVS scale factor (**b**)

4.3 Wavelet Analysis of the GCM and Scale Factor Variation

In this paper, continuous Wavelet Transform (CWT) was employed to analysis the nonlinear changes of GCM and scale factor variation. The results are shown in Figs. 9 and 10.

The results indicated that all of the annual movements in the X, Y and Z directions are especially significant, and the cycle detected by the wavelet variance analysis is 356 days. The strength of the geocenter annual movement changes significantly year by year, where the significant changes in the X, Y and Z-directions occur before the year of 2001, from 1994 to 2006 and after the year of 2010, and during 1996–2002, respectively. The geocenter annual movement in Z direction is stronger than those of the X and Y directions, yet the cycles in the X and Y directions are longer than that in the Z direction. For the scale change, ILRSA and IVS scales show an obvious annual variation, the cycles of which detected by wavelet variance analysis are 356 and 359 days, respectively, and there are some changes in their strengths year after year. The ILRSA scale shows strong long-period changes. In conclusion, the results obtained from the wavelet analysis are similar to Fourier power spectrum analysis.

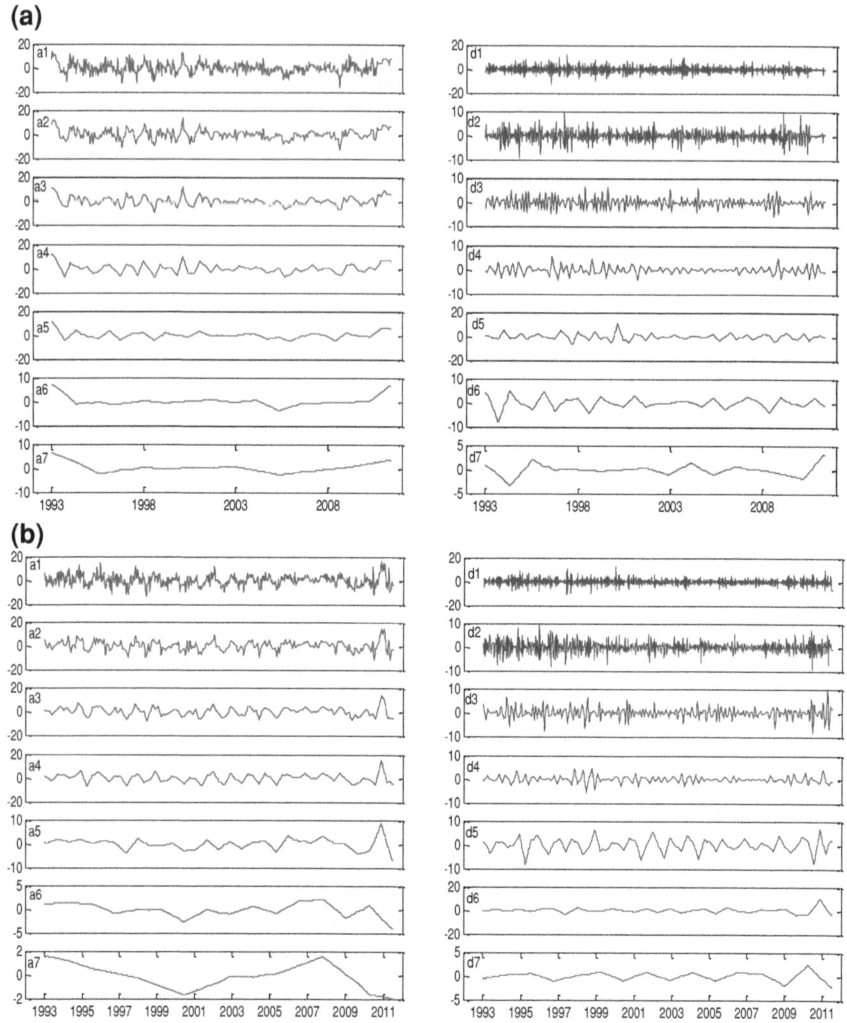

Fig. 11 Proximations and details information of ILRSA GCM in X, Y and Z directions from a seven level multi-resolution wavelet transformation using db3 wavelet. **a** X-direction **b** Y-direction. **c** Z-direction

Wavelet Multi-Resolution Transform (WRA) was further employed to refine the analysis of GCM and scale factor changes, where the long period and seasonal items of the time series were decomposed. In this paper, db3 wavelet was used to give the time series seven layers decomposition, and the approximation signal a_1–a_7 and detailed signal d_1–d_7 of GCM and scale factor changes were obtained. Results are shown in Figs. 11 and 12.

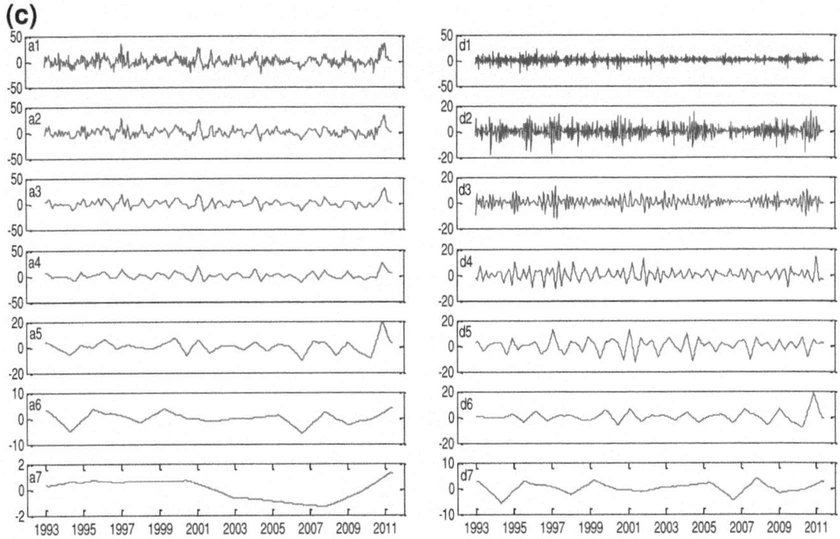

Fig. 11 (continued)

Figures 11 and 12a indicate that a_5 proves the annual movements of GCM and scale factor. d_5 proves that ILRSA GCM and scale factor have an obvious annual cycle, which shows that the annual cycle changes year by year. The GCM in the Z direction is significantly stronger than those in the X and Y directions, which is consistent with the results obtained by CWT. Figure 12b indicates that d_7 accurately shows the annual cycle and the annual variation of the scale factor.

4.4 Least Square Analysis of the GCM and Scale Factor Variation

It is confirmed that the main component of the GCM and scale factor is the seasonal variation, dominated by the annual movement. The long periodic component of more than a decade appearing in the spectrum analysis will be absorbed by the velocity component when constructing the ITRF combination solutions based on a 14 parameter Helmert transformation [2, 17]. From the geophysical mechanism of the GCM, the GCM are mainly produced by the mass load variation of the whole earth. The variations of the GCM and scale are dominated by seasonal variations, which are annual and semiannual movements. The time series of the GCM and scale factor are fitted with a polynomial. The equation is given by,

$$y_i = y_0 + \dot{y}_0(t_{i'} - t_0) + \sum_{j=1}^{m} [c_j \cos(2\pi f_j(t_i - t_0)) + s_j \sin(2\pi f_j(t_i - t_0))] \quad (4.1)$$

$$i = 1, 2, \cdots n$$

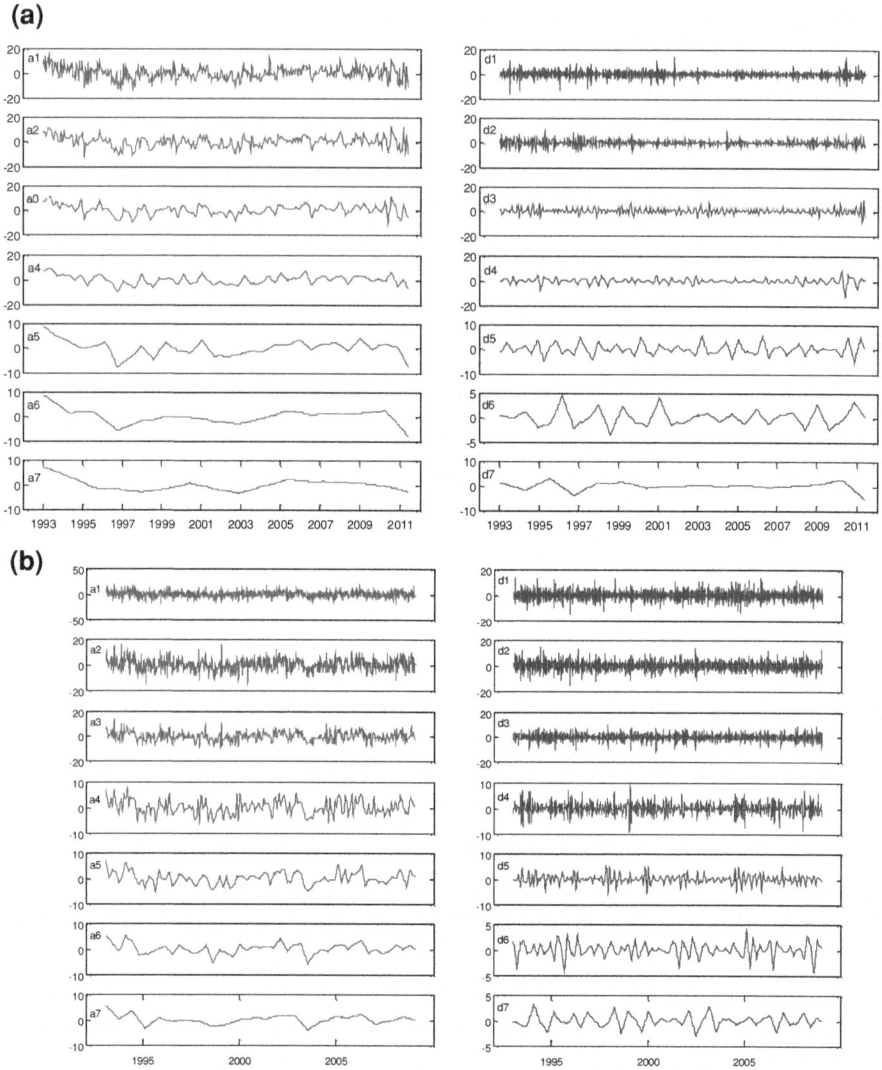

Fig. 12 Proximations and details information of ILRSA and IVS scales factor from a seven level multi-resolution wavelet transformation using db3 wavelet. **a** ILRSA scale. **b** IVS scale

where y_0 is constant; $\dot{y}_0$ is the secular rate; f_j is the frequency for the j-periodic term; the amplitude and phase of the j-periodic term are, $A_j = \sqrt{(c_j^2 + s_j^2)}$ and $\phi_j = \arctan(c_j/s_j)$, respectively. Suppose m equals two, we can obtain the annual and semiannual terms in terms of Eq. (4.1).

Tables 1 and 2 show the observed and predicted amplitude and phase of the GCM and scale factor, respectively, where the time series used in the first line is

Table 1 The observed and predicted amplitude and phase of the GCM

Source		Annual component		Semiannual component	
		Amplitude /mm	Phase /deg	Amplitude /mm	Phase /deg
ILRSA1(1993–2011) in this paper	tx	2.5 ± 0.3	216.5 ± 5.2	0.3 ± 0.3	170.5 ± 328.1
	ty	3.2 ± 0.2	310.5 ± 2.2	0.1 ± 0.2	63.2 ± 155.8
	tz	5.2 ± 0.5	253.2 ± 1.7	1.3 ± 0.5	38.4 ± 13.7
ILRSA2(1993–2011) in this paper	tx	2.5 ± 0.2	225.7 ± 3.1	0.6 ± 0.2	164.4 ± 47.4
	ty	3.0 ± 0.2	317.7 ± 2.2	0.4 ± 0.2	22.3 ± 26.9
	tz	5.2 ± 0.4	243.1 ± 1.6	1.5 ± 0.4	34.8 ± 9.6
ILRSB(1993–2008) in this paper	tx	2.8 ± 0.3	238.2 ± 2.3	0.9 ± 0.3	164.4 ± 38.2
	ty	3.2 ± 0.2	313.2 ± 2.1	0.5 ± 0.2	43.7 ± 12.9
	tz	5.8 ± 0.4	249.3 ± 1.5	1.0 ± 0.4	37.6 ± 16.5
ILRS [4]	tx	2.6 ± 0.1	222 ± 3	–	–
	ty	3.1 ± 0.1	135 ± 2	–	–
	tz	5.5 ± 0.3	202 ± 10	–	–
ILRS [8]	tx	3.0 ± 0.2	216 ± 4	–	–
	ty	3.3 ± 0.2	147 ± 3	–	–
	tz	2.5 ± 0.4	186 ± 11	–	–
SLR [9]	tx	2.7 ± 0.6	220 ± 14	–	–
	ty	3.8 ± 0.6	321 ± 10	–	–
	tz	5.4 ± 1.4	265 ± 15	–	–
GPS [21]	tx	3.3 ± 0.3	184 ± 3	–	–
	ty	4.8 ± 0.3	285 ± 3	–	–
	tz	11.0 ± 0.2	214 ± 1	–	–
GPS [17]	tx	2.1 ± 0.3	224 ± 3	–	–
	ty	3.3 ± 0.3	297 ± 3	–	–
	tz	7.1 ± 0.3	232 ± 3	–	–
Predicted results					
Dong et al. [7]	tx	4.2	224	0.83	30
	ty	3.2	339	0.43	26
	tz	3.5	235	1.10	313
Chen et al. [6]	tx	2.4	244	0.75	181
	ty	2.0	270	0.89	221
	tz	4.1	228	0.50	238
Bouille et al. [5]	tx	1.6	236	–	–
	ty	1.8	309	–	–
	tz	3.1	254	–	–

from our computation, and those in the second and third lines are from the seven parameter transformations between ILRSA/ILRSB combination solutions and SLRF2005 solutions.

Table 1 indicates that there are good agreements in the first three groups of the annual and semiannual amplitude and phase. Our annual amplitude and phase are similar to those of Angermann et al. [9]. The annual amplitudes and phases are similar to those of Altamimi et al. [4], except for the phase in Y direction with a

Table 2 The observed amplitude and phase of the scale factor

Source	Annual component		Semiannual component	
	Amplitude /mm	Phase /deg	Amplitude /mm	Phase /deg
ILRSA(1993–2011) in this paper	1.5 ± 0.3	221.0 ± 7.4	0.2 ± 0.3	48.1 ± 40.1
ILRSA(1993–2011) in this paper	1.5 ± 0.2	206.2 ± 3.2	0.2 ± 0.2	43.1 ± 37.8
ILRSB(1993–2008) in this paper	1.3 ± 0.2	200.9 ± 22.2	0.4 ± 0.2	38.9 ± 17.5
IVS(1980–2009) in this paper	2.1 ± 0.3	217.5 ± 7.7	1.1 ± 0.3	175.6 ± 43.9
ILRS [4]	0.6 ± 0.1	255 ± 10	–	–
ILRS [8]	1.7 ± 0.2	231 ± 7	–	–
IVS [4]	2.2 ± 0.1	241 ± 3	–	–
IVS [8]	2.7 ± 0.1	220 ± 3	–	–
SLR [9]	16.6 ± 0.9	205 ± 28	–	–

180° difference. Compared with the results of other researchers, we can find similar results. Hence, the phase in Y direction of Altamimi et al. [4] might be wrong.

Table 2 indicates that our annual amplitude and phase of the scale factor are similar to those of Altamimi et al. [4, 8], but compared with the results of Angermann et al. [9], there is a huge difference. Hence, we might conclude that the results of Angermann et al. [9] are unreasonable.

5 Conclusions

The main results in this paper are as follows.

(1) The time series of GCM and scale factor analyzed by different spectrum analysis are similar, where there exits a strong annual item and a weak semiannual item. Furthermore, we also found a long periodic item longer than 10 years. The amplitude and phase of the GCM annual change annually, where the significant changes in the X, Y and Z-directions occur before the year of 2001, from 1994 to 2006 and after the year of 2010, and during 1996–2002, respectively. Generally speaking, the amplitude of GCM in Z direction is wider than those of the X and Y directions. The annual change of the amplitude and phase of scale factor is minor. The geophysical mechanism of the long period of a decade or more in scale may come from crustal rebound after the ice age, ice and snow melt, and sea level change, or from the core-mantle coupling in the deep Earth. The long periodic component of more than a decade appearing in the spectrum analysis will be absorbed by the velocity component when constructing the ITRF combination solutions based on a 14 parameter Helmert transformation.
(2) Our results confirm to the latest international results very well. The results in this paper confirm that the scale of ITRF is not a uniform system in the sense of the relativistic theory of gravitation, and there is an obvious annual change.

(3) The status of the poor distribution of the ITRF fiducial stations in the Northern and Southern Hemispheres would have an important effect on the monitoring and analysis of the definition and variation of the GCM and scale factor. The seasonal change produced by the mass load variation of the whole earth is difficult to explain the obvious annual change of the scale of ITRF, which is most likely from the poor distribution of the ILRS and IVS fiducial stations in the Northern and Southern Hemisphere. The poor distribution of the ILRS fiducial stations may lead to the fact that the annual component of the GCM in Z direction is obviously larger than those in X and Y directions. Hence, our further researches will be focused on the effect of the poor distribution of the ITRF fiducial stations in the Northern and Southern Hemispheres on GCM and scale factor.

References

1. Boucher, C. (1990). Definition and realization of terrestrial reference systems for monitoring earth rotation. *Variations in Earth Rotation, 1*, 197.
2. Zhu, W., Song, S., & He, L. (2011). Recommendations for construction of a nonlinear international terrestrial reference frame. *Science China Physics, Mechanics and Astronomy, 54*(1), 164–171.
3. Song, S., Zhu, W., & Xiong, F. (2009). Construction of mm-level terrestrial reference frame. *Chinese Journal of Geophysics, 52*(11), 2704.
4. Altamimi, Z., Collilieux, X., & Metivier, L. (2011). ITRF2008: An improved solution of the international terrestrial reference frame. *Journal of Geodesy*, 1–17.
5. Bouille, F., Cazenave, A., Lemoine, J., & Cretaux, J. (2000). Geocentre motion from the DORIS space system and laser data to the Lageos satellites: Comparison with surface loading data. *Geophysical Journal International, 143*(1), 71–82.
6. Chen, J., Wilson, C., Eanes, R., & Nerem, R. (1999). Geophysical interpretation of observed geocenter variations. *Journal of Geophysical Research, 104*(B2), 2683–2690.
7. Dong, D., Dickey, J., Chao, Y., & Cheng, M. (1997). Geocenter variations caused by atmosphere, ocean and surface ground water. *Geophysical Research Letters, 24*(15), 1867–1870.
8. Altamim, Z., Collilieux, X., Legrand, J., Garayt, B., & Boucher, C., (2007). ITRF2005. A new release of the international terrestrial reference frame based on time series of station positions and Earth Orientation Parameters, *Journal of Geophysical Research, 112*(B9), B09401.
9. Angermann, D., & Muller, H. (2008). On the strength of SLR observations to realize the scale and origin of the terrestrial reference system. *Observing our Changing Earth*, 21–29.
10. Cretaux, J. K., Soudarin, L., & Dabidson, F. J. M. (2002). Seasonal and interannual geocenter motion from SLR and DORIS measurements: Comparison with surface loading data. *Journal of Geophysical Research, 107*(B12), 2374.
11. Eanes, R., Kar, S., Bettadapur, S., & Watkins, M. (1997). Low-frequency geocenter motion determined from SLR tracking. *Eos Transactions AGU, 78*, 46.
12. Feissel-Vernier, M., Bail, K. L., Berio, P., Coulot, D., Ramillien, G., & Valette, J. J. (2006). Geocentre motion measured with DORIS and SLR, and predicted by geophysical models. *Journal of Geodesy, 80*(8), 637–648.
13. Kang, Z., Tapley, B., Chen, J., Ries, J., & Bettadpur, S. (2009). Geocenter variations derived from GPS tracking of the GRACE satellites. *Journal of Geodesy, 83*(10), 895–901.

14. Moore, P., & Wang, J. (2003). Geocentre variation from laser tracking of LAGEOS1/2 and loading data. *Advances in Space Research, 31*(8), 1927–1933.
15. Guo, J., Han, Y., & Huang, J. (2008). Analysis of geocenter motion monitored by CHAMP satellite. *Science China Physics, 38*(G10), 1404.
16. Altamimi, Z., Boucher, C., Drewes, H., Ferland, R., Larson, K., Ray, J., & Rothacher, M. (2002). Combination of station positions and velocities.
17. Dong, D., Yunck, T., & Heflin, M. (2003). Origin of the international terrestrial reference frame. *Journal of Geophysical Research, 108*(B4), 2200.
18. Watkins, M. M., & Eanes, R. J. (1993). Long term changes in the Earth's shape, rotation, and geocenter. *Advances in Space Research, 13*(11), 251–255.
19. Zhu, W., & Xiong Fand Song, S. (2008). Notes and commentary on the ITRF2005. *Progress in Astronomy, 26*(1).
20. Zhu, W., & Song, S. (2010). Origin and no net rotation of the international terrestrial reference frame. *Progress in Astronomy, 28*, 321.
21. Blewitt, G., Lavallee, D., Clarke, P., & Nurutdinov, K. (2001). A new global mode of earth deformation: seasonal cycle detected. *Science, 294*(5550), 2342.

A New Positioning Algorithm with Elevation-Dependent Data Weighting

Ranran Su, Lei Zhang, Li Liu, Guifen Tang and Guangming Hu

Abstract As important and fundamental space establishment for global economy and national defence, satellite navigation systems (GNSS) can provide users with high precision navigation, positioning and timing services that are available without interruption of weather conditions. Same to all GNSS systems such as GPS, GALILEO and GLONASS, COMPASS positioning precision varies with a user's viewing geometry of the GNSS constellation. In case of un-optimal viewing geometries, simple positioning algorithm would lead to deteriorated positioning that is unable to satisfy the user's requirements. This paper proposes a new algorithm to improve positioning precision under un-optimal viewing geometry conditions by properly weighting psudo-range measurements based on their elevations. Application experiments with COMPASS data show significant improvement over the traditional positioning algorithm which treats all measurements with equal weight.

1 Introduction

The satellite navigation system can provide high precision, all-weather's navigation, position and timing service for users. Now, in satellite navigation system, for example, GPS, GALILEO and COMPASS. Positioning precision of navigation system is used to measure the performance of the satellite navigation system [1].

Based on the existing navigation and positioning system, positioning accuracy are greatly depends on the quality of the quality of the data, when the satellite elevation angle is low, poor data quality, the use of conventional positioning

R. Su (✉) · L. Zhang · L. Liu · G. Tang · G. Hu
Beijing Global Information Center of Application and Exploration, 100094 Beijing, China
e-mail: shuyy_2003@yahoo.com.cn

J. Sun et al. (eds.), *China Satellite Navigation Conference (CSNC) 2012 Proceedings*, Lecture Notes in Electrical Engineering 160, DOI: 10.1007/978-3-642-29175-3_24,

algorithm will inevitably lead to the positioning accuracy drops considerably, unable to meet the needs of users. Considering geometric satellite real-time changes, put forward a kind of weighted improved under the condition of low elevation angle positioning algorithm, and the weighted matrix selection are analyzed. The results of computer simulation show that, at low elevation angle conditions, using the weight matrix of the poor data quality and data quality better satellite satellite differently, thus greatly improving the low altitude conditions user positioning accuracy.

2 The Basic Principle of RNSS Positioning

According to the analytic geometry of space, in three-dimensional space, if the user can be precisely measured signal from the spatial location of known satellite transmission to the user's delay, the speed of light, can be used to determine the satellite to the user's distance, i.e. the pseudorange [2]. Considering the user clock synchronization, when the satellite number more than four, can simultaneously calculate the user of three-dimensional space coordinates and clock error, position location.

For user i observation satellite j pseudo-range observation equation:

$$\begin{aligned} P_i^j(t_k) = & \rho + C\delta t_r - C\delta t_s + \Delta D_{trop} + \Delta D_{ion} + \Delta D_{rel} \\ & + \Delta D_{ant}^r + \Delta D_{ant}^s + \Delta D_{tide} + \Delta D_{mult} + \varepsilon \end{aligned} \tag{1}$$

On type, P_i^j is emission time satellite to receive the signal time among users of theoretical distance; ρ is satellite to user communication distance; ΔD_{trop} is troposphere delay correction; ΔD_{iono} is ionospheric delay correction; ΔD_{ant}^r is station eccentricity correction and antenna phase center corrections effect; ΔD_{ant}^s is Eccentricity correction and satellite antenna phase center correction effect; ΔD_{tide} is tidal effect; ΔD_{mult} is the impact of multi-path effect; ε is pseudo range measurement noise.

$$P_i^j(t_k) = \sqrt{(x_k - x_i)^2 + (y_k - y_i)^2 + (z_k + z_i)^2} \tag{2}$$

On type, (x_k, y_k, z_k) is t_k satellite position, (x_i, y_i, z_i) is user location.

In order to solve the equations, the formula (2) are linearized processing.

$$\begin{bmatrix} \Delta oc_1 \\ \Delta oc_2 \\ \vdots \\ \Delta oc_n \end{bmatrix} = \begin{bmatrix} a_{x1} & a_{y1} & a_{z1} & 1 \\ a_{x2} & a_{y2} & a_{z2} & 1 \\ \vdots & \vdots & \vdots & \vdots \\ a_{xn} & a_{yn} & a_{zn} & 1 \end{bmatrix} \begin{bmatrix} \Delta x \\ \Delta y \\ \Delta z \\ \Delta t \end{bmatrix} \tag{3}$$

On type $(\Delta x, \Delta y, \Delta z)$ is the iterative process of receiver positioning error, Δt is clock error.

$$\Delta oc = [\Delta oc_1\ \Delta oc_2 \cdots \Delta oc_n]^T$$

$$\Delta x = (\Delta x\ \Delta y\ \Delta z\ \Delta t)^T$$

$$\mathrm{H} = \begin{bmatrix} a_{x1} & a_{y1} & a_{z1} & 1 \\ a_{x2} & a_{y2} & a_{z2} & 1 \\ M & M & M & M \\ a_{xn} & a_{yn} & a_{zn} & 1 \end{bmatrix}$$

Then type (2) can be written as the following simple form

$$\Delta oc = H\Delta x, \quad \Delta x = H^{-1}\Delta oc \tag{4}$$

The type that, when the satellite is larger than 4, matrix H is irreversible. Through repeated iterations, first from the user to approximate the location and rough clock start, end users get accurate location.

3 Satellite Elevation Angle Weighted Quality Control Positioning Method

For conventional positioning algorithm, usually considered the user for each satellite ranging accuracy equal, but in the actual system, altitude angle lower satellite observation data quality variation in elevation angle high satellite, i.e. low altitude satellite will affect the data quality. In order to solve the problem of unequal precision measurement, using the weighted least squares method, through different weights to different satellite elevation angle difference in elevation, weakening of lower importance while improving satellite elevation angle of satellite in high position resolution of important degree, so as to improve the calculation accuracy [3].

First of all, on the satellite elevation angle a calculation. To determine the visible star height angle (Fig. 1).

$$a = \arctan\left(\frac{x_P}{\sqrt{y_P^2 + z_P^2}}\right) \tag{5}$$

In which:

$$x_P = x_U + \overline{SU} \cdot (x_S - x_U)$$

$$y_P = y_U + \overline{SU} \cdot (y_S - y_U)$$

$$z_P = z_U + \overline{SU} \cdot (z_S - z_U)$$

On type, $(x_s\ y_s\ z_s)$ is the coordinates of satellites, $(x_u\ y_u\ z_u)$ is the coordinates of users.

Fig. 1 Satellite elevation

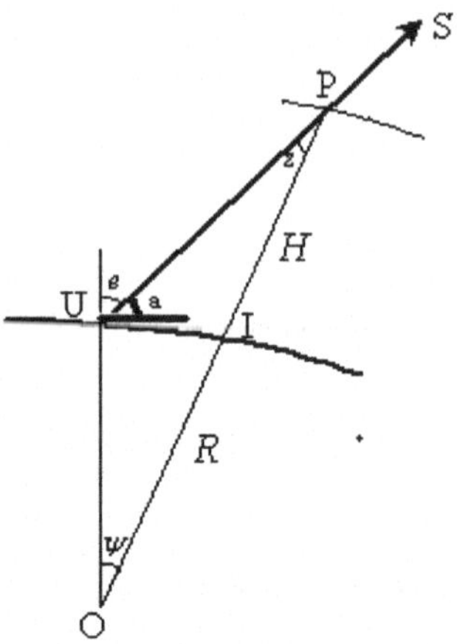

Considering that each satellite to the user of the observation equation is independent, therefore, to optimize the accuracy, weighted equation as follows:

$$\begin{cases} W_1 \cdot \sum_{m1,m2}^{m1+m2\le M} c_{m1m2}(\theta_1-\theta_o)^{m1}(\varphi_1-\varphi_o)^{m2}\cos^{m1}(\varphi_1-\varphi_o) = W_1\cdot\Delta\tau^{ij}(\theta_1,\varphi_1) \\ W_2 \cdot \sum_{m1,m2}^{m1+m2\le M} c_{m1m2}(\theta_2-\theta_o)^{m1}(\varphi_2-\varphi_o)^{m2}\cos^{m1}(\varphi_2-\varphi_o) = W_2\cdot\Delta\tau^{ij}(\theta_2,\varphi_2) \\ \vdots \\ W_K \cdot \sum_{m1,m2}^{m1+m2\le M} c_{m1m2}(\theta_K-\theta_o)^{m1}(\varphi_K-\varphi_o)^{m2}\cos^{m1}(\varphi_K-\varphi_o) = W_K\cdot\Delta\tau^{ij}(\theta_K,\varphi_K) \end{cases} \tag{6}$$

On type, W is the weight coefficient matrix. c_{m1m2} is a constant coefficients.

Through the satellite and the user position angle calculation, the observation equations are weighted according to the type.

$$W = \begin{bmatrix} 1/\Gamma_1^2 & 0 & L & 0 \\ 0 & 1/\Gamma_2^2 & L & 0 \\ 0 & 0 & L & 1/\Gamma_3^2 \end{bmatrix} \tag{7}$$

In the formula, if the weight coefficient matrix is known, it can be used as a weighted least squares of weighted matrix. How to choose the weight coefficient matrix Γ values related to weighted algorithm is effective, the Γ selection and the actual situation about compliance, weighted algorithm more effective. Among them, the measurement error variance according to the satellite elevation angle setting.

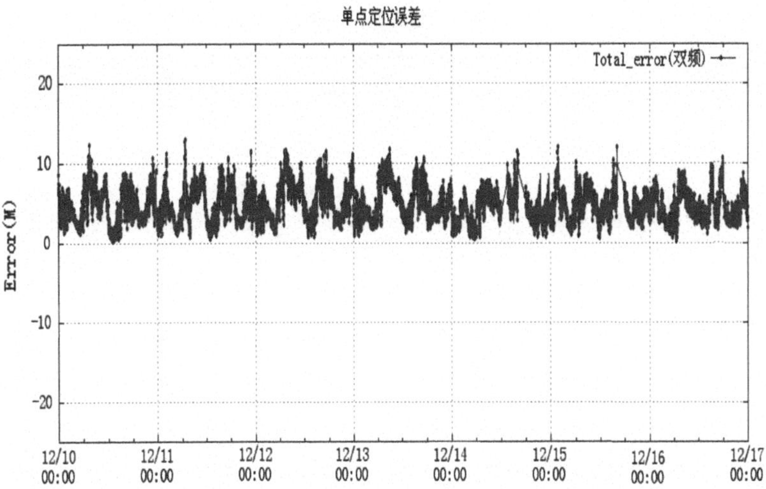

Fig. 2 Positioning result without using bad geometry arithmatic

Angle weighting method can be used to highly specific, angle cosine $\Gamma = \cos\alpha$. The weight coefficient matrix into the observation equation:

$$\Delta x = [H^T WH]^{-1} H^T W \Delta oc \tag{8}$$

On type, Δoc, Δx, H is as previously shown matrix, W is the weight coefficient matrix.

By using the method of least squares to the observation equations are solved, can the user position coordinates.

4 Precision Analysis

Using the measured data of seven stars, conditions, namely, three GEO and four IGSO satellites for conventional positioning and the altitude of the weighted condition analysis of positioning accuracy, the result is as follows:

(1) Do not use high angle weighted positioning algorithm of positioning error, as shown in Fig. 2. Below are given in 2011 December seven consecutive days Beijing area on orbit seven satellites use highly angular weighting algorithm positioning result.
(2) The use of high angle weighted positioning algorithm of positioning error, as shown in Fig. 3. Below are given in 2011 December seven consecutive days in orbit seven satellites in Beijing using highly angular weighting algorithm positioning result.

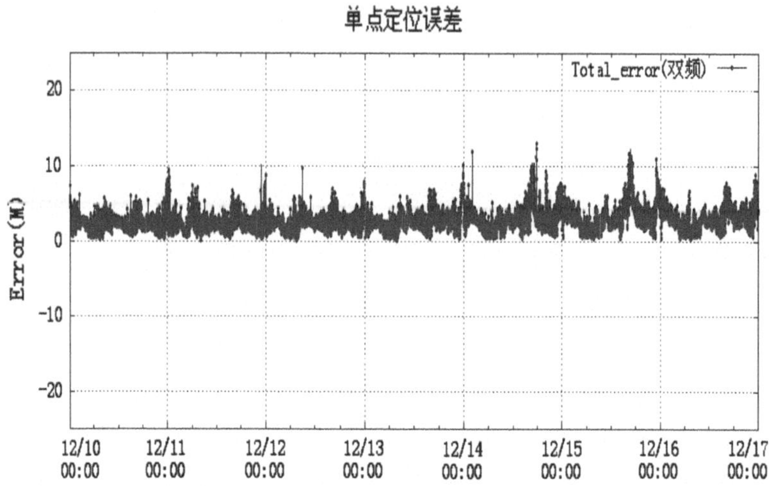

Fig. 3 Positioning result using bad geometry arithmatic

Table 1 Change of DOP value

Date	DOP do not use high angle weighted algorithm	DOP the use of highly angular weighting algorithm
12.10	4.778	4.313
12.11	4.693	4.302
12.12	4.752	4.298
12.13	4.997	4.387
12.14	4.838	4.391
12.15	4.761	4.277
12.16	4.958	4.404
Average	4.825	4.339

From the above chart can be seen, without the use of high angle weighted localization algorithm, positioning error for seven consecutive days of 3D RMS is 5.788 m. The use of high angle weighted localization algorithm, positioning error for seven consecutive days of three-dimensional improved to 3.936 m. The average DOP value changes shown in Table 1.

From the table we can see, without the use of highly angular weighting algorithm, DOP for seven consecutive days average 4.825. The use of high angle weighted location algorithm, for seven consecutive days to enhance to 4.339.

5 The End

To improve positioning accuracy is guaranteed to satellite navigation system index is one of the important factors, in this paper, based on the positioning of the thorough analysis, proposed one kind based on the satellite elevation angle weighted quality control positioning method. The simulation results show that: the positioning accuracy of the premise, the elevation angle weighted localization algorithm can effectively improve the positioning accuracy, to make up for the altitude is low with poor data quality deficiency.

References

1. Wu, T., & Li, G. (1998). *The earth synchronization satellite position*. The PLA Publishing Company.
2. Shunsen, T. (1997). *The satellite navigation engineering*. Beijing: The PLA Publishing company.
3. Jianming, T., & Li, W. (1989). *The application of augmentation technology in quick position communication system*. Beijing: Chinese Journal of Space Science.

Analysis of Real Valued Ambiguity Variance for Long Baselines Between GNSS Reference Stations

Feng Zhang, Hui Ren, Chunyang Han, Teng Li and Man Sun

Abstract The technology of virtual reference station can achieve real-time and kinematic positioning of high-precision for medium and long baseline, which is a kind of positioning method integrating the internet, wireless communication technique, data base and GNSS positioning, and represents a novel developing trend of positioning. The ambiguities resolution between reference stations of network is most important to the technology of Virtual reference station. Because the baselines between reference stations are too long, the relativity of space-related error becomes weaker, which results in that residual errors of tropospheric delay and ionospheric delay are still large after double difference, and then the ambiguities can not be fixed easily. Based on dual-frequency observation of GPS, the feature of the real valued ambiguity variance is analyzed for different linear combinations. The impact on variance of ambiguity evaluation is derived after pseudo range is smoothed by carrier phase. The optimal linear combination is obtained for ambiguity resolution. On this basis, the ambiguity resolution method and quality control program are proposed between reference stations. The calculation and analysis of experimental data shows that it is not affected by the length of the baseline. For low elevation satellites, it can also fix ambiguities successfully within dozens of epochs. It can also provide a reference value for ambiguity resolution of COMPASS multi-frequency observation.

F. Zhang (✉) · H. Ren · C. Han · T. Li · M. Sun
Beijing Application and Development Center of Round-the-World Information,
P.O. Box 5128, 100094 Beijing, China
e-mail: zhangfengchxy@163.com

J. Sun et al. (eds.), *China Satellite Navigation Conference (CSNC) 2012 Proceedings*,
Lecture Notes in Electrical Engineering 160, DOI: 10.1007/978-3-642-29175-3_25,

1 Introduction

For Network of continuously operating reference stations, because the baselines between reference stations are too long, which are tens to hundreds of kilometers typically, the relativity of space-related error becomes weaker, which results in that residual errors of tropospheric delay and ionospheric delay are still large after double difference, sometimes up to several L1 wavelengths, and then the ambiguities can not be fixed easily [1–3]. However, between reference stations have their own characteristics. First, the coordinates of reference stations are known accurately, and they are observed continuously, so the parameters to be estimated do not include the coordinates; Secondly, the precise forecast ephemeris can be obtained to reference stations in real-time, thus the orbit error is regardless. Even so, the ambiguities resolution between reference stations are difficult, especially for satellite that is new or that is tracking after unlocking long time. Because the elevation of satellite is low, which results in noise of measurement, multipath, ionospheric delay and tropospheric delay are great, and which is disadvantage for ambiguities resolution. However, long baseline causes that the number of common satellites becomes small. In order to obtain more available satellites, the low elevation satellites have to be taken into account, so we need to improve the efficiency and reliability of ambiguity resolution for low elevation satellites [4].

Whether the ambiguities resolution between reference stations is correct, differential corrections and positioning result of rover can be affected directly. Ambiguity resolution is based on generally, so the matrix of plays an important role in ambiguity resolution. The covariance matrix is the smaller, the more accurate. On the contrary, the real valued ambiguity is the worse. Through analyzing real valued variance–covariance matrix of ambiguity of different linear combination, the purpose of this paper is to find the optimal combination for ambiguity resolution.

2 Model of Variance Analysis for Ambiguity Estimation

As the length of baseline increases, the correlation of atmospheric delay becomes weak, so ionospheric delay through double difference can not be ignored, and it becomes the major error in ambiguity resolution. For VRS positioning, the distances between reference stations are up to tens to hundreds of kilometers, so ionospheric delay must be taken into account for ambiguity resolution. For the combination of dual-frequency observation, equations can be written as [5, 6],

$$
\begin{aligned}
p_1 &= \rho + I_1 + \varepsilon_{p_1} \\
p_2 &= \rho + \frac{f_1^2}{f_2^2} I_1 + \varepsilon_{p_2} \\
\Phi_{ij} &= \rho - \lambda_{ij} N_{ij} - \lambda_{ij} \frac{if_2 + jf_1}{f_2 \lambda_1} I_1 + \varepsilon_{\Phi_{ij}} \\
\Phi_{ab} &= \rho - \lambda_{ab} N_{ab} - \lambda_{ab} \frac{af_2 + bf_1}{f_2 \lambda_1} I_1 + \varepsilon_{\Phi_{ab}} \\
R &= \rho + \varepsilon_R
\end{aligned} \tag{1}
$$

where,

p and Φ are code and phase observables,
R is the calculated value of geometrical distance between receiver and satellite using precise coordinates of reference stations,
λ and f are wavelength and frequency,
ρ is the geometrical distance between receiver and satellite,
$\lambda_j$is the wavelength of the carrier of the carrier B_i (meter),
N is integer ambiguity,
I is ionospheric delay,
ε are random observation errors,
1 and 2 represent L1 and L2 signals,
i, j, a, $b$are coefficients of combination.

Equation (1) is the equation for each observed satellite (except for the one selected as reference) in an epoch, and the double-difference operator $\Delta\nabla$ is omitted. It can be expressed in matrix notation as [7],

$$
L_c + v_c = A_c X_c \tag{2}
$$

where

$L_c = FL - [p_1\, p_2\, \Phi_{ij}\, \Phi_{ab}\, R]'$, $v_c = [v_{p_1}\, v_{p_2}\, v_{\Phi_{ij}}\, v_{\Phi_{ab}}\, v_R]'$, $X_c = [\rho\, N_{ij}\, N_{ab}\, I_1]'$,
$L = [p_1\, p_2\, \Phi_1\, \Phi_2\, R]'$,

$$
A_c = \begin{bmatrix} 1 & 0 & 0 & 1 \\ 1 & 0 & 0 & \frac{f_1^2}{f_2^2} \\ 1 & -\lambda_{ij} & 0 & -\lambda_{ij}\frac{if_2+jf_1}{f_2\lambda_1} \\ 1 & 0 & -\lambda_{ab} & -\lambda_{ab}\frac{af_2+bf_1}{f_2\lambda_1} \\ 1 & 0 & 0 & 0 \end{bmatrix}, \quad F = \begin{bmatrix} 1 & 0 & 0 & 0 & 0 \\ 0 & 1 & 0 & 0 & 0 \\ 0 & 0 & \frac{i\lambda_{ij}}{\lambda_1} & \frac{j\lambda_{ij}}{\lambda_2} & 0 \\ 0 & 0 & \frac{a\lambda_{ab}}{\lambda_1} & \frac{b\lambda_{ab}}{\lambda_2} & 0 \\ 0 & 0 & 0 & 0 & 1 \end{bmatrix}.
$$

Based on model that the ratio of the standard errors of the pseudoranges and the carrier phase for each frequency is a constant, and it is assumed that the relation $\sigma_{\Phi_1}/\sigma_{\Phi_2} = f_2/f_1$ is valid for standard errors scaled to meters. With the law of transformed error, the covariance matrix of combinatorial observations is [8]

Table 1 Comparison of standard errors of different combination of ambiguities

i	j	$\lambda_{ij}\ (m)$	σ_{Nij}	σ_{N1}	Correlation coefficient $N_{ij} - N_1$
1	−1	0.862	0.565	1.992	−0.9971
4	−3	0.115	0.330		0.9195
5	−4	0.101	0.312		−0.8322
9	−7	0.054	0.306		0.13204

$$\Sigma_{L_c} = F\Sigma_L F^T = 4\sigma_{\Phi_1}^2 \begin{bmatrix} k^2 & 0 & 0 & 0 & 0 \\ 0 & \frac{f_1^2}{f_2^2}\cdot k^2 & 0 & 0 & 0 \\ 0 & 0 & \frac{\lambda_{ij}^2}{\lambda_1^2}(i^2+j^2) & \frac{\lambda_{ij}\lambda_{ab}}{\lambda_1^2}(ia+jb) & 0 \\ 0 & 0 & \frac{\lambda_{ab}\lambda_{ij}}{\lambda_1^2}(ia+jb) & \frac{\lambda_{ab}^2}{\lambda_1^2}(a^2+b^2) & 0 \\ 0 & 0 & 0 & 0 & \frac{\sigma_R^2}{4\sigma_{\Phi_1}^2} \end{bmatrix}$$

Equation (2) can be computed from the cofactor matrix

$$Q_{X_c} = (A_c^T \Sigma_{L_c}^{-1} A_c)^{-1} \tag{3}$$

We use the numerical values of $\sigma_{\Phi_1} = 0.002\,m$, $k = 154$, $f_1 = 1575.42$ MHz and $f_2 = 1227.60$ MHz. First, let us compare different linear combinations by changing i and j in area [−10, 10] and keeping $a = 1, b = 0$. This means N_{ij} and N_1 are estimated directly. The goal is to analyze the standard errors in resolving N_{ij} and N_1 directly and the impact of fixing N_{ij} to N_1 estimation. The results are listed in Table 1, but it lists only combinations that produce reasonable results.

In Table 1, the standard error of N_1 remains the same for all combinations: $\sigma_{N_1} = 1.992$, which make it impossible to yield correct integer values of ambiguities. They are easy to fix the integer value of $\hat{N}_{9,-7}$, $\hat{N}_{5,-4}$ and $\hat{N}_{4,-3}$, as they have small standard errors. However, they do not help much to estimate $\hat{N}_1$ with fixed $\hat{N}_{9,-7}$, $\hat{N}_{5,-4}$ and $\hat{N}_{4,-3}$, because with low correlations they can not reduce the standard error of N1 markedly. Once $N_{4,-3}$ is fixed to an integer, we consider it as a known parameter, and we can move it to Eq. (2), then $\sigma_{N_{1,-1}}$ becomes 0.261, which is less than half of the value of 0.565 with direct estimation, so we can estimate wide-lane ambiguity easily. Finally, using fixed wide-lane ambiguity, we can estimate $\hat{N}_1$ with the small standard error: $\sigma_{N_1} = 0.153$, which allows immediate fixing to an integer.

The pseudoranges are original observations in analysis. Because original pseudoranges have large random observation errors, and it is the major difficulty for fixing integer ambiguities, we should analyze real valued ambiguity variance after smoothing of pseudoranges with phases. If we use smoothed pseudoranges with phases, the covariance matrix at the kth epoch can be written as,

Table 2 Standard errors of real valued ambiguities with smoothed pseudoranges

Epochs Standard error Ambiguities	1	2	3	4	5	10	20	100
$N_{9,-7}$	0.31	0.31	0.31	0.31	0.31	0.31	0.301	0.30
$N_{4,-3}$	0.33	0.26	0.23	0.21	0.20	0.18	0.16	0.15
$N_{1,-1}$	0.56	0.40	0.33	0.28	0.25	0.18	0.13	0.06
$N_{1,-1}$ based *MW*	0.56	0.40	0.33	0.28	0.25	0.18	0.13	0.06
$N_{1,-1}$ with fixed $N_{4,-3}$	0.26	0.24	0.22	0.20	0.19	0.15	0.12	0.06
N_1	1.99	1.41	1.16	1.0	0.90	0.65	0.47	0.25
N_1 with fixed $N_{1,-1}$	0.15	0.15	0.15	0.15	0.15	0.15	0.15	0.14

$$\Sigma_{\bar{L}_k} = \frac{4}{k+1}\begin{bmatrix} \sigma_{P_1} + k\left(A^2\sigma^2_{\Phi_1} + B^2\sigma^2_{\Phi_2}\right) & kA\left(C\sigma^2_{\Phi_1} + B\sigma^2_{\Phi_2}\right) & kA\sigma^2_{\Phi_1} & -kB\sigma^2_{\Phi_2} & 0 \\ kA\left(C\sigma^2_{\Phi_1} + B\sigma^2_{\Phi_2}\right) & \sigma_{P_2} + k\left(C^2\sigma^2_{\Phi_1} + A^2\sigma^2_{\Phi_2}\right) & kC\sigma^2_{\Phi_1} & -kA\sigma^2_{\Phi_2} & 0 \\ kA\sigma^2_{\Phi_1} & kC\sigma^2_{\Phi_1} & (k+1)\sigma^2_{\Phi_1} & 0 & 0 \\ -kB\sigma^2_{\Phi_2} & -kA\sigma^2_{\Phi_2} & 0 & (k+1)\sigma^2_{\Phi_2} & 0 \\ 0 & 0 & 0 & 0 & k+1\sigma^2_R \end{bmatrix} \tag{4}$$

The smoothing procedure reduces the standard errors of pseudoranges, however, it introduces correlations between code and phase observations. If number of epochs is sufficiently large, then the smoothing also reduces the multipath effect on pseudoranges. Obviously, we must repair cycle slips before the smoothing procedure. The results of smoothing on different linear combinations of observations are summarised in Table 2.

As shown in Table 2, with the k increasing, the standard errors of ambiguities reduce, but this does not hold for $N_{9,-7}$. The standard error of ambiguity $N_{1,-1}$ has reduced to 0.18 at the tenth epoch, so it can be fixed to integer easily. It is helpful to improve $N_{1,-1}$ estimation with fixed $N_{4,-3}$ to integer for anterior epochs, but the help is not marked after the tenth epoch.

3 The Optimum Linear Combination of Ambiguity Resolution for Long Baseline

For ambiguities resolution of long baselines between reference stations, whether exists the optimum linear combination of L_1 and L_2 phase observations. That is the linear combination $\hat{N}_1 + x\hat{N}_2$ of the primary estimates $\hat{N}_1$ and $\hat{N}_2$ which has the smallest standard error. The arbitrary estimate of linear combination is

$$N_c = \hat{N}_1 + x\hat{N}_2 \tag{5}$$

Following the law of transformed error,

$$\sigma^2_{N_c} = \sigma^2_{N_1} + x^2\sigma^2_{N_2} + 2x\sigma_{N_1N_2} \tag{6}$$

The minimum variance of $\sigma^2_{N_c}$ is attained at $\hat{x} = -\sigma_{N_1N_2}\Big/\sigma^2_{N_2}$

$$\upsilon^2_{\hat{N}_c} = \sigma^2_{N_1} - \sigma^2_{N_1N_2}\Big/\sigma^2_{N_2} \tag{7}$$

As this solution for $\hat{x}$ is not an integer, however, if $-n/m \approx \hat{x}$, where m and n are integers, the linear combination $m\hat{N}_1 - n\hat{N}_2$ may be the optimum combination, provided that m and n are the smallest number of all.

From Eq. (3),

$$\begin{aligned} \sigma^2_{N_1} &= \frac{\sigma^2_p}{\lambda^2_2}\left(g^2 + h^2k^2 + k^2\upsilon\right) \\ \sigma^2_{N_2} &= \frac{\sigma^2_p}{\lambda^2_2}\left(g^2k^2 + h^2 + \upsilon\right) \\ \sigma_{N_1N_2} &= \frac{\sigma^2_p}{\lambda^2_2}hg\left(k^2 + 1\right) \end{aligned} \tag{8}$$

where,
$\upsilon = \left(\sigma_\Phi/\sigma_p\right)^2 \approx 10^{-6}$, $g = 2k/(k^2-1)$, $h = (k^2+1)/(k^2-1)$, $k = \lambda_2/\lambda_1$. Omitting the small terms with υ we thus obtain,

$$\begin{aligned} \sigma^2_{N_1} &= \frac{\sigma^2_p}{\lambda^2_2}\frac{4k^2 + k^2(k^2+1)^2}{(k^2-1)^2} \\ \sigma^2_{N_2} &= \frac{\sigma^2_p}{\lambda^2_2}\frac{4k^4 + (k^2+1)^2}{(k^2-1)^2} \\ \sigma_{N_1N_2} &= \frac{\sigma^2_p}{\lambda^2_2}\frac{2k(k^2+1)^2}{(k^2-1)^2} \end{aligned} \tag{9}$$

Moving Eq. (9) to $\hat{x} = -\sigma_{N_1N_2}\Big/\sigma^2_{N_2}$,

$$\hat{x} = -\frac{2k(k^2+1)^2}{4k^4 + (k^2+1)^2} \tag{10}$$

For the actual wavelengths of L_1 and L_2 phases, $\hat{x} \approx 1.007$, which is very close to the wide-lane combination, so we consider it is the optimum combination for ambiguities resolution of long baselines.

4 Algorithm and Quality Control of Ambiguity Resolution for Long Baseline

Concluding the results from previous sections, we propose the following steps for ambiguity resolution of long baselines.

(1) Using $i = 1, j = 0$, $a = 1, b = -1$ in Eq. (2), compute $N_{1,-1}$ and fix it to the correct integer.
(2) Using $i = 1, j = 0$, $a = 1, b = -1$ and the integer value of $N_{1,-1}$ in Eq. (2), compute N_1 and fix it to the correct integer.
(3) Repeat steps (1) to (2) for each satellite and test the validity of estimated ambiguities using the test procedure described in follows. If the test pass, then the integer ambiguities are correct, if not, take the next epoch of observations.

When ambiguities of L_1 and L_2 phases are resolved, the phase observations have higher accuracy than pseudorange observations. We remove the pseudorange observations in the observation equation,

$$\Delta\nabla\Phi_i^j - \Delta\nabla\rho - \Delta\nabla T + \lambda_i \Delta\nabla N_i^j = \Delta\nabla I_i^j + \varepsilon_i^j \tag{11}$$

Solving Eq. (11) through least-squares algorithm we can obtain $\hat{\sigma}_0^2$ [9]

$$\hat{\sigma}_0^2 = \frac{\hat{\varepsilon}^T \sum_L^{-1} \hat{\varepsilon}}{(n-1)} \tag{12}$$

Based on the posterior variance factor, the ambiguities group of passing the two tests is fixed to the correct integer ambiguity.

(1) The a priori and a posterior variance factor are consistent: $\chi^2_{\alpha/2}(n) \le \hat{\sigma}_0^2 / \sigma_0^2 \le \chi^2_{1-\alpha/2}(n)$.
where σ_0^2 is priori variance factor, $\hat{\sigma}_0^2/\sigma_0^2$ has a χ^2 (Chi-square) distribution, α is the significance level.
(2) There is a significant difference between smallest and the second smallest posteriori variance factors: $\hat{\sigma}_0^2/\hat{\sigma}_0'^2 \ge F_{1-\alpha/2}(n, n)$
where $\hat{\sigma}_0^2$ is smallest posteriori variance factor, $\hat{\sigma}'_0 2$ is the second smallest posteriori variance factors, $\hat{\sigma}_0^2/\hat{\sigma}'_0 2$ has an Fdistribution.

In order to ensure high reliability of ambiguity resolution, we consider the ambiguities group as the correct ones only when they pass all the tests in a continuous period of time, because incorrect ambiguities group can pass the test occasionally, especially for the low elevation satellites.

Furthermore, when all ambiguities of double difference are fixed in the network of reference stations after, all ambiguities should be closed to each other.

Fig. 1 Real valued estimates of wide-lane ambiguities for PRN04

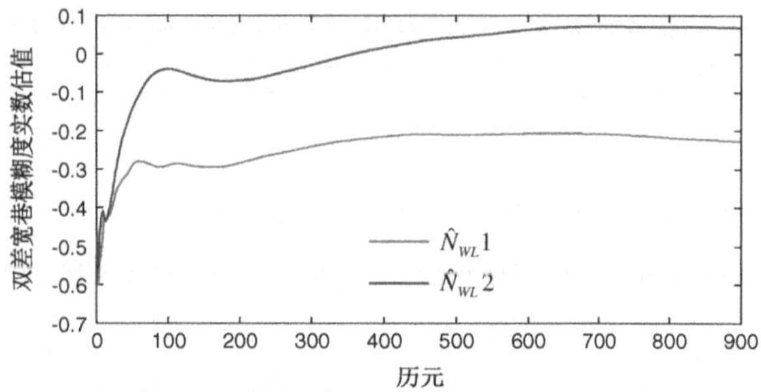

Fig. 2 Real valued estimates of wide-lane ambiguities for PRN23

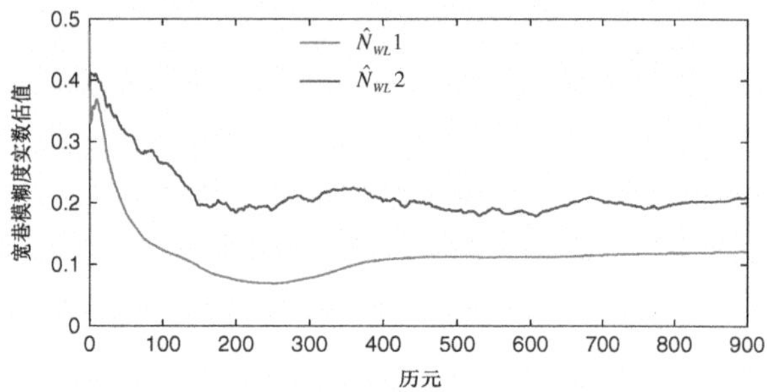

Fig. 3 Fixing process of wide-lane ambiguities for PRN04

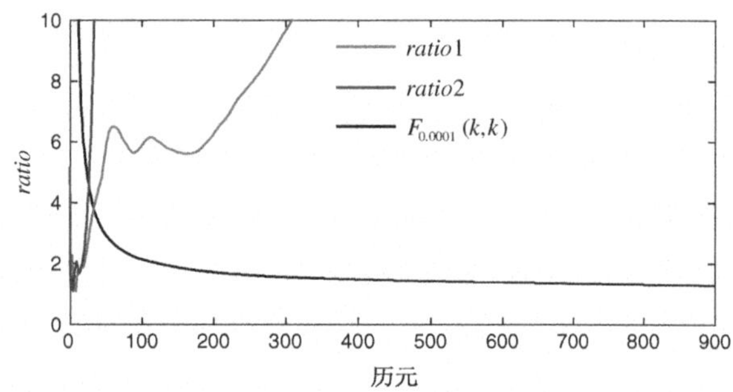

5 Practical Experiments

Experimental data is obtained from a VRS network which is continuously operating. The time span of data is 7:00–9:00, 23/12/2005 in GPS time, and the interval is one second. During this period, the reference satellite is the PRN17, and the PRN04 and PRN23 satellites are just rising, so their elevation angles are rising from 10° to 22.5 and 53° respectively. In order to verify the validity of algorithm proposed in the paper, the two different lengths of baselines, A1–A3 (27 km) and A1–A5 (116 km) are selected to test. Based on the algorithm of ambiguity resolution proposed in the paper, the double difference ambiguities of PRN04 and PRN23 are estimated for the two baselines respectively. The results are shown in Figs. 1, 2, 3 and 4.

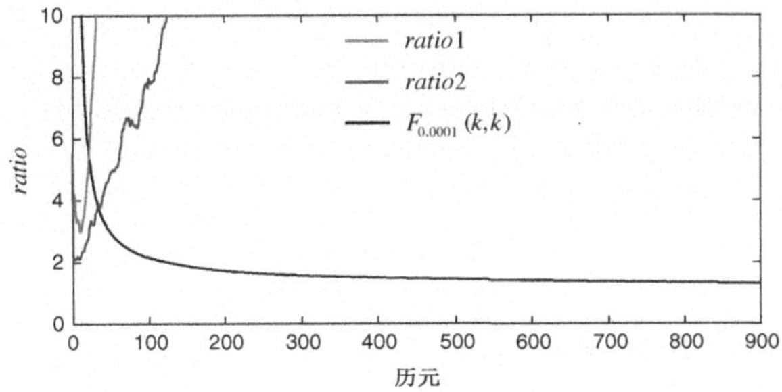

Fig. 4 Fixing process of wide-lane ambiguities for PRN23

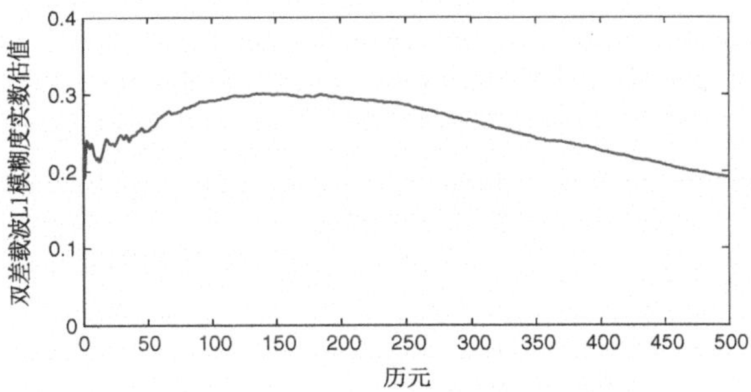

Fig. 5 Real valued estimation of L1 ambiguities for PRN04

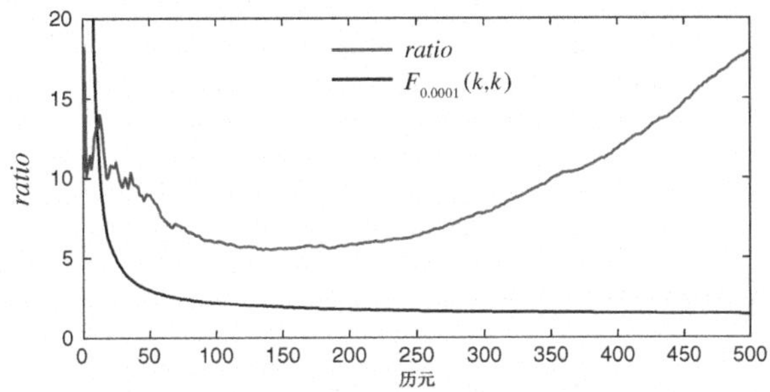

Fig. 6 Fixing process of L1 ambiguities for PRN04

In the four figures, $\hat{N}_{WL}1$ and *ratio*1 are real valued estimation and ratio of wide-lane ambiguities for baseline A1–A3, and $\hat{N}_{WL}2$ and *ratio*2 are for baseline A1–A5. The results show that, for baseline A1-A3, the times of resolving ambiguities of PRN04 and PRN23 are 44 epochs and 22 epochs, and for baseline A1–A5, the times of resolving ambiguities of PRN04 and PRN23 are 36 epochs and 34 epochs.

As can be seen from results of PRN04, however, the efficiency of wide-lane ambiguities resolution for long baseline is better than the short baseline, which indicates that the algorithm is affected weakly by length of baseline, and it can be affected mainly by the satellite elevation and pseudoranges noise, so it fits to ambiguities resolution for long baselines of GNSS.

Moving integer ambiguities of wide-lane combination to Eq. (2), we can fix integer ambiguities of L1 phase. The results of PRN04 for baseline A1–A5 are shown in Figs. 5 and 6.

As can be seen from the figure, when the wide-lane ambiguities of double difference are fixed, ambiguities of double difference L1 phase can be resolved only through 12 epochs. The fixed value of 6391182 is the same as the computing result of commercial software, which indicates that the algorithm is quite efficient for ambiguity resolution.

6 Conclusion

In order to resolve ambiguities for long baselines, based on the characteristics of baselines between reference stations network, the model of variance analysis is proposed, and the optimum linear combination is obtained, which proves to be correct by experimental data. The conclusions are as follows.

(1) Wide-lane combination is the optimum linear combination for ambiguities resolution of long baselines.
(2) For ambiguities resolution of long baselines, too large standard errors of direct estimates make it impossible to yield correct integer values, however, it is easy to fix wide-lane ambiguities, which can be resolved in tens of epochs by use of smoothed pseudoranges.
(3) Using fixed wide-lane ambiguities, we can estimate L1 ambiguities with very small standard errors, which allow immediate fixing to an integer after dozen of epochs.

References

1. He, H. B. (2002). *Precise kinematic GPS surveying and quality control*. Zheng Zhou: Doctor Paper of Institute of Surveying and Mapping, Information Engineering University.
2. Han, S., & Rizos, C. (1997). An instantaneous ambiguity resolution techniques for medium-range GPS kinematic positioning, *10th International Technical Meeting of the Satellite Division of the U.S. Institute of Navigation*. Kansas City, Missouri, 16–19 Sept 1997
3. Gao, X. W., Liu, J. N., et al. (2002). An ambiguity searching method for network RTK baselines between base stations at single epoch. *Acta Geodaetica et Cartographica Sinica, 31*(4), 305–309.
4. Zhang, F. (2010). *Algorithm researching of VRS kinematic positioning and implementing based on the regional reference station network*. Zheng Zhou: Doctor Paper of Institute of Surveying and Mapping, Information Engineering University.
5. Kim, D., & Langley R. B. (2000). A reliable approach for ambiguity resolution in real-time long-baseline kinematic GPS applications. *Proceedings of ION GPS 2000*. Salt Lake City, Utah, 19–22 September, p. 12.
6. Wei, Z. Q. (1998). *The mathematic model of relative positioning for GPS* (pp. 33–62). Beijing: Publication of Surveying and Mapping.

7. Teunissen, P. J. G. (1993). Least-squares estimation of the integer GPS ambiguities. *Invited lecture. Section IV Theory and Methodology, IAG General Meeting*, Beijing, China, p. 16, Also in Delft Geodetic Computing Centre LGR series no. 6.
8. Hatch, R. R. (1982). The synergism of GPS code and carrier measurements. In *Proceeding Third International Symposium Satellite Doppler Positioning* Las Cruces New Mexico State University, 8–12 February, vol 2, pp. 1213–1231.
9. Lu, G., & Lachapelle, G. (1992). Statistical quality control for kinematic GPS positioning. *Manuscripta Geodaetica, 17*, 270–281.

The Correction Method of Overall Pseudo-Rotation on Autonomous Navigation of Navigation Constellation

Haihong Wang, Xingyuan Han, Shanbao He, Haibin Chu and Xiangjun Wu

Abstract This paper takes research on the reason and elimination method of overall pseudo-rotation of navigation constellation. The inter-satellite ranging system errors can cause the overall pseudo-rotation of navigation constellation. In this paper, the reasons for overall pseudo-rotation of navigation constellation have been analyzed for four different systematic errors. Studies show that, short-cycle system errors and long-cycle system errors do not cause the overall pseudo-rotation, constant system errors may cause the overall pseudo-rotation, and linear system errors usually can cause the overall pseudo-rotation. When constant system errors or linear system errors of all satellites have the same direction, the overall pseudo-rotation of navigation constellation will occur. On this basis, this paper uses the overall pseudo-rotation correction algorithm to eliminate the overall pseudo-rotation of navigation constellation, simulation analysis shows that the algorithm can effectively eliminate the overall pseudo-rotation of navigation constellation. Finally, combining with the engineering practice of autonomous navigation, the paper gives the conclusions that on-board data processing units should adopt the overall pseudo-rotation correction algorithm to eliminate the overall pseudo-rotation of navigation constellation.

1 Introduction

High-precision autonomous navigation is the development trend of the new generation of Global Navigation Satellite System, which aims to make the navigation satellite with no ground master control station intervention to continue to provide

H. Wang (✉) · X. Han · S. He · H. Chu · X. Wu
Beijing Institute of Spacecraft System Engineering, 100094 Beijing, China
e-mail: jiayouhai@hotmail.com

J. Sun et al. (eds.), *China Satellite Navigation Conference (CSNC) 2012 Proceedings*, Lecture Notes in Electrical Engineering 160, DOI: 10.1007/978-3-642-29175-3_26,

accurate navigation service. The ground station measurements involved in navigation, there are some drawbacks: first, the ground station can easily become the target of wartime precision strike long-range weapons, and thus cannot meet the requirements of navigation warfare; due to the number of ground stations and distribution constraints, they cannot continuously track space navigation, orbit determination can only intermittent tracking data processing, limiting the accuracy of orbit determination. Multi-satellite constellation can take advantage of the ISLs in the pseudo range measurement information, to some extent with the conditions of the autonomous navigation [1].

The GPS system is the only constellation of autonomous navigation engineering projects. By inter-satellite ranging to achieve through in-orbit validation of the Block IIR and Block IIF satellite constellation of autonomous navigation, and precision index Block IIR autonomous operation of 180 days, the URE is less than 6 m [2], is expected to Block IIF independent run for 60 days, URE 3 m [3].

China is setting up a new generation of satellite navigation systems, in order to ensure the system safe and reliable operation, to avoid the paralysis caused by natural disasters, human error or war ground systems damaged or destroyed the entire space system [4], a new generation of satellite navigation systems considering the construction of inter-satellite links, inter-satellite ranging and data transfer functions, to achieve autonomous navigation. Autonomous navigation in the loss of support of the ground system can operate independently long time, and enhance the viability of the system, ground system support, can further improve system performance, improve navigation and positioning accuracy [5].

Navigation satellite autonomous navigation based on the inter-satellite links, in theory, exist rank deficient [6], Orientation Parameters i and Ω cannot be corrected, which is called overall rotation errors. First, because of the long-term ephemeris satellite dynamics modeling errors and the Earth Orientation Parameters (EOP) prediction error caused by the rigid rotation; The second is due to inter-satellite ranging system errors may cause the filtering algorithm to a miscarriage of justice, causing the constellation as a whole pseudo-rotation. This paper studies the overall pseudo-rotation of the constellation, to explore the reasons that lead to the overall pseudo-rotation and eliminate, given the overall pseudo-rotation correction algorithm to eliminate the constellation as a whole pseudo-rotation, and a simulation analysis results show that the method can effectively eliminate the pseudo-rotation of the constellation as a whole.

2 Analysis of the Constellation Rotation Error Sources

According to the overall rotation of the constellation of different incentives, the overall rotation is divided into two categories:

1. Rigid rotation: Whole rigid rotation of the ground uploaded long-term ephemeris caused by orbital dynamics model error; EOP prediction error, the

performance of the overall constellation rotation in the Earth-fixed coordinate system.
2. Pseudo-rotation: Overall rotation of systematic errors caused by inter-satellite ranging, this rotation is introduced in the process of filtering data, rather than kinetic factors led, also known as the overall pseudo-rotation.

For the first rotation, the rigid rotation, only by filtering based on inter-satellite ranging data processing cannot be amended, shall be with the other measurements to eliminate, Such as increasing the satellite position measurement information, and laying on the ground anchor station [7]. Anchor station can address the constellation of EOP prediction error and the orbital dynamics model error caused by the rigid rotation, Reference [0] and reference [0] give the method to eliminate the overall rigid rotation with anchor station point positioning. The second type of rotation, can be corrected through improved filtering data processing algorithms, and this paper study the causes of pseudo-rotation and eliminate the overall pseudo-rotation correction algorithm.

3 Analysis and Modeling of the Inter-Satellite Ranging Error

Inter-satellite distance measurement is achieved by inter-satellite link distance measuring equipment, measurement accuracy is mainly affected by the delay of the transceiver and antenna calibration accuracy constraints, inter-satellite ranging error, including random errors and systematic errors, random errors by the ranging the random measurement noise of the equipment, has little effect on the autonomous orbit determination results.

Zero value of the distance measuring equipment calibration residuals and aging equipment will cause the inter-satellite ranging system error, the systematic errors on the autonomous orbit determination results may be caused by a whole constellation of pseudo-rotation. Inter-satellite ranging system error modeling are as follows,

$$\Delta\rho = v + a\cos(\frac{2\pi}{T_S}t + \varphi_S) + b\cos(\frac{2\pi}{T_L}t + \varphi_L) + c + dt \tag{1}$$

Here, $\Delta\rho$ is inter-satellite pseudo-range measurement error, v is random white noise, $a\cos(\frac{2\pi}{T_S}t)$, $b\cos(\frac{2\pi}{T_L}t)$, c, dt are Systematic errors of the short-period and long-period system error constant systematic error, the linear system error, T_S is orbital period, T_L is long period of the cycle, about five to ten days, φ_S is the initial phase of short-period, φ_L is the initial phase of long-period, *a, b, c, d* are constant coefficients.

To study how each of the four kinds of systematic errors affect the constellation as a whole pseudo-rotation, the simulation analysis of the configuration for wide-beam inter-satellite links, wide-beam beam angle range is 15–50 degrees.

3.1 Analysis of Short-Period System Error

Inter-satellite ranging only contain systematic errors of short-period, there is no long-period system error, the constant systematic error, linear systematic errors, that is $a \neq 0, b = 0, c = 0, d = 0$, a take random between 0.6 and 0.9 m, φ_S take random between $-\pi$ and π, and different satellite has difference value of a and φ_S. The whole constellation of autonomous orbit determination simulation results show that: short-cycle system error did not affect the results of the autonomous orbit determination and the end of 90 days, URE accuracy of better than 1 m, the short-period system error does not cause the overall pseudo-rotation of the constellation.

3.2 Analysis of Long-Period System Error

Inter-satellite ranging only contain systematic errors of long-period short-period system error, constant error and linear systematic errors, that is $b \neq 0, a = 0$, $c = 0, d = 0$, b take random between 0.6 and 0.9 m, $T_L = 10$ days, φ_L take random between $-\pi$ and π, and different satellite has difference value of b and φ_L. The whole constellation of autonomous orbit determination simulation results show that: long-cycle system error did not affect the results of the autonomous orbit determination and the end of 90 days, URE accuracy of better than 1 m, the Long-period system error does not cause the overall pseudo-rotation of the constellation.

3.3 Analysis of Constant Systematic Error

1. Inter-satellite ranging only contains the constant systematic error, there is no short-period system error, system error and linear systematic errors of long period, that is $c \neq 0, a = 0, b = 0, d = 0$, take $c = 1.0$ m (corresponding 3 ns delay calibration system residuals), the value of c for all satellites are the same, the whole constellation of autonomous orbit determination URE and RTN three-dimensional position error shown in Fig. 5.

Further analysis of the orbital inclination and ascending node errors in the filter results are as follows:

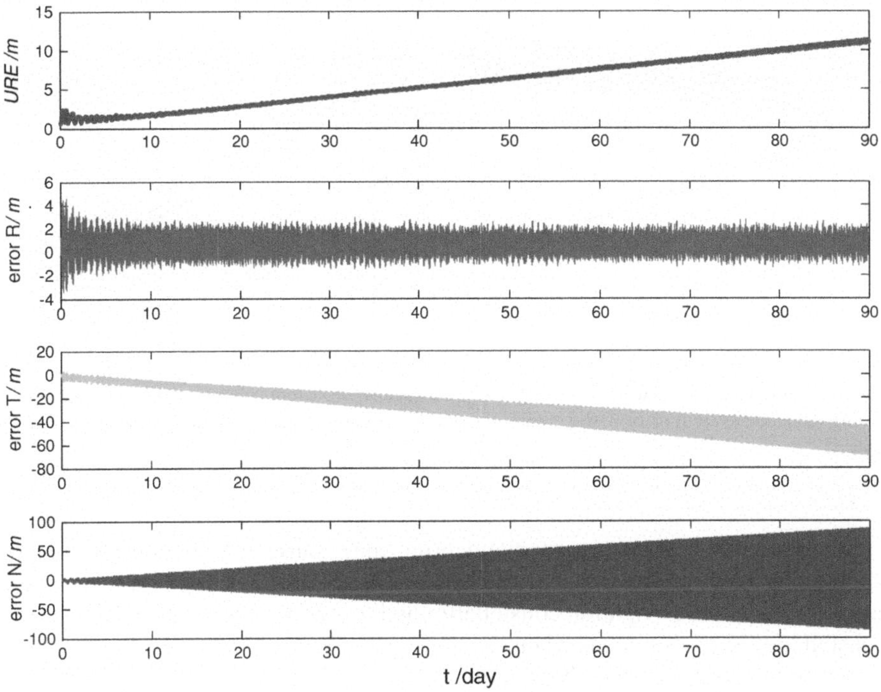

Fig. 1 Constellation URE and three-dimensional RTN position error statistics of autonomous orbit determination: inter-satellite ranging only contains constant system errors, and all satellites have the same system errors

Each cluster curve in Fig. 2 corresponds to a constellation of three orbital planes. We can see from Figs. 1 and 2, systematic errors has a great influence on the autonomous orbit determination results. Autonomous orbit determination the the URE precision and RTN three-dimensional position errors continue to diverge, mainly along the track and normal to the error increasing, which is consistent with the error characteristics of the overall constellation rotation,that is when the constellation of satellites ranging devices have exactly the same constant systematic error, the overall pseudo rotation of constellation occur.

2. Random number (corresponding to c taken (−0.3, 0.3 m) between the residuals of ±1 ns delay calibration system), and different satellite c take different values. Simulation results show that: when the constant systematic error of satellite distance measuring equipment with better randomness, and different satellite has different error values, Autonomous orbit URE accuracy and RTN three-dimensional position errors remain stable, there is no obvious trend of divergence in orbital inclination and ascending node, the constellation did not occur as a whole pseudo-rotation.

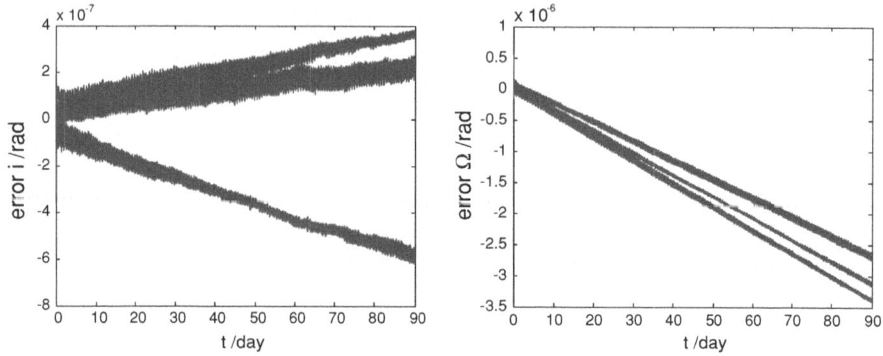

Fig. 2 When the inter-satellite ranging only containing constant system errors (all satellites having the same system errors), the errors of orbital inclination i and right ascension of ascending node Ω

Therefore, the constant systematic error may cause the pseudo-rotation of whole constellation, When the constant systematic error for each satellite in the same direction, the navigation constellation will occur the overall pseudo-rotation; Constellation does not occur when the constant systematic error for each satellite randomness better, the overall pseudo-rotation.

3.4 Linear Term Systematic Error Analysis

1. Inter-satellite ranging contain only the linear systematic error, there is no systematic errors of short-period, long-period system error and constant systematic errors, that is $d \neq 0, a = 0, b = 0, c = 0$. Take d = 0.01 m/day, and all satellites have the same values of d. Simulation results show that: the error of linear systems results a great influence on autonomous orbit determination, URE accuracy and RTN three-dimensional position errors continue to diverge, that is, The constellation will occur as a whole pseudo-rotation.
2. d take the −0.01 m/day or 0.01 m/day with 20% of the random error, and the different satellites d values, the whole constellation of autonomous orbit determination URE and RTN three-dimensional position error shown in Fig. 3.

Filtering results of the orbital inclination and ascending node error is shown in Fig. 4.

Therefore, the linear system error is usually caused by the overall pseudo-rotation of the constellation of more serious when the linear systematic error of each satellite in the same direction, the overall pseudo-rotation.

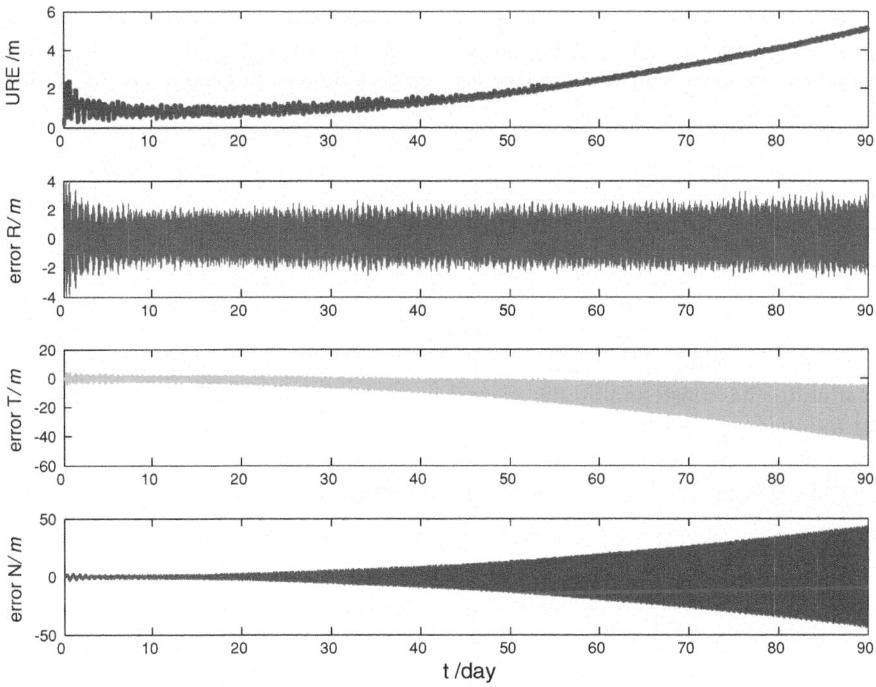

Fig. 3 Constellation URE and three-dimensional RTN position error statistics of autonomous orbit determination: inter-satellite ranging only contains linear system errors, and different satellites have different system errors

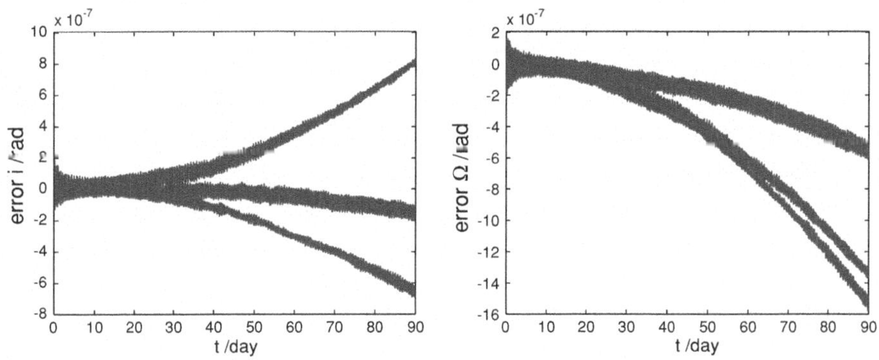

Fig. 4 When the inter-satellite ranging only containing linear system errors (different satellites having different system errors), the errors of orbital inclination *i* and right ascension of ascending node node

In short, the systematic errors of the systematic errors of short-cycle and long cycle would not cause the overall pseudo-rotation of the constellation and constant systematic errors may cause the overall pseudo-rotation, linear systematic error is usually caused by the overall pseudo-rotation.

4 Pseudo-Rotation Correction Algorithm

In the filtering process on the satellite, the general process of filtering on the orbital inclination and ascending node by filtering [5], and use the overall pseudo-rotation correction algorithm [8] to correct the i and Ω, to eliminate the overall pseudo-rotation of the constellation, the overall pseudo-rotation correction algorithm see the following:

$$\Omega^{+} = \hat{\Omega} - \overline{\delta\Omega} \tag{1}$$

$$i^{+} = \hat{i} - \overline{\delta i} \tag{2}$$

$$\overline{\delta\Omega} = \frac{\sin\Omega\cos i}{\sin i}\delta\theta_x + \frac{\sin\Omega\cos i}{\sin i}\delta\theta_y + \delta\theta_z \tag{3}$$

$$\overline{\delta i} = \sin\Omega\delta\theta_x + \cos\Omega\delta\theta_y \tag{4}$$

$$\delta\theta_x = \frac{\sum_{k=1}^{n}\sin\Omega_k\delta i_k\sum_{k=1}^{n}\sin\Omega_k\cos\Omega_k}{\sum_{k=1}^{n}\cos^2\Omega_k + (\sum_{k=1}^{n}\cos\Omega_k\sin\Omega_k)^2} \tag{5}$$

$$\delta\theta_y = \frac{\sum_{k=1}^{n}\cos\Omega_k\delta i_k\sum_{k=1}^{n}\sin\Omega_k\cos\Omega_k}{\sum_{k=1}^{n}\sin^2\Omega_k + (\sum_{k=1}^{n}\cos\Omega_k\sin\Omega_k)^2} \tag{6}$$

$$\delta\theta_z = \frac{\sum_{k=1}^{n}\delta\Omega_k + \delta\theta_y\sum_{k=1}^{n}\cos\Omega_k + \delta\theta_x\sum_{k=1}^{n}\sin\Omega_k}{n} \tag{7}$$

Using the pseudo-rotation correction algorithm simulation, Inter-satellite Ranging System Error is set to $a \neq 0, b \neq 0, c \neq 0, d \neq 0$, a and b are random between 0.6 and 0.9 m, φ_S and $\varphi_L$are random between $-\pi$ and π, $T_L = 10$ days; c = 0.5 m, d = 0.01 m/day. Each satellite with same constant systematic error or linear systematic errors, the constellation will have the overall pseudo-rotation, the use of the above pseudo-rotation correction algorithm to the constellation as a whole pseudo-rotation correction, the whole constellation of autonomous orbit determination URE and RTN three-dimensional position error shown in Fig. 5.

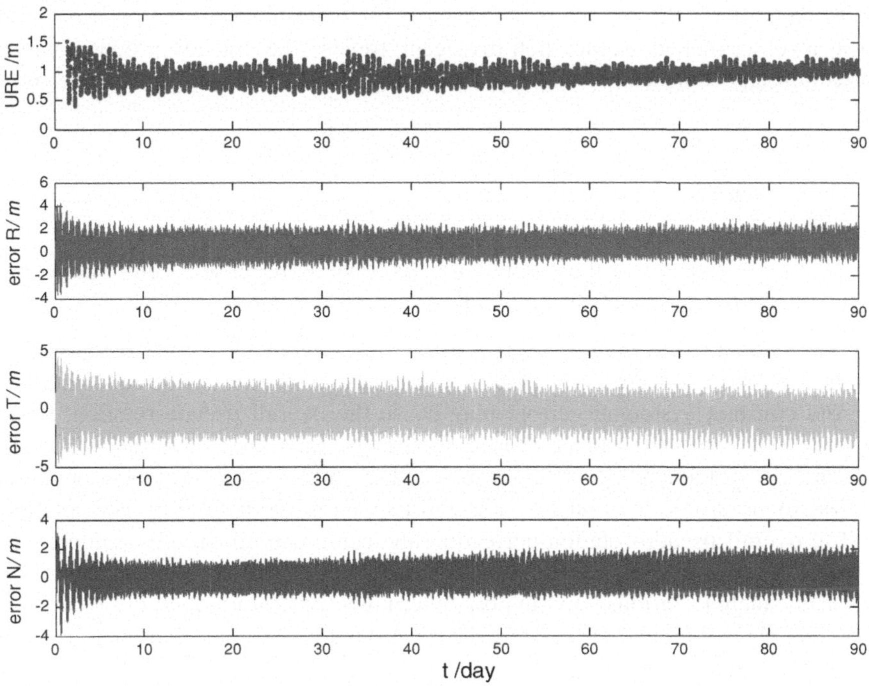

Fig. 5 Constellation URE and three-dimensional RTN position error statistics of autonomous orbit determination after using the overall pseudo-rotation correction algorithm to eliminate the overall pseudo-rotation of navigation constellation

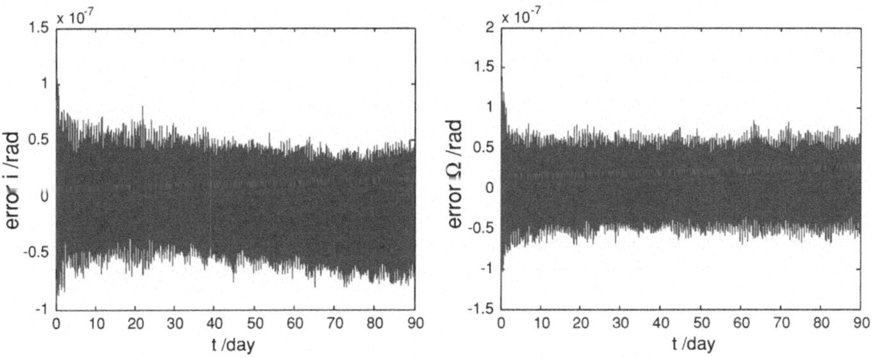

Fig. 6 The errors of orbital inclination i and right ascension of ascending node Ω after the overall pseudo-rotation correction

Revised the orbital inclination and ascending node right ascension error in Fig. 6.

Figures 5 and 6 shows, the overall pseudo-rotation correction algorithm can effectively eliminate the overall pseudo-rotation of the constellation of the end of 90 days, the URE accuracy of 1.4 m, RTN three-dimensional position error error level remained stable, non-divergent trends; the overall pseudo-rotation correction algorithm can effectively eliminate the overall pseudo-rotation errors.

5 Conclusions

Cause pseudo-rotation of the constellation as a whole because its elimination has been studied through analysis and simulation, the following conclusions:

1. Four inter-satellite ranging system error, the error of the systematic errors of short-period and long-cycle system will not cause the overall pseudo-rotation. And constant systematic errors may cause the overall pseudo-rotation, linear systematic error is usually caused by the overall pseudo-rotation, when the constant systematic error for each satellite in the same direction or linear systematic errors, the constellation will happen to the overall pseudo-rotation.
2. The overall pseudo-rotation correction algorithm can effectively eliminate the pseudo-rotation of the constellation as a whole, this method is also beneficial to have "anchor stations" to support inheritance of autonomous orbit determination algorithm, and to facilitate the processing of the constellation as a whole rigid rotation.

Based on the above conclusion, this study suggests the implementation process of the autonomous navigation project, the overall pseudo-rotation correction algorithm to eliminate a whole constellation of pseudo-rotation, on the one hand, to improve the accuracy of the ground uploaded long-term ephemeris, in particular, the accuracy i and Ω [4]. On the other hand, use a whole pseudo-rotation correction algorithm for correction, and ultimately the effective elimination of the overall pseudo-rotation of the constellation.

References

1. Abusali, P. A. M., et al. (1998). Autonomous navigation of global positioning system satellites using crossing-link measurements. *Journal of Guidance, Control, and Dynamics, 21*(2), 321–327.
2. Rajan, J. A. (2002). Highlights of GPS II-R autonomous navigation. *ION 58th Annual Meeting/ CIGTF 21th Guidance Test Symposium, Albuquerque, NM, 24–26, June 2002.*
3. Fisher, S. C., & Kamran, G. (1999). GPS IIF—The next generation. *Proceeding of the IEEE, 87*(1), 24–47.
4. Chen, J., Jiao, W., Ma, J. et al. (2005). Autonav of navigation satellite constellation based on crosslink range and orientation parameters constraining. *Geomatics and Information Science of Wuhan University, 30*(5), 439–443.
5. Wang, H., Chen, Z., Zheng, J., et al. (2011). A new algorithm for on-board autonomous orbit determination of navigation satellites. *Journal of Navigation, 64*, S162–S179.

6. Lin, L., & Yingchun, L. (2000). The deficient-rank problem of autonomous orbit determination based on satellite-to-satellite relative measurement. *Journal of Spacecraft TT and C Technology*, *19*(3), 13–16.
7. Zhu, J., Liao, Y., & Wen, Y. (2009). The integrated autonomous orbit determination of the navigation constellation based on crosslink range and ground-based emitter. *Journal of National University of Defense Technology*, *31*(2), 15–19.
8. Mean, M. D., & Bernstein, H. (1994). Ephemeris Observability Issues in the Global Positioning System Autonomous Navigation[C]. IEEE Position Location and Navigation Symposium, New York, 677–680.

Analysis and Modeling of PPP Residuals from GPS and GLONASS

Qianxin Wang, Yingyan Chen and Jing Zhao

Abstract The PPP results are different for the same station at the same time when using different GNSS observations. The reason is that there are some different system errors in every navigation system, such as satellite orbit error, satellite clock error, frequency drift, etc. In this paper, a lot of data from GPS and GLONASS mixed observation stations are processed using CSRS-PPP software. The results show that the positioning precision of GPS is better than that of GLONASS and the positioning precision of the N component is better than those of the E and U component. There are some regular varieties in the residuals, which includes the trend items and periodic items. Therefore, an error compensation model is constructed. The experiment results show that the PPP precision of GPS and GLONASS can be improved by using the proposed model.

1 Introduction

Following the rapid development of GNSS (such as GPS, GLONASS, COMPASS, GALILEO, QZSS, etc.), the study of GNSS is moving from only GPS into multi-GNSS. Due to the different system errors in every navigation system, the

Q. Wang (✉) · Y. Chen
Institute of Geodesy and Geodynamics, Chinese Academy of Surveying and Mapping, 100830 Beijing, People's Republic of China
e-mail: wangqianxin8012@163.com

Q. Wang
School of Geosciences and Info-Physics, Central South University, Changsha 410083, People's Republic of China

J. Zhao
National Remote Sensing Center of China, 100036 Beijing, People's Republic of China

J. Sun et al. (eds.), *China Satellite Navigation Conference (CSNC) 2012 Proceedings*, Lecture Notes in Electrical Engineering 160, DOI: 10.1007/978-3-642-29175-3_27,

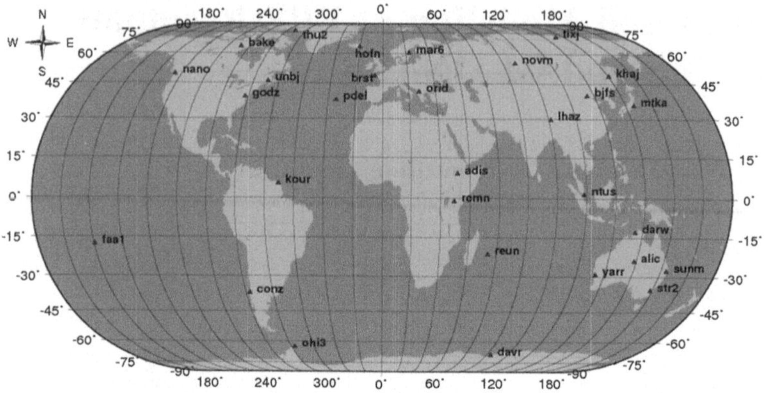

Fig. 1 Locations of the chosen GPS/GLONASS mixed stations

positioning results will be different for the same station at the same time, using different GNSS observation [1–5]. What are the characters of residuals from the different GNSS? How to model these residuals? And what are the physical significations of above phenomenon? It is very helpful for us to comprehend and apply the multi-GNSS in the future that making clear above questions. In this paper, one month observation data of 30 mixed GPS/GLONASS stations are processed using CSRS-PPP software, and the residuals are analyzed carefully. Some regulations of residual variety are concluded based on the statistical results. And an error compensation model is constructed for removing these systematic errors. The results show that the PPP precision of GPS and GLONASS can be improved by using this model.

2 Data Processing and Analysis

For analysis of the PPP residuals from GPS and GLONASS, 30 mixed GPS/GLONASS observation stations are chosen from about 200 IGS stations, which are well-distributed on the Earth. Figure 1 shows that the locations of these chosen stations. And 30 days observation data of these stations are collected for resolution. The data sample is 30 s. The CSRS-PPP software is used to processing these data. CSRS-PPP is a high-precision GNSS point positioning software, which is developed by Natural Resources Canada [6, 7]. In the data processing, every 6 h data is solved once. Figure 2 is the PPP results of BJFS station using 6 h GPS and GLONASS data. The results show the PPP solution has been convergence and the internal precision is better than 1 cm. Therefore, it can represent the positioning precision of each navigation system objectively. The precise ephemeris are used, where the precisions of satellite orbit and clock error of GPS and GLONASS are better than 2 cm, 15 cm, 0.1 ns, 1.5 ns, respectively [8] .

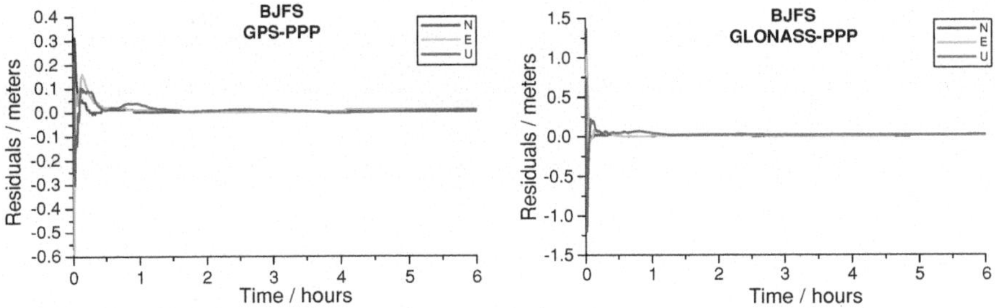

Fig. 2 The PPP results of BJFS station using 6 h data

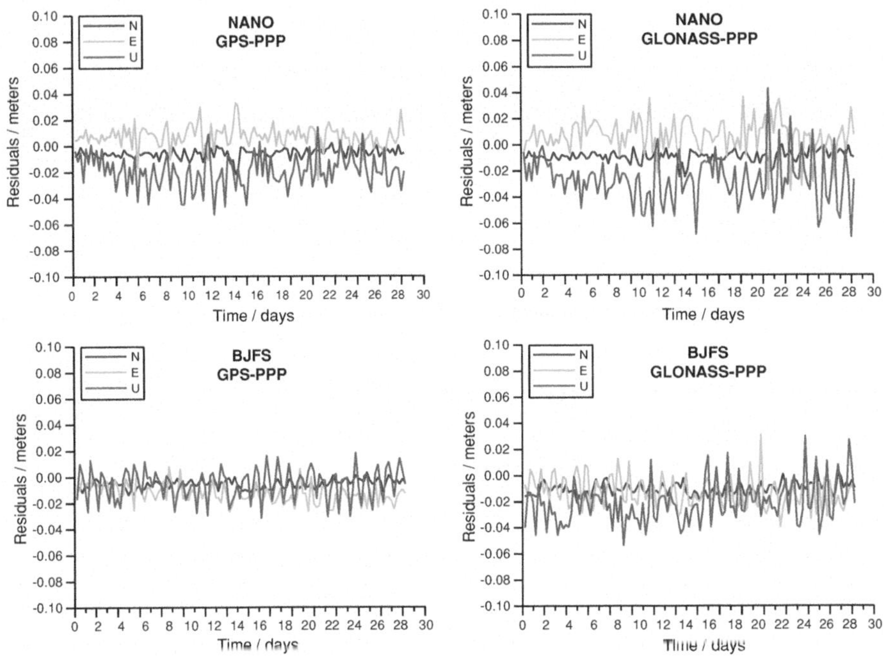

Fig. 3 PPP residuals of two representative stations on the Northern Hemisphere

All calculation results are compare with the IGS published coordinates. And the solution residuals of all stations are obtained. Figure 3 shows the residual variety of two representative sites on the Northern Hemisphere in 30 days. Figure 4 is the residual variety of two representative sites on the Southern Hemisphere in 30 days. Table 1 is the average value of PPP precision of all stations.

Based on the above experiments and results, some conclusion can be obtained. In general, the PPP precision of GPS is better than that of GLONASS, and the positioning precision of the N component is better than those of the E and U components. Secondly, there are some regular varieties in the residuals of all

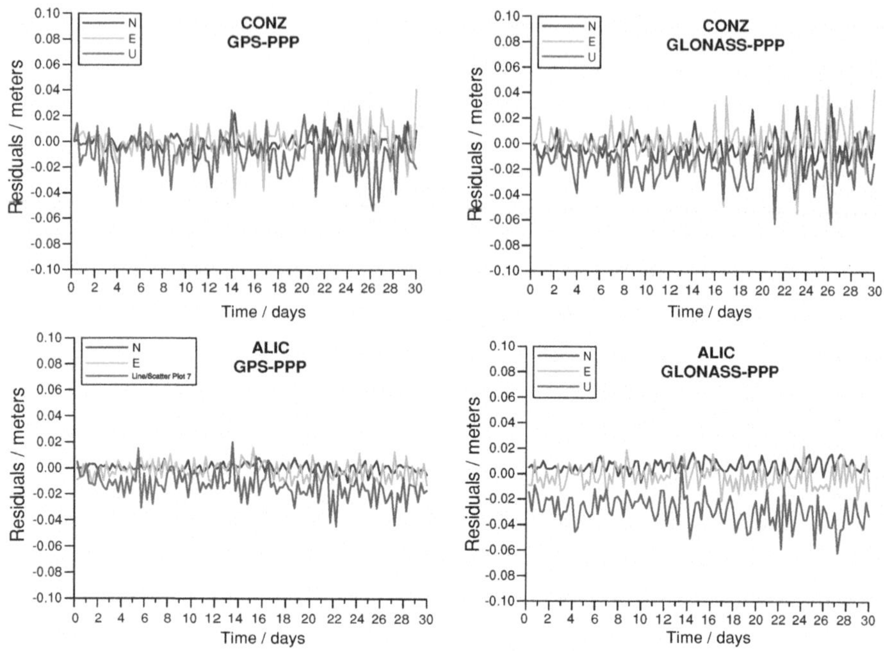

Fig. 4 PPP residuals of two representative stations on the Southern Hemisphere

Table 1 The average value of PPP precision of 30 stations (unit: meters)

RMS	N	E	U
GPS	0.008	0.015	0.027
GLONASS	0.013	0.018	0.039

stations, which include some trend and periodic variety. And there is about 2 cm bias in the positioning results of the U component. Additionally, there is not obvious relationship between the location of station and the residual character. The residual variety of 30 stations on the N, E, and U components are similar.

3 Model of PPP Residuals

Since there are some regular varieties in the PPP residuals of GPS and GLONASS, one suitable model can be constructed for modeling the residual variety. This model includes two parts: one is the trend items and the other is the periodic items, see Eq. (1).

$$v(t) = a + bt + \sum_{n=1}^{m} (A_n \sin(\omega_n t) + B_n \cos(\omega_n t)) \quad (1)$$

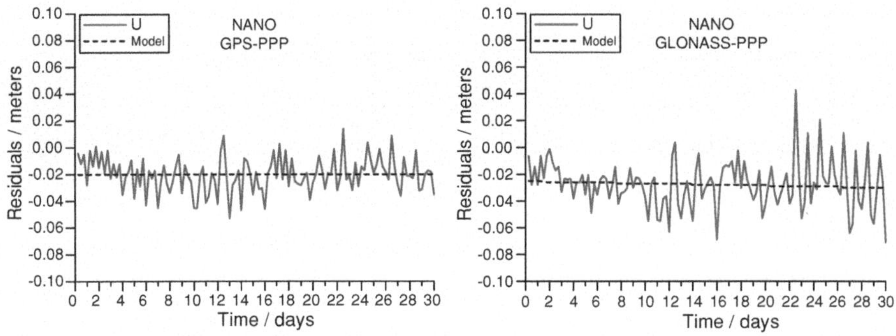

Fig. 5 The fitting results of residual on the U component using trend items

Where, $v(t)$ is the correction value at the t moment; a, b are the factors of linear function; A_n, B_n, are the amplitudes; $\omega_n = 2\pi f_n$, $1/f_n$ notes the period, m = 1, 2,…n. The calculation method as follows: firstly, the factors a, b are obtained by the least square adjustment; then the main periods $1/f_n$ are obtained by the spectrum analysis; finally, the amplitudes A_n, B_n are obtained based on Eq. (1). Due to the limited space, the fitting results of NANO station on the U component are listed. Figure 5 shows the fitting result of residuals based on the trend items.

It can be see that there is about 2 cm bias in the residuals of GPS and GLONASS on the U component from the fitting results of Fig. 5. And the bias can be found in the residuals of all stations. In general, there are three factors which affect GNSS positioning precision on the U component. The 1st factor is the GDOP value; the 2nd factor is the tropospheric delay error; the 3rd factor is the antenna phase center bias. In our experiments, the bias probably is come from the tropospheric delay error correct model of CSRS-PPP software.

Figure 6 shows the fitting results of residuals after the trend items are removed based on the periodic items. The results of spectrum analysis show that the main periods of GPS and GLONASS are 12 and 24 h, respectively. The time is similar with the period of satellite orbit. Therefore, the errors probably are from the errors of GNSS satellite orbit. The fitting precision of Eq. (1) is about 8 mm for the residuals of NANO station on the U component. And the fitting precision of the N and E components are better than 1 and 3 mm, respectively.

4 Validation of Model

In order to verify the validation of proposed model, the next two days data of NANO station are used. The method of data processing is same with that of above section. Figure 7 shows the positioning residuals from CSRS-PPP and the correction values from the proposed model. Table 2 shows the statistical results of positioning precision using CSRS-PPP and adding the correction values of proposed model.

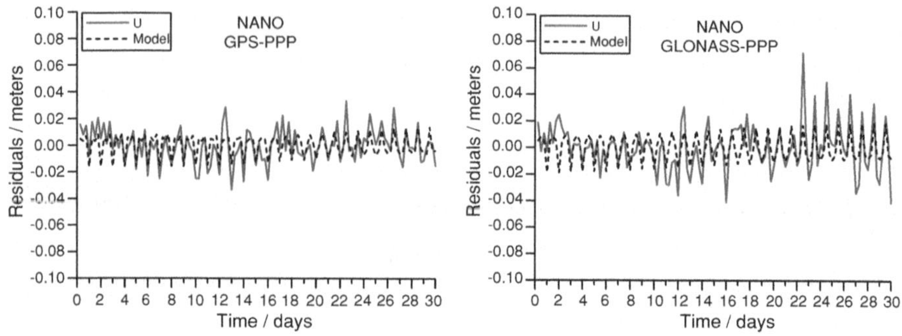

Fig. 6 The fitting results of residual on the U component using periodic items

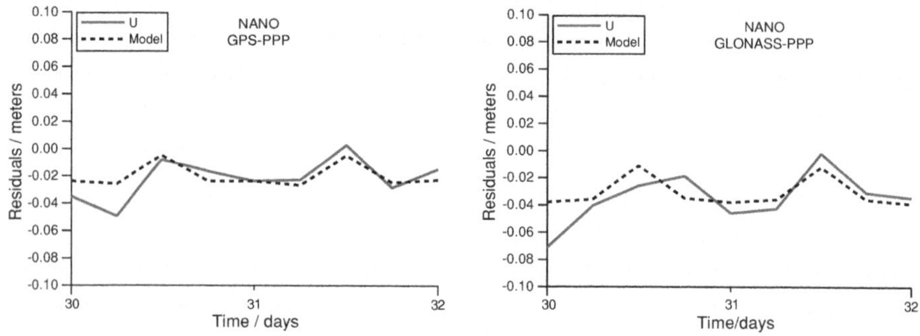

Fig. 7 The residuals of CSRS-PPP and correction values of proposed model

Table 2 The statistical results of PPP precision of NANO using CSRS-PPP and proposed model (unit: meters)

	CSRS-PPP			Adding correction model		
RMS	N	E	U	N	E	U
GPS	0.003	0.012	0.035	0.001	0.006	0.011
GLONASS	0.008	0.014	0.043	0.005	0.007	0.013

The experiment results show the PPP precision of GPS and GLONASS can be improved after using the proposed model. Certainly, this conclusion is drawn based on the results of CSRS-PPP. It should be further verified whether there is the similar regulation in the other software.

5 Conclusions

There are many advantages using multi-GNSS to navigation and positioning relative to the single GNSS. However, it brings some new challenges and problems at the same time. It is very important that study of the similarities and differences in the different GNSS and summarizing the characters and regulations of each GNSS for utilizing them better. In this paper, the characters of PPP residuals of GPS and GLONASS are analyzed and compared carefully. Some conclusions can be drawn:

(1) In general, the PPP precision of GPS is better than that of GLONASS and the positioning precision of the N component is better than those of the E and U component.
(2) There is not obvious relationship between the location of station and the residual character. The residual characters of GPS and GLONASS are similar on the N, E, and U components.
(3) There are some regulations in PPP residuals of GPS and GLONASS based on the results of CSRS-PPP. Therefore, the correction model can be constructed, which include the trend and periodic items.
(4) The results of spectrum analysis show that the main periods of residuals from GPS and GLONASS are 12 h and 24 h, respectively. The time is similar with the period of satellite orbit. Therefore, it is very important that improving the precision of GNSS orbit for PPP.
(5) The experiment results show the PPP precision of GPS and GLONASS can be improved by using the proposed correction model.

Acknowledgments This work was supported by the Key Laboratory of Mapping from Space, National Administration of Survey Mapping and Geo-information (K201104), the National Science and Technology Pillar Program (2012BAB16B01), the National High Technology Research and Development of China (2011AA120503).

References

1. Wang, J. (2001). *Modeling and quality control for GPS and GLONASS satellite positioning* (Vol 61). University of New South Wales. UNISURV Report S.
2. Wang, J., Rizos, C., Stewart, M. P., & Leick, A. (2001). GPS and GLONASS integration: modeling and ambiguity resolution issues. *GPS Solutions, 5*(1), 55–64.
3. Wang, J. (1999). Stochastic modeling for real-time kinematic GPS/GLONASS positioning. *Navigation, 46*(4), 297–305.

4. Chen, J., Dang, Y., & Cheng, P. (2007). Development and progress in GNSS. *Journal of Geodesy and Geodynamics, 27*(5), 1–4.
5. Wang, Q., & Xu, T. (2010). Combining GPS carrier phase and Doppler observations for precise velocity determination. *Science China, Series G, 54*(6), 1022–1028.
6. Tsakiri, M. (2008). GPS processing using online services. *Journal of Surveying Engineering, 134*(4), 115–125.
7. Pierre, T., Jan, K., Pierre, H., & Patrick, L. (2005). CSRS-PPP: An internet service for GPS user access to the Canadian spatial reference frame. *Geomatica, 59*(1), 17–28.
8. Griffiths, J., & Ray, J. R. (2008). On the precision and accuracy of IGS orbits. *Journal of Geodesy, 83*(3), 277–287.

Improved Pseudorange Smoothing Method for Standing Multipath Mitigation

Bo Chen, Xuanying Zhou, Dechen Yin, Xiaoxiao Ma, Han Yu, Xiaojun Duan and Jiying Liu

Abstract In satellite-based augmentation system (SBAS), because of limited satellite-station geometry change, some multipath interference varies very slowly and repeats at a time shift from several hours to one day. To mitigate the interference in real time, which is needed in Augmentation System, the code noise and multipath (CNMP) method was proposed, where the smoothed pseudorange is used to estimate the "standing" multipath. But some experiments show that the left error still looks with the same periods as original multipath. An improved pseudorange smoothing method is proposed to eliminate the left error. Firstly, the differential equation about the left error and the standing multipath is deduced, which can be solved to get the exact expression of the standing multipath. Following the formula, the corresponding real time filtering procedure is designed by modifying the CNMP method. And the experiments show that the proposed method can actually cancel the periodic component in the standing multipath.

1 Introduction

Multipath is a significant and sometimes dominant contributor to error in current Global Navigation Satellite System (GNSS). Multipath is corresponding to characterize of signal, the processing method in the receiver, the antenna and receiving scenario, which makes it quite difficult to eliminate multipath using difference or a common model. Furthermore, it is noticed that multipath of geostationary (GEO)

B. Chen (✉) · X. Zhou · D. Yin · X. Ma · H. Yu · X. Duan · J. Liu
College of Science, National University of Defense Technology, Changsha, Hunan province, China
e-mail: chenbo0354@yahoo.com.cn

J. Sun et al. (eds.), *China Satellite Navigation Conference (CSNC) 2012 Proceedings*, Lecture Notes in Electrical Engineering 160, DOI: 10.1007/978-3-642-29175-3_28,

satellites in SBAS has some "standing" components, the periods of which are from several hours to one day and much longer than a few minutes of the medium Earth orbit (MEO) satellites. And it is because of limited GEO satellite-station geometry change [1, 2].

The combination of radio frequency and software approaches are pursued to reduce multipath. The software methods could be classified into time domain processing and frequency domain processing. Bartone and Rizos' frequency methods are based on the separability of different errors of GNSS data in frequency domain and multipath is mitigated by suppressing the corresponding frequency in Fourier or wavelet domain [3–5]. While the frequency of multipath of the GEO satellite is much more lower, which may lap over those of ionosphere, troposphere and so on, and it is difficult to separate them in frequency domain. Furthermore, it needs the data in one period at least to calculate its spectrum. For the multipath of one day period, the frequency domain method needs a whole day to initialization.

The time domain processing can be done in two different ways. The first one is developed by exploiting the repeatability of multipath errors between successive sidereal days for a fixed-location receiver, and the multipath is estimated and cancelled with that one or several periods before [6, 7]. But because the repeatability is just some approximate, it is not entirely satisfy. The other one, code noise and multipath (CNMP), is proposed for real time processing in Wide Area Augmentation System (WAAS) [8]. Because multipath in the carrier-phase measurement is quite smaller, the multipath in pseudorange can be cancelled in the similar way as pseudorange smoothing. But unfortunately, in our experiments (See Sect. 3) it is found that the smoothed code in CNMP still has some error left, which has the same periods as standing multipath.

In this paper we proposed an improved method to eliminate the left error. Firstly, CNMP is introduced in Sect. 2. Then the formula of the left error is deduced and its periodicity is explained in Sect. 3. Based on the formula, a modified filtering method is given and the experiment results are shown in Sect. 4.

2 CNMP Method

The double frequency code and carrier-phase measurement are used in CNMP. When multipath in carrier is ignored, the ionosphere delay (with ambiguity) can be calculated using double frequency carrier-phase. Then the code and carrier-phase pseudorange are got, the difference of which mainly contains code multipath, noise and ambiguity. The ambiguity can be estimated using smoothed code and carrier-phase by filtering. At last the multipath can be calculated and subtracted from the code pseudorange. The code and carrier-phase pseudorange can be expressed as:

$$C_i = r + \frac{I}{f_i^2} + \delta_{trop} + C(\tau_r - \tau_s) + MP_i + \varepsilon_{pi},$$

$$P_i = r - \frac{I}{f_i^2} + \delta_{trop} + C(\tau_r - \tau_s) + N_i\lambda_i + \varepsilon_{Li}, \quad i = 1, 2.$$

C_i and P_i are the code and carrier-phase pseudorange at the ith frequency. r is the real distance between the satellite and the receiver. I is the parameter of the ionosphere delay. δ_{trop} is the troposphere delay. τ_r and τ_s are the receiver clock error and the satellite clock error. MP_i is multipath. ε_{pi} is noise in code pseudorange. ε_{Li} is noise in carrier-phase pseudorange. N_i is the ambiguity.

Using the dual-frequency code and carrier, multipath can be calculated as

$$\begin{aligned} MPN_i &= C_i + \frac{f_j^2 + f_i^2}{f_j^2 - f_i^2} P_i + \frac{2f_j^2}{f_i^2 - f_j^2} P_j = MP_i + N + \varepsilon \\ &= \left(C_i - \frac{I}{f_i^2} \right) - \left(P_i + \frac{I}{f_i^2} \right) = C_i^* - P_i^*, \\ I &= \frac{f_i^2 f_j^2}{f_i^2 - f_j^2} (P_i - P_j), \quad i, j = 1, 2, i \neq j. \end{aligned} \tag{1}$$

From formula (1), it can be seen that in the difference (MPN_i) of code pseudorange (C_i^*, no ionosphere delay) and carrier-phase pseudorange (P_i^*, no ionosphere delay), clock error and troposphere error are canceled, but the ambiguity (N) and noise (ε) are still left, among which N is the main one, and should be estimated and reduced.

The ambiguity can be taken as the mean of MPN_i. While in real time processing, N needs to be calculated using exponential filter as in (2).

$$\begin{aligned} \overline{N}(0) &= C_i^*(0) - P_i^*(0), \\ \overline{N}(t) &= \frac{t-1}{t} \overline{N}(t-1) + \frac{1}{t} \left(C_i^*(t) - P_i^*(t) \right), \\ \overline{MP}_i &= MPN_i - \overline{N}. \end{aligned} \tag{2}$$

In formula (2) the bar on the top of $\overline{N}$ and $\overline{MP}_i$ means they are only the estimation of real value. And actually $\overline{N}$ approximates the mean of MPN_i gradually. And it should be noted that the cycle slip detection and repair must be done before filtering.

3 The Improved Pseudorange Smoothing Method

When using CNMP to mitigate multipath in pseudorange of GEO satellite,we find that the left multipath in the processed pseudorange still looks as a periodical function, while its amplitude decreases. (See blue dashed curve in Fig. 2). And we find the reason of the periodicity of the left multipath is that in CNMP the multipath must can be looked as white noise, but the standing multipath in GEO is

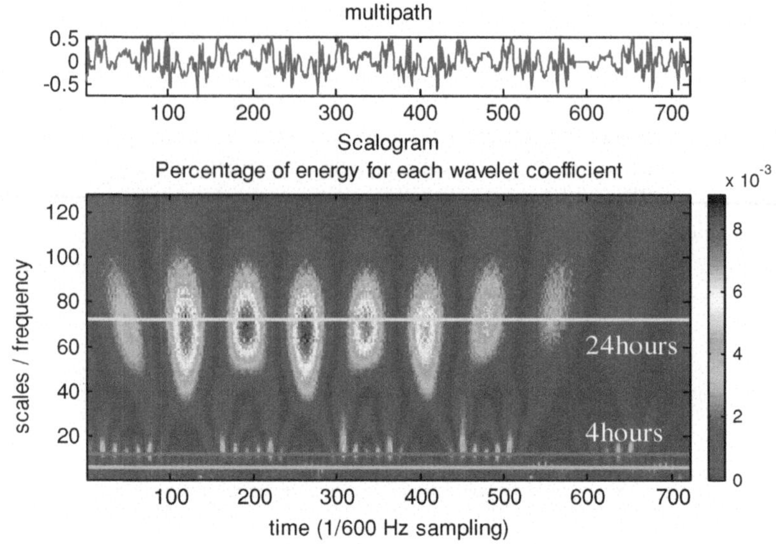

Fig. 1 Multipath of GEO and its scalogram

absolutely not (See Fig. 1, in the wavelet scalogram of standing multipath, components with periods about 24 and 4 h are shown clearly. Because the Fourier transform can not process the time variable characterize that may exist in multipath (different frequencies will be aliased using global Fourier transform), Wavelet transform is used here.) Now, we will explain the periodicity of the left multipath by deducing its expression.

Using formula (2) repeatedly, we can get

$$\overline{N}(t) = \frac{\sum_{s=0}^{t}\left(C_i^*(s) - P_i^*(s)\right)}{t} = \frac{\sum_{s=0}^{t} MPN_i(s)}{t}. \tag{3}$$

Considering the continuous/integral form of formula (3), and substitute (1) in, the result is

$$\begin{aligned}\overline{N}(t) &= \frac{\int_{s=0}^{t} MPN_i(s)ds}{t} \\ &= \frac{\int_{s=0}^{t} (MP_i(s) + N + \varepsilon)ds}{t} \\ &= \frac{\int_{s=0}^{t} MP_i(s)ds}{t} + N,\end{aligned} \tag{4}$$

In formula (4) ε is white noise, so its integral equals 0. The estimation error of ambiguity (difference between real value N and asymptotic estimation $\overline{N}$) can be expressed as the mean of the multipath in integral form. When MP_i is periodical but not white noise, its mean, the error, is just a periodical function with decreasing amplitude (considering the mean of sine function).

Furthermore, combining formula (1), (2) and (4), we can deduce the integral function about the real multipath and its estimation value.

$$MP_i(t) - \overline{MP_i}(t) = \frac{\int_{s=0}^{t} MP_i(s)\mathrm{d}s}{t}, \tag{5}$$

where $\overline{MP_i}$ can be calculated using CNMP method. And the analytic solution of MP_i is

$$MP_i(t) - MP_i(0) = \overline{MP_i}(t) + \int_{s=0}^{t} \frac{\overline{MP_i}(s)}{s} ds. \tag{6}$$

But because the initial value $MP_i(0)$ of multipath is unknown, formula (6) cannot be used to correct $\overline{MP_i}$ directly.

Noticing that $\overline{MP_i}$ is already an approximation to multipath, it can be used to substitute MP_i in the integral in formula (5), then we can get the explicit expression of the corrected multipath $\left(\widetilde{MP}_i(t)\right)$

$$\widetilde{MP}_i(t) = \overline{MP}_i(t) + \frac{\int_{s=0}^{t} \overline{MP_i}(s)\mathrm{d}s}{t}, \tag{7}$$

And the corresponding discrete form is

$$\widetilde{MP}_i(t) = \overline{MP_i}(t) + \frac{\sum_{s=0}^{t} \overline{MP_i}(s)}{t}. \tag{8}$$

According formula (8) about the modified multipath estimation, $\widetilde{MP}_i(t)$ can be calculated only by adding the asymptotic mean of $\overline{MP_i}$ to itself, which can be done after the formula (2) in CNMP procedure. So $\widetilde{MP}_i(t)$ can also be calculated by filtering procedure slightly different form CNMP, and the increasing of the amount of computation is quite limited. As CNMP, the improved multipath estimation method can also be used in real time processing.

4 Experiments and Results

The improved method proposed here and CNMP are both used to reduce the multipath in the pseudorange of GEO satellite. In Fig. 2, the typical processing results are shown. The green continuous curve is the multipath (ambiguity is taken as the mean of the cycle slip repaired MPN_i), whose component with periods of one day looks obvious. The blue dashed line is the left multipath processed by CNMP method. The periodicity keeps, and its amplitude decreases. The red dotted curve is processed using our improved filtering, the periodicity is eliminated.

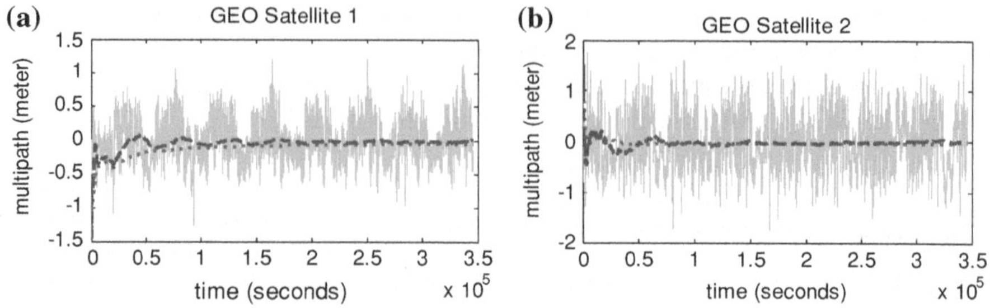

Fig. 2 Mitigated multipath using CNMP and the improved method. The *green continuous curve* is the multipath. The *blue dashed curve* is the one mitigate using CNMP, and the *red dotted curve* is mitigate using our improved method

The multipath mitigation method proposed in this paper can be implemented in real time as CNMP, and can further eliminated the left periodicity in CNMP, which ban be used in GEO-satellite based Navigation or Augmentation system, especially the real time processing scenario.

Acknowledgments At last, we must express our acknowledgment to Professor Hu Xiaogong, Doctor Chao Yueling in SHAO for their help on the analysis and implementation about CNMP. Professor Cui Xiaowei in Tsinghua University also inspires us on the nature of standing multi-path.and we thank him at the same time. This research is supported by National Natural Science Foundation of China, No. 60974124 and by Program for NCET.

References

1. Schempp, T., Bruke, J., & Rubin, A. (2008). WAAS benefits of GEO ranging. In G. A. Savannah(Ed.), *Proceedings of the 21st International Technical Meeting of the Satellite Division of The Institute of Navigation* (*ION GNSS 2008*), September 2008, pp. 1903–1910.
2. Wanninger, L., & Stephan, W. -F. (2007). Combined processing of GPS, GLONASS, and SBAS code phase and carrier phase measurements. In T. X. Fort Worth (Ed.), *Proceedings of the 20th International Technical Meeting of the Satellite Division of The Institute of Navigation* (*ION GNSS 2007*), September 2007, pp. 866–875.
3. Zhang, Y., & Bartone, C. (2004). *Multipath Mitigation in the Frequency Domain, Position Location and Navigation Symposium*, pp. 486–495.
4. Bartone, C. G., & Zhang,Y. (2005). Real-time code multipath mitigation in the frequency domain using Fdsmooth for global navigation satellite systems, US Patent Application 2005021621.
5. Satirapod, C., & Rizos, C., (2003). Multipath mitigation by wavelet analysis for GPS base station applications. *Survey Review*, *38*(295).
6. Axelrad, P., Larson, K., & Jones, B. (2005). Use of the correct satellite repeat period to characterize and reduce site-specific multipath errors. In *Proceedings of ION GNSS 2005* pp. 2638–2648.
7. Miceli, R. J., Psiaki, M. L., O'Hanlon, B W., & Chiang, K. Q. Z. (2011). Real-time multipath estimation for dual frequency GPS ionospheric delay measurements. In *Proceedings of the*

24th International Technical Meeting of The Satellite Division of the Institute of Navigation (*ION GNSS 2011*), Portland, OR, September 2011, p. 1173.

8. Shallberg, K. et al. (2001). WAAS measurement. In *Processing of the Reducing the Effects of Multipath, Proceedings of the 14th International Technical Meeting of the Satellite Division of The Institute of Navigation* (*ION GPS 2001*), Salt Lake City, UT, September 2001, pp. 2334–234.

Bayesian Methods for Cycle Slips Detection Based on Autoregressive Model

Qianqian Zhang, Qingming Gui, Jianwen Li, Yisong Gong and Songhui Han

Abstract A new approach, Bayesian method for single frequent cycle slips detection based on autoregressive model, is presented by exploiting modern Bayesian statistical theory. Besides, this paper deals with the problem of masking and swamping about cycle slips detection in a thorough new conception and gives the corresponding solution. First of all, considering the character of cycle slips in phase data and the relations between cycle slips and outliers, this paper proposes a Bayesian method for cycle slips detection based on the posterior probabilities of classification variables in the respective of Bayesian hypothesis. Secondly, an adaptive Gibbs sampling algorithm is designed through analyzing the reasons of masking and swamping about cycle slips detection. Then an unmasking Bayesian method for cycle slips detection is proposed. Thirdly, accurate estimation of cycle slips is given based on Bayesian point estimation. Finally, the method is applied to real phase data to test its correction and efficiency.

Q. Zhang (✉) Q. Gui · S. Han
Institute of Science, Information Engineering University,
62 Kexue Road, 450001 Zhengzhou, China
e-mail: zhangqianqian0216@163.com

Q. Gui · J. Li · S. Han
Institute of Surveying and Mapping, Information Engineering University,
66 Middle Longhai Road, 450052 Zhengzhou, China
e-mail: guiqm@public.zz.ha.cn

J. Li
e-mail: zzljw@126.com

S. Han
e-mail: hansonghui@126.com

Y. Gong
61618 Troops, 102102 Beijing, China

J. Sun et al. (eds.), *China Satellite Navigation Conference (CSNC) 2012 Proceedings*, Lecture Notes in Electrical Engineering 160, DOI: 10.1007/978-3-642-29175-3_29,

1 Introduction

Cycle slips is a common phenomenon in carrier phase data, and it is also an important session that must be handled in highly accurate positioning and navigation solutions [16, 22]. Currently, some approaches concerning cycle slips detection include: High-order differences [16], polynomial fitness [16], combination of pseudorange and carrier phase [22], Kalman filter method [11] and so on in domestic. Besides, there are many methods in abroad, such as methods utilizing double difference observations, methods exploiting geometry-free combination and ionosphere-free combination [7, 18] and methods of read-time cycle slips detection [1, 3] in GPS dynamic geodesy.

It is noticeable that although these cycle slips detection methods can be able to solve problems in the corresponding practical fields and are different in characters, they have two common disadvantages. Firstly, prior information is not considered. Ignoring the exploitation of prior information is not only profligacy but also it can lead to unreasonable conclusions [13, 14]. Bayes methods which connect prior information with sampler information have already successfully been used into other aspects of GNSS data processing, such as ambiguity resolution [2, 8], construction of confidence regions for GPS baselines [10] and so on. However, it is seldom discussed that Bayes methods can be utilized to solve the problem of GNSS cycle slips detection in surveying field. Secondly, the characters of cycle slips and their relations with outliers are not analyzed deeply. Not to mention to notice that when cycle slips create patches of additive outliers, masking and swamping often occur by utilizing unreasonable detection methods, so that conclusions will be wrong. What's worse, it can yield bad results. Masking [9, 20, 21] is that some certain outliers cause other outliers not to be detected, so that not all of the outliers can be detected or even all of the outliers can not be detected. Swamping [20] which can also be called amplification phenomenon means that because of the appearance of some unusual outliers, some correct observations are misspecified into outliers. Thus the positions of outliers are not in accord with the actual locations of outliers. In the detection process of patches of additive outliers, either many scholars are not fully acquainted with the existence and seriousness of masking and swamping or they do not analyzed clearly about the cause of it. Thereby, the corresponding measures which can prevent the occurrence of masking and swamping effectively are not adopted in the process of designing, constructing and implementing cycle slips detection methods.

Therefore, this article proposes a new method for cycle slips detection, Bayes methods for cycle slips detection based on autoregressive model, by means of analyzing the relations of cycle slips and outliers. Section 2 describes the characteristic analysis of cycle slips. In Sect. 3, a Bayesian method for cycle slips detection based on the posterior probabilities of classification variables is proposed in the respective of Bayesian hypothesis. The calculation method of the posterior probabilities of classification variables and the estimation approach of outliers are given by Gibbs sampling. In Sect. 4, this paper deals with the problems of masking and swamping about cycle slips detection in a thorough new conception and give the corresponding solution. An adaptive Gibbs sampling algorithm is designed through analyzing the reasons of masking and swamping in cycle slips detection. Besides, accurate estimation of cycle slips is given based on Bayesian point estimation. In Sect. 5, the numerical experiments made on real data in order to test the correction and efficiency of the method are presented.

2 Characteristic Analyses of Cycle Slips

2.1 Phase Date and its Preprocessing

This paper adopt the following observation equation [16]

$$L_i^k(t) = D_i^k(t) + \lambda N_i^k - I_i^k(t) + (a_i^k)_L(t) \tag{1}$$

where, $L_i^k(t)$ is phase pseudorange between satellite k and receiver i at epoch t, $D_i^k(t)$ is the pseudo geometric distance which contains satellite clock error, receiver clock error and tropospheric delay, N_i^k is ambiguity, λ is wavelength, $I_i^k(t)$ is ionospheric delay, $(a_i^k)_L(t)$ is white noise.

Generally speaking, differential or dual-differential carrier phase data between satellites and receivers are utilized to weaken errors [16]

$$\begin{aligned} L_i^k(t) - L_j^k(t) =& \lambda(N_i^k - N_j^k) + D_i^k(t) - D_j^k(t) - (I_i^k(t) - I_j^k(t)) \\ &+ (a_i^k)_L(t) - (a_j^k)_L(t) \\ L_i^k(t) - L_i^l(t) =& \lambda(N_i^k - N_i^l) + D_i^k(t) - D_i^l(t) - (I_i^k(t) - I_i^l(t)) \\ &+ (a_i^k)_L(t) - (a_i^l)_L(t) \\ L_{ij}^{kl} =& (L_i^k(t) - L_j^k(t)) - (L_i^k(t) - L_i^l(t)) \end{aligned} \tag{2}$$

It can be known from (2) that satellite clock error, receiver clock error, tropospheric delay and ionospheric delay can be weakened largely by using differential or dual-differential carrier phase data especially in the situation of high sampling rate. Only multipath effect and measurement noise have been increased. Since the measurement noise of phase data is very low, the observations that have been preprocessed above contain a low noise level except the cycle slips under a well surveying environment.

What need to be stated is that this paper only concerns the characters of cycle slips based on the preprocessed differential or dual-differential phase data, because baseline solution and network adjustment usually utilize differential or dual-differential phase data. That is to say, when detecting the locations of cycle slips, the positions of cycle slips in differential or dual-differential phase data are given in this paper and the repairs of cycle slips are also based on differential or dual-differential phase data.

2.2 The Relations Between Cycle Slips and Outliers

(1) Outliers caused by cycle slips are isolated additive outliers

Generally speaking, differential or dual-differential phase data between satellites and receivers are non-stationary time series. According to the theories of time series [4], if we want to fit the time series by autoregressive model, non-stationary time series should be changed into stationary time series. The ordinary approach is differencing the time series between consecutive epochs. If the one-differential time series become stationary, the outliers in the epoch differential phase data caused by cycle slips are isolated additive outliers [19]. The reason is that if a cycle slip occurs between epoch i and epoch $i+1$ and the size of this cycle slip is δN cycles, the error of the observation at epoch $i+1$ is δN cycles. Therefore, the error in the one-differential observation between epoch i and epoch $i+1$ is δN cycles. Since cycle slips are inherited, the next and the following one-differential observations are normal. That is to say, a cycle slip in differential or dual-differential phase data between satellites and receivers only makes one of the one-differential observations between epochs wrong. The wrong observation above is isolated additive outlier.

(2) Outliers caused by cycle slips are patches of additive outliers

If the stationary process of the differential or dual-differential phase data between satellites and receivers need multiple differences between consecutive epochs, then a cycle slip leads to a patch of additive outliers in multi-differential observations between consecutive epochs according to the analysis of (1). Of course, continuous cycle slips are more likely to cause patches of additive outliers.

3 Bayes Method of Cycle Slips Detection

3.1 Bayes Model and Principle of Cycle Slips Detection

It is assumed that the differential observations between epochs are stationary, and they are fit for the following $AR(p)$ model:

$$\begin{cases} x_t = \phi_1 x_{t-1} + \cdots + \phi_p x_{t-p} + a_t \\ a_t \quad i.i.d. \quad N(0, \sigma^2) \end{cases} \tag{3}$$

where, $\phi = (\phi_1, \ldots, \phi_p)^T$ and σ^2 are unknown parameters, assuming that the prior distributions [4] of them are $\phi \sim N_P(\phi_0, V^{-1})$, $\sigma^2 \sim IG(\upsilon/2, \upsilon\lambda/2)$ and $\phi_0, V, \upsilon, \lambda$ are hyper-parameters. a_t is white noise and $E[a_s x_t] = 0$ while $s > t$.

In order to detect and repair the additive outliers caused by cycle slips in $\{x_1, \ldots, x_n\}$, additive classification variables are introduced corresponding to each observation x_t [15]

$$\delta_t = \begin{cases} 1 & x_t \text{ is an additive outlier} \\ 0 & x_t \text{ is normal} \end{cases} \tag{4}$$

It is assumed that each observation has a small probability α of being aberrant [19], $P(\delta_t^{AO} = 1) = \alpha$. In this case, the model for additive outlier detection is as follows

$$\begin{cases} z_t = \phi_1 z_{t-1} + \cdots + \phi_p z_{t-p} + a_t \\ x_t = z_t + w_t \delta_t \\ a_t \quad i.i.d. \quad N(0, \sigma^2), \quad t = 1, \ldots, n \end{cases} \tag{5}$$

where, z_t denote normal observations. w_t denotes the magnitudes of additive outliers and assuming that $w_t i.i.d. N(\mu, \xi^2), t = 1, \ldots, n$, μ, $\xi^2$are hyper-parameters.

In order to judge that whether the observation is an additive outlier or not and give the threshold, the following Bayes hypothesis is constructed.

The null hypothesis H_0: x_j is normal, namely $\delta_j = 0$;

The alternative hypothesis H_1: x_j is additive outlier, namely $\delta_j = 1$

According to the idea of Bayes hypothesis testing [13], when the posterior probabilities $P(\delta_j = 1|X)$ is larger than $P(\delta_j = 0|X)$, namely $P(\delta_j = 1|X) > 0.5$, we identify that x_j is an additive outlier. Otherwise, x_j is normal. Where, $X = (x_1, \ldots, x_n)^T$.

3.2 Calculation of the Posterior Probabilities of the Classification Variables and Estimation of the Sizes of Outliers Based on Gibbs Sampling

As mentioned above, cycle slips detection has been changed into the calculation of the posterior probabilities of classification variables $q_j = P(\delta_j = 1|X)$, $j = 1, \ldots, n$. Considering the complexity of distributions related to the posterior probabilities, we introduce Gibbs sampling [5] into the computation of posterior probabilities of classification variables and the sizes of outliers. Let $\delta = (\delta_1, \ldots, \delta_n)$, $W = (w_1, \ldots, w_n)^T$, we give full conditional posterior distributions of unknown parameters according to Bayesian theorem [14]:

$$\phi | X, \sigma^2, \delta, W \sim N_p(a, b) \tag{6}$$

$$\sigma^2 | X, \phi, \delta, W \sim IG\left(\frac{\upsilon_1}{2}, \frac{\upsilon_1 \lambda_1}{2}\right) \tag{7}$$

$$\delta_j | X, \phi, \sigma^2, \delta_{(-j)}, W \sim b(1, \ p_j) \tag{8}$$

$$w_j | X, \phi, \sigma^2, \delta, W_{(-j)} \sim N(\hat{w}_j, \xi_j^2) \tag{9}$$

where,

$$a = \left(\frac{1}{\sigma^2} \sum_{t=p+1}^{n} X_{t-1} X_{t-1}^T + V \right)^{-1} \cdot \left[\frac{1}{\sigma^2} \sum_{t=p+1}^{n} X_{t-1}(x_t - w_t \delta_t) + V \phi_0 \right]$$

$$b = \left(\frac{1}{\sigma^2} \sum_{t=p+1}^{n} X_{t-1} X_{t-1}^T + V \right)^{-1}, \quad X_{t-1} = (x_{t-1}, \ldots, x_{t-p})^T$$

$$\upsilon_1 = n - p + \upsilon, \quad \lambda_1 = \frac{1}{n - p + \upsilon} \cdot \left[\sum_{t=p+1}^{n} \left(x_t - \sum_{i=1}^{p} \phi_i z_{t-i} - w_t \delta_t \right)^2 + \upsilon \lambda \right]$$

$$\delta_{(-j)} = (\delta_1, \ldots, \delta_{j-1}, \delta_{j+1}, \ldots, \delta_n)^T$$

$$p_j = P(\delta_j = 1 | X, \phi, \sigma^2, \delta_{(-j)}, W) = \frac{q_1}{q_1 + q_2} \tag{10}$$

$$q_1 = \alpha \exp\left\{ -\frac{1}{2\sigma^2} \sum_{t=j}^{T} (x_t^* - \sum_{i=1}^{p} \phi_i x_{t-i}^* + C_{t-j} w_j)^2 \right\}$$

$$q_2 = (1 - \alpha) \exp\left\{ -\frac{1}{2\sigma^2} \sum_{t=j}^{T} \left(x_t^* - \sum_{i=1}^{p} \phi_i x_{t-i}^* \right)^2 \right\}$$

$$x_t^* = \begin{cases} x_t, & t = j \\ z_t, & t \neq j \end{cases}, \quad C_{t-j} = \begin{cases} -1, & t = j \\ \phi_j, & j = 1, \ldots p, \\ 0, & j > p \end{cases} \quad T = \min(n, \ p + j)$$

$$\hat{w}_j = \xi_j^2 \cdot \left\{ \delta_j \left[\left(x_j - \sum_{i=1}^{p} \phi_i z_{j-i} \right) + \sum_{t=j+1}^{T} \phi_{t-j} \left(\sum_{i=1}^{p} \phi_i x_{t-i}^* - x_t^* \right) \right] \right\} \tag{11}$$

$$\xi_j^2 = \left[\frac{(\delta_j)^2}{\sigma^2} \left(1 + \sum_{i=1}^{p} \phi_i^2 \right) + \frac{1}{\xi^2} \right]^{-1}$$

Then we can implement Gibbs sampling according to the full conditional posterior distributions above. It is assumed that $\Phi^{(r)}, (\sigma^2)^{(r)}, (\delta^{AO})^{(r)}, (\delta^{IO})^{(r)}, (W^{AO})^{(r)}, (W^{IO})^{(r)}$ is a sample of size R drawn from the full conditions. Then, the formula of the posterior probabilities of classification variables is (12) according to (10)

$$P(\delta_j = 1 | X) \approx \frac{1}{R} \sum_{r=1}^{R} \frac{(q_1)^{(r)}}{(q_1)^{(r)} + (q_2)^{(r)}} \tag{12}$$

where,

$$(q_1)^{(r)} = \alpha \exp\left\{-\frac{1}{2(\sigma^2)^{(r)}}\sum_{t=j}^{T}(x_t^* - \sum_{i=1}^{p}\phi_i^{(r)}x_{t-i}^* + C_{t-j}w_j^{(r)})^2\right\}$$

$$(q_2)^{(r)} = (1-\alpha)\exp\left\{-\frac{1}{2(\sigma^2)^{(r)}}\sum_{t=j}^{T}\left(x_t^* - \sum_{i=1}^{p}\phi_i^{(r)}x_{t-i}^*\right)^2\right\}$$

The formula of the estimation of the sizes of outliers is (13) according to Bayes point estimation and (9)

$$\hat{w}_j = \xi_j^2 \cdot \left\{\delta_j\left[\left(x_j - \sum_{i=1}^{p}\phi_i z_{j-i}\right) + \sum_{t=j+1}^{T}\phi_{t-j}\left(\sum_{i=1}^{p}\phi_i x_{t-i}^* - x_t^*\right)\right] + \mu\right\} \quad (13)$$

3.3 The Implementation of Bayes Method for Cycle Slips Detection (AlgorithmI)

Step 1: Choice for hyper-parameters. For example, we assumed that

$$\phi_0 = \begin{pmatrix} 0 \\ \cdots \\ 0 \end{pmatrix}_{p\times 1}, V = \begin{pmatrix} 10 & & \\ & \cdots & \\ & & 10 \end{pmatrix}_{p\times p}, \upsilon = 3,\ \lambda = 0.5,\ \alpha = 0.05,$$

$$\xi^2 = 2,\ \mu = 0$$

in our experiments.

Step 2: Choose initial conditions of Gibbs sampling according to Bayesian estimation method and the values of hyper-parameters above.

Step 3: Gibbs sampling. Suppose that the current configuration is $\left(\phi^{(s-1)}, (\sigma^2)^{(s-1)}, \delta^{(s-1)}, W^{(s-1)}\right)$, then the next sample is produced by the Gibbs algorithm

(1) draw $\phi^{(s)}$ from $p\left(\phi|X, (\sigma^2)^{(s-1)}, \delta^{(s-1)}, W^{(s-1)}\right)$

(2) draw $(\sigma^2)^{(s)}$ from $p\left(\sigma^2|X, \phi^{(s)}, \delta^{(s-1)}, W^{(s-1)}\right)$

(3) draw $(\delta_j)^{(s)}$ from $p\left(\delta_j = 1|X, \phi^{(s)}, (\sigma^2)^{(s)}, \delta_{(-j)}^{(s-1,j+1)}, W^{(s-1,j)}\right)$

(4) draw $(w_j)^{(s)}$ from $p\left(w_j|X, \phi^{(s)}, (\sigma^2)^{(s)}, \delta^{(s-1,j+1)}, W_{(-j)}^{(s-1,j+1)}\right)$

The above step is repeated for R times and we guarantees that the Markov chain is convergent in each time. Then we obtain the following samples with size R:

$$(\phi^{(1)}, (\sigma^2)^{(1)}, \delta^{(1)}, W^{(1)}), \ldots, (\phi^{(R)}, (\sigma^2)^{(R)}, \delta^{(R)}, W^{(R)})$$

Step 4: Cycle slips detection. Compute the posterior probabilities of classification variables from the formula (12) and identify outliers by the principle in Sect. 3.1. Then we can obtain the positions of cycle slips according to the positions of outliers and the differential numbers between consecutive epochs.

Step 5: Cycle slips estimation. Estimation of the magnitudes of $\hat{w}_j$, then we can obtain the cycles of the cycle slips m by m = round$(\hat{w}_j/\lambda)$, and then the magnitudes of cycle slips λm, where round(x) denotes the function of rounding.

4 Unmasking Bayes Method for Cycle Slips Detection

4.1 Reasons for Masking and Swamping

Algorithm I is indeed feasible in many situations. However, through many experiments and calculations, the author find that when the outliers caused by cycle slips are patches of additive outliers, cycle slips detection will be suffering from masking and swamping while we utilize Algorithm I directly. So cycle slips detection fail. (For example: see the experiments and analysis in Sect. 5).

In view of this, this paper analyze the reasons of masking and swamping and take some measures to ensure the correctness of the values of the posterior probabilities of classification variables when designing Gibbs sampling algorithm. This will be a feasible way for researching unmasking Bayes method for cycle slips detection.

When we implement Gibbs sampling, the estimation of w_j is its posterior mean

$$\hat{w}_j = \xi_j^2 \cdot \left\{ \delta_j \left[\left(x_j - \sum_{i=1}^{p} \phi_i z_{j-i} \right) + \sum_{t=j+1}^{T} \phi_{t-j} \left(\sum_{i=1}^{p} \phi_i x_{t-i}^* - x_t^* \right) \right] + \mu \right\}$$

According to the Bayes theorem, posterior mean is the weighed average mean of sample information and prior information.

$$\hat{w}_j = C_1 \times \text{ sample information } + C_2 \times \mu \tag{14}$$

where, C_1, C_2 are constants that are greater than zero and $C_1 + C_2 = 1$. We quantitatively depict the influence of sample information on $\hat{w}_j$ by $x_j - x_{j|n}^*$ and this paper will analyze the reasons of masking and swamping through a simple example. Where, $x^*_{j|n}$ is an estimation of x_j.

It is assumed that the epoch differential phase data is fit for AR(1): $x_t = \phi x_{t-1} + a_t$, and the observations at epochs $j-1, j, j+1$ contain three consecutive additive outliers with the sizes of w_{j-1}, w_j, w_{j+1} respectively. That is to say, there is a patch of additive outliers at epochs $j-1, j, j+1$. In order to simplify the analysis, we assume that $w_{j-1} = w_j = w_{j+1} = w > 0$. Let $x^*_{j|n} = \frac{\phi}{1+\phi^2}(x_{j-1} + x_{j+1})$ to be an interpolation of x_j. Then

$$x^*_{j-1|n} = \frac{\phi}{1+\phi^2}(z_{j-2} + w_{j-2} + z_j + w_j) = z^*_{j-1|n} + \frac{\phi}{1+\phi^2}w$$

$$x^*_{j|n} = \frac{\phi}{1+\phi^2}(z_{j-1} + w_{j-1} + z_{j+1} + w_{j+1}) = z^*_{j|n} + \frac{2\phi}{1+\phi^2}w$$

$$x^*_{j+1|n} = \frac{\phi}{1+\phi^2}(z_j + w_j + z_{j+2}) = z^*_{j+1|n} + \frac{\phi}{1+\phi^2}w$$

And then

$$x_{j-1} - x^*_{j-1|n} = z_{j-1} + w - z^*_{j-1|n} - \frac{\phi}{1+\phi^2}w = z_{j-1} - z^*_{j-1|n} + \left(1 - \frac{\phi}{1+\phi^2}\right)w$$

$$x_j - x^*_{j|n} = z_j + w - z^*_{j|n} - \frac{2\phi}{1+\phi^2}w = z_j - z^*_{j|n} + \frac{(1-\phi)^2}{(1+\phi)^2}w$$

$$x_{j+1} - x^*_{j+1|n} = z_{j+1} + w - z^*_{j+1|n} - \frac{\phi}{1+\phi^2}w = z_{j+1} - z^*_{j+1|n} + \left(1 - \frac{\phi}{1+\phi^2}\right)w$$

Because the estimation of ϕ will approach the unity as the outlier sizes and numbers increase [12] and z_j denote the normal observation, then $z_j - z^*_{j|n} \approx 0$. Therefore, $x_{j-1} - x^*_{j-1|n} \approx \frac{1}{2}w$, $x_j - x^*_{j|n} \approx 0$, $x_{j+1} - x^*_{j+1|n} \approx \frac{1}{2}w$. If we let the prior mean to be 0, then we can obtain $\hat{w}_{j-1} \approx C_1 \times \frac{1}{2}w \leq \frac{1}{2}w$, $\hat{w}_{j+1} \approx C_1 \times \frac{1}{2}w \leq \frac{1}{2}w$, $\hat{w}_j \approx 0$ according to (14). It is state that when we implement the standard Gibbs sampling the calculation of $\hat{w}_j$ according to (11) has utilized the observations which contain outliers before and after x_j. This caused $\hat{w}_j$ deviate from its realistic value. Since Gibbs sampling is very sensitive to the initial conditions [5], so it may convergent to a false posterior distribution. Then masking and swamping will occur and finally cycle slips detection will fail. It is found by a further study that when we use standard Gibbs sampling to detect patches of additive outliers either the phenomenon that two ends observations are masking and middle observations are swamping occurs or the opposite phenomenon occurs.

4.2 Measures for Overcoming Masking and Swamping

According to the reasons of masking and swamping analyzed above, we propose a measure to overcome masking and swamping. The measure is that if there is a patch of additive outliers we utilize the normal observations before and after this patch of additive outliers to obtain an accurate estimation of the outlier sizes, assign the initial conditions of Gibbs sampling by the accurate estimation above and resample the Gibbs sampling.

4.2.1 Searching for the Positions of Potential Patches of Additive Outliers

It is assumed that $\{x_{i_k}, k = 1, 2, \ldots, m\}$ are additive outliers detected by the standard Gibbs sampling. Then

(1) If the epoch distance of two consecutive additive outliers is smaller than $0.05n$ (n is the total numbers of observations), specify step length h and threshold C. Considering one of the outliers x_{i_k} that is fit for the standard, we check $x_{i_k}, x_{i_k+1}, \ldots, x_{i_k+h}$ and compare the posterior probabilities of them with C. Any point within the window with the posterior probabilities greater than C is regarded as a possible ending point of this outlier patch associated with x_{i_k}. We select the farthest epoch $i_k + m_{k1}$ from i_k as the ending epoch of the outlier patch [12]. Do the same for $x_{i_k}, x_{i_k-1}, \ldots, x_{i_k-h}$, then we can obtain the beginning epoch $i_k - m_{k2}$ of the outlier patch. Therefore, the potential outlier patch is $\{x_{i_k-m_{k2}}, \ldots, x_{i_k+m_{k1}}, k = 1, 2, \ldots, m\}$. Any overlapping or consecutive patches should be merged to form a large patch [12]. If the total number of outliers is greater than $0.5n$, then we should increase C and reduce h.

(2) If the epoch distance of two consecutive additive outliers is greater than $0.05n$, then we assume that the $0.05n$ observations before and after the outliers are potential outlier patch. That is to say, if x_i is an outlier that is fit for the demands, then we denote $\{x_{i-0.05n}, \ldots, x_i, \ldots, x_{i+0.05n}\}$ as the potential outlier patch. Any overlapping or consecutive patches should be merged to form a large patch [12]. If the total number of outliers is greater than $0.5n$, then we reduce $0.05n$.

Let $\{x_j, x_{j+1}, \ldots, x_{j+k-1}\}$ to be the potential outlier patch and let $W_{j,k} = (w_j, w_{j+1}, \ldots, w_{j+k-1})^T$ to be the sizes of the outliers.

4.2.2 Estimate the Magnitudes of Potential Additive Outlier Patches

Considering the correlation of time series, this paper utilizes least squares collocation method and the normal observations before and after the potential additive outlier patch $\{x_j, x_{j+1}, \ldots, x_{j+k-1}\}$ to give an accurate estimation of $\{x_j, x_{j+1}, \ldots, x_{j+k-1}\}$. Then we can obtain an accurate estimation of $W_{j,k} = (w_j, w_{j+1}, \ldots, w_{j+k-1})^T$, where $k > p$.

From model (3), we can obtain the following equations

$$\begin{cases} x_{j+k} = \phi_1 x_{j+k-1} + \cdots + \phi_p x_{j+k-p} + a_{j+k} \\ x_{j+k+1} - \phi_1 x_{j+k} = \phi_2 x_{j+k-1} + \cdots + \phi_p x_{j+k-(p-1)} + a_{j+k+1} \\ \quad\cdots \qquad \cdots \\ x_{j+k+p-1} - \phi_1 x_{j+k+p-2} - \phi_{p-1} x_{j+k} = \phi_p x_{j+k-1} + a_{j+k+p-1} \end{cases}$$

Namely

$$L = B\begin{pmatrix} S \\ S' \end{pmatrix} + \Delta$$

where,

$$L = (x_{j+k}, x_{j+k+1} - \phi_1 x_{j+k}, \ldots, x_{j+k+p-1} - \phi_1 x_{j+k+p-2} - \phi_{p-1} x_{j+k})^T,$$

$$\Delta = \begin{pmatrix} a_{j+k} \\ a_{j+k+1} \\ \cdots \\ a_{j+k+p-1} \end{pmatrix} \quad F = \begin{pmatrix} \phi_p & \cdots & & \phi_1 \\ 0 & \phi_p & \cdots & \phi_2 \\ & \cdots & \cdots & \\ 0 & \cdots & 0 & \phi_p \end{pmatrix}, \quad S = \begin{pmatrix} x_j \\ \cdots \\ x_{j+k-(p+2)} \\ x_{j+k-(p+1)} \end{pmatrix},$$

$$S' = \begin{pmatrix} x_{j+k-p} \\ x_{j+k-(p-1)} \\ \cdots \\ x_{j+k-1} \end{pmatrix}, \quad B = (0, F)$$

and it is assumed that the estimations of $\phi_1, \ldots, \phi_p$ is obtained by least squares method.

According to the theories of least squares collocation [6]

$$S^* = \Sigma_{SS'} F \Sigma_L^{-1} L, \quad (S')^* = \Sigma_{S'} F \Sigma_L^{-1} L \tag{15}$$

Then the estimation of $W_{j,k} = (w_j, w_{j+1}, \ldots, w_{j+k-1})^T$ is

$$W_{j,k}^* = \begin{pmatrix} x_j \\ x_{j+1} \\ \cdots \\ x_{j+k-1} \end{pmatrix} - \begin{pmatrix} S^* \\ (S')^* \end{pmatrix} \tag{16}$$

The determining method of the (cross) covariance Σ_s, $\Sigma_{SS'}$, Σ_L follows the idea that the longer distance between epochs i, j, the smaller correlation of x_i, x_j. Generally, an ordinary formula [17] is $\mathrm{Cov}(x_i, x_j) = e^{-2|i-j|}$.

4.3 The Implementation of Unmasking Bayes Method for Cycle Slips Detection (Algorithm II)

An adaptive Gibbs sampling algorithm is designed for the calculation of the posterior probabilities of classification variables by the ideas above. Then an unmasking Bayes method for cycle slips detection is given. The specific implementation process is as follows:

Step 1: Standard Gibbs sampling according to the method of Algorithm I from step 1 to step 3.

Step 2: Searching for the locations of the potential additive outlier patch according to the method in Sect. 4.2.1. Let $\{x_j, x_{j+1}, \ldots, x_{j+k-1}\}$ to be the potential outlier patch and let $W_{j,k} = (w_j, w_{j+1}, \ldots, w_{j+k-1})^T$ to be the magnitudes of outliers.

Step 3: Estimate the magnitudes of potential additive outlier patch according to the method in Sect. 4.2.2. Let $W^*_{j,k}$ to be the estimation of the magnitudes of outlier patch.

Step 4: Adaptive Gibbs sampling.

Firstly, redefine the initial conditions of Gibbs sampling. The initial conditions of $\delta = (\delta_1, \cdots, \delta_n)$: the classification variables corresponding to the potential outlier patch and the isolated outliers detected by standard Gibbs sampling assign 1 and the others assign 0.

The initial conditions of $w_1, \ldots, w_n$: the sizes corresponding to the potential outlier patch assign $W^*_{j,k}$, the sizes of the isolated outliers detected by the standard Gibbs sampling are determined by (13) and the others assign 0. Secondly, the determining method of the initial conditions of other parameters is the same as standard Gibbs sampling. Finally, running step 3 of Algorithm I again, then we obtain the following samplers:

$$(\phi^{(1)*}, (\sigma^2)^{(1)*}, \delta^{(1)*}, W^{(1)*}),$$
$$\cdots$$
$$(\phi^{(m)*}, (\sigma^2)^{(m)*}, \delta^{(m)*}, W^{(m)*})$$

Step 5: Cycle slips detection according to step 4 of Algorithm I.

Step 6: Cycle slips estimation according to step 5 of Algorithm I.

5 Examples and Analysis

5.1 Experiment 1 and Analysis

The simulation data is collected by two JAVAD LEGACY receivers with dual frequencies to observe six GPS satellites (SV11, 27, 8, 31, 28, 7). The length of the baseline is 3220.340 m. The time of collection is 2003-02-18, at 11:00–12:00. The sampling rate is 1 s with 15°. The data of this experiment is from epoch 2434 to epoch 2634 on L1 wave with the wavelength 19.03 cm. The dual-differential phase data between receivers and satellites (SV11-27) is obtained by the IGG GPS data processing software. According to the theories of time series modeling analysis, the one-differential of the data between consecutive epochs will be stationary as is shown by Fig. 1

Scheme 1: introduce cycle slips of size equal to 1 cycle at epoch 31, 96 and 161 in the dual-differential phase data between receivers and satellites. Then outliers in the stationary time series caused by these cycle slips are three isolated additive outliers at $t = 30$, $t = 95$ and $t = 160$ with the sizes of 19.03 cm respectively. Detecting cycle slips by Algorithm I. The result is shown by Fig. 2

As can be seen from the result, the posterior probabilities of the classification variables corresponding to $t = 35$, $t = 95$ and $t = 160$ at stationary time series are 1.0000, 1.0000, and 0.9977 respectively. The estimation of the outlier sizes are 18.3109 cm, 19.5576 cm and 15.6109 cm respectively. According to the positions and sizes of the outliers, we can obtain that the locations of the cycle slips are at epoch 36, 96 and 161 in the dual-differential phase data. Then according to round $(18.3109/19.03) = 1$, round $(19.5576/19.03) = 1$, round $(15.6109/19.03) = 1$, we obtain that all of the sizes of these cycle slips are 1 cycle. Namely, all of the estimation of the sizes of the cycle slips are 19.03 cm.

Scheme 2: introduce cycle slips of size equal to 1 cycle at epoch 100, 101 and 102 in the dual-differential phase data between receivers and satellites. Then outliers in the stationary time series caused by these cycle slips are an patch of additive outliers at $t = 99$, $t = 100$ and $t = 101$ with the sizes of 19.03 cm respectively. Detect outliers by Algorithm I. The result is shown by Fig. 3.

As can be seen from the result, the outlier at $t = 101$ in the stationary time series is masked and the estimation of it biases from the realistic value. Besides, the normal observations at $t = 98$ and $t = 102$ are swamped.

Scheme 3: introduce cycle slips of size equal to 1 cycle at epoch 100, 101 and 102 in the dual-differential phase data between receivers and satellites. Then outliers in the stationary time series caused by these cycle slips are a patch of additive outliers at $t = 99$, $t = 100$ and $t = 101$ with the sizes of 19.03 cm respectively. Detect outliers by Algorithm II. The result is shown by Fig. 4

As can be seen from the result, the posterior probabilities of the classification variables corresponding to $t = 99$, $t = 100$ and $t = 101$ at stationary time series are 0.9899, 1.0000, and 1.0000 respectively. The estimation of the outlier sizes are 14.5582,

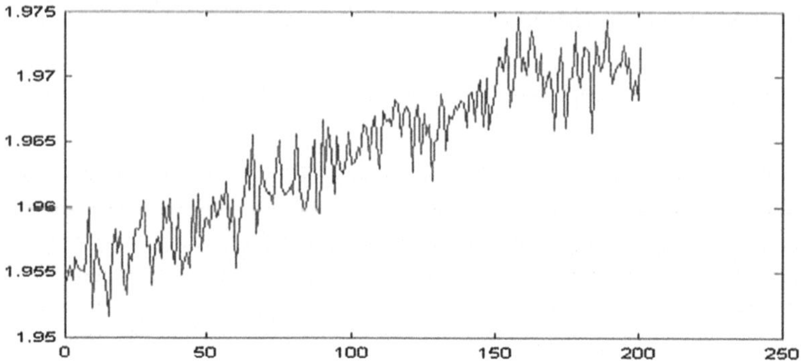

Fig. 1 Dual-differential carrier phase between receivers and satellites of 11–27 satellites

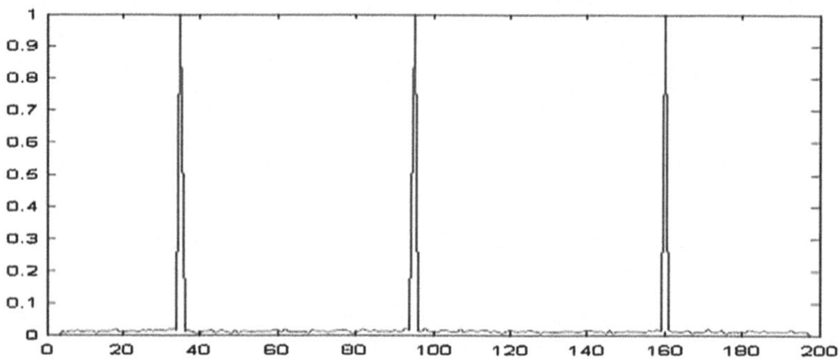

Fig. 2 Posterior probabilities of classification variables of scheme 1

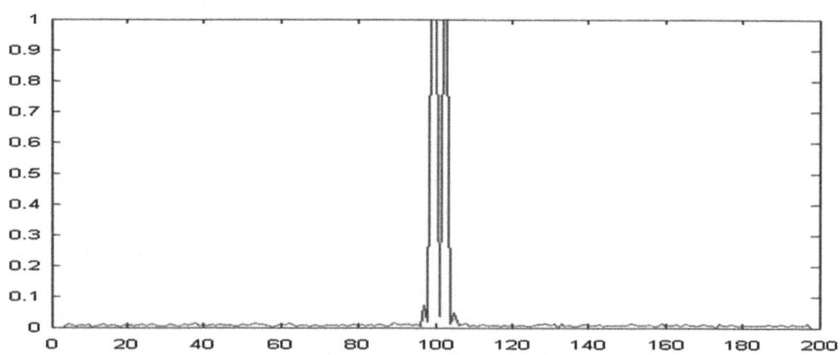

Fig. 3 Posterior probabilities of classification variables of scheme 2

22.6014 and 18.7860 cm respectively. According to the positions and sizes of the outliers, we can obtain that the locations of the cycle slips are at epoch 100, 101 and 102 in the dual-differential phase data. Then according to round $(14.5582/19.03) = 1$,

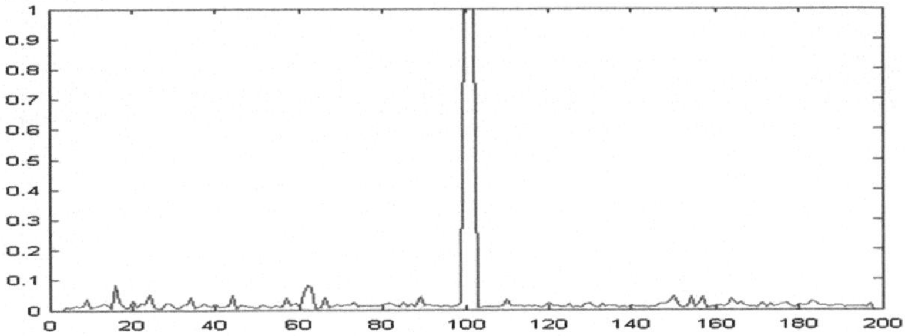

Fig. 4 Posterior probabilities of classification variables of scheme 3

round (22.6014/19.03) = 1, round (18.7860 /19.03) = 1, we obtain that all of the sizes of these cycle slips are 1 cycle. Namely, all of the estimation of the sizes of the cycle slips is 19.03 cm.

5.2 Experiment 2 and Analysis

The simulation data is collected at 2001-03-04, 00:37:54 to observe ten satellites (SV28, 03, 25, 29, 20, 22, 13, 01, 31, 19). The sampling rate is 5 s. The data of this experiment is from epoch 100 to epoch 171 on L1 wave. The one-differential phase data between satellites (SV28-03) is obtained. It can be seen in Fig. 5. According to the theories of time series modeling analysis, three-differential of the data between consecutive epochs will be stationary.

Scheme 1: introduce cycle slips of size equal to 1 cycle at epoch 20 and 50 in the one-differential phase data between satellites. Then outliers in the stationary time series caused by these cycle slips are two patches of additive outliers. One is at t = 17, t = 18 and t = 19 with the sizes of 19.03 cm, −38.06 and 19.03 cm, and another one is at t = 47, t = 48 and t = 49 with the sizes of 19.03, −38.06 and 19.03 cm respectively. The result is shown by Fig. 6. Detect outliers by Algorithm II. The result is shown by Fig. 7.

As can be seen from the result, outliers at t = 17, 19, 47 and 49 are masked and the estimation of the sizes of them are biased from their realistic values.

Scheme 2: introduce a cycle slip of size equal to 1 cycle at epoch 20 in the one-differential phase data between satellites. Then outliers in the stationary time series caused by these cycle slips is an patch of additive outliers at t = 17, t = 18 and t = 19 with the sizes of 19.03, −38.06 and 19.03 cm respectively. Detect outliers by Algorithm II. The result is shown by Fig. 8

As can be seen from the result, all of the posterior probabilities of the classification variables corresponding to t = 17, t = 18 and t = 19 at stationary time series are 1.0000 respectively. The estimation of the outlier sizes are 15.7353 cm, −38.9815 cm and 13.3974 cm respectively. According to the positions and sizes

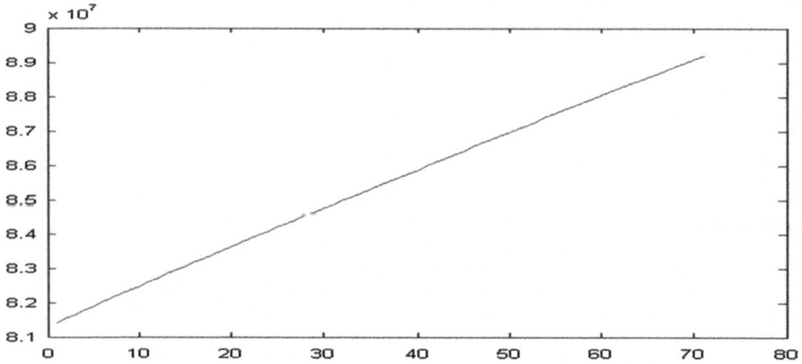

Fig. 5 Differential carrier phase between satellites of 28–03 satellites

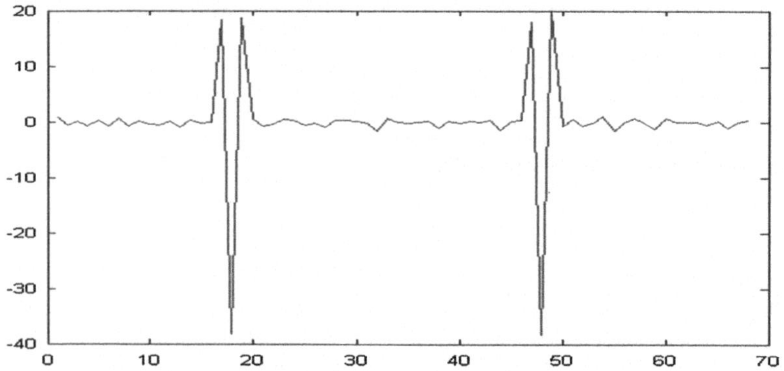

Fig. 6 Three time-differential of satellites differential carrier phase data

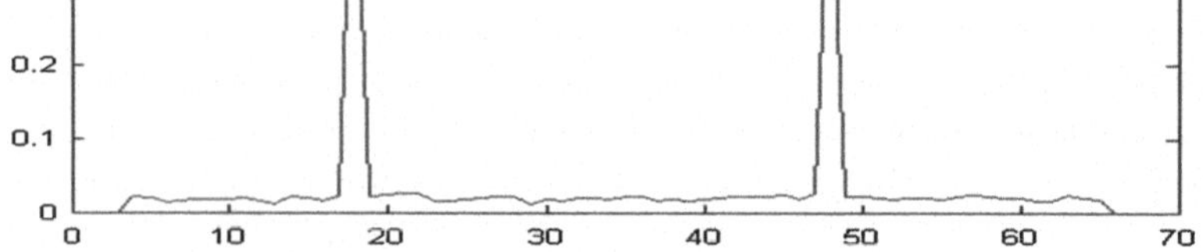

Fig. 7 Posterior probabilities of classification variables of scheme 1

of the outliers, we can obtain that the location of the cycle slip is at epoch 20 in the one-differential phase data. Then according to round $(15.7353/19.03) = 1$, round $(-38.9815/19.03) = -2$, round $(13.3974/19.03) = 1$, we obtain that the

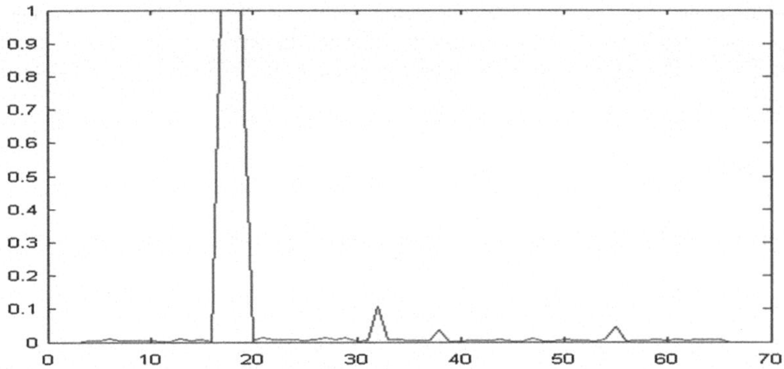

Fig. 8 Posterior probabilities of classification variables of scheme 2

size of the cycle slip is 1 cycle. Namely, the estimation of the size of the cycle slip is 19.03 cm.

6 Conclusions

(1) When we solve the problems of baseline solution and network adjustment based on the differential or dual-differential phase data, cycle slips in the differential or dual-differential phase data with epoch difference yield either isolated additive outliers or patches of additive outliers.
(2) A Bayesian method for cycle slips detection based on the posterior probabilities of classification variables is proposed in the respective of Bayesian hypothesis. The calculation method of the posterior probabilities of classification variables and the estimation approach of abnormal disturbance is give by Gibbs sampling
(3) Reasons for masking and swamping are analyzed. Accurate estimation of patches of additive outliers is given by the least squares collocation method and we utilize this accurate estimation to redefine the initial conditions of Gibbs sampling. Then an adaptive Gibbs sampling algorithm is designed. And then Bayes method for overcoming masking and swamping in cycle slips detection is given. It is a feasible way for researching unmasking Bayes method of cycle slips detection. That is to say, take some measures to ensure the correctness of the posterior probabilities of classification variables when designing Gibbs sampling algorithm to overcome the masking and swamping phenomenon.
(4) The experiments show that the new method is not only efficient for the detection of isolated additive outliers but also efficient for frequent cycle slips detection.

Acknowledgments This research was supported jointly by National Science Foundation of China (No. 40974009, No. 41174005), Planned Research Project of Technology of Zhengzhou City, and Funded Project with youth of Annual Meeting of China's satellite navigation.

References

1. Banville, S., & Langley, R. B. (2010). Instantaneous cycle slip correction for real time ppp applications. *Journal of Navigation, 57*(4), 325–334.
2. Betti, B., Crespi, M., & Sanso, F. (1993). A geometric illustration of ambiguity resolution in GPS theory and a Bayesian approach. *Manus Geodesy, 18*, 317–330.
3. Bisnath, S. B. & Langley, R. B. (2000). Efficient automated cycle-slip correction of dual-frequency kinematic GPS data. In *Proceeding of 47th conference of Canadian Aeronautics and Space Institute* (pp. 121–125), Ottawa.
4. Box, G. E. P., & Jenkins, G. M. (1970). *Time series analysis, forecasting and control.* San Francisco: Holden-Day.
5. Christian, P. R. (2004). *Monte Carlo statistical methods*. Berlin: Springer.
6. Cui, X., Yu, Z., & Tao, B. (2001). *Generalized surveying adjustment* (New ed.). Wuhan: Publishing House of Wuhan Technical University of Surveying and Mapping.
7. De Lacy, M. C., Reguzzoni, M., Sanso, F., & Venuti, G. (2008). The Bayesian detection of discontinuities in a polynomial regression and its application to the cycle slip problem. *Journal of Geodesy, 82*(9), 527–542.
8. De Lacy, M. C., Sanso, F., Rodriguez-Caderot, G., & Gil, A. J. (2002). The Bayesian approach applied to GPS ambiguity resolution, a mixture model for the discreet-real ambiguities alternative. *Journal of Geodesy, 76*, 82–94.
9. Gui, Q., Li, X., Gong, Y., Li, B., & Li, G. (2011). A Bayesian unmasking method for locating multiple gross errors based on posterior probabilities of classification variables. *Journal of Geodesy, 85*, 191–203.
10. Gundlich, B., & Koch, K. R. (2002). Confidence regions for GPS baselines by Bayesian statistics. *Journal of Geodesy, 76*, 55–62.
11. He, H., & Yang, Y. (1999). Detection of successive cycle slip for GPS kinematic positioning. *Acta Geodaetica et Cartographica Sinica, 28*(3), 199–203.
12. Justel, A., Pena, D., & Tsay, R. S. (2001). Detection of outlier patches in autoregressive time series. *Statistical Sinica, 3*(11), 651–673.
13. Koch, K. R. (1990). *Bayesian inference with geodetic applications*. Berlin, Springer.
14. Koch, K.R (2000) Einfuhrung in Die Bayes-Statistic. Springer, Berlin
15. Li, X., Gui, Q., & Xu, A. (2008). Bayesian method for detection of gross errors based on classification variables. *Acta Geodaetica et Cartographica Sinica, 37*(3), 355–360.
16. Li, Z., & Huang, J. (2005). *GPS surveying and data processing*. Wuhan: Wuhan University Press.
17. Mira, J., & Sanchez, M. J. (2004). Prediction of deterministic functions: an application of a Gaussian kriging model to a time series outlier problem. *Computational Statistics & Data Analysis, 44*, 477–491.
18. Sanso, F., & Venuti, G. (1997). Integer variable estimation problems: The Bayesian approach. *Ann Geofisica, 5*(XL), 1415–1431.
19. Tsay, R. S. (1986). Time series model specification in the presence of outliers. *Journal of American Statistical Association, 393*(81), 132–141.
20. Zhang Q., & Gui, Q. (2011). Bayesian methods for simultaneous detection of additive and innovation outliers in ARMA model. *In Proceedings of 2011 International Symposium on Statistics & Management Science*. Paseo Segovia, Irvine, CA: Scientific Research Publishing, Inc.

21. Zhang, Q., Gui, Q., & Wang, Y. (2012). Bayesian methods for outliers detection in autoregressive model based on different types of classification variables. *Acta Geodaetica et Cartographica Sinica, 37*(4), 482–488.
22. Zhang, C., Xu, Q., & Li, Z. (2009). Improving method of cycle slip detection and correction based on combination of GPS pseudo range and carrier phase observation. *Acta Geodaetica et Cartographica Sinica, 38*(5), 402–407.

Influence of Satellite-to-Ground Link on the Autonomous Navigation of Navigation Constellation

Wei Wang, Xurong Dong, Wanke Liu, Ying Liu, Sihui Liu and Chengeng Su

Abstract Autonomous navigation is a significant element of satellite navigation systems. Current generation autonomous navigation of navigation satellite constellation typically employs the inter-satellites information only, which shows vulnerabilities due to the restriction of deficiencies and the on board processing units of inter-satellites links. In this paper, we describe an autonomous navigation of constellation method in virtue of satellite-ground links, along with aspects of the operating mode of autonomous navigation and the autonomous navigation of constellation with the aid of satellite-ground links. The proposed method is tested by simulations, proving its effectiveness for the precision improvement of autonomous navigation of constellation.

1 Introduction

Autonomous navigation is the capability and process to ensure navigation system running independently and steadily by constellation autonomous management, measurement and data exchange among satellites or between satellites and ground and satellite autonomous data processor while the satellites can not obtain normal support from ground system for a long time [1].

W. Wang (✉) · X. Dong
Academy of Equipment Command & Technology, Beijing, China
e-mail: weiwei101977@163.com

Y. Liu · S. Liu · C. Su
Beijing Institute of Tracking and Telecommunication Technology, Beijing, China

W. Liu
Wuhan University, Wuhan, China

J. Sun et al. (eds.), *China Satellite Navigation Conference (CSNC) 2012 Proceedings*, Lecture Notes in Electrical Engineering 160, DOI: 10.1007/978-3-642-29175-3_30,

As early as 1990, Ananda M.P. has pointed out that the relative position of satellites in constellation could be determined exactly by using inter-satellites ranging data. But a question that the constellation as a rigid body relative to the earth exists "unobservable" the satellite constellation rotation problem [2]. In 2008, Liu Wanke analysis pointed out that the constellation rotation "unobservable" is inevitable from disturbing force. If the autonomous orbit determination algorithm cannot reduce the constellation of the overall rotation error, it will seriously affect the accuracy of the satellite ephemeris, thereby reducing the majority of the user's navigation and positioning accuracy [3].

When using inter-satellite ranging data to complete Autonomous orbit determination, due to the constellation rotation cannot be observed, orbit determination is lack of adequate direction of baseline information and induces the rank deficient problems. Aiming at this problem, In the case of considering only the Earth's center of gravity, China's Liu Lin has proved theoretically that the rank deficient problems is due to the estimation of the orbital plane of the longitude ascending node Ω to the Autonomous orbit determination with same orbit and different orbit observations. He also pointed out that if the number of autonomous orbital elements could be fixed on, it can solve the problem of rank defect, but the autonomous orbit determination accuracy is heavily dependent on a priori ephemeris in Ω of accuracy [4, 5]. In 2005, Zhang Yan has researched the problem more detailedly theoretically, and also reached a similar conclusion [6].

In order to eliminate the rank deficient in the autonomous orbit determination, it generally fixed or constrained part or all of the satellites reference orbit orientation parameters to solve the satellite orbit. Constellation rotation error is influence by the reference orbit accuracy, so if the reference orbit forecast accurately, constellation rotation errors can be well controlled or even eliminate, if the reference orbit prediction forecast inaccurately, the rotation of the constellation rotation errors may increase. The paper analyze the GPS satellite, the results show that the forecast of 180 days with the broadcast ephemeris as the initial value, the maximum prediction error of the ascending node right ascension is about 60mas, the equivalent of 7.5 m to the track error. Therefore, in accordance with the above ideas completely, orbit determination error and time synchronization accuracy influenced by the reference orbit badly. In general it is shown the trends that accuracy is getting worse with the increase in autonomous navigation of time.

In order to eliminate the navigation satellites constellation rotation, in 2003, Rajan, JA, et al. studied autonomous orbit determination Increase of the ground anchor station (ground stations). The essence of this method is not different from combined orbit determination. In his design, the anchor station does not transmit to all visible GPS satellite ranging signal, but only receives and measures the GPS satellite signals. The results showed that: the increase an "anchor stations" substantially increase the accuracy of the satellite time synchronization and orbit estimation, it has solved the constellation rotation; increase several "anchor stations" will not only improve the accuracy of satellite navigation systems, but also enhance system redundancy performance [7].

To the non-polar ground anchor station, under the control of getting four navigation satellites distance observation simultaneity, the constellation will not be able to rotate, then circle the equator, otherwise it is difficult to meet the conditions of satellite and ground distance of the observed values. In other words, the ground anchor station could fix the constellation by getting multi-epoch observations of multiple navigation satellites to prevent it from rotating.

On these research groundwork, The paper study the manne, the advantages and disadvantages, and the function of the autonomous orbit determination with the supporting of satellite-ground links. The paper also takes the simulation method to discuss the influence of the satellite-ground links on the accuracy of autonomous orbit determination, and ground monitoring station layout problems. However, the presence of ground anchor station is undoubtedly reduces the autonomy of the navigation satellite systems.

2 Influence Analysis of Satellite-Ground Links to Autonomous Navigation

2.1 Generation of Simulated Observations

(1) **Simulation of GNSS constellation configuration**

According to characteristics of COMPASS-2 in building, the constellation configuration is set Walker24/3/2. 24 MEO satellites with orbital altitude of 21700 km, dip 55°; 3 GEO satellites, positioned at 80°, 100°, 140°; 3 IGSO satellites with orbital inclination of 55°. The software STK generates the simulated constellation orbit initial value under the above parameters, this value serves as the initial value of the orbit integration and simulation of satellite ephemeris.

(2) **Simulation of the satellite ephemeris**

According to the initial orbit state vector of the simulated constellation of satellites at a given moment t_0, to consider the necessary perturbation, using numerical integration of orbital integral, we could obtain instantaneous state vector in each sampling interval. At the same time according to the GPS clock error model published by IGS and other organizations, the we could generate navigation satellite clock error information in each sampling interval. On this basis, we could generate the orbital position and clock error, Furthermore, generate the final satellite ephemeris, to do the reference by the subsequent numerical simulation and the orbit determination accuracy assessment.

(3) **Generation data of intersatellite and satellite-ground observation**

Taking into account the factors such as the signal propagation time, the satellite clock error, relativistic effects, the observation noise and residual systematic errors, we generate inter-satellite distance measurement using a TDMA wide beam with emission signal time of each satellite distributed 1.0 s. The noises of Inter-satellite and satellite-ground observations are all assumed to be 30 cm.

There are three types of satellites in COMPASS-2, so we design three forms of inter-satellite topological configuration including the MEO to MEO, MEO to GEO, MEO to IGSO, with one-way and two-way ranging. Transmitting antenna beam angle of MEO is 15–60°, the angle of GEO and IGSO is 10–45°, and the high cut-off angle of simulation of ground observations is 10° [5].

2.2 The Resolution Strategy of Orbit Determination

In order to assess correctness of combined orbit determination method, to test the reliability of the software developed, and to analyze the influence of ground stations to combined orbit determination accuracy, the paper designs variety of programs to do the combined orbit determination of GPS satellites in 2007 from 112th to 118th day, and compare the results with the IGS precise ephemeris. The Resolution Strategy are as follows in Table 1.

2.3 The Design of Orbit Determination Strategy

2.3.1 Ground Station Layout Schemes

In the subsequent orbit determination calculations, the paper select 17 stations in the world and 5 stations in China respectively. Its geographical distribution is as follows as shown in Figs. 1 and 2 [8].

2.3.2 Orbit Determination Schemes

In order to easily analyze the influence of satellite-ground links to Autonomous Navigation, the paper designed 20 kinds of orbit determination as follows in Table 2.

2.4 Analysis of Results

2.4.1 Influence Analysis of Satellite-Ground Links to Autonomous Orbit Determination

The paper completed the autonomous orbit determination calculation with and without the supporting of satellite-ground links respectively based on 2011 the 32nd day to 91st day Inter-satellite and satellite-ground Ranging observations, and compared the results with the IGS precise ephemeris.

Table 1 The resolution strategy used in a■tonomous orbit determination

Project	Centralized autonomous orbit determination	Distributed autonomous orbit determination
Observations	Satellite-ground, inter-satellite	Inter-satellite
Sampling interval	Ground: 1 min, inter-satellite: 15 min	Inter-satellite: 15 min
Three body gravitational	Sun and moon gravitational	Sun and moon gravitational
Earth gravity field model	12 order EGM96	12 order EGM96
Tide Correction	Solid tide, tide	Solid tide
Solar radiation pressure model	ROCK4/42 (2 Parameters to be estimated)	Priori model
Earth orientation parameters	EOP C04 support by IERS	EOP C04 support by IERS
Estimator	Least squares collocation	Extended Kalman filter with a priori constraint conditions
State parameters	Each arc (24 h) of each satellite to estimate the state vector (fixed ground tracking station coordinates)	Each epoch of each satellite orbital elements of a set of orbital elements; constrain the orbital plane orientation parameters
Clock error parameter	Each satellite two hours to set up a set of parameters, each receiver for each epoch to set up a clock error parameters (estimated at a ground station clock error)	Each epoch of each satellite is set for two differential parameters; select the prediction of a satellite clock difference as the benchmark bell

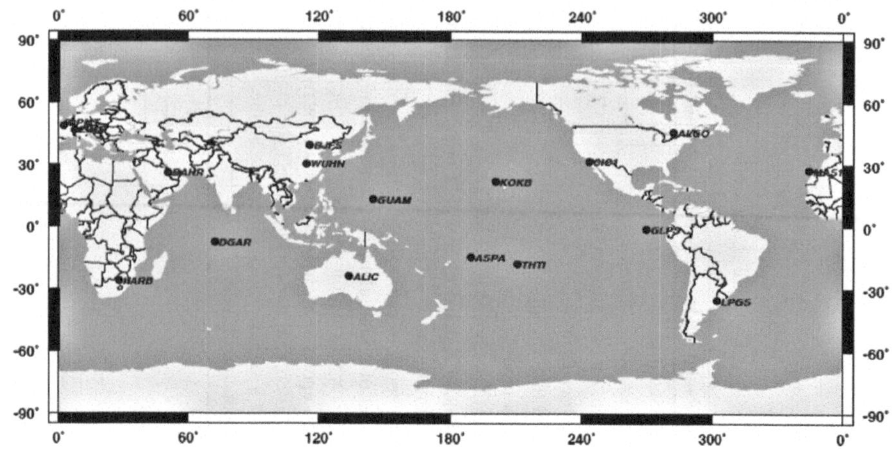

Fig. 1 Geographic distribution map of the 17 ground tracking stations in the world

Fig. 2 Geographic distribution map of the 5 ground tracking stations in China

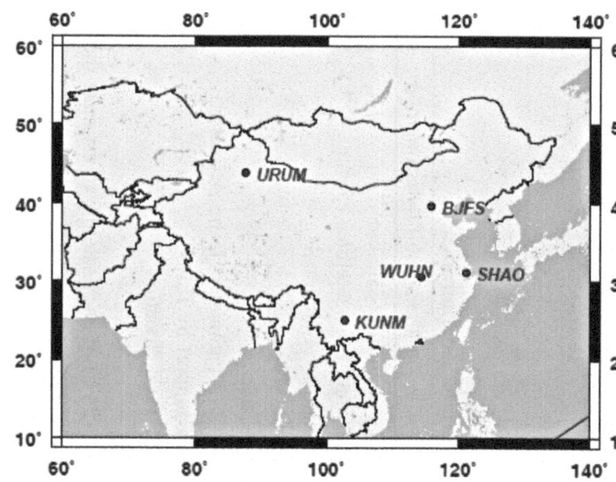

In order to facilitate the analysis, we did a survey the entire constellation in the radial direction of the orbital plane, clock error and the user ranging error index, expressed as the RERR, PERR, CERR and URE.

(1) **Distributed autonomous orbit determination accuracy without satellite-ground links**

The following Fig. 3 shows the 60 days of distributed autonomous orbit determination results. We could clearly see that with increasing time of autonomous orbit determination, the error of radial and orbital plane increases gradually, and ultimately affects the URE results.

(2) **Centralized autonomous orbit determination accuracy with satellite-ground links**

Table 2 Schemes of autonomous orbit determination

Scheme	Number of ground station	Ground station
1	1	BJFS
2	1	KUNM
3	1	SHAO
4	1	WUHN
5	1	KUNM
6	2	BJFS + KUNM
7	2	BJFS + URUM
8	2	BJFS + WUHN
9	3	SHAO + URUM
10	2	SHAO + KUNM
11	3	BJFS + SHAO + KUNM
12	3	BJFS + WUHN + URUM
13	3	BJFS + KUNM + URUM
14	3	SHAO + KUNM + URUM
15	4	BJFS + SHAO + URUM + KUNM
16	4	BJFS + WUHN + URUM + KUNM
17	5	BJFS + WUHN + URUM + SHAO
18	4	BJFS + WUHN + SHAO + KUNM
19	5	BJFS + WUHN + KUNM + URUM + SHAO
20	17	as Fig. 1

The following Fig. 4 shows the centralized autonomous orbit determination results with the supporting of BJFS station in 60 days. We could clearly see that with increasing time of autonomous orbit determination, the error of radial and orbital plane does not increase.

Compare to Figs. 3 and 4, we could reduce the constellation of the overall rotation and the absolute clock error, improve the autonomous navigation accuracy, and ensure the long-term precision and effectiveness of autonomous orbit determination with the method of the inter-satellite links distributed autonomous orbit determination adding the BJFS ground stations centralized autonomous orbit determination results.

2.4.2 Analysis of the Number of Satellite-Ground Links

In order to analyze the influence of the number and geographic distribution of ground stations to the centralized autonomous orbit determination, the paper completed 20 kinds calculation of autonomous orbit determination in accordance with the ground monitoring stations layout on the Table 2.

(1) **Analysis the number of ground stations to autonomous orbit determination**

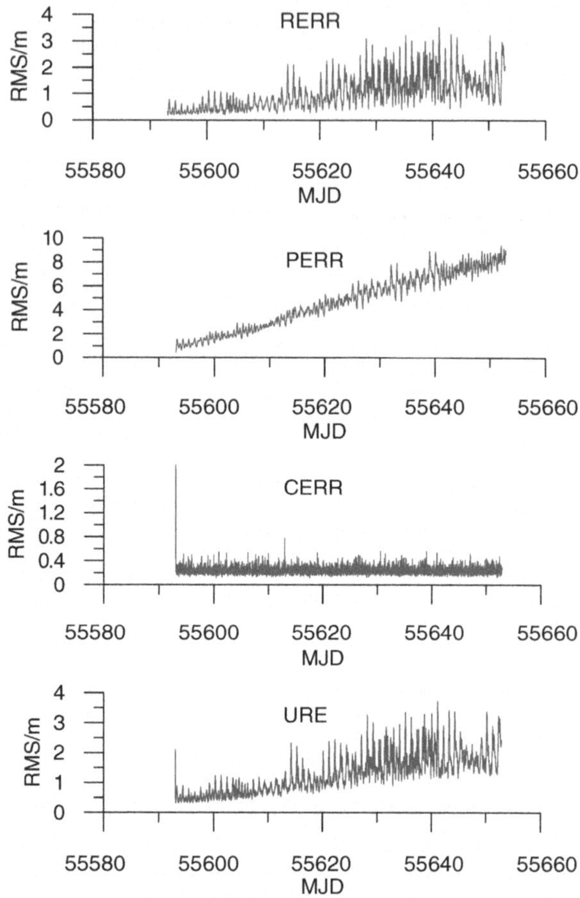

Fig. 3 Distributed autonomous orbit determination result without satellite-ground links

In order to analyze the impact of the number of the stations to autonomous orbit determination, the paper selected domestic 1 station, 2 Stations, 3 Stations, 4 Stations, 5 Stations and 17 stations in the world to complete autonomous orbit determination calculation of 7 days, the results are as Table 3 and Fig. 5.

From the Table 3 and Fig. 5, centralized autonomous orbit determination with the ground station supporting, more stations could not improve the accuracy of orbit determination, using only the single station, orbit determination orbital accuracy will be able to reach 1 m, clock error is better than 1 ns. The first reason is that the number of the inter-satellite distance observations is much larger than the satellite and ground observations, the second is the inter-satellite observations did not significantly improve the graphical structure, its role is limited to solve the autonomous orbit determination constellation rotation problem.

(2) **The influence of different geographic distribution of station to the combined orbit determination results**

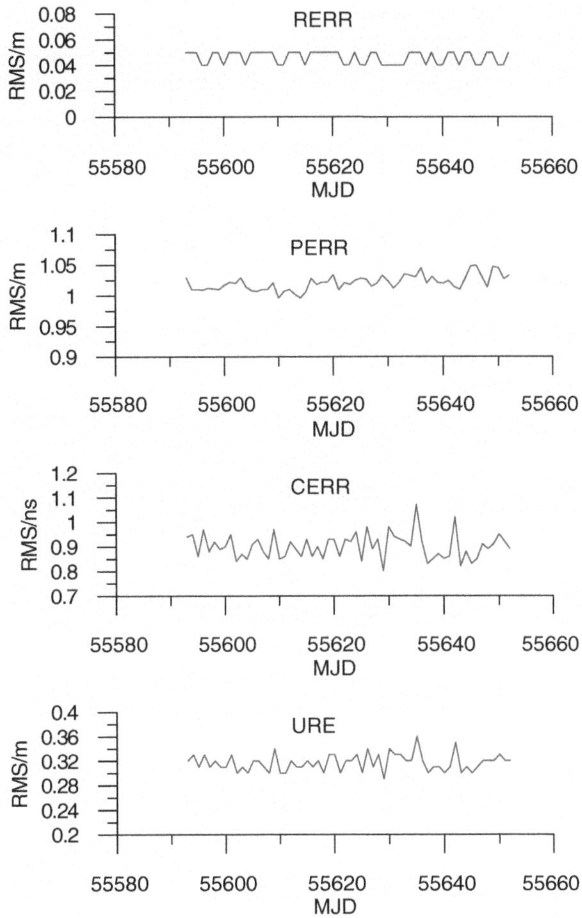

Fig. 4 Centralized autonomous orbit determination result with satellite-ground links

Table 3 Result of autonomous orbit determination with different ground stations

Scheme	RMS of 7 day average				
	Radial/m	Along/m	Cross/m	CLK/ns	URE/m
1	0.05	0.93	0.32	0.91	0.32
6	0.05	0.93	0.31	0.92	0.32
11	0.05	0.92	0.33	0.93	0.32
15	0.05	0.92	0.33	0.94	0.32
19	0.05	0.92	0.33	0.94	0.33
20	0.04	0.88	0.32	0.94	0.32

In order to analyze the impact of the geographical distribution of the stations to combined orbit determination, the paper selected domestic 1 station, 2 Stations, 3 Stations, 4 Stations to complete orbit determination calculation, the results are as follow Table 4 and Figs. 6, 7, 8 and 9.

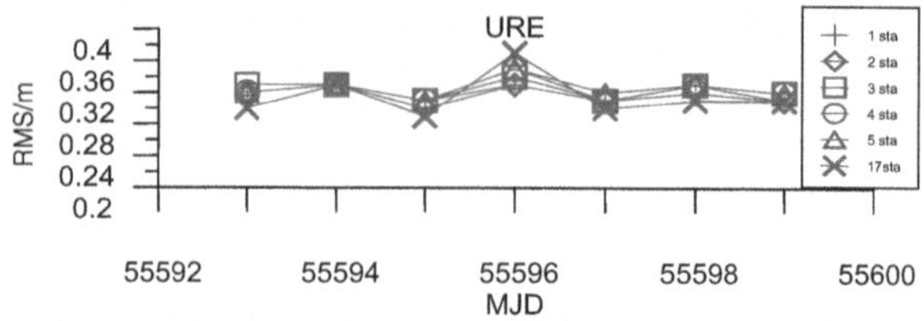

Fig. 5 Autonomous orbit determination result of 7 days on different stations

Table 4 Result of autonomous orbit determination with different geographic distribution ground stations

Scheme	RMS of 7 day average				
	Radial/m	Along/m	Cross/m	CLK/ns	URE/m
1	0.05	0.92	0.32	0.91	0.32
2	0.05	0.93	0.30	0.92	0.32
3	0.05	0.92	0.39	0.92	0.32
4	0.05	0.93	0.30	0.92	0.32
5	0.05	0.92	0.35	0.92	0.32
6	0.05	0.93	0.31	0.92	0.32
7	0.05	0.92	0.33	0.92	0.32
8	0.05	0.93	0.31	0.92	0.32
9	0.05	0.92	0.36	0.92	0.32
10	0.05	0.93	0.34	0.92	0.32
11	0.05	0.92	0.33	0.93	0.32
12	0.04	0.92	0.32	0.92	0.32
13	0.04	0.92	0.32	0.92	0.32
14	0.05	0.92	0.34	0.93	0.32
15	0.05	0.92	0.33	0.94	0.32
16	0.04	0.92	0.32	0.93	0.32
17	0.05	0.92	0.33	0.93	0.32
18	0.05	0.92	0.32	0.94	0.32

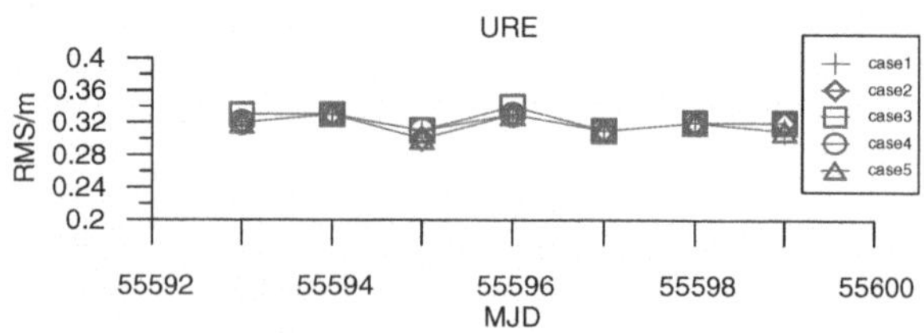

Fig. 6 Autonomous orbit determination URE result on different domestic stations

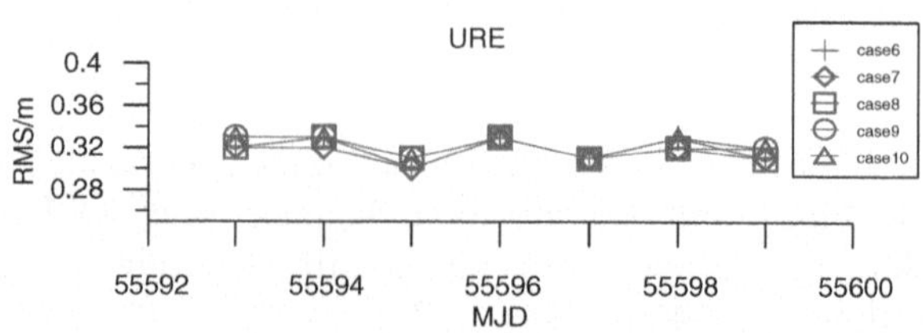

Fig. 7 Autonomous orbit determination URE result on different two domestic stations

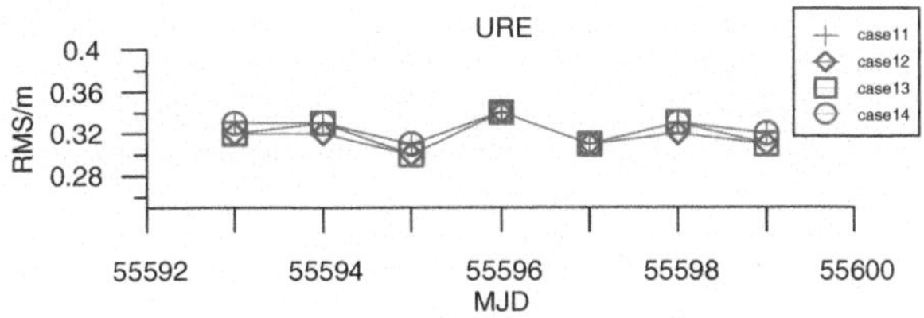

Fig. 8 Autonomous orbit determination URE result on different three domestic stations

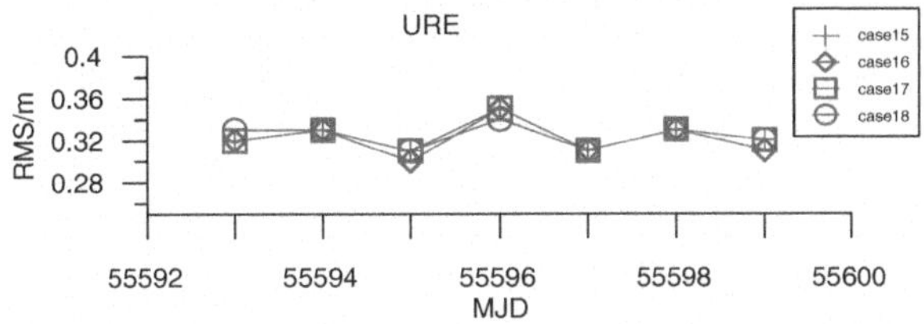

Fig. 9 Autonomous orbit determination URE result on different four domestic stations

On summary, given number of ground stations, the effect of different geographical distribution of the stations to orbit determination is very limited. Regardless of the orbit and clock error, the differences are very small.

3 Conclusions

In this context, we proposed the use of satellite-ground links for autonomous navigation of constellation based on the traditional methods and their demerits, its operating mode is also analyzed especially for the effects of satellite-ground links on the autonomous navigation of constellation. The preliminary results which can be concluded from the above analysis are 1) when a ground station is added, the constellation rotation and absolute clock bias increment can be decreased leading to the improvement of autonomous navigation precision; 2) when the increment of the number and the improvement of geographical distribution of ground stations cannot enhance the centralized autonomous time synchronization significantly.

References

1. Rajan, J. A. (2002). Highlights of GPS II-R autonomous navigation. In: *Proceedings of ION 58th Annual Meeting and CIGTF 21st Guidance Test Symposium, June 24–26* (pp. 354–363). New Mexico, VA: Institute of Navigation.
2. Liu, W. K. (1990). Research and simulation on autonomous. In M. P. Ananada, H. Bernstein & K. E. Cunningham, et al. (Eds.), Global positioning system autonomous navigation. *IEEE Position Location and Navigation Symposium*, Las Vegas, Nevada.
3. Liu, W. K. (2008). *Research and simulation on autonomous orbit determination and combined orbit determination of navigation satellites.* Dissertation for the doctoral degree, Wuhan University, Wuhan

4. Liu, L., & Liu, Y. (2000). The problem of rank defect about inter-satellite relative measurements. *Aerocraft Monitor Transaction*, *29*(3), 80–83.
5. Liu, Y. C., Liu, L., & Wang, C. B. (2000). On the orbit determination combined the satellite–satellite tracking and the ground based tracking. *Publication of Purple Mountain Observatory, 19*(2), 17–120 (in Chinese).
6. Zhang, Y. (2005). *Study on autonomous navigation of constellation using inter-satellite measurement*. Dissertation for the doctoral degree (in Chinese), National University of Defense Technology, Changsha.
7. Rajan, J. A., Brodie, P., Rawicz, H., et al. (2003). Modernizing GPS autonomous navigation with anchor capability. *Proceedings of the Institute of Navigation, GNSS 2003, Portland.*
8. Liu, W. K., Li, Z. H., Gong, X. Y., & Wang, F. H. (2010). Combined orbit determination of navigation satellites with crosslink ranging observations and ground tracking observations. *Geomatics and Information Science of Wuhan University, 35*(7), 811–815 (in Chinese).

Research on Integrated Orbit Determination Combined Satellite-Ground and Inter-Satellite Observation Based on Helmert Method of Variance Components Estimate

Xing Su, Tao Geng, Qile Zhao, Lizhong Qu and Xingkai Li

Abstract Because of the different precision of the satellite-ground and inter-satellite observations, how to correctly determine the weight of the two kinds of observations is still a key technique in integrated orbit determination. According to the ground stations and different kinds of the inter-satellite cross-link of the Chinese COMPASS satellite navigation system, the weight of the two kinds of observations was determined by the Helmert method of variance components estimate, and then the new weight was used in the integrated orbit determination. The results show that this method could reasonably determine the weight of satellite-ground and inter-satellite observations and improve the precision of integrated orbit determination. For the integrated orbit determination based on the Satellite-Ground and Inter-Satellite cross-link fixed antenna data, the orbit accuracies improve from 0.15 to 0.04 m after use the Helmert estimation weight to replace the priori. And the result is 0.06–0.03 m for the Satellite-Ground and Inter-Satellite cross-link UHF antenna data.

X. Su (✉) · T. Geng · Q. Zhao · L. Qu · X. Li
Research Center of GNSS, Wuhan University, Wuhan, China
e-mail: longtian00@163.com

T. Geng
e-mail: gt_gengtao@whu.edu.cn

Q. Zhao
e-mail: zhaoql@whu.edu.cn

L. Qu
e-mail: winnerzz@126.com

X. Li
e-mail: lxky2005@126.com

J. Sun et al. (eds.), *China Satellite Navigation Conference (CSNC) 2012 Proceedings*,
Lecture Notes in Electrical Engineering 160, DOI: 10.1007/978-3-642-29175-3_31,

1 Introduction

The technology of orbit determination and time synchronization of navigation satellites is a significant subject in the construction and operation of navigation system, and impacts directly on the positioning accuracy of the whole system. There are some problems in orbit determination of the GEO such that the ground observation structure is bad and its system error is difficult to detect [1]. The elevation of the IGSO changes slowly, and its observation structure is not very well. Only with the continuous observation data from the evenly-distributed global stations, the orbit determination accuracies of the MEO could be very precise and the navigation system could operate well. But if the tracking stations are distributed regionally, the requirements of the high-precision positioning service are difficult to satisfy because of the limited satellites tracking arc, bad ground observation structure and low precision of the orbit determination of the navigation system [2, 3]. In order to overcome the shortcomings of the discontinuous tracking arc and improve the precision of the navigation constellation orbit determination, the cross-link technology is adopted in the navigation system orbit determination [4, 5].

Because of the differences of observation noise and system error between satellite-ground and inter-satellite observation data [6], the weight determination of two kinds of observations is an important technology in integrated orbit determination. Zhang Feipeng proposed that Helmert variance component estimation could be used in the weight determining of the different kinds of the satellites observation data [7], and this method had been used in the orbit determination experiments of ERS-2 and Lageos2 to improve the accuracy [8, 9].

With the construction of China COMPASS system [10], the study of integrated orbit determination by combining the satellite-ground with inter-satellite data based on the COMPASS system is very urgent.

This paper introduces the mathematical principle of the Helmert variance component estimation in Sect. 2. Section 3 introduces the COMPASS navigation system and simulates the satellite-ground, inter-satellite fixed link and UHF antenna link data, meanwhile the satellite-ground data and satellite-ground together with inter-satellite data are compared by using the Helmert method in the integrated orbit determination. Section 4 illustrates the conclusions.

2 Helmert Method of Variance Components Estimate Mathematic Principle

$\underset{n_1\times 1}{L_1}$, $\underset{n_2\times 1}{L_2}$, …, $\underset{n_m\times 1}{L_m}$ etc. are m kinds of independent observations in L, and their weight matrixes are $\underset{n_1\times n_1}{P_1}$, $\underset{n_2\times n_2}{P_2}$, …, $\underset{n_m\times n_m}{P_m}$ etc. So the error equation is

$$\underset{n_i\times 1}{V_i} = \underset{n_i\times t}{B_i}\ \underset{t\times 1}{\hat{X}} - \underset{n_i\times 1}{L_i},\ (i = 1, 2, \cdots, m) \tag{2.1}$$

$$D(L_i) = \sigma_{0_i}^2 P_i^{-1} \tag{2.2}$$

Let the

$$\begin{aligned} N &= B^T PB = B_1^T P_1 B_1 + B_2^T P_2 B_2 + \cdots + B_m^T P_m B_m \\ &= N_1 + N_2 + \cdots + N_m \end{aligned} \tag{2.3}$$

$$\begin{aligned} W &= B^T PL = B_1^T P_1 L_1 + B_2^T P_2 L_2 + \cdots + B_m^T P_m L_m \\ &= W_1 + W_2 + \cdots + W_m \end{aligned} \tag{2.4}$$

Then the equation

$$\hat{X} = N^{-1} W = N^{-1} \sum_{j=1}^{m} W_j \tag{2.5}$$

is substituted in Eq. (2.1), and the Eq. (2.1) is sorted out to be

$$\begin{aligned} V_i &= B_i \hat{X} - L_i = B_i N^{-1} \sum_{j=1}^{m} W_j \\ &= (B_i N^{-1} B_i^T P_i - 3) L_i + B_i N^{-1} \sum_{j=1, j\neq i}^{m} B_j^T P_j L_j \end{aligned} \tag{2.6}$$

Obeyed the covariance propagation law, the covariance of the errors is obtained as

$$\begin{aligned} D(V_i) &= (P_i^{-1} + B_i N^{-1} N_i N^{-1} B_i^T - 2B_i N^{-1} B_i^T) \sigma_{0_i}^2 \\ &\quad + \sum_{j=1, j\neq i}^{m} \left\{ (B_i N^{-1} N_j N^{-1} B_i^T) \sigma_{0_j}^2 \right\} \end{aligned} \tag{2.7}$$

On the basis of the quadratic form expectation law:

$$E(Y^T MY) = tr(M\Sigma) + \eta^T M\eta \tag{2.8}$$

and $Y = V_i$, $M = P_i$, $\Sigma = D(V_i)$, $\eta = E(V_i) = 0$ are taken into account, the expectation of the observation variance is gotten as

$$\begin{aligned} E(V_i^T P_i V_i) &= tr(P_i P_i^{-1} + P_i B_i N^{-1} N_i N^{-1} B_i^T - 2P_i B_i N^{-1} B_i^T) \sigma_{0_i}^2 \\ &\quad + \sum_{j=1, j\neq i}^{m} \left\{ tr\left(P_i B_i N^{-1} N_j N^{-1} B_i^T\right) \sigma_{0_j}^2 \right\}, \end{aligned} \tag{2.9}$$

Then the Eq. (2.9) could change to

$$E(V_i^T P_i V_i) = (n_i - 2tr(N^{-1}N_i) + tr(N^{-1}N_i)^2)\sigma_{0_i}^2 + \sum_{j=1, j \neq i}^{m} \left\{ tr(N^{-1}N_i N^{-1}N_j)\sigma_{0_j}^2 \right\}. \tag{2.10}$$

There are m parameters needed to be estimated in the Eq. (2.10). It could be written in matrix form, so it is the Helmert variance component estimation equation of m kinds of observations:

$$\underset{m\times m}{S}\ \underset{m\times 1}{\hat{\theta}} = \underset{m\times 1}{W_\theta} \tag{2.11}$$

where

$$S = \begin{bmatrix} \begin{matrix} n_1 - 2tr(N^{-1}N_1) \\ + tr(N^{-1}N_1)^2 \end{matrix} & tr(N^{-1}N_1N^{-1}N_2) & \cdots & tr(N^{-1}N_1N^{-1}N_m) \\ \cdots & \begin{matrix} n_2 - 2tr(N^{-1}N_2) \\ + tr(N^{-1}N_2)^2 \end{matrix} & \cdots & tr(N^{-1}N_2N^{-1}N_m) \\ \vdots & \vdots & \ddots & \vdots \\ \text{(symmetrical)} & \cdots & \cdots & \begin{matrix} n_m - 2tr(N^{-1}N_m) \\ + tr(N^{-1}N_m)^2 \end{matrix} \end{bmatrix},$$

$$\hat{\theta} = \begin{bmatrix} \hat{\sigma}_{0_1}^2 & \hat{\sigma}_{0_2}^2 & \cdots & \hat{\sigma}_{0_m}^2 \end{bmatrix}^T,$$

$$W_\theta = \begin{bmatrix} V_1^T P_1 V_1 & V_2^T P_2 V_2 & \cdots & V_m^T P_m V_m \end{bmatrix}^T.$$

The resolution of the Eq. (2.11) is

$$\hat{\theta} = S^{-1} W_\theta. \tag{2.12}$$

What has been discussed above is the strict formula of Helmert variance component estimation [11]. But the simplified one is widely used in practical application for the faster speed and saving operation space.

The part of computation of the matrix trace in the strict Eq. (2.11) is omitted, so the simplified one is described as $V_i^T P_i V_i = n_i \hat{\sigma}_{0_i}^2$. The estimation of unit weight variance of the different kinds of observations is $\hat{\sigma}_{0_i}^2 = (V_i^T P_i V_i)/n_i$.

The iteration process of Helmert variance component estimation is described as follows:

(1) All the observations are sorted into different kinds depending on the accuracy grade or the observation type. The initial weights of all kinds of observations (P_1, P_2, …, P_m) are determined by the priori variances.
(2) The parameter estimation is done to get the $V_i^T P_i V_i$ for the first time.

(3) Then the variance component estimation is done following the Eq. (2.12) for the first time to get the unit weight variances of all kinds of observations ($\hat{\sigma}_{0_i}^2$). The weights are determined by the estimated variances again as follows:

$$\hat{P}_i = c/(\hat{\sigma}_{0_i}^2 P_i^{-1}), \tag{2.13}$$

where c is an arbitrary constant, generally it is chosen from $\hat{\sigma}_{0_i}^2$.

(4) Repeating the step (2) and (3), which means: parameter estimation—Helmert variance component estimation—weight determination and new parameter estimation, until $\hat{\sigma}_{0_1}^2 = \hat{\sigma}_{0_2}^2 = \cdots = \hat{\sigma}_{0_m}^2$ or the ratio of all kinds of observations equal 1.

3 Helmert Method used in Integrated Orbit Determination Combined Satellite-Ground and Inter-Satellite Observation Data

3.1 *Observation Data Simulation*

3.1.1 COMPASS Constellation Simulation

The COMPASS constellation orbits were used the target of the system by 2020, which are 3GEOs + 3IGSOs + 24MEOs. The PRN number 01, 02, and 03 satellites are GEO and launched above the east longitude 80°, 110.5°, and 140°, respectively. The PRN number 04, 05, and 06 satellites are IGSO, which are distributed on 3 different orbit planes but the same sub-point track. The PRN number 07–30 are MEO, each of them has a 27878 km semi-major axis and 55° inclination orbit, and they are well-distributed on 3 different orbit planes. The sub-point tracks of the constellation are presented in Fig. 1.

3.1.2 Ground Tracking Stations

The six ground tracking stations used for simulating satellite-ground observation data are distributed in China region, as presented in Fig. 2.

3.1.3 Satellite-Ground Observation Data Simulation

For the satellite-ground observation data, the pseudo-range measurement $\Re_1$ from the transmitting antenna k on the satellite j to the receiving antenna n on station m is computed at the receive time t_i as follow:

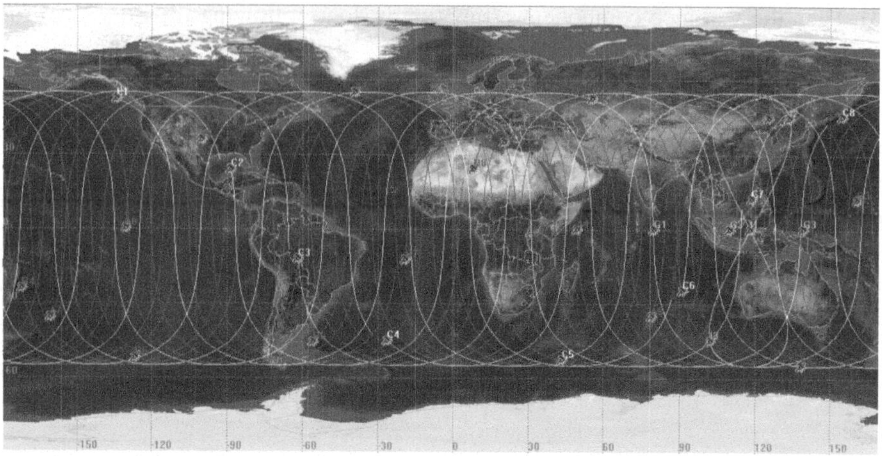

Fig. 1 COMPASS Navigation satellite sub-point track

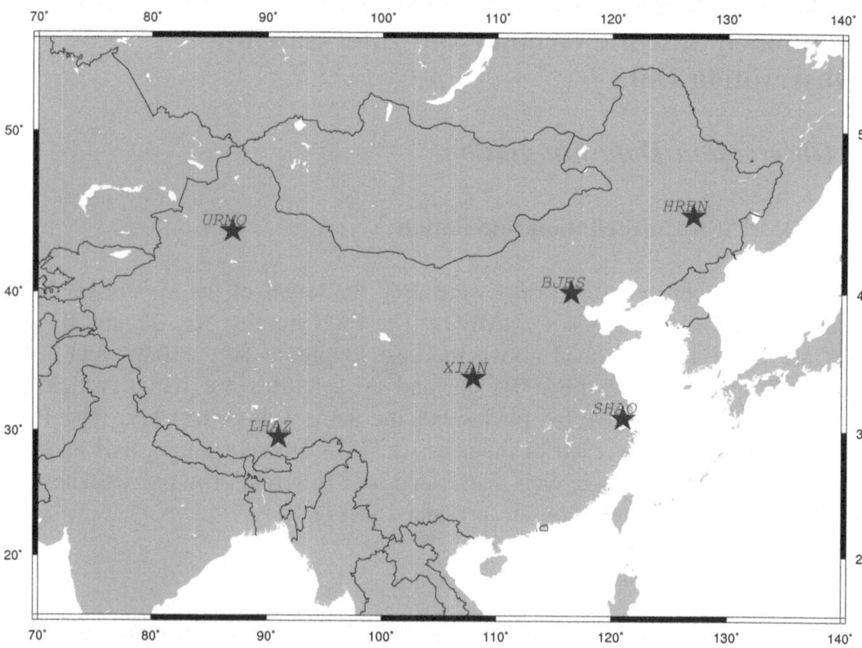

Fig. 2 Distribution of ground tracking stations

$$\Re_1^{(j)}(t_i) = \rho_1^{(m,j)}(t_i) + (b_r^{(m)}(t_i) - b_t^{(j)}(t_t)) \cdot c + \Delta\rho_{iono}^{(j)}(t_i) + \Delta\rho_{tropo}^{(j)}(t_i) + \sigma_n^{(j)}(t_i) + b_m(t_i)$$

where $\rho_1^{(m,j)}$ is the geometrical distance between the transmitting and receiving antennas, b_r and b_t are the clock bias parameters for the transmitter and receiver, t_t

is the transmission time, c is the speed of light, $\Delta\rho_{iono}$ and $\Delta\rho_{tropo}$ are the ionospheric and tropospheric refraction delay respectively, σ_n is the random measurement noise, and b_m is the pseudo-range measurement bias.

The carrier-phase measurement C at the receive time t_i using transmitter j is given by:

$$C^{(j)}(t_i) = \frac{\Re_c^{(j)}(t_i)}{\lambda_C} - N^{(j)}$$
$$\Re_c^{(j)}(t_i) = \rho_1^{(m,j)}(t_i) + (b_r^{(m)}(t_i) - b_t^{(j)}(t_t)) \cdot c + \Delta\rho_{iono}^{(j)}(t_i) + \Delta\rho_{tropo}^{(j)}(t_i) + \sigma_{n,c}^{(j)}(t_i) + b_{m,c}(t_i)$$

where $\sigma_{n,c}$ is the random measurement noise, $b_{m,c}$ is the measurement bias, λ_C is the carrier signal wavelength.

In this experiment, the pseudo-range error was set as 1 m, the carrier phase error is 0.01 cycles, and the system error is 3 cm. Three days observation data was simulated.

3.1.4 Inter-Satellite Observation Data Simulation

For the inter-satellite observation data, two different kinds of cross-link, fixed link and UHF antenna link, were used. The fixed link is that every MEO links with a MEO on the neighboring orbit plane and two MEO on the same orbit plane, the three GEO and three IGSO link with one MEO on each orbit plane. The UHF antenna link is that the transmitting signal angle is 15°–50°. In the data simulation, the range measure error was set as 0.1 m, and the system error is 0.1 m too.

3.2 Satellite-Ground Data Result Analysis

Using the six ground stations only and regarding the pseudo-range and carrier phase data as two different kinds of observation types, the estimate results of priori weight and Helmert variance components estimate weight were analyzed respectively. The priori weights were determined by the priori standard deviations, which of the pseudo-range is $\sigma_{0_{SR}} = 1.0$ m and the carrier phase is $\sigma_{0_{SP}} = 1.903$ mm.

After 3 time iterations, the unit weight variances of pseudo-range and carrier phase of the ground stations were approach to 1 to 1, then the progress of the iteration was stopped. The standard deviations were changed to $\sigma_{0_{SR}} = 2.96$ m and $\sigma_{0_{SP}} = 57.5$ mm.

There are only two kinds of observation types in ground stations data, and all the six stations are located in China region. The results of the orbit determination are not very well because of its short continuous tracking curve and a little number

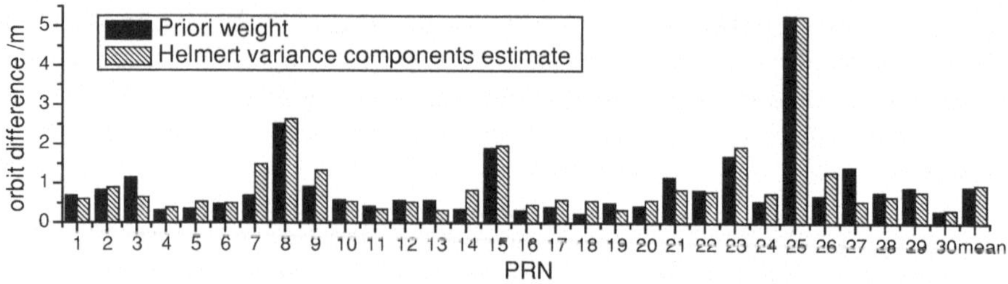

Fig. 3 The comparison of Orbit determination precision between priori and Helmert estimation based on SatGround data

of observation data. The mean orbit determination accuracy of using priori and Helmert method weight is 0.97 and 0.93 m, respectively (Fig. 3).

3.3 Combined Satellite-Ground and Inter-Satellite Fixed Link Data Integrated Orbit Determination Analysis

The integrated orbit determination was analyzed by combining the satellite-ground pseudo-range and carrier phase data with the inter-satellite fixed link data, so there will be three kinds of observation data in the Helmert variance estimation.

(1) The priori standard variances of the pseudo-range and carrier in satellite-ground are $\sigma_{0_{SR}} = 1.0\,\text{m}$, and $\sigma_{0_{SP}} = 1.903$ mm. The priori standard variance of inter-satellite range measurements is $\sigma_{0_N} = 0.15$ m. The three priori standard variances are used to determine the weights for the first time, which are

$$P_{SR} = \sigma^2_{0_{SR}}/\sigma^2_{0_{SR}} = 1, P_{SP} = \sigma^2_{0_{SR}}/\sigma^2_{0_{SP}} = 276135.6, P_N = \sigma^2_{0_{SR}}/\sigma^2_{0_N} = 44.44$$

The unit weight variances of the three kinds of observation data ($\hat{\sigma}^2_{0_{SR}}$, $\hat{\sigma}^2_{0_{SP}}$ and $\hat{\sigma}^2_{0_N}$) in the first parameter estimation are used to compute the estimated variances of the satellite-ground and inter-satellite observation data, as the following equations.

$$\hat{\sigma}^2_{SR} = \hat{\sigma}^2_{0_{SR}} P^{-1}_{SR} = 8.78\ \text{m}^2, \hat{\sigma}^2_{SP} = \hat{\sigma}^2_{0_{SP}} P^{-1}_{SP} = 0.00317\ \text{m}^2, \hat{\sigma}^2_N = \hat{\sigma}^2_{0_N} P^{-1}_N = 0.01299\ \text{m}^2$$

(2) The estimated variances in the first time are used to determine the weights for the second time.

$$P_{SR} = \hat{\sigma}^2_{0_{SR}}/\hat{\sigma}^2_{0_{SR}} = 1, P_{SP} = \hat{\sigma}^2_{0_{SR}}/\hat{\sigma}^2_{0_{SP}} = 2768.8, P_N = \hat{\sigma}^2_{0_{SR}}/\hat{\sigma}^2_{0_N} = 676$$

Table 1 The iteration progress of Helmert variance component estimation

Iteration times	P_i (weight)	$\hat{\sigma}_i^2$ (observation variance)	$\hat{\sigma}_{0_i}^2$ (unit weight variance)	$\sigma_{0_{SR}}^2 : \sigma_{0_{SP}}^2 : \sigma_{0_N}^2$
1	1,276135,44.44	8.7,0.0031,0.012	8.78,876.1,0.5776	1:99.7:0.0657
2	1,2768,676	8.7,0.0040,0.008	0.99,1.2,0.62	1:1.289:0.6236
3	1,2147,1083	8.7,0.0047,0.007	1,1.17,0.95	1:1.174:0.956
4	1,1829,1133	8.7,0.0050,0.007	1,1.06,0.97	1:1.061:0.977
5	1,1723,1159	8.7,0.0052,0.007	1,1.02,1	1:1.021:1.000
6	1,1687,1159	8.7,0.0052,0.007	1,1.005,1	1:1.005:1.000
7	1,1678,1159	8.7,0.0052,0.007	1,1,1	1:1:1

And then the estimated variances of the observations are computed by the unit weight variances which are calculated in the second parameter estimation.

$$\hat{\sigma}_{SR}^2 = \hat{\sigma}_{0_{SR}}^2 P_{SR}^{-1} = 8.779 \text{ m}^2, \hat{\sigma}_{SP}^2 = \hat{\sigma}_{0_{SP}}^2 P_{SP}^{-1} = 0.00408 \text{ m}^2, \hat{\sigma}_N^2 = \hat{\sigma}_{0_N}^2 P_N^{-1} = 0.0081 \text{ m}^2$$

(3) The computation processes of the step (2) are repeated for weight determination, parameter estimation, Helmert estimation the third time. The weights are $P_{SR} = 1$, $P_{SP} = 2147.4$, and $P_N = 1083$.
(4) The processes of weight determination, parameter estimation, Helmert estimation are reiterated until the unit weight variances of the three kinds of observation data are approach 1 to 1 to 1. The progresses of the iterations are collected in the Table 1.

After the progresses of the Helmert variance estimation are completed, the orbit differences of the navigation constellation are computed and then compared to the results of the first iteration used the priori weight in the parameter estimation.

As is depicted in the Fig. 4, the mean orbit determination error reduces from 0.15 to 0.04 m after using the Helmert weights in the parameter estimation. Comparing to the 0.97 m in satellite-ground data orbit determination, the orbit error improves obviously after adding the inter-satellite observation data.

3.4 Combined Satellite-Ground and UHF Antenna Link Data Integrated Orbit Determination Analysis

The results of integrated orbit determination combined satellite-ground with UHF antenna link data are shown in Fig. 5. The mean orbit determination error reduces from 0.06 to 0.03 m after using the Helmert weights in the parameter estimation.

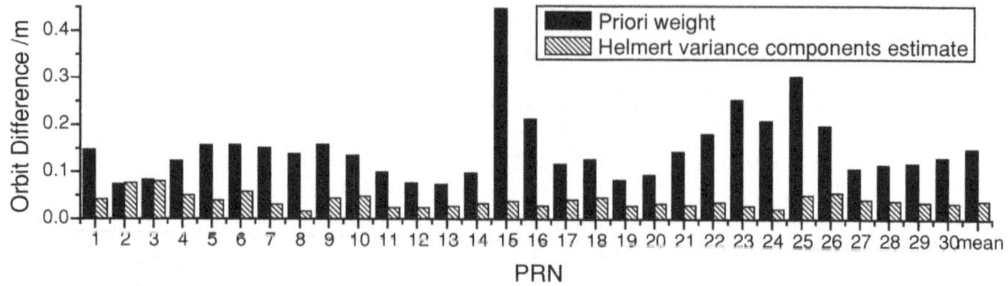

Fig. 4 The comparison of Orbit determination precision between priori and Helmert estimation based on SatGround and cross-link fix data

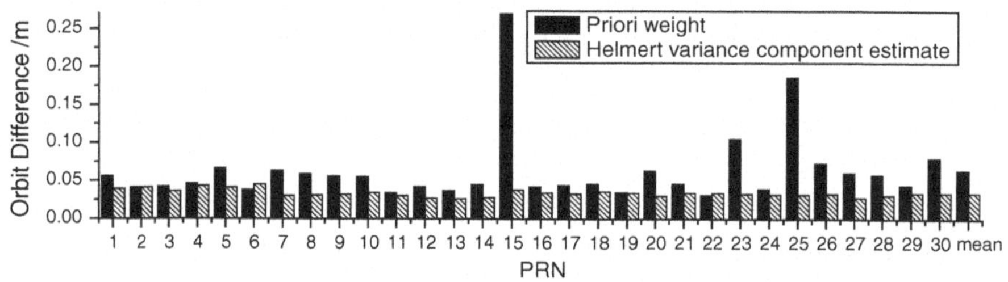

Fig. 5 The comparison of Orbit determination precision between priori and Helmert estimation based on SatGround and cross-link UHF antenna data

4 Summaries

Based on the target of satellite constellation of China COMPASS navigation system in 2020, the integrated orbit determinations using satellite-ground data only, satellite-ground and inter-satellite fixed link data, satellite-ground and UHF antenna link data are done by means of priori and Helmert weight. The results illustrate:

(1) Comparing to using the satellite-ground observation data only in the integrated orbit determination, the accuracy of the orbit could be improved obviously by using the satellite-ground and inter-satellite data.
(2) Helmert variance component estimation is an effective method in reasonably determining the weights of different kinds in satellite-ground and inter-satellite data. In this paper, for the satellite-ground and inter-satellite fixed link data, the mean orbit accuracy improves from 0.15 to 0.04 m, and so does 0.06–0.03 m in using satellite-ground and inter-satellite fixed link data.

Acknowledgments This study is sponsored by the Fundamental Research Fund for the Central Universities, Specialized Research Fund for the Doctoral Program of Higher Education (No. 20110141120076), and Open Research Fund Program of the Key Laboratory of Geospace Environment and Geodesy, Ministry of Education, China (No. 10-02-01).

References

1. Ou, J., Liu, J., & Sun, B. (2007). Precision orbit determination of a geostationary satellite geo with mirror surface projection method. *Geomatics and Information Science of Wuhan University, 11*, 975–979, 997.
2. Wei, Z., & Ge, M. (1998). *GPS relative positioning mathematical model*. Beijing: Survey and Mapping Press.
3. Geng, T. (2009). *Real-time precise orbit determination theory for navigation satellite and its experimental application based on regional reference stations*. Dissertation, Wuhan University, Wuhan.
4. Rajan, J. A., Brodie, P., & Rawicz, H. (2003). Modernizing GPS autonomous navigation with anchor capability. *ION GPS/GNSS* (pp. 1534–1543). Portland: [s. n.].
5. Shuai, P. (2009). *X-ray pulsar-based navigation system theory and methods*. Beijing: China Space Press.
6. Liu, W. (2008). *Research and simulation on autonomous orbit determination and combined orbit determination of navigation satellite*. Dissertation, Wuhan University, Wuhan.
7. Zhang, F. (2000). The A posteriori estimation of stochastic model in statistical orbit determination. *ACTA Astronomica Sinica, 41*(2), 129–137.
8. Zhang, F. (2000). Application of the improved Helmert method for variance component estimation in precise orbit determination. *ACTA Geodaetica et Cartographica Sinica, 29*(3), 216–233.
9. Zhang, F. (2001). Application of variance component estimation to precision orbit determination for ERS-2. *ACTA Astronomica Sinica, 42*(2), 173–183.
10. Ran, C. (2010). Plan of compass development. Beijing: Chinese Compass Satellite Navigation Systems Management Office.
11. Cui, X. (2009). *Generalized surveying adjustment*. Wuhan: Wuhan University Press.

Combined Prediction of Earth Orientation Parameters

X. Q. Xu, L. Zotov and Y. H. Zhou

Abstract Earth orientation parameters (EOP) are essential for transformation between the celestial and terrestrial coordinate systems, which has important applications in the Earth sciences, astronomy and satellite navigation. The latter cannot be considered as self-reliant without accurate EOP predictions, which, in particular, are required for real-time precise orbit determination. In this paper we firstly describe the principles and analyze the characteristics of several EOP prediction methods. Then, according to the forecast accuracy, the weights are assigned to EOP predictions to get a combined solution. Results show that no single forecasting method is suitable for all the parameters at different time spans. The combined solution integrates advantages of different prediction methods, which avoid the limitations and instability of any of them. The resultant combined prediction is issued to the JPL Earth Orientation Parameters Combination of Prediction Pilot Project (EOPC PPP).

X. Q. Xu (✉) · L. Zotov · Y. H. Zhou
Shanghai Astronomical Observatory, Chinese Academy of Sciences,
200030 Shanghai, China
e-mail: xqxu@shao.ac.cn

X. Q. Xu
Science and Technology on Aerospace Flight Dynamics Laboratory,
100076 Beijing, China

L. Zotov
Sternberg Astronomical Institute of Moscow State University,
119992 Moscow, Russia

J. Sun et al. (eds.), *China Satellite Navigation Conference (CSNC) 2012 Proceedings*, Lecture Notes in Electrical Engineering 160, DOI: 10.1007/978-3-642-29175-3_32,

1 Introduction

Earth rotation (speed and axis orientation) has been monitored in recent decades with increasing accuracy by advanced space-geodetic techniques, including Lunar and Satellite Laser Ranging, Very Long Baseline Interferometry and Global Positioning System [11, 13]. The observational precision now reaches $\sim 100\ \mu as$ for X and Y coordinates of the pole (PMX and PMY), ~ 10 μs for the length of day (LOD) variations (10 μs corresponds to 150 μas and ~ 5 mm on the Earth sphere). But due to the complex data processing, Earth Orientation Parameters (EOP) are usually available with a delay of hours to days. The growing demands by spacecraft tracking and navigation requires real-time EOP data and the questions of safety and stability requires the EOP predictions at least two weeks and even longer in the future, what prompts the respective research.

Earth rotation is sensitive to global processes in the atmosphere, ocean, crust and Earth interior. Changes of pole position and LOD can be considered as a general response to these processes in the earth systems. The complexity of input factors, presence of nonlinearities and nonstationarities make EOP prediction quite complex. The state of the art here is much verse than the quality of observations [18].

Nevertheless, the results of the Earth Orientation Parameters Prediction Comparison Campaign (EOP PCC) took place in 2005–2009 has shown, that the quality of predictions can be improved, by incorporating of the contemporary physical models and forecasts of angular momentum changes. EOP PCC was initiated as an international competition, in purpose to estimate the accuracy of the EOP predictions and provoke the improvement [8]. It attracted 13 participants from various countries and institutes, and collected almost 6,500 submissions by a number of techniques, e.g., (1) the least squares extrapolation (LS) of a harmonic model and autoregressive (AR) prediction [8, 19], (2) Spectral analysis and least squares extrapolation [2], (3) Artificial Neural Networks (ANN) [12, 15, 19], (4) Least Squares Collocation (LSC) [19], (5) Wavelet decomposition and auto-covariance prediction [9, 10], (6) Adaptive transformation from the atmospheric angular momentum to length-of-day (LOD) change [8], (7) Kalman filter with atmospheric angular momentum forecasts [5, 6, 17].

The major conclusion reached by the EOP PCC was that there is no particular prediction technique superior to the others for all EOP components and all prediction intervals. While some techniques (1–3) work well for polar motion (PMX, PMY) prediction, the other (5–7) show better results for UT1-UTC and LOD forecasts. And the combined solution performs better than all individual predictions techniques.

Based on the EOP PCC results, International Earth Rotation and Reference Systems Service (IERS) and Jet Propulsion Laboratory (JPL) initiated the Earth Orientation Parameters Combination of Prediction Pilot Project (EOPC PPP) in 2010. The main goal is to develop a strategy for predictions combinations. Authors of this study are participants of EOPC PPP, the methods of EOP prediction presented here are under submission to this competition.

In this study, we investigate the statistics of EOP predictions by different methods such as LS, AR, LSC and ANN, based on mathematical approximation models, and then develop the combined prediction strategy. Section 2 describes the principle of LS, AR, LSC, and ANN models, which are the methods of prediction used. Section 3 explains errors estimation and describes the combination method and shows the examples. We finally summarize the results in Sect. 4.

2 Mathematical Approximation Models

2.1 Least-Squares Method

Polynomial trend and harmonic oscillations often present in the time series. EOP of our interest also includes them, which can be captured by spectral analysis [7]. In this study, we use the following model to fit the trend and periodic terms of EOP, which parameters can be estimated by the least-squares method, and the residuals are then processed by the AR and other models.

$$X_t = A + Bt + C\cos\left(\frac{2\pi t}{p_1}\right) + D\sin\left(\frac{2\pi t}{p_1}\right) + E\cos\left(\frac{2\pi t}{p_2}\right) + F\sin\left(\frac{2\pi t}{p_2}\right), \quad (2.1)$$

where t is time, and A, B, C, D, E, F are unknown amplitudes to be estimated. In the case of UT1-UTC and LOD, the main period terms is annual and semiannual oscillations, $p_1 = 365.24$, $p_2 = 182.62$. For polar motion (PMX, PMY) Chandler and annual harmonics with periods $p_1 = 435.00$, $p_2 = 365.24$ are under estimation.

2.2 AR Model

After deterministic part of the time series is modeled and subtracted, a stochastic process model can be used for the residuals, where the AR (p) model [3] is used. For a stationary random sequence $z_t(t = 1, 2, \ldots, N)$, it is expressed as follows,

$$z_t = \sum_{i=1}^{p} \varphi_i z_{t-i} + a_t. \quad (2.2)$$

where a is zero-mean white noise, p is the order, and $\varphi_1, \varphi_2, \ldots, \varphi_p$ are auto regressive coefficients, obtained by solving the Yule-Walker equations, which can be done through Levinson-Durbin recursion [4].

The optimum order p can be determined by Akaike's Final Prediction Error (FPE) criterion that corresponds to the smallest FPE [1],

$$\mathrm{FPE}_p = \mathrm{P}_p(N + p + 1)/(N - p - 1), \tag{2.3}$$

$$\mathrm{P}_p = 1/(N - p) \sum_{t=p+1}^{N} \left(z_t - \sum_{j=1}^{p} \varphi_j z_{t-j} \right)^2. \tag{2.4}$$

2.3 Least Squares Collocation

LSC is a matrix regression method, and wildly used in geodesy surveying and combining of gravity measurements [7]. Let the observational model be written in form $l = t + n$, where t is useful signal, n is noise. If we know the covariance matrixes of observations Q_{ll}, and noise Q_{nn}, the cross-covariance matrix of useful signal can be obtained as $Q_{tt} = Q_{ll} - Q_{nn}$. Let's subdivide the vector of useful signal t into the up to now part x and the future values f, then the matrix Q_{fx} can be obtained as the left lower part of Q_{tt} and the vector of future signal can be estimated as

$$\hat{f} = Q_{fx} Q_{ll}^{-1} l. \tag{2.5}$$

Auto- covariance function can be estimated from observations or modeled.

2.4 Artificial Neural Networks

ANN are powerful mathematical tools, which came to birth in 1960-th with the development of brain studies [12, 20]. Simple neuron can be represented by the equation $y = f(\sum_{i=1}^{n} w_i x_i)$, where w_i are weights, ($w_0 = 1$ corresponds to the polarization $x_0 = b$), $x_1 \ldots x_n$ are the input vector values, f is a transfer function. One neuron produces one scalar output y. The neurons can be organized in layers. Every neuron in the layer has the same input vector and transfer function; it produces one element (coordinate) of the layer output, formed by the neurons of this layer. The dimensionality of the layer output is equal to the number of its neurons. The output of one layer can sequentially become the input of the next layer, etc. Thus, 3-layer network can be represented by equation

$$y = f_3(W_3 f_2(W_2 f_1(W_1 x))). \tag{2.6}$$

Unknown parameters here are weight matrixes W_i. They are to be adjusted in the iterative learning procedure from the comparison of the selected known answers with the responses to the corresponding inputs. The error is propagated backward in the network and changes its weights. This learning process is called back propagation of error, it requires all the transfer functions to be differentiable. In such way the network is trained to provide desirable responses for the learning sequence, and then used on the real data. In this study the ANN we used has 15 neurons, organized in 3 layers as (5, 7, 1) [14, 19].

3 Combined Solution of EOP Prediction

Among many estimation indicators of the prediction accuracy, the mean error (ME) and mean squared error (MSE) were selected.

$$ME_i = \frac{1}{n}\sum_{j=1}^{n}(p_j^i - o_j^i) \tag{3.1}$$

$$MSE_i = \sqrt{\frac{1}{n}\sum_{j=1}^{n}(p_j^i - o_j^i)^2} \tag{3.2}$$

where, o is the EOP observation, p is the EOP prediction, i is the prediction interval, n is the number of predictions used to calculate statistics.

We used ME and MSE to characterize the discrepancy between the predictions and real data. It should be mentioned that ME and MSE represents different characteristics and sometimes even contradicts to each other. For example, it could be fund that some changes in the prediction method reduce the ME, while MSE may increase. Thus, it should be chosen what characteristic is more important for the particular consumer.

IERS C04 bulletin was used as an input for all the predictions. All the predictions were made for 90 days in the future. Statistics was calculated from comparison of predictions with data from this bulletin.

To obtain the combined solution, we selected two LS + AR predictions (made in SHAO and SAI), LSC and ANN methods. The weights of particular predictions in combination were determined from the mean ME and mean squared MSE error calculated on the test interval. There are different ways to calculate the weights w. After some experiments, supposing that both ME and MSE are important for us, we adopted time-dependent weights for the t_i moment in the future and for the lth forecast technique

$$w_l(t_i) = \frac{k}{ME(t_i)^2 + MSE(t_i)^2} \tag{3.3}$$

where k was selected to satisfy the criteria$\sum_l w_l(t_i) = 1$.

Figure 1 represents the overall statistics for different predictions and statistics for the combined solution calculated on the selected test interval since MJD 55659 till 55803 with daily step. The weights calculated from the obtained statistics are shown on bottom of Fig. 1. It can be seen, that the combination reduces the mean error ME, and MSE becomes at the level of the best prediction involved.

The statistics can change if the interval of estimation and number of predictions will change. For comparison we estimated the predictions by LS + AR (SAI) method on the 6.4-year interval of beating between Chandler and annual periods since MJD 53371 till 55791 with 20-day step. Discrepancies of PMX, PMY, UT1-UTC and LOD for about 100 of predictions are represented on Fig. 2. The plots on the top show the deviations for the short interval MJD 55659-55803. It is seen, that

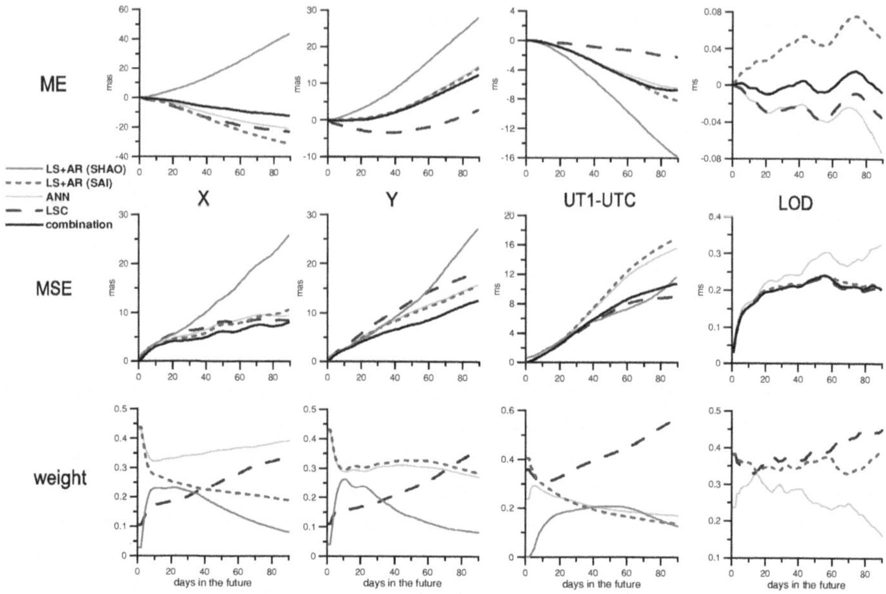

Fig. 1 Predictions statistics and weights obtained from MJD 55659 till 55803 with daily step

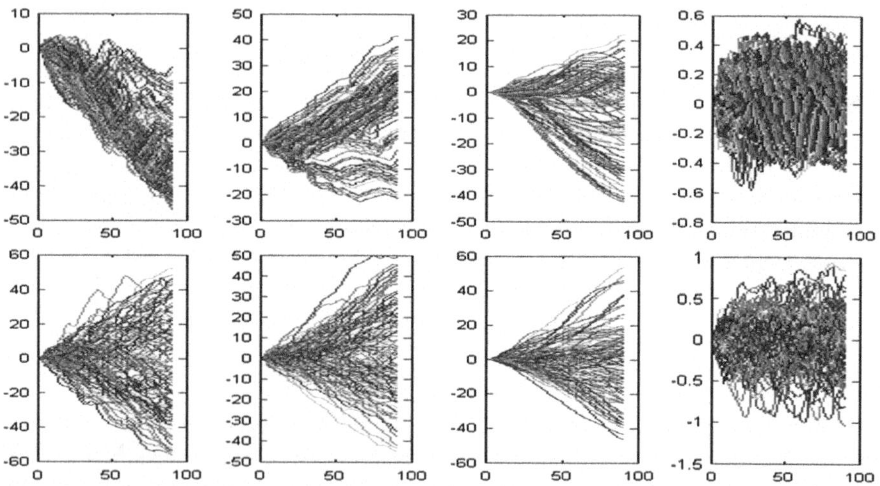

Fig. 2 Discrepancies for the LS + AR (SAI) predictions on short MJD 55659-55803 (top) and long MJD 53371-55791 (bottom) time spans, from left to right is PMX, PMY (mas), UT1-UTC, LOD (ms). Along the abscissa is days in the future

for PMX there is general underestimation, which is also seen on left top of Fig. 1. While the statistics of predictions for the longer time span MJD 53371-55791, the bottom plot of Fig. 2 shows more or less symmetric distribution for all the components.

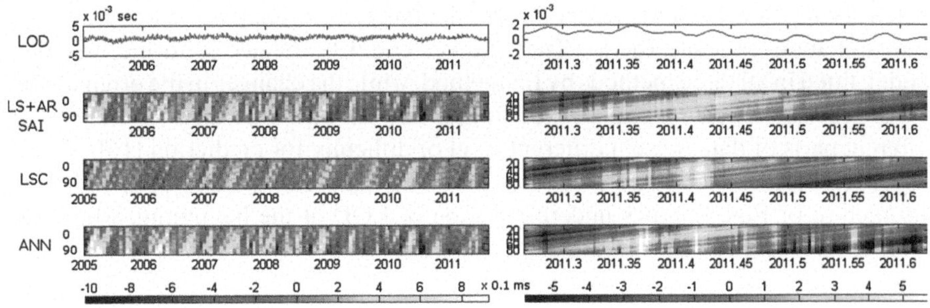

Fig. 3 Error maps of LOD for 6-year MJD 53371-55791 (left) and last half-year MJD 55659-55803 time span (right)

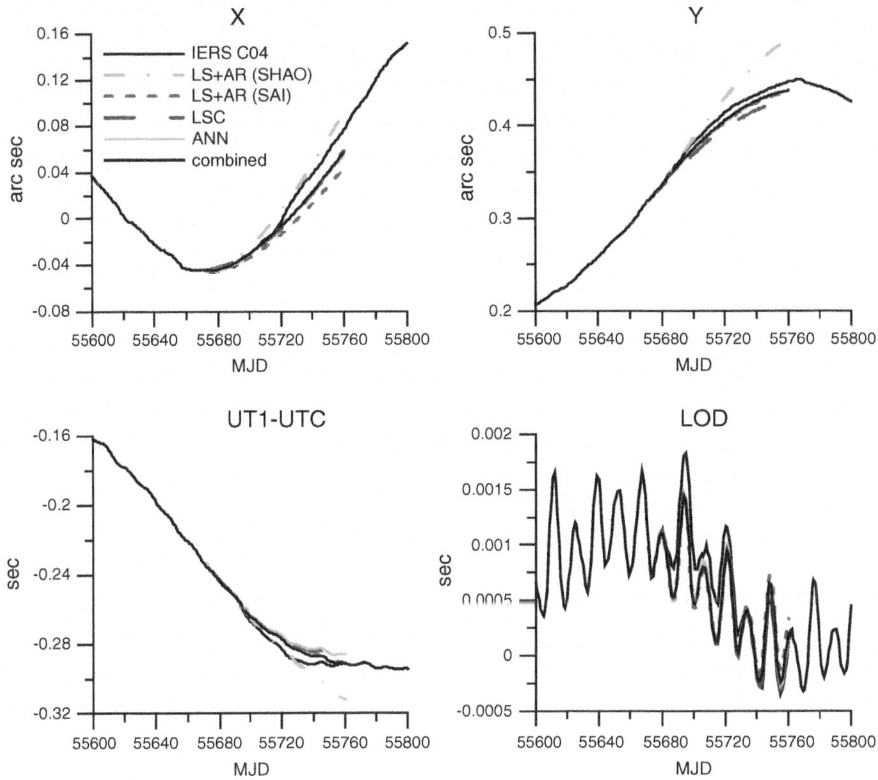

Fig. 4 Results of prediction starting from MJD 55671

On Fig. 3 the time scans of the discrepancy between the real data and LOD predictions obtained by LS + AR (SAI), LSC and ANN methods are represented in form of color maps. The color shows the deviation. The plot on top represents the LOD time series from IERS C04 bulletin, it can be seen that there are temporal waves, which means that on some intervals, predictions have a tendency to overestimate (light) or underestimate (dark). These waves could be related with the harmonic model, fitted in all three methods by LS method, while the changes in the original time series are nonstationary and do not follow any model. This also illustrates that different parts of data present different level of difficulty for prediction [16].

We can also look at one particular prediction started on MJD 55671.The last sub-picture of Fig. 4 shows underestimation of LOD at the beginning, which can be also seen on Fig. 3 as a dark region (at 2011.3).

4 Conclusions

In this article we briefly described the principles of LS, AR, LSC and ANN models, which we use to generate EOP predictions for EOPC PPP. We developed our method of their combination, based on weights, obtained as a result of some kind of trade-off between the mean ME and mean squared MSE error. These weights and the combined predictions were calculated based on the statistics of predictions accuracy on the selected time span. Comparing with the EOP predictions by single models, combination has reduced ME and MSE. No single forecasting method works as good for all the parameters and all the time spans, as a combined solution, which integrates the advantages of different models and shows higher accuracy and stability. The result of the presented work, SAI-SHAO combined solution, is to be submitted to the EOPC PPP competition, which lasts till the end of 2012 year.

The presented forecast strategies have many degrees of freedom, in sense of possible options, like the number of AR parameters, time span of the base period, inclusion of different harmonics and number of iterations in nonlinear LSM. But, generally speaking, the power of numerical approximations is limited. We need to make more investigation about dynamic modeling and introduce the knowledge about input processes. In the future works we plan to involve atmospheric and oceanic excitation functions (AAM and OAM) to obtain better EOP Predictions.

Acknowledgments The research is supported by the NSFC grant (11073045, 11133004), STCSM (06DZ22101), CAS (KJCX2-YW-T13), the Open Research Foundation of Science and Technology on Aerospace Flight Dynamics Laboratory grant (2012afdl011) and Project for Excellent Young Scientists (CSNC2010-QY-017). Second author is supported by the Chinese Academy of Sciences Fellowship for Young International Scientists grant. We thank IERS and NASA JPL for providing the EOP data and organizing EOPC PPP competition.

References

1. Akaike, H. (1971). Autoregressive model fitting for control. *Annals of the Institute of Statistical Mathematics*, *23*, 163–180.
2. Akulenko, L. D., Kumakshev, S. A., & Markov, Y. G. (2002). Motion of the Earth's pole. *Doklady Physics*, *47*, 78–84.

3. Box, G. E. P., & Jenkins, G. M. (1976). *Time series analysis: forecasting and control* (pp. 24–78). San Francisco: Holden Day.
4. Brock well, P. J., & Davis, R. A. (1996). *Introduction to time series and forecasting* (p. 420). New York: Springer.
5. Freedman, A. P., Steppe, J. A., Freedman, A. P., Dickey, J. O., Eubanks, T. M., & Sung, L. Y. (1994). The short-term prediction of universal time and length of day using atmospheric angular momentum. *Journal of Geophysical Research, 99*, 6981–6996.
6. Gross, R. S., Eubanks, T. M., Steppe, J. A., Freedman, Dickey J. O., & Runge, T. F. (1998). A Kalman filter-based approach to combining independent Earth-orientation series. *Geodesy, 72*, 215–235.
7. Gubanov, V. S. (1997). *Generalized Least Squares Method Theory And Application To Astrometry*, St. Petersburg.
8. Kalarus, M., Schuh, H., Kosek, W., Akyilmaz, O., Bizouard, Ch., Gambis, D., et al. (2010). Achievements of the Earth orientation parameters prediction comparison campaign. *Geodesy, 84*, 587–596.
9. Kosek, W., Kalarus, M., Johnson, T. J., Wooden, W. H., McCarthy, D. D., & Popinski, W. A. (2005a). Comparison of LOD and UT1–UTC forecasts by different combination prediction techniques. *Artificial Satellites, 40*, 119–125.
10. Kosek, W., Kalarus, M., Johnson, T. J., Wooden, W. H., McCarthy, D. D., & Popinski, W. A. (2005b). Comparison of UT1–UTC forecasts by different prediction techniques. In: N. Captained (Ed.), Proceedings Journeys Systems de Reference Spatiotemporal 2004, Observatoire deParis, Systems de Reference Temps-Espace, pp. 140–141.
11. Ratcliff, J. T., & Gross, R. S. (2010). Combinations of Earth orientation measurements: SPACE2008, COMB2008, and POLE2008. *Jet Propulsion Laboratory Publications, 10–4*, pp. 1–27.
12. Schuh, H., Ulrich, M., Egger, D., Muller, J., & Schwegmann, W. (2002). Prediction of Earth orientation parameters by artificial neural networks. *Geodesy, 76*, 247–258.
13. Schuh, H., Johannes, B., Heinkelmann, R., Hobiger, T., Mendes Cerveira, P. J., Pany, A., Tanir, E., Teke, K., Todorova, S., & Wresnik, J. (2007). *International VLBI Service for Geodesy and Astrometry 2006 Annual Report, no. NASA/TP-2007-214151*, pp. 208–211.
14. Wang, Q. J., Liao, D. C., Zhou, Y. H., & Liao, X. H. (2007). Determination of the topology of the neural networks in the prediction of LOD. *Annals of Shanghai Astronomical Observatory, 28*, pp. 75–80.
15. Wang, Q. J., Liao, D. C., & Zhou, Y. H. (2008). Real-time rapid prediction of variations of Earth's rotational rate. *Chinese Science Bulletin, 53*, 969–973.
16. Xu, X. Q., & Zhou, Y. H. (2010). High Precision Prediction Method of Earth Orientation Parameters. *Journal of Spacecraft TT & C Technology, 29*, 70–76.
17. Xu, X. Q., Zhou, Y. H., & Liao X. H. (2011). *Short-term earth orientation parameters predictions by combination of the least-squares*, AR model and Kalman filter. Geodyn. doi:10.1016/j.jog.2011.12.001.
18. Zotov, L.V. (2010). Dynamical modeling and excitation reconstruction as fundamental of earth rotation prediction, Artificial Satellites, Warsaw, *Proceedings of IERS Workshop on EOP Combination and Prediction*. Warsaw, 19–21 October 2009, *45*(Number 2/2010), 95–106.
19. Zotov, L. V. (2005). Regression methods of Earth rotation prediction. *Moscow University Physics Bulletin, 5*, 36–50.
20. Zotov, L. V. (2011). *Theory of filtering and time series processing*. Moscow State University. pp. 78–85.

Analysis of the Impact of Satellite Payload's Channel Characteristic on the Performance of Carrier Tracking

Caihua Li, Lei Chen, Xiangyu Wu and Fan Chen

Abstract As the effect of analog device on Satellite Payload, such as amplifier, mixer and filter, there will be distortion for the constant envelope signals, which would affect the performance of carrier tracking. To address this issue, this paper established the generic model for analyzing navigation signals channel characteristic and non-linear distortion, and conducted the relation between channel characteristic and PLL tracking error. Then, we analyzed the impact of channel characteristic on performance of Navigation Signals carrier tracking quantificationally. At last, the analysis result is verified in a software receiver for Galileo signals.

1 Introduction

With the development of Galileo and modernized GPS and the increase in performance of the associated signals (ALTBOC, MBOC, Cos-BOC…), channel characteristic, often neglected, may become essential factors in the Satellite's payload design to better characterize the navigation signals and to asses their best achievable performance.

Most considerations of the code tracking performance for Satellite Payload are based on the ideal rectangular shaped spread spectrum signals up to now. In paper [1] the degradation of code tracking accuracy due to generic linear signal distortions and for BOC (m, n)-type random signals is investigated. In paper [2] the

C. Li (✉) · L. Chen · X. Wu · F. Chen
Satellite Navigation Research and Development Center,
National University of Defense Technology, Changsha, China
e-mail: alexfo14@126.com

J. Sun et al. (eds.), *China Satellite Navigation Conference (CSNC) 2012 Proceedings*, Lecture Notes in Electrical Engineering 160, DOI: 10.1007/978-3-642-29175-3_33,

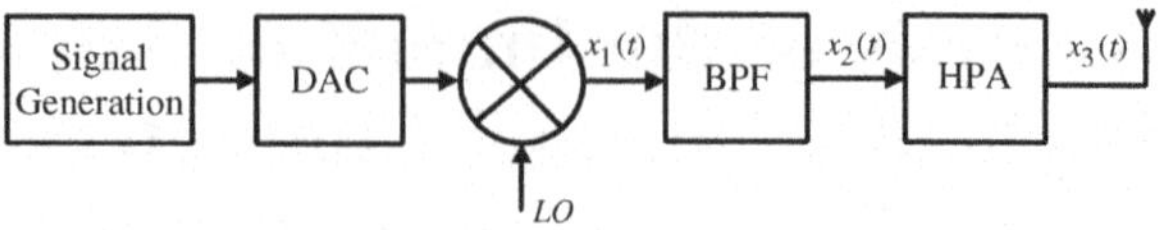

Fig. 1 Analysis mode of navigation signal

bandwidth restrictions and nonlinear effects impacted on Rectangular pulse shaped, SRC pulse shaped, BPSK modulation and BOC modulation is analyzed. The aim of those papers is to evaluate the code tracking performance due to the Traveling wave tube amplifier (TWTA) model, neglecting the impact to the carrier tracking performance and the Solid state power amplifier (SSPA) model.

The aim of this paper is to evaluate the impact of the channel characteristic on the receiver carrier tracking performance due to the SSPA or TWTA in the Satellite payload.

The HPA non-linearization model are described in Sects. 2 and 3, the performance of carrier tracking error are simply presented. Some types of channel characteristic are investigated by means of a generalised approach in Sect. 4. The analysis and computer simulations of channel characteristic effects on PLL tracking error for Galileo signals are given in Sect. 5 with some conclusions drawn in Sect. 6.

2 Signal Model of Satellite Payload

2.1 Model Summarize

The non-linear mode of navigation signal is showed in Fig. 1. The navigation signal is modulated to high frequency constant envelope signals $x_1(t)$ after passing DAC and mixer. Then it becomes unconstant envelope signals $x_2(t)$ after passing band-pass filter (BPF). And it becomes $x_3(t)$ which consist of phase jitter by passing non-linear High power amplifier (HPA). At last the signal is transmitted by antenna. $x_1(t)$ in Fig. 1 showed as

$$x_1(t) = s(t)\cos[2\pi f_0 t + \varphi(t)] \tag{1}$$

$s(t)$ is the modulated envelope which valued at ± 1 for navigation signals, f_0 is the frequency of carrier, $\varphi(t)$ is the instantaneous phase.

It becomes $x_2(t)$ after passing BPF,

$$x_2(t) = s'(t)\cos[2\pi f_0 t + \varphi(t)] \tag{2}$$

$s'(t)$ is the changed envelope signals after filter.

$x_2(t)$ passes HPA and generate an amplified and distorted output signal $x_3(t)$. It could be expressed as

$$x_3(t) = A[s'(t)]\cos\{2\pi f_0 t + \varphi(t) + \varphi[s'(t)]\} \tag{3}$$

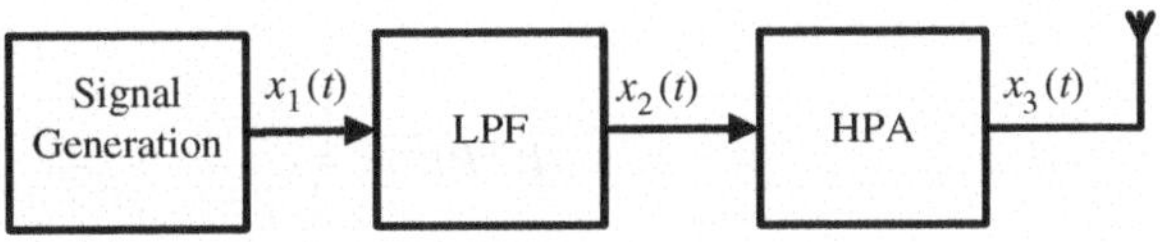

Fig. 2 Equivalent baseband mode of non-linear distortion on satellite

$A[s(t)]$ and $\varphi[s(t)]$ is the gain and phase character of the HPA. Make nonlinear weighting for the amplitude and phase of output signal.

For predigestion, suppose that (which are satisfied with the character of signal on satellite and the demands for navigation performance)

(1) HPA is wideband memory-less equipment, whose non-linear response could reflect the change of carrier envelope rapidly;
(2) The transform function of BPF is $H_{BP}(f)$ and the equivalent Fourier transform $H_{LP}(f)$ of LPF is:

$$H_{LP}(f - f_0) = \begin{cases} H(f) & (f > 0) \\ 0 & (f < 0) \end{cases} \tag{4}$$

(3) Without regard to the influence of phase noise from local oscillator

The baseband mode could be set according the supposition above, showed as Fig. 2.

Before passing LPF, the signal could be expressed as

$$x_1(t) = s(t)\cos[\varphi(t)] \tag{5}$$

After passing LPF

$$x_2(t) = s'(t)\cos[\varphi(t)] \tag{6}$$

$x_2(t)$ generates an amplified and distortion output signal $x_3(t)$ after passing HPA. The expression is

$$x_3(t) = A[s'(t)]\cos\{\varphi(t) + \varphi[s'(t)]\} \tag{7}$$

What need to be emphasized is that the passing is just for liner phase filter since it's hard to define the verify comes from of non-linear phase filter. The effect of non-linear phase filter will be analyzed in Chap. 5 by the dissimulation of software receiver.

2.2 The Envelope's Change of Navigation Signal After Passing the Liner Phase Filter

The traditional parameter to scale verify of envelope is ratio of peak/average power (PAPR), which is defined as [3]

$$PAPR = \frac{\left[\max_{t\in[0,T_c]}[p_c(t)]\right]^2}{\frac{1}{T_c}\int_0^{T_c}[p_c(t)]^2 dt} \tag{8}$$

When $p_c(t)$ is the average power during $[0,T_c]$.

PAPR is consisted of the ratio of peak and average power, from which the probability information could not be got and is hard to build corresponding relation when analyse the effect on carrier tracking from envelope verify. So that the power spectrum density with varying envelope is used for analysing the effect from envelope verify in this paper, which makes many advantages for analysing on condition of different bandwidth and signal system.

For analysing the envelope verify of navigation signal, we regard the navigation signal s(t) during a period as an constant signal, and define it s′(t) after passing ideal lowpass filter. The varified value by passing liner phase filter is $s_s(t)$, and define the phase response of liner phase filter is $\Phi_{LP}(f)$ satisfied with the function below

$$\phi_{LP}(f) = -2\pi f\tau_0 \tag{9}$$

τ_0 is the absoluteness group-delay of liner phase filter. Then we have

$$s_s(t) = s(t) - s'(t + \tau_0) \tag{10}$$

SI(f) and SLP(f) is Fourier transform of s(t) and s′(t), then $S_s(f)$ the Fourier transform of $S_s(t)$, could be expresses as

$$S_S(f) = S_I(f) - S_{LP}(f)e^{-jw\tau_0} \tag{11}$$

$$S_S(f) = \left[1 - H_{LP}(f)\cdot e^{-jw\tau_0}\right]S_I(f) \tag{12}$$

By considering the random character of C/A code. We define $G_I(f)$ to be unitary power spectrum density of s(t). While the *e* unitary power spectrum density with varying envelope is $G_s(f)$. According the define of random process and (12), we get

$$G_S(f) = S_S(f)S_S^*(f) = \left|1 - H_{LP}(f)\cdot e^{-jw\tau_0}\right|^2 G_I(f) \tag{13}$$

When $S_s^*(f)$ is conjugation of $S_s(f)$.

2.3 Equivalent Model of HPA

There are two kinds of HPA in navigation system: TWTA and SSPA [4].

Saleh model [5] is is always used for TWTA, the AM–AM and AM–PM function could be expressed as:

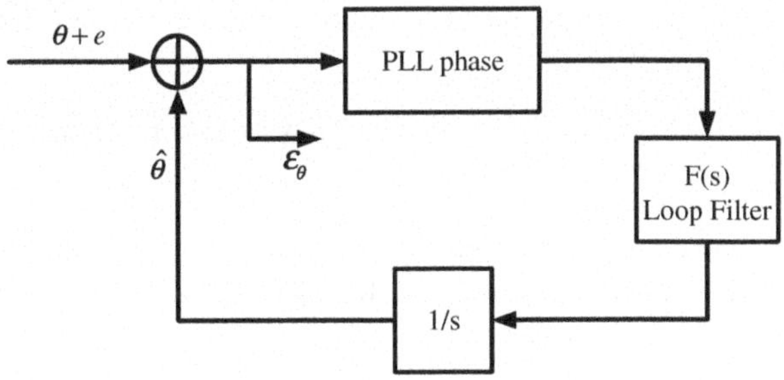

Fig. 3 The ideal structure of PLL tracking

$$A[|s(t)|] = \frac{\alpha_a |s(t)|}{1 + \beta_a |s(t)|^2} \tag{14}$$

$$\Phi[|s(t)|] = \frac{\alpha_\phi |s(t)|^2}{1 + \beta_\phi |s(t)|^2} \tag{15}$$

A[|s(t)|] and Φ[|s(t)|] is the weighting results of nonlinear amplifiers' gain and phase character. α_a is amplitude gain of little signal. $1/(\beta a)^{1/2}$ is saturated input voltage of TWTA, α_Φ is the phase gain of little signal, $1/(\beta a)^{1/2}$ is saturated input phase of TWTA.

The modified Rapp model [6] is always used for SSPA. The AM–AM and AM–PM function could be expressed as the functions below:

$$A[s(t)] = \frac{g|s(t)|}{[1 + (|s(t)|/A_{sat})^{2p}]^{1/2p}} \tag{16}$$

$$\Phi[|s(t)|] = \frac{\alpha |s(t)|^{q1}}{1 + \left(\frac{|s(t)|}{\beta}\right)^{q2}} \tag{17}$$

g is the amplitude gain of little signal. p is smoothness parameter, which decides the smoothness condition of amplifier from liner area to saturated area. A_{sat} is saturated input voltage. a, β, q1, q2 can be adjusted for different amplifier.

2.4 Assessment Parameter

To assess the impact of the resulting signal distortions on the carrier tracking accuracy, the PLL tracking error as a generalised measure more independent of the type of receiver are evaluated here [7].

The ideal structure of PLL tracking is illustrated in Fig. 3.

No concerning the phase noise of clock, the PLL tracking error is determined by

$$\sigma_{\varepsilon_\theta}{}^2 = \int_0^\infty S_{N_e}(f) \cdot |1 - H_{PLL}(jf)|^2 df \tag{18}$$

When S_{Ne} is the power spectrum density of input signal phase noise, HPLL(*jf*) is the transfer function of the PLL. Considering a 3rd order PLL, the following model can be used [8]:

$$|1 - H_{PLL}(jf)|^2 = \frac{f^6}{f_L^6 + f^6} \tag{19}$$

where $f_L = 2\pi \times 1.2 \times B_L$, B_L is the loop noise bandwidth.

3 Analytical Expression of PLL Tracking Error

As the phase noise is small signal, the phase jitter assuming linear relationship with input amplitude $s_0(t)$.

$$\Delta\phi = k_{S_0} \cdot \Delta s(t) \tag{20}$$

where $ks_0 = \Phi'[|s_0(t)|]$ is derivative of AM–PM curve at $|s_0(t)|$ point.

Considering unit impedance, the relationship of input amplitude $|s_0(t)|$ with power p is: $p = |s_0(t)|^2/2$.

Linking (18–20)

$$\sigma^2 = p \cdot k_{S_0} \cdot \int_0^\infty (1 - |H_{LP}(f)|)^2 \cdot G_I(f) \cdot |1 - H_{PLL}(jf)|^2 df \tag{21}$$

4 Analysis of Simulation

4.1 Software Receiver

The generic structure of software receiver used for base-band simulations is illustrated in Fig. 4. There are five special channel Characteristic in paper [2].

4.2 Parameter

This simulation Parameter is from paper [3] and paper [6], TWTA Parameter: $\alpha_a = 128.32$, $\beta_a = 13.83$, $\alpha_\Phi = 10.99$, $\beta_\Phi = 10.49$, $P_{in} = 16.78$ dBm, $P_{out} = 51.85$ dBm. SSPA Parameter: $g = 19$, $A_{sat} = 1.4$, $p = 0.81$, $a = 1$, $\beta = 0.123$, $q_1 = 3.8$, $q_2 = 3.7$, $P_{in} = 31.45$ dB m, $P_{out} = 53.14$ dBm.

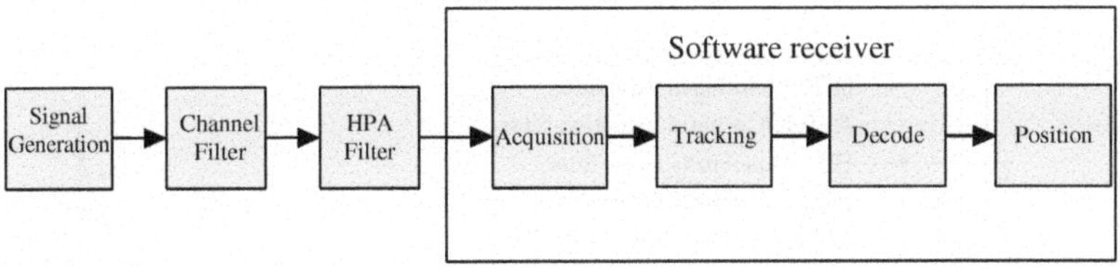

Fig. 4 The structure of software receiver

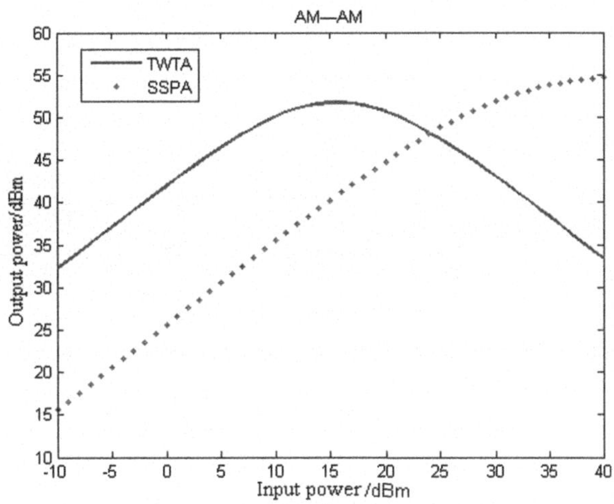

Fig. 5 AM–AM curve for TWTA and SSPA

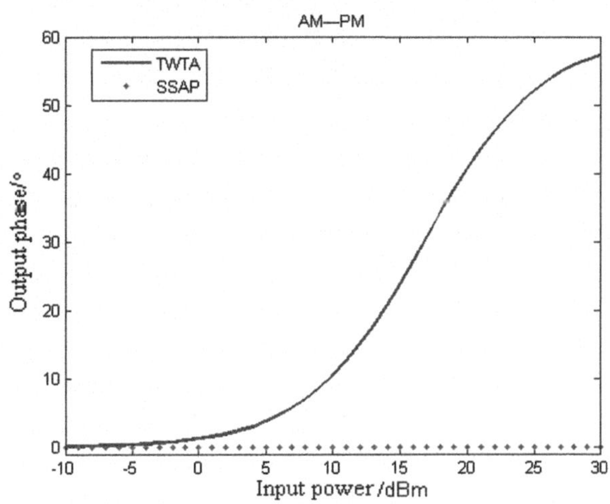

Fig. 6 AM–PM curve for TWTA and SSPA

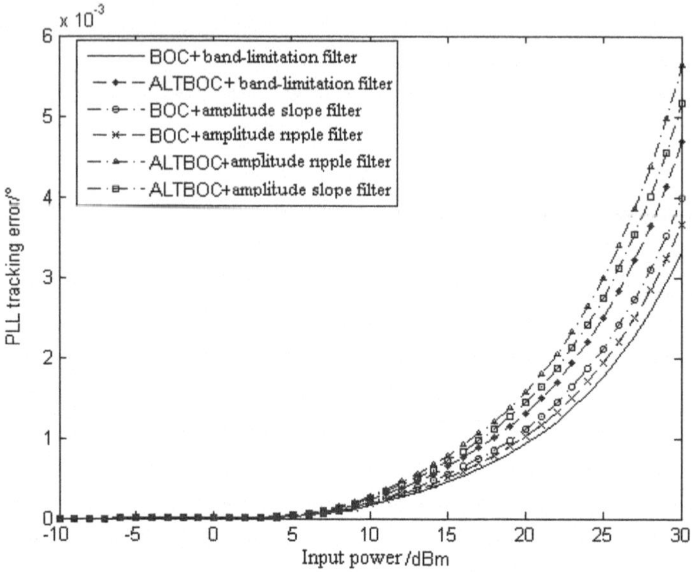

Fig. 7 PLL tracking error with amplitude distortions for SSPA

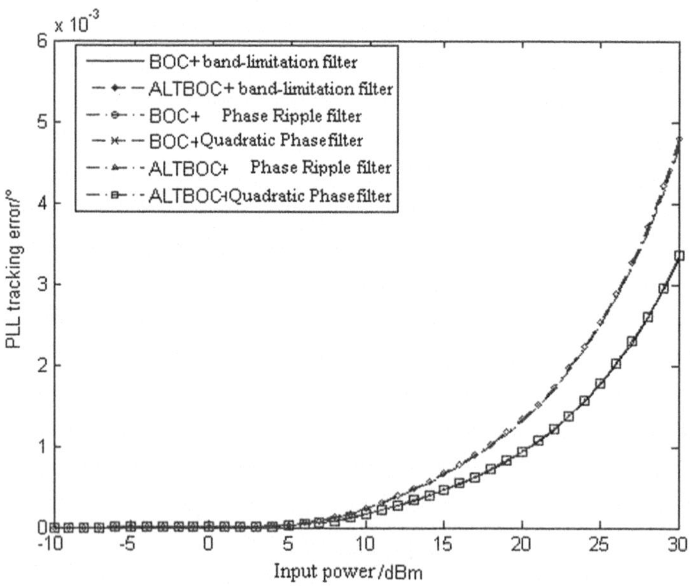

Fig. 8 PLL tracking error with phase distortions for SSPA

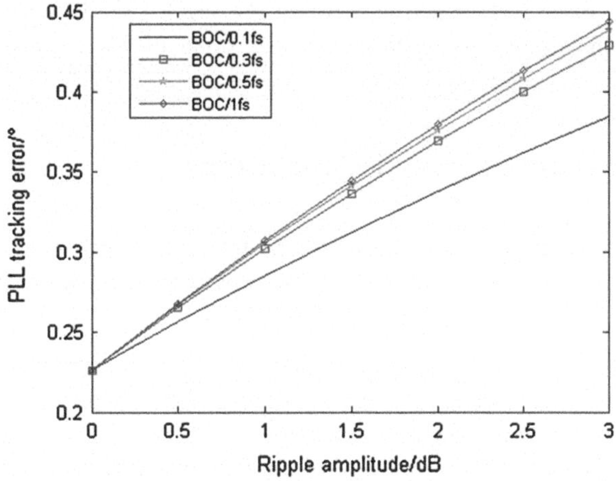

Fig. 9 PLL tracking error with amplitude distortions for TWTA

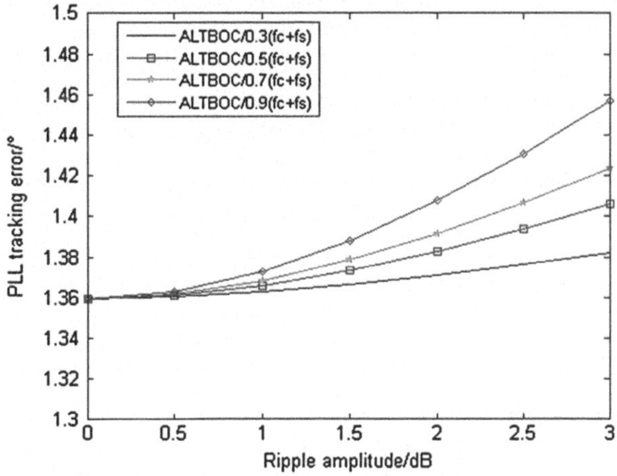

Fig. 10 PLL tracking error with amplitude Slope Filter on ALTBOC(15,10) for TWTA

The AM–AM curve and AM PM curve is showed as (Figs. 5 and 6):

5 Results

For SSPA, The PLL tracking error can be showed in Fig 7, when Sin-BOC (10,5) and ALTBOC(15,10) going through amplitude-distortion filters, with $B_n = 30$ MHz for Sin-BOC(10,5) and $B_n = 50$ MHz for ALTBOC(15,10), A = 1 dB, $\Delta d = 0.5f_{c+s}$, $\Delta f = 0.3f_{c+s}$. When going through phase-distortion filters, the PLL tracking error can be showed in Fig 8.

For TWTA, The PLL tracking error of Sin-BOC(10,5) can be showed below, when the input power is on the saturation point of TWTA($p = 16$ dBm) (Fig. 9).

The PLL tracking error of ALTBOC(15,10) can be showed below, when the input power is on the saturation point of TWTA(p = 16 dBm) (Fig. 10).

6 Summary and Conclusions

This paper has presented a detailed analysis of the Channel Characteristic appearing in Satellite Payload. a summary of the main distortion effects from the generic investigations is given below.

(1) For SSPA, no matter band-limitation or amplitude and phase, the PLL tracking error is quantity of 10^{-3} degree. The reason is that the AM–PM curve of SSPA is approximately constant.
(2) For TWTA, Band-limitation is assessed to be the most critical type of distortion on the carrier-tracking performance, then the amplitude distortions will be less important within small ripple amplitude($A < 3$ dB), when phase distortions can be neglected.
(3) For TWTA, typical variations of the Amplitude Ripple Filter seem to have generally stronger impact on the signal distortion parameters than Amplitude Slope Filter variations with only small impact for these parameters.

The SSPA will be used more comprehensive in the next few years for navigation application, due to mass, volume and performance advantages. These results may also be the main reason for Galileo experimental phase choosing SSPA to meet very stringent space standards.

References

1. Soellner, M., Kohl, R., & Luetke, W., et al. (2002). *The impact of linear and non-linear signal distortions on Galileo code tracking accuracy* (pp. 1270–1285). Portland: ION GPS.
2. Hu, X., Hu, X., & Tang, Z. (2009). Impact of satellite's high power amplifiers on spectrum of navigation signals and PRN tracking accuracy. *Acta Electronica Sinica 37*(3), P640–P645.
3. Li, X. (2007). Research on key techniques of earthstation time synchronization network in satellite navigation system. Ph.D. thesis National University of Defense Technology. National University of Defense Technology.
4. Jan, M. W., & Mangus, B. J. (2005). TWTA Versus SSPA: A comparison of on-orbit reliability data. *IEEE Transactions on Communications Electron Devices, 52*(5), 650–652.
5. Saleh, A.A.M. (1981) Frequency-independent and frequency-dependent nonlinear models of TWT amplifiers. *IEEE Transactions on Communications, 29*(11), 1715–1720.
6. Gil Jiménez, V. P., & Jabrane, Y., et al. (2011). High power Amplifier pre-distorter based on neural-fuzzy systems for OFDM signals. *IEEE Transactions on Broadcasting, 57*(1), 313–317.
7. Rebeyrol, E., & Christophe, M., et al. (2006) Phase noise in GNSS transmission/reception system (pp. 698–708). Monterey, CA: ION NTM.
8. Irsigler, Makus, & Eissfeller, Bernd. (2002). PLL tracking performance in the presence of oscillator phase noise. *GPS Solutions, 5*(4), 45–57.

Application of Inter-system Hardware Delay Bias in GPS/GLONASS PPP

Xiao Pei, Junping Chen, Jiexian Wang, Yize Zhang and Haojun Li

Abstract GPS applies Code Division Multiple Access technique in signal coding, while GLONASS's signal is produced with Frequency Division Multiple Access technique. The differences in signal frequency results in inter-system hardware delay bias for GPS/GLONASS receivers. Subjecting to these hardware delays, strategies and models of GPS/GLONASS PPP based positioning needs to be modified. We derived a GPS/GLONASS combined PPP positioning models by introducing inter-system hardware delay biases. Several scenarios were simulated to test the introduced models using the observation of IGS stations. The results show that: ① Adding a couple of GLONASS satellites can improve the positioning accuracy in the environment without enough GPS satellites being tracked; ② The inter-system hardware delay bias is stable on daily base, and it could be predicted and be fixed in the GPS/GLONASS combined PPP.

X. Pei (✉) · J. Wang · Y. Zhang
Department of Surveying and Geo-Informatics, Tongji University, Shanghai, People's Republic of China
e-mail: pxtoday@hotmail.com

X. Pei · J. Chen (✉) · Y. Zhang · H. Li
Shanghai Astronomical Observatory (SHAO), Chinese Academy of Sciences (CAS), Shanghai, People's Republic of China
e-mail: junping.chen@shao.ac.cn

J. Wang
Key Lab of Advanced Surveying Engineering of SBSM, Shanghai, People's Republic of China

J. Sun et al. (eds.), *China Satellite Navigation Conference (CSNC) 2012 Proceedings*, Lecture Notes in Electrical Engineering 160, DOI: 10.1007/978-3-642-29175-3_34,

1 Introduction

Precise Point Positioning (PPP) is based on un-differenced dual-frequency pseudo-range and carrier phase observations along with precise GNSS orbits and satellite clocks. Precise models of error correction also need to be considered [1]. Although PPP is mostly implemented for GPS system, combined GPS/GLONASS PPP has been recently investigated [2–6]. At present, there are several GNSS Analysis Centers (IAC, ESA, GFZ, CODE, SHA) provide GLONASS products. Their products are different in accuracy and precision because of different data processing approaches and data set were used. GPS is based on code division multiple access (CDMA), while GLONASS is based on frequency division multiple access (FDMA). Carriers with different frequencies have different delays while propagating in both satellite and receiver. For the combined GPS/GLONASS receivers, the difference of GPS and GLONASS carrier frequency produces the inter-system hardware delay bias (ISDB). In this paper, the ISDB is defined as the difference between the receiver clock under GPS system and GLONASS system. For combined GPS/GLONASS PPP, ISDB parameters could be defined in the model, which requires that 4 GPS satellites at minimum must be tracked with additional GLONASS observations for the ISDBs estimation. Based on the routine results of the GNSS data analysis center at SHAO (SHA) [7], we found the ISDBs are stable on daily base and could be well modeled for short period. With the ISDBs be precisely modeled and fixed, combined GPS/GLONASS PPP could be derived with minimum 4 GNSS satellites.

2 Inter-System Hardware Delay Bias

Un-differenced pseudo-range and carrier phase functions can be defined as:

$$P_i = \rho + c(dt - dT) + d_{orb} + d_{trop} + d_{ion} + \varepsilon(P_i) \tag{1}$$

$$\Phi_i = \rho + c(dt - dT) + d_{orb} + d_{trop} - d_{ion} + \lambda_i N_i + \varepsilon(\Phi_i) \tag{2}$$

Where ρ is the true geometric range; P is the measured pseudo-range; Φ is the measured carrier phase; c is the speed of light; dt is the receiver clock offset; d_{orb} is the satellite orbit error; d_{trop} is the tropospheric delay error; d_{ion} is the ionospheric delay error; $\varepsilon(P_i)$ is the pseudo-range measurement error; $\varepsilon(\Phi_i)$ is the phase measurement error.

In PPP data processing, d_{ion} could be ignored by forming the ionosphere-free combination using dual-frequency pseudo-range and carrier phase observations.

dT and d_{orb} can be corrected by precise ephemeris and precise satellite clocks. In this paper, we use products from Shanghai Observatory GNSS Analysis Center (SHA) [7], where its orbit precision is about 2 cm for GPS and 5 cm for GLONASS. By conducting integrated GPS/GLONASS orbit determination, SHA obtains precise ephemeris and precise satellite clock of GPS and GLONASS simultaneously under the same coordinate and time reference frame.

Applying precise ephemeris and precise satellite clock to eliminate satellite orbit error and satellite clock offset, the ionosphere-free code and phase combinations can be expressed as:

$$\begin{aligned} P_{IF}^{g} &= \rho + (cdt + b^{g}) + d_{trop} + \varepsilon \\ \Phi_{IF}^{g} &= \rho + (cdt + b^{g}) + d_{trop} + N_{IF}^{g} + \varepsilon \\ P_{IF}^{r} &= \rho + (cdt + b^{r}) + d_{trop} + \varepsilon \\ \Phi_{IF}^{r} &= \rho + (cdt + b^{r}) + d_{trop} + N_{IF}^{r} + \varepsilon \end{aligned} \tag{3}$$

Where superscript g and r stand for GPS and GLONASS satellite; P_{IF} is ionosphere-free code combinations; Φ_{IF} is ionosphere-free phase combinations; c is the speed of light; dt is the receiver clock offset; b^{g} is the hardware delay for GPS satellites; b^{r} is the hardware delay for GLONASS satellite; d_{trop} is the tropospheric delay error;

Hardware delay b^{g} b^{r} cannot be separated from station clock cdt, therefore they will be estimated as one single parameter as:

$$\begin{aligned} cdt^{g} &= cdt + b^{g} \\ cdt^{r} &= cdt + b^{r} \end{aligned} \tag{4}$$

We can define the inter-system hardware delay bias by taking the difference between cdt^{g} and cdt^{r} :

$$diff\_HD = b^{g} - b^{r} = cdt^{g} - cdt^{r} \tag{5}$$

3 Application of Inter-System Hardware Delay Bias

ISDB parameters are one of the routine products of SHA, which could benefit users for combined GPS/GLONASS PPP positioning. Two groups of experiments with several different scenarios were conducted.

1. Comparison between static PPP result using GPS observation and GPS/GLONASS combined observations. Urban canyon environment were simulated and the results are compared to normal tracking conditions.
2. Pseudo-range is used in real-time navigation and positioning. Considering the inter-system hardware delay bias, a fifth satellite is required to process parameter estimation. In order to overcome this limit, we estimate inter-system hardware delay bias when satellites are adequate, and use it as a known value in case no sufficient satellites are tracked.

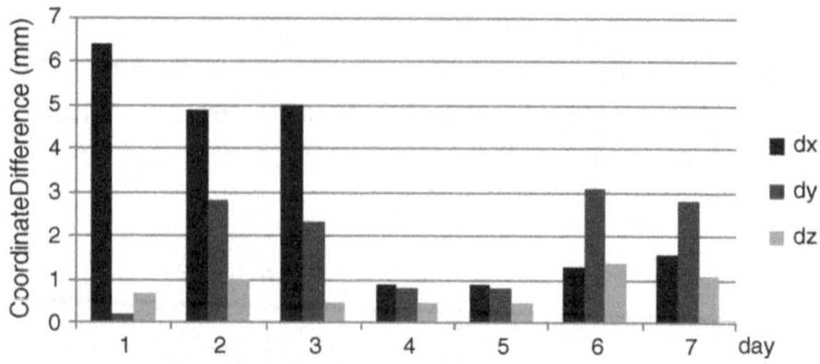

Fig. 1 Comparison between PPP using GPS observations and combined GPS/CLONASS observations

3.1 Combination GPS/GLONASS PPP

In order to compare the precision of combined GPS/GLONASS PPP, we conduct static PPP using GPS observations and combined GPS/GLONASS observations at IGS station BJCO. One week data (from March 18, 2011 to March 24, 2011) was used and Fig. 1 shows the results.

Figure 1 shows that the positioning result from combined GPS/GLONASS PPP is consistent with PPP using GPS system only. Their coordinate differences are within 1 cm.

Given the statistics of Fig. 1, adding GLONASS satellite to PPP will not reduce the precision, so combined GPS/GLONASS PPP can bring many advantages, as enhancing system reliability, improving the accuracy and integrity in positioning. Combined GPS/GLONASS PPP has many important applications especially in certain environment, such as urban canyon, mining areas, gorges and ravines. We simulate a circumstance with limited satellite visibility. Artificial selection is conducted to make sure there are 6 GPS satellites and 2 GLONASS satellites at each epoch. Data of IGS station COCO on August 21, 2011 is processed to compare PPP using GPS only and combined GPS/GLONASS observations. Positioning result is shown in Fig. 2:

Figure 2 shows that when there are few GPS satellites at each epoch, PPP using GPS only is seriously influenced and even failed at some epochs. The accuracy and stability are significantly improved by adding 2 GLONASS satellites.

3.2 Application of Inter-System Hardware Delay Bias

Combined GPS/GLONASS navigation and positioning have great applications for navigation users, especially in limited satellite conditions. However, introducing inter-system hardware delay bias will add in new parameters following the traditional combined GPS/GLONASS PPP, which needs at least 4 GPS satellite together with a fifth satellite to fulfill the parameter estimation. Checking the long-time integrated GPS/GLONASS routine of Shanghai Observatory GNSS Analysis Center [7], we found that the inter-system hardware delay bias is stable, especially

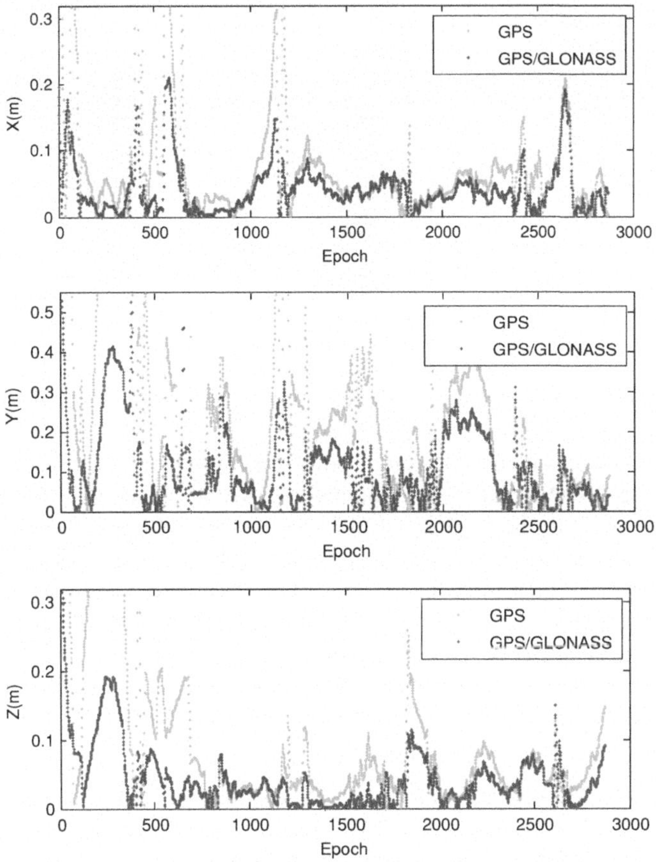

Fig. 2 Positioning results at COCO with 6 GPS satellites and 2 GLONASS satellites at each epoch

for the same satellite in one day. We can carry out a strategy to reduce unknown parameters: we estimate inter-system hardware delay bias when satellites are adequate, and use it as a known value in case no sufficient satellites are tracked. To validate this idea, we make some tests of Kinematic navigation and positioning with the pseudo-range observation data at the station POTS on August 21, 2011. These tests are applied in 4 scenarios:

1. GPS system
2. GLONASS system
3. Combined GPS/CLONASS navigation
4. Based on the third strategy, introduce the inter-system hardware delay bias, which is provided by Shanghai Observatory GNSS Analysis Center [7].

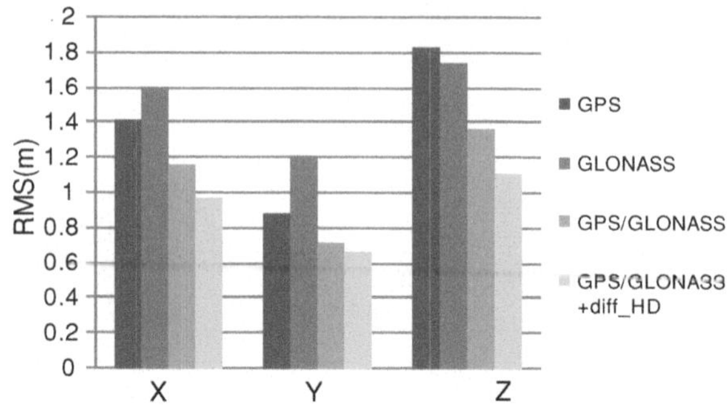

Fig. 3 RMS of coordinate parameters by different strategies

RMS of coordinate parameters calculated by all of the strategies above is shown in Fig. 3:

Figure 3 shows that, pseudo-range point positioning has a precision of 2 m. PPP using GLONASS data only obtains the worst precision; PPP using GPS system only is slightly better than GLONASS system only; precision of combined GPS/GLONASS navigation is obviously better than using data of single system; The forth strategy which introduce inter-system hardware delay bias, provided by Shanghai Observatory GNSS Analysis Center (SHA), obtains the best precision

Based on the above tests, we simulate observation conditions like 3.1 and analyze the effect of inter-system hardware delay bias in combined GPS/GLONASS navigation. Pseudo-range data at IGS station POTS on August 21, 2011 is used.

Two simulated observation conditions are:

1. R1 + G4: 4 GPS satellites and 1 GLONASS satellites at each epoch.
2. R2 + G3: 3 GPS satellites and 2 GLONASS satellites at each epoch. In this test, only one ISDB parameter was set up for the two GLONASS satellites.

Two tests are conducted to ensure the reliability of the conclusion. RMSs of coordinate parameters calculated under both conditions are shown in Table 1.

In GPS/GLONASS combined navigation without known value of inter-system hardware delay bias, there will be 5 unknown parameters at each epoch (3 coordinate parameters, 1 receiver clock offset, 1 inter-system hardware delay bias). Using 5 observations will bring no redundancy, consequently the performance will be affected. As it shows in Table 1, no matter using the R1 + G4 combination or the R2 + G3 combination, the positioning result is rather bad Among the 2880 epochs in the whole day, respectively 27 and 32 epochs has no solution due to the bad GDOP(>200).

The situation is much improved after introducing known value of inter-system hardware delay bias. Unknown parameter numbers declined to 4 (3 coordinate parameters, 1 receiver clock offset). With redundant observations, the numbers of unsolvable epochs both reduce to zero. RMS in all directions (X, Y, Z) is obviously smaller than the traditional strategy. The comparison shows that introducing

Table 1 RMS of coordinate under different conditions

Strategy	Method	RMS X/m	RMS Y/m	RMS Z/m	Invalid epochs
R1 + G4	Introduce diff_ HD	6.38	3.48	7.67	0
	Without diff_ HD	16.08	7.68	16.68	27
R2 + G3	Introduce diff_ HD	4.99	2.96	5.55	0
	Without diff_ HD	10.52	10.56	25.66	32

empirical value of inter-system hardware delay bias can improve the accuracy and validity of positioning, especially when the valid satellite number is low.

4 Conclusion

Several experiments with different scenarios were performed to test the introduced model by using observation data of several IGS stations. Results show that: ① adding a couple of GLONASS satellites can improve the positioning when the number of GPS satellites is not enough. There is a precondition that new parameters need to be added, which is the inter-system hardware delay biases. ② The inter-system hardware delay bias is comparatively stable on daily base, and it could be predicted and be used as a known parameter. By adopting this strategy, the number of necessary observation reduced from 5 to 4, which is efficient to deal with inevitable limited satellite visibility.

Acknowledgments This research is supported by 100 Talents Programme of the Chinese Academy of Sciences. IGS community is acknowledged for providing Rinex data.

References

1. Zumberge, J. F., Heflin, M. B., Jefferson, D. C., Watkins, M. M., & Webb, F. H. (1997). Precise point positioning for the efficient and robust analysis of GPS data from large networks. *Journal of Geophysical Research, 102*(B3), 5005–5017. doi:10.1029/96JB03860.
2. Cai, C., & Gao, Y. (2007). Precise point positioning using combined GPS and GLONASS observations. *Journal of Global Positioning System, 6*(1), 13–22.
3. Habrichm, H. (1999). *Geodetic applications of the Global Navigation Satellite System (GLONASS) and of GLONASS/GPS combinations*. Ph.D. dissertation, Astronomical Institute, University of Berne, Switzerland.
4. Meng, X., & Guo, J. (2010). GPS/GLONASS and their combined precise point positioning. *Geomatics and Information Science of Wuhan University, 35*(12), 1409–1413.
5. Zhang, X., Guo, W., & Li, X. (2010). Study on precise point positioning based on combined GPS and GLONASS. *Geomatics and Information Science of Wuhan University, 35*(1), 9–12.
6. Zhang, Y., Xu, S., & Wang, Z. (2001). Ambiguity processing approach in combined GPS/GLONASS positioning. *Geomatics and Information Science of Wuhan University, 26*(1), 58–63.
7. Shanghai Observatory GNSS Analysis Center. www.shao.ac.cn/shao_gnss_ac.

A New Method of Satellite Link Antenna Pointing Error Analysis for the Mixed Constellations

Zheng Song, Qinghua Wang, Lifang Yuan and Wenyu Hao

Abstract Satellite link communication is a rising technique problem in global satellite navigation systems of each country in nowadays. Spot antenna can be used to deal with satellite link communication. As the designed beam angle of spot antenna is very small, satellite link antenna is very sensitive with the satellite platform attitude error. In this paper, a new method of satellite link antenna pointing error modeling and analysis is presented for a kind of mixed constellations navigation satellite systems. The rotation matrix method is used for the integrated error analysis, each kind of error sources which affect the satellite link antenna pointing precision in the mixed constellation is considered synthetically. The analysis results are the conclusions of the whole satellite level, which can exactly reflect that whether the selected spot antenna can be able to satisfy the geometric basic condition of building the satellite link. Numerical example is given to illustrate the effectiveness of the analysis method.

1 Introduction

Satellite link communication is a rising technique problem in global satellite navigation systems of each country in nowadays. However, only the GPS III system uses the satellite link technology in the existing satellite navigation system. Consider a class of satellite navigation system composed of mixed constellations, due to the satellite link topology characteristics, the spot antenna can be used to deal with satellite link communication. As the designed beam angle of spot

Z. Song (✉) · Q. Wang · L. Yuan · W. Hao
Beijing Institute of Spacecraft System Engineering, 100094 Beijing, China
e-mail: skarlz@163.com

J. Sun et al. (eds.), *China Satellite Navigation Conference (CSNC) 2012 Proceedings*, Lecture Notes in Electrical Engineering 160, DOI: 10.1007/978-3-642-29175-3_35,

antenna is very small, satellite link antenna is very sensitive with the satellite platform attitude error. Thus, the research of satellite link antenna pointing error analysis is of great significance in practical engineering application.

The satellite two-axis antenna pointing mechanism was taken as the study object in Ref. [1], the influence factors on the pointing accuracy were analyzed, but the modeling was only fit for the antenna itself. The problem of antenna beam center did not superpose upon rotation center was analyzed in Ref. [2], but the error factors were not all things considered. Although the error sources such as satellite attitude control error, boom spread mechanism error, drive mechanism error and antenna reflecting surface error etc. were taken into account in Ref. [3, 4], the terminal goal of the arithmetic methods was to analyze the error effect on the accuracy of antenna itself. These methods are not fit for solving the problem of antenna pointing error as viewed from the whole satellite level. Besides, the methods given in those papers cannot measure the integrated satellite link antenna pointing error of all kinds of error sources influenced by the satellite roll axis, pitch axis and yaw axis total error effect. Hence, the criterion conditions of building the satellite link of mixed constellations cannot be established. In conclusion, the existing antenna pointing error analysis methods are not suitable for solving the problem of satellite link antenna pointing in mixed constellation navigation systems. It is necessary to provide a novel integrated analysis method to deal with the new problem.

A new method of satellite link antenna pointing error modeling and calculation analysis is presented in this paper for a class of mixed constellation navigation satellite systems. The creative works of this paper are stated as follows. Firstly, the rotation matrix method is used for the integrated error analysis as viewed from the whole satellite level for the first time, all kinds of error sources which affect the satellite link antenna pointing precision in the mixed constellations are considered synthetically. The analysis results are the whole satellite level conclusions, which can exactly reflect that whether the selected spot antenna can be able to satisfy the geometric basic condition of building the satellite link. Secondly, this method can calculate the exact maximum error value in the effective antenna angle range of the link topology structure. Thirdly, this method can exactly calculate the maximum error value of any given initial link direction, such value is closer to the actual error level. Finally, a numerical example is given to illustrate the effectiveness of the analysis method.

2 Mixed Constellation Satellite Link Analysis

Mixed constellation usually means that a constellation system composed of satellites with different orbit characteristics, especially different orbit altitudes and inclinations. GPS system and Galileo system are constellation systems composed of satellites with the same orbit altitudes and inclinations, respectively. Hence the link types in the satellite link topology are the same. In this paper, the mixed

Fig. 1 Link topology map

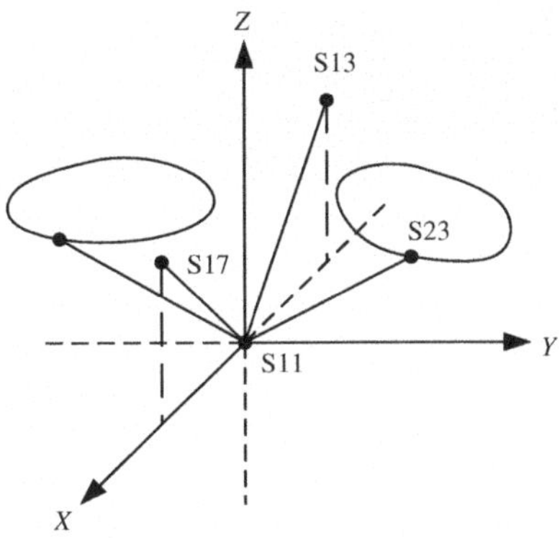

constellation is composed of satellites with different orbit altitudes, the satellite link include the low–low link type and high-low link type. Despite of the antenna installation point position coordinates in the satellite coordinate system, as shown in Fig. 1, set up a satellite coordinate system for a given reference satellite, the reference satellite is located at the origin. Consider the vector of other satellite relatively to the origin, thus for the two satellites in the same orbit the coordinates are invariable, for the satellite in different orbits the coordinate trajectory is a closed curve. As for the high-low link the satellite direction angle relations are within the spot antenna scanning range, hence the considered pointing error analysis covers both the case of low–low link and high-low link.

3 Definitions

As shown in Fig. 2, the definitions are listed as follows.

3.1 Antenna Coordination System

Define the satellite coordination system OXYZ: The origin is the center of satellite mass. +Z axis is the satellite yaw axis pointing to the geocenter. +X axis is the satellite roll axis pointing to the satellite +X side and is normal to +Z axis. Y axis is normal to X and Z axis, satisfies the right-handed system.

Define the antenna coordination system OaXaYaZa: The origin is the center of antenna beam. The coordinate of the origin in the satellite coordinate system is (x0, y0, z0). The +Za axis is the antenna beam centric line, pointing towards the earth. +Xa axis is pointing towards the satellite +X axis and is normal to +Za axis. Ya axis is normal to Xa and Za axis, satisfies the right-handed system.

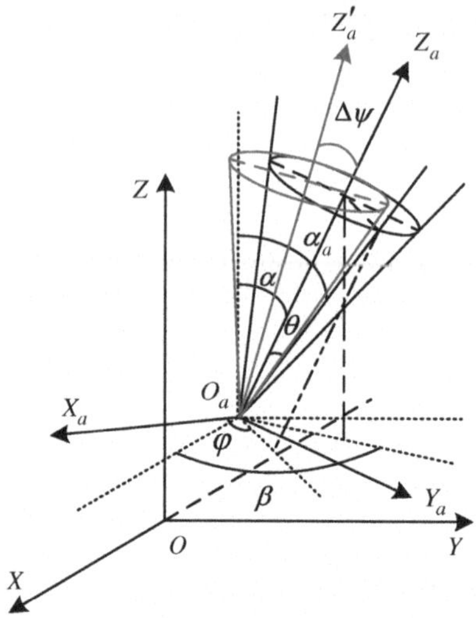

Fig. 2 Definitions of antenna pointing angle and pointing error

3.2 Antenna Pointing Angle

Antenna pointing angles include antenna center angle and antenna azimuth.

Antenna center angle α: Angle between antenna beam centric line (+Za axis) and satellite +Z axis.

Antenna azimuth β: Angle between the projection of the antenna beam centric line in XOY plane and the satellite +X axis.

Antenna beam half angle θ: Angle between the antenna beam edge line and antenna beam centric line satisfying a definite gain index, that is the center angle of the antenna beam edge line in the antenna coordinate system.

Consider the coordinate positioning of the antenna beam edge line, the edge is the conic line that has the center angle value θ.

Antenna beam edge azimuth in the antenna coordinate system φ: Angle between the projection of the antenna beam edge line in XaOaYa plane and the satellite +Xa axis, the positive direction is the counterclockwise.

Define the angle between antenna beam edge line and satellite +Z axis as α_a.

3.3 Antenna Pointing Error

Antenna pointing error is the angle of antenna beam centric line pointing vector and the object pointing vector, as shown in Fig. 2, the error angle $\Delta\psi$ between Za' axis and Za axis.

4 Error Sources Analysis

The error sources which affect the antenna pointing accuracy include three types such as antenna error, control system error and installation error.

As viewed from the error effect mode, the antenna pointing error can be divided into three parts: constant error, orbit period error and short term error.

Because of the navigation satellite character and the demands of satellite link, the antenna pointing error analysis considers the satellite roll, pitch and yaw axes at the same time in this paper.

4.1 Antenna Error

4.1.1 Antenna Thermal Deformation Error

For the phased array antenna, influenced by the sunshine in the space and thermodynamics factors, the thermal deformation affects the antenna beam electric axis pointing accuracy. For the reflecting surface antenna, the theoretical beam center is superposed with the geometric beam center, when the thermal deformation happens due to the thermal environment in the space, the actual beam center will change and influence the antenna pointing accuracy further. Simultaneously, the sustentation structure thermal deformation, the feed source installation structure thermal deformation will lead to the deviation of the beam center. This error source belongs to the orbit period error.

4.1.2 Antenna Measure Error

Antenna measure error is caused by the measure system and is determined by the accuracy of the ground adjust equipment. Usually the accuracy of the ground adjust equipment is considered as the antenna measure accuracy. This error source belongs to the constant error.

4.1.3 Antenna Control Error

For the phased array antenna, the actual exciting current amplitude and phase of each array feed source cannot be consistent with the theoretical value, the amplitude and phase error exist. For the reflecting surface antenna, the error is caused by the drive mechanism rotation angle measure error and controller error. This error source belongs to the constant error.

4.1.4 Antenna Installation Error

Antenna installation error includes antenna assembly error and installation error, caused by the gravity, installation measure accuracy and satellite public cabin fixing and tuning effect. This error source belongs to the constant error.

4.2 Control System Error

4.2.1 Sensor Error

Sensors used in control system normal mode include infrared earth sensor and gyro.

Infrared Earth Sensor Error

Infrared earth sensor electronics error: The measure error caused by the earth sensor electronics short term pulse (e.g. white noise). This error source belongs to the short term error.

Uncertainty of zero axis adjustment: Error caused by the uncertainty of the zero axis relative to the optical collimator position. This error source belongs to the constant error.

Unrepeatability of parts calibration: Parts' absolute precision error, defined as the bias of the output transfer function obtained by ground testing, influenced by ground calibration, launch environment and other conditions. This error source belongs to the constant error.

Error caused by the measure object characteristics: measure error caused by the variation of the infrared radiation due to the earth temperature variation and the difference temperature between the space and the earth. This error source belongs to the orbit period error.

Gyro Error

Gyro constant drift: During the invisible term of the sun, the measure error caused by the gyro constant drift. This error source belongs to the orbit period error.

Gyro integration initial value error: As the sun sensor is used to calibrate the gyro, the gyro integration initial value error include the sun sensor installation error. This error source belongs to the constant error.

Gyro random drift: This error source belongs to the orbit period error.

Gyro installation error: The installation bias may cause the gyro influenced by the orbit angular velocity on the yaw axis. As the sun sensor is used to calibrate the gyro integration initial value, this error source belongs to the orbit period error.

4.2.2 Controller Error

Controller error can be compensated by the control law, thus this error is not taken account.

4.2.3 Orbit error

Inclined orbit error is the earth sensor measure error caused by the earth oblateness. This error source belongs to the orbit period error.

4.3 Installation Error

Satellite accurate measurement basic: This error source belongs to the constant error.

Satellite structure deformation: This error source belongs to the constant error.

5 Antenna Pointing Error Calculation Methods

Next, the satellite link antenna pointing error calculation method is presented, including single axis error calculation and three axes integrated error. The result of the three axes integrated calculation is used as the satellite link antenna pointing error accuracy and as the geometric basic condition of whether or not the satellite link can be built.

5.1 Single Axis Error Calculation

The method is shown in Table 1.

5.2 Three Axes Integration Error Calculation

The rotation matrix method is used in the three axes integration error calculation. Define the satellite roll axis, pitch axis and yaw axis error are $\Delta\psi_x$, $\Delta\psi_y$ and $\Delta\psi_z$, respectively, the three axes integration error is $\Delta\psi$. As the antenna pointing is affected by the satellite three axes attitudes, the maximum integrated error is directly related to the antenna initial vector pointing angle. In this paper different ways of three axes rotation order are considered in the method, such as 2-1-3,

Table 1 Single axis error calculation

Error sources	Constant error	Orbit period error	Short term error
Errors (3σ)	ε_{11}	ε_{21}	ε_{31}
	ε_{12}	ε_{22}	ε_{32}
	……	……	……
	ε_{1n}	ε_{2n}	ε_{3n}
RSS	$\varepsilon_1 = \sqrt{\sum_{i=1}^{n} \varepsilon_{1i}^2}$	$\varepsilon_2 = \sqrt{\sum_{i=1}^{n} \varepsilon_{2i}^2}$	$\varepsilon_3 = \sqrt{\sum_{i=1}^{n} \varepsilon_{3i}^2}$
Single axis pointing error		$\varepsilon = \sum_{i=1}^{3} \varepsilon_i$	

1-2-3, 3-1-2,…, according to the Euler angle formula, the maximum pointing error can be obtained.

Take 3-1-2 way as an example, the rotation is shown in Fig. 3. Let the initial antenna pointing angle are α and β without error. When the attitudes error exists, let the actual terminal deviation pointing angle are α_3 and β_3. Consider the coordinate transformation, after the transformation, the maximum error angle is calculated.

For different initial pointing angles, same error may obtain different integrated error angle, thus the initial pointing angles need to be ergodic. Select $\alpha \in [0°, 180°]$, $\beta \in [0°, 360°]$ the maximum error angle of all directions can be obtained. For the satellite link antenna, considering the spot antenna effective scanning range and earth shelter effect, this method can calculate the maximum error within a certain angle range.

The calculation formula is listed as follows.

Consider some initial antenna pointing unit vector:

$$\begin{bmatrix} x \\ y \\ z \end{bmatrix} = \begin{bmatrix} \sin\alpha\cos\beta \\ \sin\alpha\sin\beta \\ \cos\alpha \end{bmatrix}. \tag{1}$$

Rotation matrices are:

$$\begin{gathered} R_x(\Delta\psi_x) = \begin{bmatrix} 1 & 0 & 0 \\ 0 & \cos\Delta\psi_x & \sin\Delta\psi_x \\ 0 & -\sin\Delta\psi_x & \cos\Delta\psi_x \end{bmatrix}, \\ R_y(\Delta\psi_y) = \begin{bmatrix} \cos\Delta\psi_y & 0 & -\sin\Delta\psi_y \\ 0 & 1 & 0 \\ \sin\Delta\psi_y & 0 & \cos\Delta\psi_y \end{bmatrix}, \\ R_z(\Delta\psi_z) = \begin{bmatrix} \cos\Delta\psi_z & \sin\Delta\psi_z & 0 \\ -\sin\Delta\psi_z & \cos\Delta\psi_z & 0 \\ 0 & 0 & 1 \end{bmatrix} \end{gathered} \tag{2}$$

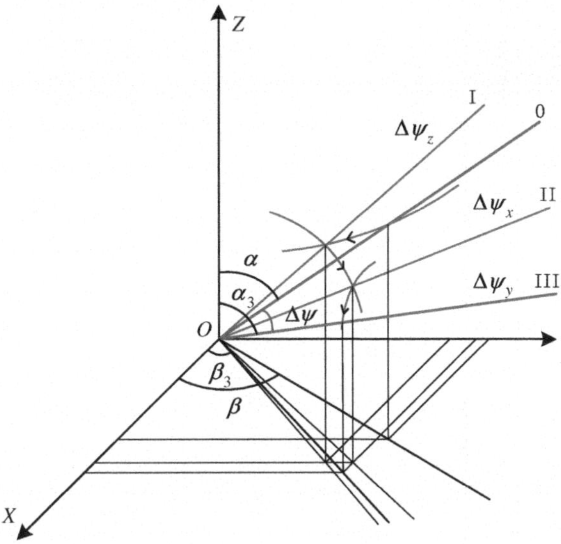

Fig. 3 Antenna pointing error integration sketch map

The terminal direction unit vector is:

$$\begin{bmatrix} x_3 \\ y_3 \\ z_3 \end{bmatrix} = \begin{bmatrix} \sin\alpha_3 \cos\beta_3 \\ \sin\alpha_3 \sin\beta_3 \\ \cos\alpha_3 \end{bmatrix} = R_y(\Delta\psi_y)R_x(\Delta\psi_x)R_z(\Delta\psi_z)\begin{bmatrix} x \\ y \\ z \end{bmatrix} = R_y(\Delta\psi_y)R_x(\Delta\psi_x)R_z(\Delta\psi_z)\begin{bmatrix} \sin\alpha\cos\beta \\ \sin\alpha\sin\beta \\ \cos\alpha \end{bmatrix} \tag{3}$$

Use the formula of intersection angle of two vectors:

$$\cos(\vec{a},\vec{b}) = (\vec{a}\cdot\vec{b})\Big/(|\vec{a}||\vec{b}|) \tag{4}$$

The error angle $\Delta\psi$ is obtained as:

$$\begin{aligned} \cos\Delta\psi &= \begin{bmatrix} x \\ y \\ z \end{bmatrix} \cdot \begin{bmatrix} x_3 \\ y_3 \\ z_3 \end{bmatrix} \\ &= \sin\alpha\cos\beta\sin\alpha_3\cos\beta_3 \\ &\quad + \sin\alpha\sin\beta\sin\alpha_3\sin\beta_3 + \cos\alpha\cos\alpha_3 \\ &= \Lambda \end{aligned} \tag{5}$$

Table 2 Antenna error sources and calculation result

Error sources			Roll			Pitch			Yaw		
			Constant error (°)	Orbit period error (°)	Short term error (°)	Constant error(°)	Orbit period error (°)	Short term error (°)	Constant error(°)	Orbit period error (°)	Short term error (°)
Antenna error	Thermal deformation			0.02			0.02			0.019	
	Antenna assembly		0.03			0.03			0.028		
	Measurement		0.03			0.03			0.028		
	Antenna control		0.02			0.02			0.019		
	Antenna installation		0.03			0.03			0.028		
RSS			0.056	0.02	0	0.056	0.02	0	0.052	0.019	0
Antenna			0.076			0.076			0.071		
Control system error	Earth sensor	Constant error	0.02	0.02		0.02	0.02				
		Earth noise			0.003			0.003			
	Gyro	Constant drift								0.1	
		Integration initial							0.1		
		Random drift								0.02	
		Gyro installation								0.01	
	Orbit	Orbit period effect		0.006			0.02			0.01	
RSS			0.02	0.021	0.003	0.02	0.028	0.003	0.1	0.103	0
Control system			0.044			0.051			0.203		
Satellite installation error	Satellite measure basic		0.01			0.01			0.01		
	Satellite structure deformation		0.08			0.08			0.07		
RSS			0.081	0	0	0.081	0	0	0.071	0	0
Installation			0.081			0.081			0.071		
RSS			0.1	0.029	0.003	0.1	0.035	0.003	0.133	0.105	0
Single axis			0.132			0.138			0.238		
Maximum error of three axex			0.305°								

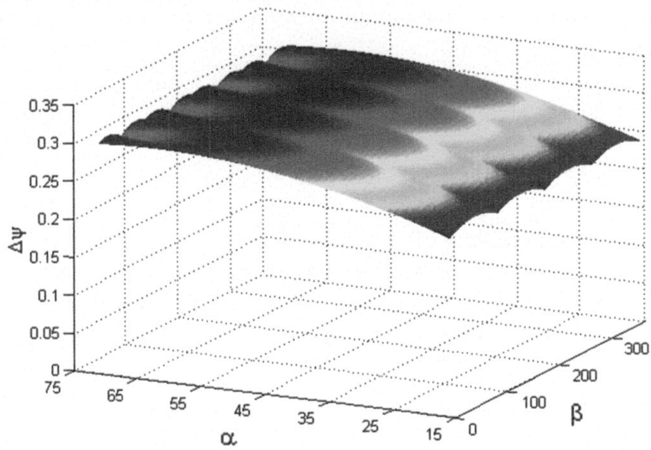

Fig. 4 Maximum antenna pointing error

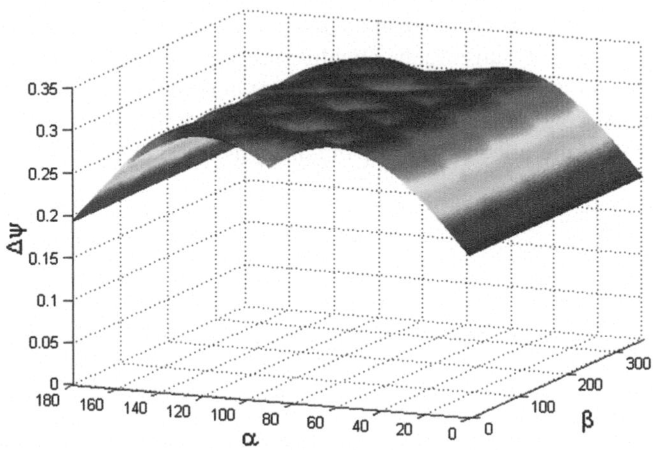

Fig. 5 Maximum antenna pointing error of all direction

$$\Delta\psi = \cos^{-1} \Lambda \tag{6}$$

The maximum error is considered as the pointing accuracy index.

The estimation of the geometric conditions whether satellite link can be built is as follows:

For the low–low satellite link, antenna pointing error needs to consider the relation of the two satellites. Thus the error value is twice the single satellite error $\Delta\psi$, that is $2\Delta\psi$ should be less than the antenna whole beam width.

For the high-low satellite link, antenna pointing error needs to consider the relation of the low orbit and the high orbit satellites. Thus the error value is the plus of the two class satellite maximum error angles, this plus should be less than the antenna whole beam width.

6 Numerical Examples

Consider the error source data of some satellite as shown in Table 2. The results and simulations are shown in Figs. 4 and 5. From the table, the error precision is 0.305°. If the selected spot antenna beam width is 1.6°, thus from the result data, the error can satisfy the geometric conditions of building the satellite link.

7 Conclusions

A new method of satellite link antenna pointing error modeling and calculation analysis is presented in this paper for a class of mixed constellation navigation satellite systems. The rotation matrix method is used for the integrated error analysis. The analysis results are the conclusions of the whole satellite level, which can exactly reflect that whether the selected spot antenna can be able to satisfy the geometric basic condition of building the satellite link. Numerical example is given to illustrate the effectiveness of the analysis method.

References

1. Sun, J., Ma, X., & Yu, D. (2007). Accuracy analysis of a satellite two-axis antenna pointing mechanism. *Journal of Astronautics, 28*(3), 545–550.
2. Tian, H., Zhao, Y., & Sun, J. (2007). Beam pointing calculation of a dual-axis oriented antenna. *Journal of Astronautics, 28*(5), 1215–1218.
3. Zhang, F., Ding, H., & Fu, T. (2010). Analysis and modeling of pointing error for antenna pointing mechanism. *Electro-Mechanical Engineering, 26*(1), 41–44.
4. Pan, B., Zhang, D., Shi, W., & Lin, T. (2011). Modeling and analysis on pointing accuracy of antenna in satellite. *Spacecraft Engineering, 20*(5), 49–54.

Present Status Analysis on the Construction and Application of CORS in China

Hui Liu, Sitong Guo, Jingnan Liu, Zongbiao Tian and Donghai Zhang

Abstract Based on the construction data of continuously operating reference stations (CORS) in China, this paper reviews the development history of CORS briefly. Then the development characteristics and existing problems of which was summarized. In the end, the construction and development trend of CORS was predicted.

1 Introduction

Continuously operating reference stations (CORS) is also called satellite positioning services system, which is an information system used to provide a precise positioning and navigation service. CORS is a kind of important information infrastructure of geospatial information, which is widely used in National economic departments such as land planning, traffic and water resource. The system user also distributed in geodynamics, environmental science of the earth and atmosphere, aerospace, etc. [1–3].

The construction of CORS in China began at the end of twentieth century, whose development expanded from city system to provincial system gradually. Now, the developed area of China which includes middle and eastern provinces has been setup provincial system. Almost all of the provincial capital city has been setup city system (maybe only 3–4 cities excluded). Base on the provincial and city systems, the national CORS platform has been setup preliminary.

H. Liu (✉) · S. Guo · J. Liu · Z. Tian · D. Zhang
GNSS Research Center, Wuhan University, Wuhan, Hubei, China
e-mail: loweliu@hotmail.com

J. Sun et al. (eds.), *China Satellite Navigation Conference (CSNC) 2012 Proceedings*,
Lecture Notes in Electrical Engineering 160, DOI: 10.1007/978-3-642-29175-3_36,

The departments and institute of city planning, national land and transportation carry on real time precise positioning based on the system.

Based on the review of the CORS development in China, the author has summarized constructions status of CORS. According to its characteristics, the application prospect is predicted to provide certain stand for construction in future.

2 Constructions and Application Status

2.1 Current Construction Status Analysis

In general, CORS in China can be classified by regional, national and professional system [4].

Regional system is the earliest system. It refers to city and provincial systems. Most of the city systems are built by the institute of city surveying and mapping and funded by the department of city planning, some are built by the department of land resource. Almost all of the provincial systems are built by the provincial department of land resource and funded by the government. The first city CORS was built and funded by Shenzhen Land Resources and City Planning Department in 1998 [5], the first provincial CORS was built by the Surveying and Mapping Institute of Land Resource of Guangdong Province and funded by the Department of Land Resource of Guangdong Province. They are the historical milestone on the CORS construction in China. At present, the sketch map for provincial and city systems is shown in Fig. 1, and the summary of each system is shown in Table 1 (only includes partial city and provincial systems).

Professional system refers to the systems which are built by the department of traffic, geological, earthquake, weather and water conservancy, whose area covered whole country as well as whole city. All these systems provide the professional service to their owner and the GNSS observation data can be shared to other departments. There are two large-scale professional systems in China. One is Crustal Movement Observation Network of China (CMONOC) that was built by several government departments; they are China Earthquake Administration (CEA), National Administration of Surveying Mapping and Geoinformation (NASMG), Ministry of Education of P.R. China, etc. There are 260 reference stations and 1000 monitor stations in this system. The goals are geodynamics science research and earthquake monitoring, etc. Another is built by China Meteorological Administration (CMA), it is used for the GPS Meteorological research and weather broadcast proposal. There are 600–1000 reference stations in this system and all of them setup in the weather station area. In addition, the Yellow River Conservancy Commission of the Ministry of Water Resources and the Changjiang Water Resource Commission committee have plan to build large system along the Yellow River basin and Yangtzi River basin.

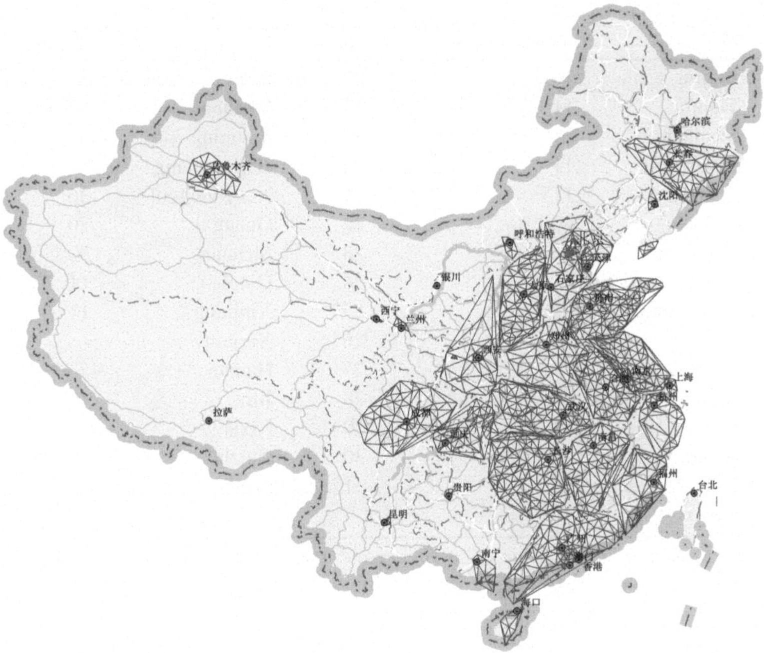

Fig. 1 Distribution graph of the reference systems in Mainland China. (By 2011, data sources from website of surveying and mapping departments of various provinces and presentation material, there may be a little different from exiting. In this figure, data of H.K, Macau and Taiwan is not included)

The country CORS built by National Administration of Surveying Mapping and Geoinformation is on the fieldwork stage. There are about 340 stations in it. It can be formed the uniform coordinate framework for other systems and can provide high precision positioning service for earthquake, weather broadcast and so on. Besides, many important projects such as western mapping, island and reef surveying and mapping have the content to build reference stations which are used to obtain high precision positioning service.

In addition, the scientific research institutions and enterprise engaged in high precision positioning services (such as South Surveying & Mapping Instrument Co, Hi-Target Co., Huace Co., etc.) built a large number of stations running in single base station model. For example, some branch company in Yunnan province has built more than 60 stations for demonstration, testing and providing the positioning services. On a conservative estimate, there are 500–600 single base stations built by company are operating in China.

In general, the total numbers of stations are more than 3000 by the end of 2011 (Table 2). Practical numbers of stations are 3500–3800 in estimate, and independently operating systems are more than 100.

Table 1 Profile of the regional reference system in Mainland China

No.	Province	Station numbers	No.	Province	Station numbers
1	Beijing	14	17	Gansu	6
2	Tianjin	11	18	Yunnan	8
3	Shanghai	14	19	Liaoning	14
4	Chongqing	11	20	Heilongjiang	9
5	Shandong	113	21	Nei Monggol	36
6	Jiangsu	80	22	Guangxi	70
7	Zhejiang	62	23	Sichuan	50
8	Fujian	81	24	Qinghai	11
9	Guangdong	95	25	Xinjiang	11
10	Jiangxi	72	26	Xinjiang	–
11	Hunan	90	27	Hainan	7
12	Anhui	86	28	Jilin	45
13	Hubei	80	29	Nixia	–
14	Henan	56	30	Guizhou	–
15	Hebei	20	31	Shanxi	26
16	Shanxi	50			

By 2011, data sources from website of surveying and mapping departments of various provinces and presentation material, there may be a little different from exiting

Table 2 Estimated number of the reference stations in China

Project	Numbers
Regional system reference stations	1228
NASMG	340
CEA	260
CMA	600
Other administrations	200
Enterprise (industry)	500
Total	3128

Summarizing the constructive condition of reference stations, there are two aspects.

One is that regional system constructs faster than country and professional system and became the main force of construction. Regional system is built by city planning and provincial departments of land resource to serve for high precision positioning. With the development of economy, the positioning needs of effectiveness, current time, accuracy and reliability are dynamic for massive investment.

The other is that all regional system can provide real time high precision positioning service. Network RTK technology (real-time carrier phase difference) is the key for system to provide real time high precision positioning service. For the reason, the average distance between contiguous reference stations is within 50–80 km to meet the network RTK technological demands. At the same time, the communication link between reference stations and data center uses optical fiber or broadband such as SDH to reduce the data delay.

Table 3 The Customer situation in provincial system

Profession	Jiangxi (%)	Hubei (%)	Jiangsu (%)
Surveying and mapping	32	18	30
Land resource	26	5	18
City planning, Urban geotechnical Investigation and surveying	19	34	19
Geology, geological mining	10	13	5
Water conservancy and hydropower engineering	6	6	15
Nuke industry	2	–	–
Earthquake research	–	–	2
Traffic	2	3	–
School, research institutions	1	4	3
Electricity	1	1	2
Instrument distributor	–	9	4
Others	1	7	2
Total	100	100	100

In general, the character of the CORS construction in China is high starting point and in strict standard.

2.2 Application Status Analysis

At present, the promotion of regional system aims at real time high precision positioning, involving surveying and mapping, land resource, territorial planning and meteorological departments. For example, the percentage of CORS registered users among various industries in the Province of Hubei, Jiangxi and Jiangsu shows in Table 3 [6–8]. The distribution of city CORS users is similar, but the user number is much more smaller than it.

As we can see in the table, there are more than 50% customers belong to the major of surveying and mapping, land resource and city planning. For those areas who built system earlier, the percentage is up to 60–70%.

This means that most of surveying and mapping engineer have already accepted the work mode base on the system. In addition, the quantity of surveying and mapping users is stable, and the numbers are 300–400 users in every province steadily.

There are almost 20% users in the departments of Geology, Geological Mining, Water Conservancy and Hydropower Engineering, Earthquake, Electricity. The quantity is probable to increase. There are 7–10% users in research institutions, instrument distributors, forestry and others, whose quantity is stable relatively.

For the traffic applications, there are only 2–3% users. In fact, the real time sub-meters precision positioning is much more useful for mobile navigation, logistics management and personal location based services. For the regional system, the user can get sub-meter precision positioning much easier than centimeter precision positing.

In general, for the regional system management, the customers in the major of surveying and mapping, land resource and city planning are their key account users. So only to provide much more high quality service for them and to develop new navigation customers, the system can create benefit and operating in a positive feedback circle.

For the these customers, it is important to improve the accuracy (precision and reliability) and availability of the regional system for real time high precision positioning service.

3 Development Characteristic Analysis

3.1 Hot Spot Analysis

Regional system provides the precise positioning service by using real time high precision technical in China. It is larger than America and Europe whether the system scale or the quantity of users. For the reason, the operation and maintenance is harder. The discussions about the operation and maintenance technology are as follow.

Firstly, the question is position standard. According to surveying and mapping law, CGCS2000 is the coordinate by law in China. Regional system should provide positioning service based on CGCS2000. But the dynamic reference frame in the basis of CORS is not formed at present in China. The positioning standard of regional system is formed by translocation of reference stations and local CGCS2000 points, and the precision of the reference station coordinates is low. The standard is not uniform throughout country. To solve the problem, it is necessary to research the theory and method of the CGCS2000 reference frame formed base on CORS. At the same time, existing data of reference stations should be integrated to handle in order to get the existing results with maximum compatibility and ensure high precision results of the overall frame.

Secondly, the problem is service compliance. According to the law and privacy regulations, the geodetic data and coordinates transform parameters are classified. Transform all these data using internet (by WLAN or wireless) is illegal. So the difficulty is to provide real time high precision positioning service to the users within the restricted conditions. To solve the problem, some security measures is indispensible to process original data, such as changing the absolute value to relative value and sending correction value only.

The third problem is due to using the centralized software architecture, the commercial network RTK software cannot process huge number of reference stations data simultaneously. When there are more than 70 stations in provincial system, the system administrators have to divided the entire network into several sub-network less than 50 stations and deploy independent software for each sub-network. In this situation, some problem happened in overlapping area, such as the positioning difference between two sub-networks was over the limitation, the services roving confused and the reliability of positioning service reduced. When two or more systems connected, those could be happened too. The way to solve the problems is to improve the algorithm model and flow design as distribute computing architecture.

Fourthly, corrected algorithm limitations of ionosphere model and troposphere model reduce the availability. Because most south area in China is located in ionosphere abnormal region, the condition is much more complex than other area in the world. With the limit of algorithm, many systems cannot work at the noon, so the availability is poor. Besides, the elevation difference between reference stations and users is too large (more than 1000 m), which lead to the user cannot get positioning result. The solution is to build model for ionosphere and research on the troposphere error interpolation model.

Fifthly, the standardization is weakness. In the recent decades, the technical standards about system operation and maintenance have been formed initially, but it is weak in standard implement. The content of most system construction standards are on the basic construction, but less in application and data processing. To solve the problem, it is necessary to absorb new technology and methods in Geology and revise the existing standard incessantly.

Sixthly, it lacks in the effective monitor measures of CORS reliability and system integrity. For regional system, many reference stations are not located on the basement rock and only setup on the roof. Lacking of data accumulation and research funding, the research on the reliability of reference stations is insufficient. For the real time systems, most systems have not the function of system integrity monitoring and the service quality maintains mechanism, they only can provide user services in the 'best I can do' mode. All the risks must be undertaken by the users themselves. To solve this problem, it is necessary to establish the roadmap of data accumulation and processing, then and develop and upgrade system in full-function operating mode.

Seventhly, the most applications concentrate in high precision positioning service limited in the professional field of geodesic survey, city planning, basic survey and cadastral survey. But the navigation and positioning service is involved less. Many administrators have recognized that application promotion is necessary in the society, and the largest user quantity will be at the field of vehicles and individual for CORS. But there are many technological difficulties for regional system, such as information security, information privacy and concurrent user capacity. It is indispensable to bring in the computer science methods of the latest cloud computing and requirements engineering, as well, to change system service process and to transform concentrated monomer services to networked services.

Eighth, the problem is the localization of key technical and products. Since 2008, the PPP-network RTK algorithm has come up, and the network-RTK and network managing software of independent intellectual property rights has been developed. However, foreign products are mostly adopted in regional system.

Besides, receiver or OEM board is from foreign manufacturers. Thus, it is significant to stimulate the Compass development. For government, academic and industry, one of the problems are how to popularize domestic equipments and spread software product.

Above, main problems in the development have been indicated. Referred to the professional technology, the author will write more detail articles professionally in future.

3.2 Development Tendency

At the current situation of the CORS construction and application development, the CORS development in China is concluded to technology progress and industrialized application.

Technologically, the first step is to develop a new generation of system management and service platform. Distributed computing architecture and service oriented architecture (SOA) is the key to implement management and service mechanism. System platform of service on demand should be set up for more than 10000 concurrent users by CORS web service standardized encapsulation to manage thousands of stations and provide the location services in region. Secondly, the problem of synchronous calculation by large-scale network stations should be solved by combination with PPP and network RTK technology; the method to decrease ionosphere delay is by multi satellite systems or combination with three or more frequency mode; the error interpolation algorithm optimized could be eliminate troposphere effects. Thirdly, the theory of reference station reliability and system integrity should be perfect. Fourthly, domestic software and hardware such as multimode Compass receivers should be supported at key system and station construction in the basic of national uniform location standard to ensure security. Fifthly, with the expansion of CORS scale, the whole system has become a complex and enormous system. It is important to research the regulation of internal alteration by the theory of complex network.

For industrialized application, the first step is to expand current service range, also ensure the information security and privacy by relevant technology, as well, spread to the field of vehicle navigation and personal positioning, then provide various accrued service for users. Secondly, improve the navigation technology up to decimeter precision by the advantage of CORS network; promote high precision navigation, timely data collection, data processing and distribution; enlarge application of industry chain and promote the development of spatial information technology. Thirdly, key equipments and software should be domestic, make the technology of geodetic OEM plate and machine manufacturing technology perfect.

In general, the development of key CORS technology will be conclude to four high, such as high reliability, high availability, high precision, high timely. The applications will involve in reliable network service on demand and coexistence with popular and professionalized service.

4 Conclusions

Human cannot live without location and time information in society. With the society development and the technological progress, various navigation technologies have been developed. CORS is the platform (infrastructure) including new technologies of geodesic survey, satellite navigation and computer network. The effects of different location technologies on platform reflect the difference of reference station distance and communication efficiency request. No matter what type of positioning technologies, they could be deployed and operated on CORS to provide the service. In the view, it is vigorous vitality for reference station network with the development of navigation and positioning technology [2, 3, 9].

With the development of Compass satellite navigation system, the alteration and update period of reference station network is coming. The schedule should be arranged to make the Compass multimode frequency system compatible. In the application, it is significant for administrator to improve the service quality and expand the application range.

Acknowledgments This paper was funded by the National high technology research and development plan (863 Program), 2007AA12Z309; The National Basic Research Program of China (973 Program) (2007CB310805); Surveying and Mapping Basic Research Fund of Earth Space Environment and Geodesic Measurement Laboratory in Wuhan University (100110).

References

1. Chen, J. (2002). New geodetic progress. International Association of Geodesy (IAG). *Advance in Earth Sciences, 3*, 6–17.
2. Liu, J., & Liu, H. (2003). Continuous operational reference system—infrastructure of urban spatial data. *Editorial Board of Geomatics and Information Science of Wuhan University, 6*(28), 259–264.
3. Liu, J., & Liu, H. (2003). Thought on establishing a continuous operational reference station network in China. *Surveying and Mapping Report in China, 3*, 10–28.
4. Liu, H. (2005). *Research on geo-spatial information grid and continuously operating reference network*. Dissertation, Wuhan University.
5. Liu, H., Shi, X., Yang, Z., Luo, H, Liu, J., & Yang, S. (2003). The establishment and experiments of Shenzhen continuous operational reference system. *Bulletin of Surveying and Mapping*, No 9, pp. 33–36.
6. Navigation and Positioning Center in Hubei Province. (2011). Introduction of HBCORS, Surveying and Mapping Bureau, 12.
7. Fei, S. (2010). *New advance and development progress about JSCORS*. Institute of Surveying and Mapping of Jiangsu Province, 12.
8. Yu, L. (2011). *Construction and application of JXCORS*. Emergency Service Center of Jiangxi Surveying and Mapping Bureau, 12.
9. Cheng, P., Yang, Y., & Li, J. (2002). The latest advances in Geodesy and GPS technique in China. *Bulletin of Surveying and Mapping 2*.

The Distributions of HDOP and VDOP in GNSS and a Corresponding New Algorithm of Fast Selecting Satellites

Haifu Ji, Lihua Ma, Guoxiang Ai and Meng Wang

Abstract In Global Navigation Satellite System (GNSS), the accuracy of position estimates depends upon the measurement geometry and pseudorange measurement errors. The Dilution of Precision (DOP) provides a simple characterization of the user-satellite geometry. An in-depth study of DOP distribution characteristics is conducted in this paper, with emphasis on performance bounds of Horizontal Dilution of Precision (HDOP) and Vertical Dilution of Precision (VDOP) for four satellites in varied hypothetical constellations. The distributions of HDOP and VDOP in 16 major world cities are analyzed by simulated calculations using the TLE sets of GPS satellites. According to the theoretical analyses about the distributions of HDOP and VDOP, this paper also proposes a new algorithm of fast selecting satellites based on HDOP and VDOP, which requires a shorter computing time than the general method and provides a feasible way to quickly achieve a high-accuracy positioning.

1 Introduction

Satellite navigation systems have burgeoned since 2000. GPS is the first GNSS, followed by Russia's GLONASS reinvigorated in recent years, Europe's Galileo and China's Compass which are under development. Position estimations are

H. Ji (✉) · L. Ma · G. Ai · M. Wang
National Astronomical Observatories, Chinese Academy of Sciences,
100012 Beijing, China
e-mail: jihaifu@nao.cas.cn

H. Ji · M. Wang
Graduate University of Chinese Academy of Sciences, 100049 Beijing, China

J. Sun et al. (eds.), *China Satellite Navigation Conference (CSNC) 2012 Proceedings*,
Lecture Notes in Electrical Engineering 160, DOI: 10.1007/978-3-642-29175-3_37,

based on measurements of pseudoranges in GNSS. Given a simple model that the errors in the measurements from different satellites are treated as zero-mean, uncorrelated and identically distributed, we can characterize the root-mean-square (rms) position error in terms of a DOP parameter and the rms measurement error. The higher the DOP value, the bigger the scatter in position error.

The distributions of DOP are first studied in this paper. Similar research has been undertaken and yielded many results of great value [1–3]. This paper gives emphasis on performance bounds of HDOP and VDOP for four satellites in varied hypothetical constellations. Simulation results about the distributions of HDOP and PDOP in 16 major world cities are then provided using the TLE sets of GPS satellites.

The minimum Geometric Dilution of Precision (GDOP) principle is generally adopted as satellite selection method [4–7]. In this general method, the minimum GDOP is searched among all four visible satellites' combinations, which inevitably consumes a relatively long time and imposes great amount of calculations on receivers. Taking into account the number of receiver channels and real-time signal processing capabilities especially for these navigation modules embedded in mobile phones, vehicles, spacecrafts, etc., as well as the fact that user is much more concerned about the horizontal positioning accuracy, a new algorithm of fast selecting satellites based on HDOP and VDOP is finally proposed in this paper, which can effectively reduce the computing time and realize fast positioning with high accuracy.

2 DOP and Position Estimation

To estimate the position and clock bias of a user, at least four satellites are observed simultaneously. The pseudorange measurement based on code phase from satellite s can be modeled as [2]

$$\rho^s = \sqrt{(x-x^s)^2 + (y-y^s)^2 + (z-z^s)^2} + b + \varepsilon^s \tag{1}$$

where ρ^s denotes the pseudorange after accounting for the satellite clock offset and compensating for the remaining errors in the measurement; b is receiver clock bias; the error term ε^s is the combined effect of the residual errors; vectors $\mathrm{x}=(x, y, z)$ and $\mathrm{x}^s = (x^s, y^s, z^s)$, for $s = 1, 2, \ldots, n$, represent the position of the user at the time of measurement and the position of the satellite at the time of signal transmission.

To characterize the quality of the estimates in terms of DOP, our model for the measurement error in Eq. (1) is simplified as [2]

$$\mathrm{E}(\varepsilon^s) = 0, \quad \mathrm{Cov}(\varepsilon^s) = \sigma^2 \mathrm{I} \tag{2}$$

where E(•) represents the mean value, Cov(•) denotes covariance, I is the identity matrix, and σ is the common standard deviation of the user range error for each of satellites.

Let $\hat{\mathrm{x}}$ and $\hat{\mathrm{b}}$ denote the estimates of the position and clock bias in the local east-north-up (ENU) coordinate frame respectively, the covariance matrix of $(\hat{\mathrm{x}}, \hat{\mathrm{b}})^T$ can been written as

$$\mathrm{Cov}\begin{bmatrix} \hat{\mathrm{x}} \\ \hat{\mathrm{b}} \end{bmatrix} = \sigma^2 \left[G^T G\right]^{-1} \tag{3}$$

G is $(n \times 4)$ matrix with each row composed of three elements of direction cosine vectors represented in ENU coordinate frame, and an entry of 1 in the last column. It is convenient to represent the unit vector of direction cosine in the ENU coordinates in terms of azimuth and elevation, we obtain

$$G = \begin{bmatrix} \cos el^{(1)} \sin az^{(1)} & \cos el^{(1)} \cos az^{(1)} & \sin el^{(1)} & 1 \\ \cos el^{(2)} \sin az^{(2)} & \cos el^{(2)} \cos az^{(2)} & \sin el^{(2)} & 1 \\ \vdots & \vdots & \vdots & \vdots \\ \cos el^{(s)} \sin az^{(s)} & \cos el^{(s)} \cos az^{(s)} & \sin el^{(s)} & 1 \\ \vdots & \vdots & \vdots & \vdots \\ \cos el^{(n)} \sin az^{(n)} & \cos el^{(n)} \cos az^{(n)} & \sin el^{(n)} & 1 \end{bmatrix} \tag{4}$$

where $az^{(s)}$ and $el^{(s)}$ are the azimuth and elevation angle of satellite s observed by user. Defining $\mathrm{H} = \left[G^T G\right]^{-1}$ for simplicity, then

$$\text{Horizontal error} = \sigma \cdot \mathrm{HDOP} \quad \text{Vertical error} = \sigma \cdot \mathrm{VDOP} \tag{5}$$

where $\mathrm{HDOP} = \sqrt{\mathrm{H}_{11} + \mathrm{H}_{22}}$ and $\mathrm{VDOP} = \sqrt{\mathrm{H}_{33}}$.

3 Distributions of HDOP and VDOP

With four satellites, DOP is lowest if three satellites are equally spaced on the horizon and a fourth satellite is at zenith [1]. Let azimuth angles of these three satellites be 0°, 120° and 240° respectively, then matrix G is as follows:

$$G = \begin{bmatrix} \cos 0^\circ \sin 0^\circ & \cos 0^\circ \cos 0^\circ & \sin 0^\circ & 1 \\ \cos 0^\circ \sin 120^\circ & \cos 0^\circ \cos 120^\circ & \sin 0^\circ & 1 \\ \cos 0^\circ \sin 240^\circ & \cos 0^\circ \cos 240^\circ & \sin 0^\circ & 1 \\ 0 & 0 & 1 & 1 \end{bmatrix} \tag{6}$$

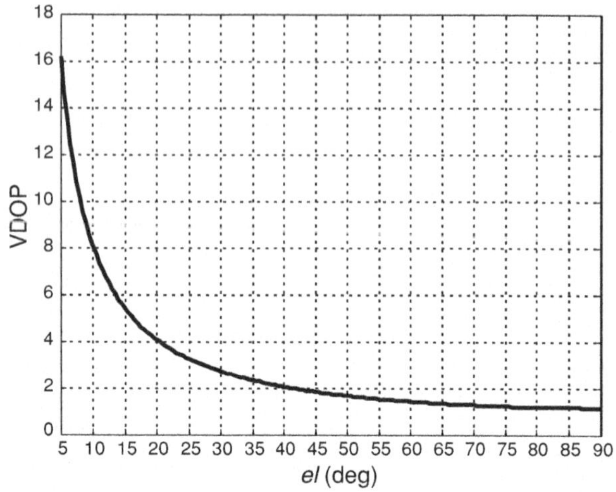

Fig. 1 VDOP versus the elevation angle of the fourth satellite

The minimums of PDOP, HDOP, and VDOP derived from Eq. (6) are

$$\mathrm{PDOP} = 2\sqrt{\frac{2}{3}} \approx 1.63; \quad \mathrm{HDOP} = \sqrt{\frac{4}{3}} \approx 1.15; \quad \mathrm{VDOP} = \sqrt{\frac{4}{3}} \approx 1.15 \tag{7}$$

In this section, except the above optimum constellation, we study the distribution characteristics of HDOP and VDOP in other hypothetical constellations with four satellites.

3.1 The First Hypothetical Constellation with Four Satellites

Suppose that three satellites are evenly spaced by 120° on the horizon and a fourth satellite has an azimuth of az and elevation of el, and the HDOP and VDOP are

$$\mathrm{HDOP} = \sqrt{\mathrm{H}_{11} + \mathrm{H}_{22}} = \sqrt{\frac{4}{3}} \approx 1.15 \quad \mathrm{VDOP} = \sqrt{\mathrm{H}_{33}} = \sqrt{\frac{2 \cdot [2 + \cos(el)^2]}{3 \cdot \sin(el)^2}} \tag{8}$$

Figure 1 gives a plot of the fourth satellite's elevation angle and the corresponding VDOP value computed from Eq. (8).

From the results, the following conclusions can be drawn:

1. The theoretically lowest HDOP = 1.15 can be obtained only if there are three satellites equally distributed by 120° on the horizon and does not involve the distribution of the fourth satellite.

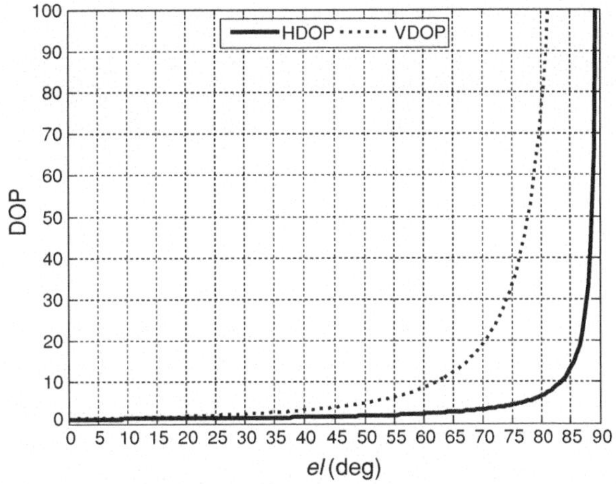

Fig. 2 HDOP and VDOP versus the elevation angle of the three satellites

2. If there are three satellites equally distributed by 120° on the horizon, VDOP value is inversely proportional to the fourth satellite's elevation and has no relation with its azimuth. When the fourth satellite has an elevation of 90°, i.e., at zenith, VDOP value decreases to 1.15 which is its theoretical minimum.

3.2 The Second Hypothetical Constellation with Four Satellites

In general, of course, we do not have one satellite at zenith and three others equally spaced on the horizon. Considering ionospheric scintillation, multipath and sheltering, etc., which could trigger abnormal observation, the observed satellite must have an elevation higher than the elevation mask angle, e.g., 5°. We thus further suppose that one satellite is at zenith and three satellites are evenly spaced in azimuth with the same elevation angle of *el*, then

$$\mathrm{HDOP} = \sqrt{\frac{4}{3}} \cdot \frac{1}{\cos(el)} \qquad \mathrm{VDOP} = \sqrt{\frac{4}{3}} \cdot \frac{1}{1 - \sin(el)} \tag{9}$$

Figure 2 illustrates the values of HDOP and VDOP and corresponding elevation angle of the three satellites *el*. As can be seen from Fig. 2, both HDOP and VDOP ascend with the increasing *el*. This conclusion provides a useful method of fast selecting satellites, i.e., first selecting a satellite near zenith and then three satellites which are most evenly distributed in azimuth and have the lowest elevation angles.

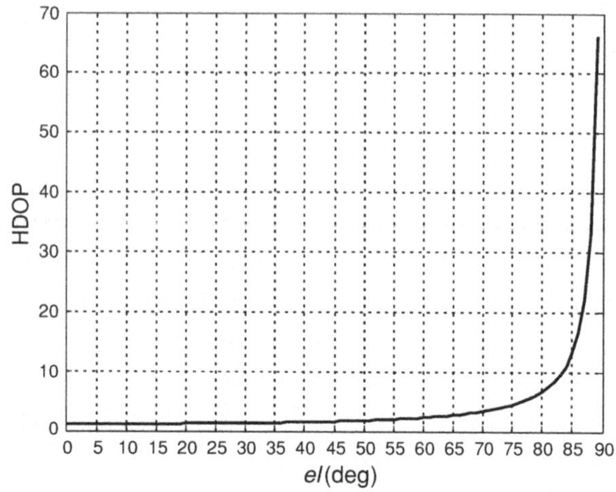

Fig. 3 HDOP versus the elevation of the three satellites

3.3 The Third Hypothetical Constellation with Four Satellites

In the following, a more practical assumption that three satellites are evenly spaced in azimuth with the same elevation angle of *el*and the fourth satellite's azimuth and elevation angle are el_4 and az_4 respectively, then

$$G = \begin{bmatrix} \cos(el)\sin 0^\circ & \cos(el)\cos 0^\circ & \sin(el) & 1 \\ \cos(el)\sin 120^\circ & \cos(el)\cos 120^\circ & \sin(el) & 1 \\ \cos(el)\sin 240^\circ & \cos(el)\cos 240^\circ & \sin(el) & 1 \\ \cos(el_4)\sin(az_4) & \cos(el_4)\cos(az_4) & \sin(el_4) & 1 \end{bmatrix} \tag{10}$$

The results will be given and discussed on two cases:

1. $el_4 \neq el$

In this case, HDOP only relates to the elevation of the three satellites *el*. As is shown in Fig. 3, HDOP decreases uniformly as we decrease *el* until $el = 0$ (horizontal) where it is at the theoretical minimum of 1.15. VDOP is not relevant with the azimuth of the fourth satellite. When *el* is given, VDOP first begins to ascend with the increasing el_4 and reaches the maximum until $el_4{=}el$ and decreases uniformly afterwards (Fig. 4).

2. $el_4{=}\,el$

The distributions of HDOP and VDOP are devoid of general characteristics in such a case. Both HDOP and DOP depend on the elevation of the three satellites and configuration of the fourth satellite, and could be an extreme value of zero or infinity.

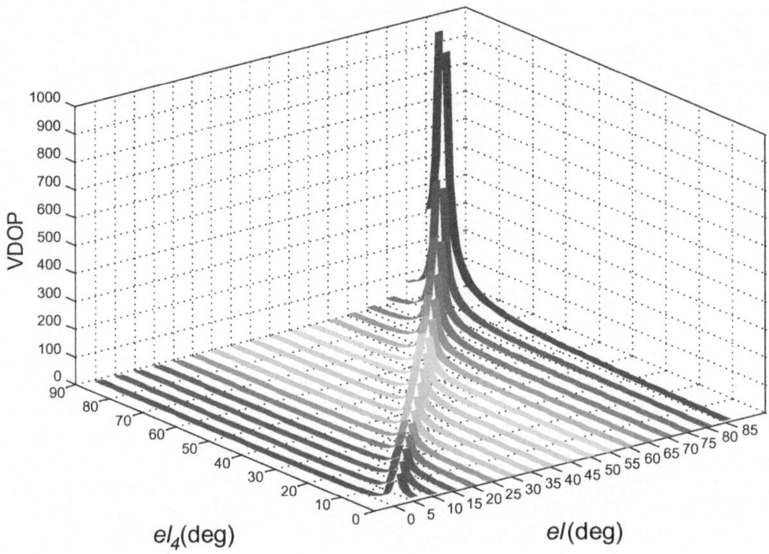

Fig. 4 VDOP versus the elevation angles of the three satellites and the fourth satellite

4 Simulating Analysis of HDOP and VDOP

In previous section, the distributions of HDOP and VDOP were analyzed in theory for four satellites in varied constellations. We will give some corresponding simulation results in this section. The two-line element (TLE) sets of GPS satellites are adopted in the simulations.

A TLE set is a data format specified by North American Air Defense Command (NORAD) and is used to convey set of orbital elements that describe the orbit of an earth satellite. The TLE can be used directly by the SGP4 model (or one of the SGP8, SDP4, and SDP8 models) to compute the precise position of a satellite at a particular time.

Sixteen major world cities are selected to calculate the distributions of HDOP and VDOP. Those cites could be considered as several of representative cities in geographical distribution, which are located in north, south, east and west hemisphere.

The ellipsoidal height is assumed to be 0-m and the elevation mask angle is 5° in the simulation. The HDOP and VDOP values are computed 1 min apart over a 24-h period with the selected four optimum satellites. A summary of the distributions of HDOP and VDOP is presented in Table 1 for the 16 cities.

The distributions of DOPs for several of the sixteen cities are shown in Fig. 5.

The GPS satellites are positioned in six Earth-centered orbital planes with four satellites in each plane. The orbits are nearly circular and equally spaced around

Table 1 The distributions of HDOP and VDOP for the sixteen cities

City	Latitude	Longitude	HDOP			VDOP		
			Max	Min	Mean	Max	Min	Mean
Helsinki	60°12′N	24°56′E	2.56	1.16	1.41	2.99	1.36	1.86
Berlin	52°31′N	13°25′E	2.18	1.16	1.42	3.02	1.34	1.86
Vancouver	49°16′N	123°07′W	2.39	1.20	1.46	2.93	1.37	1.85
Rome	41°54′N	12°29′E	2.59	1.18	1.48	3.28	1.37	1.93
New York	40°43′N	74°00′W	4.03	1.18	1.49	4.36	1.37	1.91
Beijing	39°54′N	116°23′E	2.27	1.19	1.48	3.04	1.38	1.88
Tokyo	35°41′N	139°46′E	2.07	1.17	1.42	2.78	1.34	1.86
Miami	25°47′N	80°13′W	3.64	1.17	1.44	4.32	1.40	1.94
Hong Kong	22°17′N	114°08′E	3.17	1.15	1.42	4.48	1.34	2.00
Bangkok	13°45′N	100°31′E	2.17	1.14	1.41	3.70	1.38	1.89
Singapore	1°17′ N	103°51′E	1.87	1.16	1.39	2.95	1.44	1.87
Jakarta	6°11′S	106°50′E	2.28	1.17	1.38	3.38	1.36	1.94
Lima	12°03′S	77°02′W	1.82	1.15	1.36	3.32	1.41	1.89
Rio de Janeiro	22°54′S	43°14′W	1.84	1.17	1.41	3.54	1.40	2.02
Sydney	33°52′S	151°13′E	2.07	1.17	1.45	3.54	1.31	1.94
Wellington	41°17′S	174°47′E	2.76	1.19	1.48	3.48	1.34	1.91

the equator at a 60° separation with a nominal inclination relative to the equatorial plane of 55° [8]. In such a constellation, it is easy to select three most evenly distributed satellites near horizon. However, it is less possible to have a satellite near at zenith, especially for users at higher latitudes where there are no satellites overhead for the reason that the inclination of the GPS orbits is 55°. According to the simulation results presented in Table 1, the VDOP values are about 30% larger than the VDOP values consistently, which is in accordance with the previous analyses in theory.

5 An New Algorithm of Selecting the Optimal Satellites Based on HDOP and VDOP

Considering the number of receiver channels and real-time signal processing capabilities especially for those navigation modules embedded in mobile phones, vehicles, spacecrafts, etc., satellite selection is required to fast get the best positioning results.

The general method of selecting satellites is based on the minimum GDOP obtained by searching among four visible satellites' combination, which inevitably would consume a relatively long time and bring a great amount of calculations to user receivers. According to the previous theoretical analyses about the distributions of HDOP and VDOP, as well as considering the fact that users are much more concerned about the horizontal positioning error, a new algorithm of fast

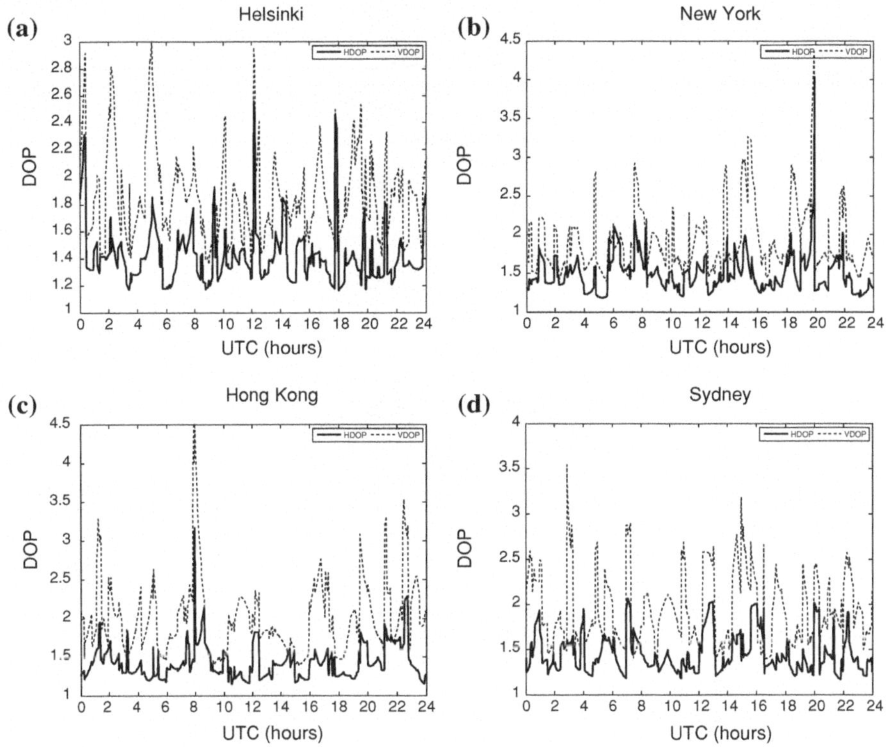

Fig. 5 Distributions of DOPs for several representative cities. **a** Helsinki, **b** New York, **c** Hong Kong, **d** Sydney

selecting satellites based on HDOP and VDOP is then proposed. The flow chart of the algorithm is presented in Fig. 6.

The main idea of this algorithm is based on the previous analysis. In order to get an optimal HDOP which means a high-accuracy horizontal positioning, only three satellites with low elevation angles and most evenly distributed in azimuth are required in this algorithm and the selection of a fourth satellite is arbitrary. If a three-dimensional positioning with high accuracy is further desirous, all we should do is further selecting the fourth satellite with the largest elevation angle. As can be inferred from the implementation process, this algorithm can effectively reduce the computing time and thus realize fast positioning.

6 Conclusion

The paper conducted a thorough study of the distributions of HDOP and VDOP in GNSS. In Sect. 2, a conclusion that the HDOP value can reach its minimum when there are there satellites equally distributed on the horizon and the VDOP value is

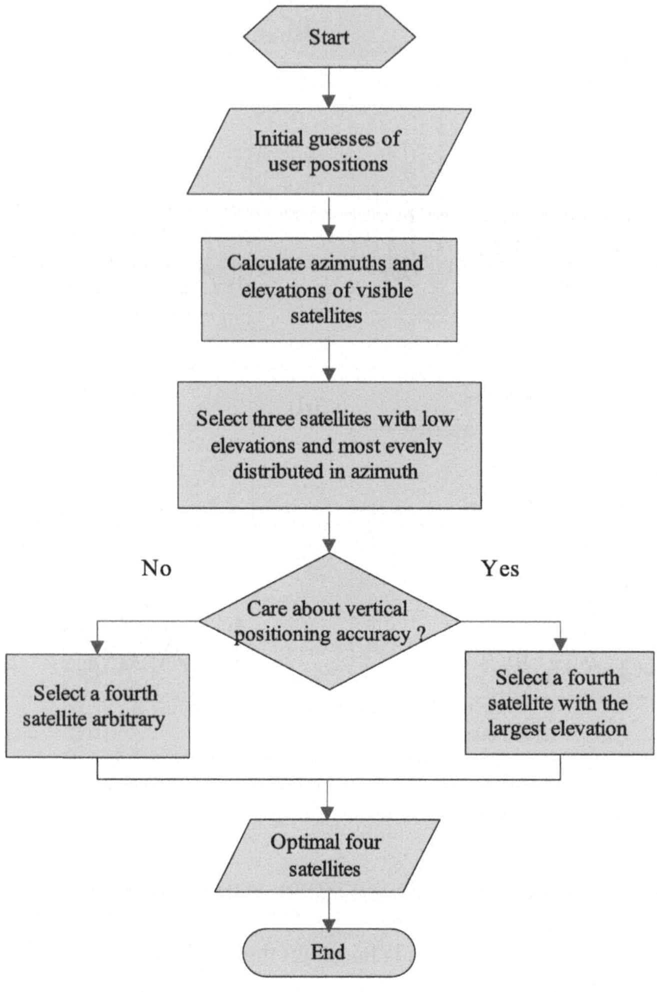

Fig. 6 An algorithm of selecting satellites based on HDOP and VDOP

inversely proportional to the elevation of the fourth satellite was first proven. We then gave an analysis of HDOP and VDOP in other varied configurations of satellites in this section. Some simulation results about the distributions of HDOP and PDOP using the TLE sets of GPS satellites were provided in Sect. 4, which are consistent with the preceding theoretical results. Finally, a new algorithm of selecting satellites based on HDOP and VDOP was proposed, which requires a shorter computing time than the general method and provides a feasible way to fast select satellites and then achieve the three dimensional positioning with high accuracy.

Acknowledgments The project was supported by the National Basic Research and Development Program of China (Grant No. 2007CB815501), the Pilot Program for the New and Interdisciplinary Subjects of the Chinese Academy of Sciences (Grant No. KJCX2-EW-J01) and the Knowledge Innovation Program of the Chinese Academy of Sciences (Grant No. KGCX2-EW-4071).

References

1. Parkinson, B. W., & Spilker, J. J, Jr. (1996). *Global positioning system: theory and applications* (Vols. 1 and 2). Washington, DC: American Institute of Aeronautics.
2. Misra, P., & Enge, P. (2006). *Global positioning system: signal, measurements, and performance*. Lincoln, MA: Ganga-Jamuna Press.
3. Zemin, W., Yang, M., et al. (2006). DOP for GPS, Galileo and combination navigation System. *Geomatics and Information Science of Wuhan University, 31*(1), 9–11.
4. Milliken, R. J., & Zoller, C. J. (1980). *Global positioning system* (pp. 12–14). Washington, DC: The Institute of Navigation.
5. Montenbruck, O., & Gill, E. (2000). *Satellite oribits: models, methods and applications*. New York: Springer.
6. Mohino, E., Gende, M., Brunini, C., & Heraiz, M. (2005). SiGOG: simulated GPS observation generator. *GPS Solution, 3*, 250–254.
7. Zong, Z. (1987). A simple and practical choice method on GPS constellation. *Navigation, 1*, 168–173.
8. Kaplan, E. D., & Hegarty, C. J. (2005). *Understanding GPS principles and applications*. London: Artech House Publishers.

Research on the Technology of Calibration of Satellite Constellation Crosslink

Xianbin Li, Chuansheng Zhang and Jianyun Chen

Abstract The method of dual-way pseudo ranging is used in navigation constellation crosslink to implement precise orbit determination and time synchronization. The accuracy of dual-way pseudo ranging is directly influenced by delay variation in transceiver. Based on the illustration of the conception and estimating model of delay, a real-time on-line closed loop calibration method of transceiver absolute delay is advanced. The experiment shows that the accuracy of this method is better than 0.3 ns.

1 Introduction

Precise orbit determination and time synchronization of navigation Constellation can be implemented by pseudo ranging between satellites. When satellite constellation crosslink is established, the space part of navigation system will not be a single part anymore. The precision of orbit determination and time synchronization and the ability of surviving of navigation system will be enhanced a lot. Because of the frequency transducer, power amplifier and low noise amplifier the transmit channel will be non-ideal. When the method of dual-way pseudo ranging is used in navigation constellation crosslink to implement precise orbit determination and time synchronization, the signal will have group time delay fluctuate and phase aberration. Meanwhile, there are error factors of multiple access interference, mismatched impedance, temperature drifting and uncertainty

X. Li (✉) · C. Zhang · J. Chen
College of Mechatronics Engineering and Automation,
National University of Defense Technology,
Changsha, 410073 Hunan, China
e-mail: lixianbincn@163.com

J. Sun et al. (eds.), *China Satellite Navigation Conference (CSNC) 2012 Proceedings*,
Lecture Notes in Electrical Engineering 160, DOI: 10.1007/978-3-642-29175-3_38,

of clock phase. These factors can bring change of transceiver fixed time delay and will influence the veracity of real range measurement. If we do nothing to the transceiver time delay, the error will reach 1 ns. So how to measure the time delay of transceiver real-timely and veraciously will be significant for the whole system of satellite constellation crosslink.

The principle of two-way satellite time and frequency transfer (TWSTFT) is introduced in paper [1]. Meanwhile, a real-time and precision Calibration system of transceiver time delay is designed. The transceiver time delay can be get through remove the symmetry part of two-way measurement. Also, group time delay of 70 MHz mid-frequency is also analyzed in this paper. Paper [2] utilizes the calibration of transceiver time delay in real-time and veraciously by designing a fixed modulate of time delay calibration. RFSR (time delay measurement method based on RF sampling and software radio) is given and has better precision in paper [3]. Paper [4] introduces a method of equipment time delay measurement based on time transfer between grand stations. This paper studies the real-time time delay calibration of transceiver of satellite constellation crosslink system based above.

2 The Influence of Time Delay Character on Inter-Satellite Range Measurement

2.1 The Principle of Dual-One-Way Measurement in Satellite Constellation Crosslink System

An important function of satellite constellation crosslink system is to measure the range and time difference between satellites and to serve for self-navigation and time synchronization. Dual-One-Way method is used wildly at present to accomplish range measurement. This method has virtues of real-time, better precision and much little influence from out surroundings and applied to international atomic time compare, radar, and TT&C [5].

Figure 1 shows the principle of dual-one way method. The process includes two parts that one is A→B and the other is B→A.

Satellite A transmits spread spectrum signal whose code phase carries the information of transition time t_1. The transmit equipment time delay is t_A and the time delay of the signal in space is τ_A. The signal is detected at moment of t_2 and the receive equipment time delay of satellite B is r_B. Because of signal is transmit at moment of t_1 and is detected at moment of t_2, so the time delay tested τ_A is a function of $\tau_A(t_1, t_2)$ which includes t_1 and t_2. $\tau_A(t_1, t_2)$ includes all of time delay of the space between two satellites. The time delay can be measured as T_B by satellite B based on self-clock.

Fig. 1 Principle of dual-way measurement in navigation constellation crosslink

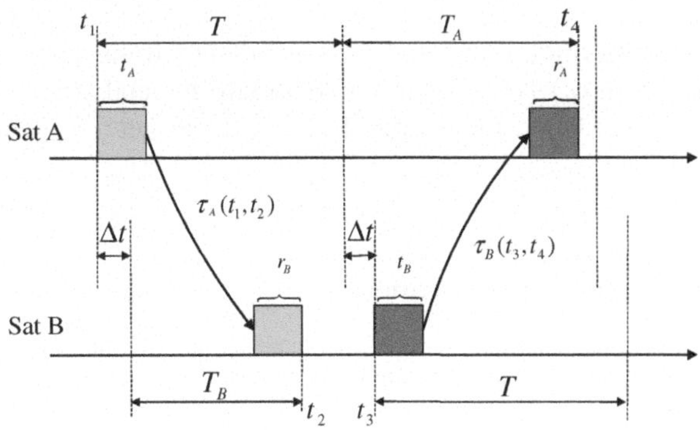

Also satellite B transmits signal at moment of t_3. The transmit equipment time delay of satellite B is t_B and the time delay of the signal in space is τ_B. The signal is detected at moment of t_4 and the receive equipment time delay of satellite B is r_A. Just like satellite A the time delay tested τ_B is a function of $\tau_B(t_3, t_4)$ and T_A is tested based on self-clock.

According to Fig. 1, we have

$$T_A = \Delta t + t_B + \tau_B(t_3, t_4) + r_A \tag{1}$$

$$T_B = \Delta t + t_A + \tau_A(t_1, t_2) + r_B \tag{2}$$

where Δt is clock difference between two satellites.

Equations 1 and 2 gives the basic relation of time and range in satellite constellation crosslink system. Time difference and pseudo range can be getting through proper change.

2.2 The Influence of Time Delay on Range Measurement

From the analysis of above, we can get that Δt is clock difference between two satellites at some moment. Based on the difference, the module of self-navigation can calculate absolute time difference between satellites.

When Eq. 1 subtracts Eq. 2, we can get

$$\Delta t = \frac{1}{2}(T_A - T_B) - \frac{1}{2}[(t_A + r_B) - (t_B + r_A)] + \frac{1}{2}[\tau_A(t_1, t_2) - \tau_B(t_3, t_4)] \tag{3}$$

where $\tau_A(t_1, t_2) - \tau_B(t_3, t_4)$ includes time delay information of two way and should be compensate according to the relatively moving between two satellites to increase precision of time synchronization.

$(t_A + r_B) - (t_B + r_A)$ reflects the time delay between satellites without taking transmit route into count. It is decided by transceiver equipment. And the influence of $(t_A + r_B) - (t_B + r_A)$ on Δt should be analyzed.

Also, when Eq. 1 adds Eq. 2, we can get

$$\frac{1}{2}[\tau_A(t_1, t_2) + \tau_B(t_3, t_4)] = \frac{1}{2}(T_A + T_B) - \frac{1}{2}[(t_A + r_B) + (t_B + r_A)] \quad (4)$$

Equation 4 can be written:

$$\frac{1}{2}[\tau_A(t_1, t_2) + \tau_B(t_3, t_4)] = \frac{1}{2}(T_A + T_B) - \frac{1}{2}[(t_A + r_A) + (t_B + r_B)] \quad (5)$$

where $(T_A + T_B)/2$ can be get through transceiver. If $(t_A + r_A)$ and $t_B + r_B$ can be measured, we will get $[\tau_A(t_1, t_2) + \tau_B(t_3, t_4)]/2$. According to the relatively moving between two satellites we can get pseudo range.

3 Analysis of Time Delay Character of Transceiver

In the basic equation of Eqs. 1 and 2, $(t_A + r_B)/2$ and $(t_B + r_A)/2$ reflect the time delay of transceiver and have not relationship of transmit path. So we should calibrate it before measurement.

According to the character of transceiver, time delay of transceiver can be classified as time delay of digital base band module and time delay of radio frequency channel.

3.1 Time Delay Character of Digital Base Band Module

Time delay of digital base band module includes jitter of high steady clock, out delay of IO, edge jitter of ADC's clock, edge jitter of DAC's clock, aperture jitter of ADC and transmitting out delay of DAC.

In the digital base band module high precision clock signal is sent to ADC and DAC through programmable IOB after synthesis by DCM in FPGA. The short stability of clock can reach 10^{-13} s/10 s, so the clock jitter can be assumed as $\sigma(T_{source}) = 0.1$ ps in short time. The clock jitter will reach 150 ps when it reaching ADC because of the character of DCM. Because there is capacitor character in PCB and pin of ADC the time delay will have an edge jitter T_{CLK} of 0.3 ps. ADC has aperture delay T_{AP1} and the value of it is less than 10 ns. Meanwhile, DAC has transmitted delay T_{AP2} and which value is also less than 10 ns. The jitter of T_{AP1} and T_{AP2} is less than 1 ps (Table 1).

The receiving or transmitting delay of digital base band module is:

Table 1 Components of digital base band delay

Items	Symbol	Average value (ns)	1σ (ps)
Clock jitter	T_{soruce}	0	0.1
DCM jitter	T_{DCM}	0	150
Time delay of IOB	T_{IOB}	2	0
Clock edge jitter	T_{CLK}	0	0.3
Aperture delay	T_{AP1}	10	1
Transmit delay	T_{AP2}	10	1

$$\tau_B = T_{source} + T_{DCM} + T_{IOB} + T_{CLK} + T_{AP} = 12\,\text{ns}$$

The averaging variance is:

$$\sigma_B = \sqrt{\sigma(T_{source})^2 + \sigma(T_{DCM})^2 + \sigma(T_{IOB})^2 + \sigma(T_{CLK})^2 + \sigma(T_{AP})^2} \approx 0.15\,\text{ns}$$

We can get that the main time jitter error is come from DCM and the whole averaging variance is about 0.15 ns.

3.2 Time Delay Character of Radio Frequency Channel

Time delay of radio frequency channel includes group time delay of radio frequency filter, time delay of amplifier, time delay of intermediate frequency AGC, time delay of frequency mixer and any other kinds of time delay of circuitry. The delay will change a lot with the change of temperature, humidity and the aging of material.

On the one hand, when signal throughout filters and transmit lines the transmit speed will not equal to the speed transmitting in vacuum. It will changed with the change of the character of transmit medium. The transmit speed is $C\sqrt{1/\mu\varepsilon}$. Where ε is the relative permittivity and μ is relative permeability of transmit medium. ε and μ are sensitive to the environment variety. For example if ε changes 20%, the time delay will change 10%.

On the other hand, analog filters are important parts of radio frequency channel. These filters are also sensitive to the environment variety especially temperature. The working frequency of intermediate frequency filters is low so the time delay of that is a main part of radio frequency channel. The group time delay of 70 MHz intermediate frequency filter can be more than 50 ns. And the time delay will get bigger as long as the filter series get more.

We can conclude that time delay error of radio frequency channel mainly caused by the change of transmit character of medium with environment variety especially temperature. So in order to increase precision of time synchronization, we should to design a system to test the time delay timely.

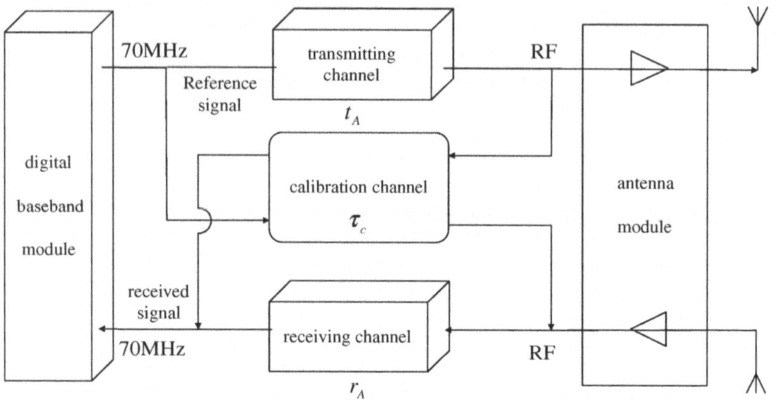

Fig. 2 Calibration system of transceiver's delay

4 Time Delay Calibration Project of Transceiver in System of Satellite Constellation Crosslink

In order to get r_A (time delay of receiving channel) and t_A (time delay of transmitting channel), a calibration channel is added between receiving channel and transmitting channel. With this module receiving channel and transmitting channel become a closed circuit and r_A and t_A can be measured just like Fig. 2.

The calibration process as follows:

1. Measurement of total time delay of receiving channel and transmitting channel. As shown in Fig. 3, calibration channel should be closed firstly. Then 70 MHz reference signal is sent from digital base band module to transmitting channel. 70 MHz reference signal is changed to radio frequency signal in transmitting channel and then access to receiving channel in which the signal changed to 70 MHz. The 70 MHz signal is processed in digital base band module and we can get $(t_A + r_A)$ (the total time delay of receiving channel and transmitting channel).
2. Measurement of total time delay of calibration channel and transmitting channel. As shown in Fig. 4, receiving channel should be closed and calibration channel should be opened. 70 MHz reference signal is sent from digital base band module to transmitting channel. 70 MHz reference signal is changed to radio frequency signal in transmitting channel. Because of the receiving channel is closed, radio frequency signal access to calibration channel in which the signal changed to 70 MHz. The 70 MHz signal is processed in digital base band module and we can get $(t_A + \tau_{cr})$ (the total time delay of calibration channel and transmitting channel).

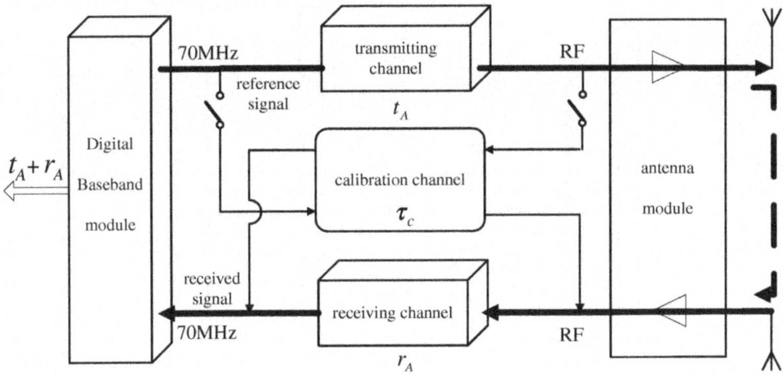

Fig. 3 Method of delay measurement between transmitter and receiver channel

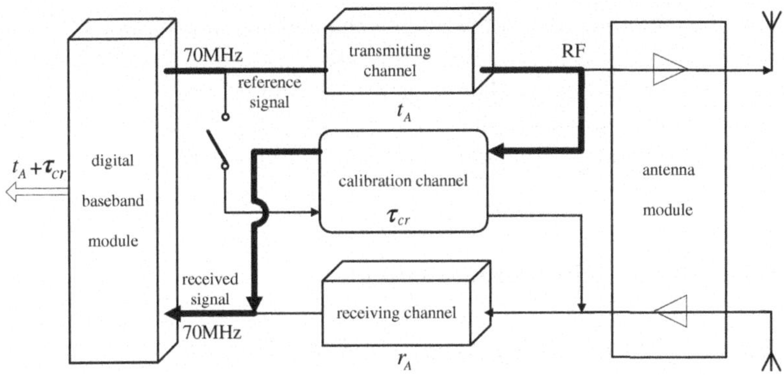

Fig. 4 Measurement method of delay between transmitter and calibration channel

3. Measurement of total time delay of receiving channel and calibration channel. As shown in Fig. 5, transmitting channel should be closed firstly. Then 70 MHz reference signal is sent from digital base band module to calibration channel. 70 MHz reference signal is changed to radio frequency signal in calibration channel and then access to receiving channel in which the signal changed to 70 MHz. The 70 MHz signal is processed in digital base band module and we can get $(t_A + \tau_{ct})$ (the total time delay of receiving channel and calibration channel).

   From the above three steps, we can get:

$$\begin{cases} t_A + r_A = \tau_1 \\ t_A + \tau_{cr} = \tau_2 \\ t_A + \tau_{ct} = \tau_3 \end{cases}$$

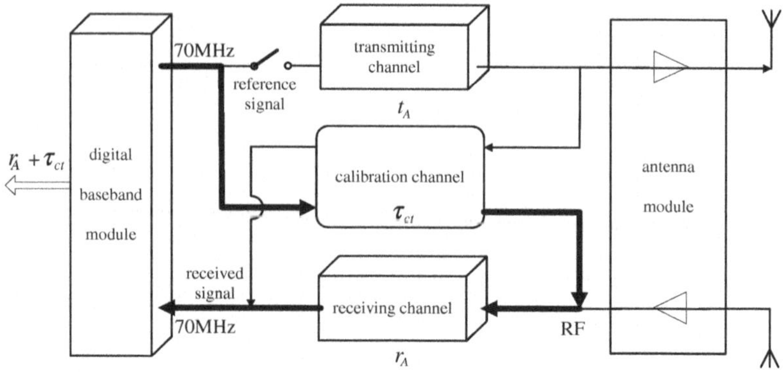

Fig. 5 Measurement method of delay between receiver and calibration channel

where τ_1 is the total time delay of receiving channel and transmitting channel, τ_2 is the total time delay of calibration channel and transmitting channel and τ_3 is the total time delay of calibration channel and receiving channel.

Because it is easy to ensure that $\tau_{ct} = \tau_{cr}$, we can calculate r_A (time delay of receiving channel) and t_A (time delay of receiving channel).

5 Validation of Calibration Project

5.1 Choice of Time Delay Measurement Method

The precise measurement of time delay of transceiver is precondition of calibration. There are many methods to measure time delay. They can classified as static method and dynamic method. The static method uses instrument such as vector network signal analyzer. The dynamic method is based on the principle of time delay estimate and uses fixed signal to compare the phase to get time delay value. Spread spectrum signal method is a dynamic method with higher precision. In this paper, spread spectrum signal method is chose to implement the calibration method.

5.2 Validation Method of Time Delay Calibration Project

In order to validate time delay calibration project, we can compare results between two separate transceivers and the unit of two transceivers. Firstly, we can get r_A, t_A, r_B and t_B according to above analysis. Secondly, we can get $(t_A + r_B)$ through connecting two transceivers together. The difference between the first step and the second step reflects the validity and veracity of the time delay calibration project.

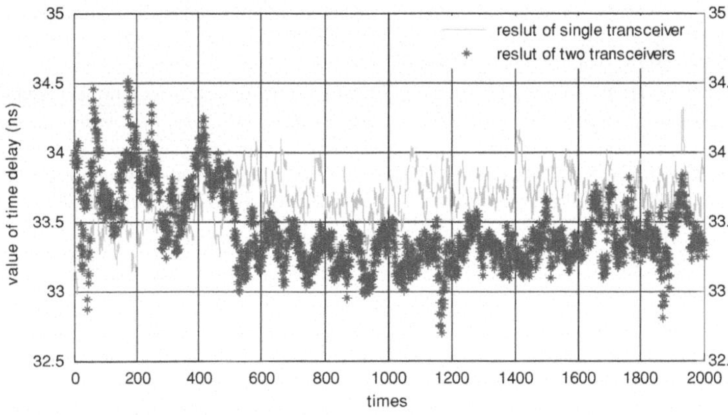

Fig. 6 Result of calibration experiment

5.3 The Testament Result

According the precious method, we can get data as Fig. 6 through 2000 times test. The average difference value between two methods is 0.1946 ns and which is less than 0.3 ns. So the time delay calibration project is validly and precision.

6 Conclusion

In the satellite constellation crosslink system, the method of dual-way pseudo ranging is used in navigation constellation crosslink to implement precise orbit determination and time synchronization. The accuracy of dual-way pseudo ranging is directly influenced by delay variation in transceiver. Consequently, the measurement and calibration of transceiver delay is significant to rising the accuracy of dual-way pseudo ranging and fulfilling the function of navigation constellation crosslink. This paper analyses the influence of time delay on precise orbit determination and time synchronization and time delay character of transceiver. Then a time delay calibration project is given. The project is testified finally and can meet the need of satellite constellation crosslink system.

References

1. Merck, P., & Achkar, J. (2005). Design of a Ku band delay difference calibration device for TWSTFT station. *IEEE Transactions on Instrumentation and Measurement*, *54*(2), 814–818.
2. Ascarrunz, F. G., Jefferts, S. R., & Parker, T. E. (1998). A delay calibration system for two-way satellite time and frequency transfer. *IEEE International Frequency Control Symposium*, 250–253.

3. Wei, H. T., Wei, B. G., Li, G., et al. (2010). The accurate method for the equipment time-delay and research on testing technology in satellite navigation system. *Science China: Physics, Mechanics & Astronomy, 40*(5), 623–P627.
4. Yuan, L., Chu, H. L., Wang, H. B., et al. (2011). The measure technique research of the satellite navigation equipment time delay. *Science China: Physics, Mechanics & Astronomy, 41*(5), 629–634.
5. Lee, A. B., Smith, A. L., & Blair, C.F. (2005). The accuracy of two-way satellite time transfer calibrations. *36th Annual Precise Time and Time Interval (PTTI) meeting*, Reston, VA, USA.

Research on Relative Navigation for Formation Flying Spacecrafts Based on Differential GNSS

Yi Li, Shancong Zhang, Changqing Wu and Wei Xu

Abstract In order to improve the precision of relative navigation for formation flying spacecrafts, a relative navigation scheme based on differential GNSS was proposed, the method of differential-pseudo-range smoothed by double-differential carrier phase was applied for the distance measurement. The measurement vector included the relative distance, azimuth and elevation between spacecrafts. The extended Kalman filter was adopted to estimate the relative position and velocity. Taking the relative dynamic formula based on non-singularity orbit elements as the state equation instead of the traditional C–W formula, the restriction of eccentricity could be ignored, the transform formula between relative non-singularity orbit elements and relative position(velocity) between spacecrafts was derived, by which the singular point in the Kepler's elements was avoided. Corresponding numerical simulation and experimental demonstration was provided to verify the precision of relative navigation, the scheme proposed by the paper was proved to be available.

1 Introduction

With the rapid development of on-orbit servicing such as formation flying and rendezvous, the precision requirements for relative navigation between spacecrafts has been achieved to a high standard, which would determine the success of space

Y. Li (✉) · S. Zhang · C. Wu · W. Xu
Beijing UCAS Space Technology CO., Ltd, 100190 Beijing, China
e-mail: liyi320303@yahoo.com.cn

Y. Li
Academy of Opto-Electronics, Chinese Academy of Science, 100094 Beijing, China

Y. Li
Graduate University of Chinese Academy of Science, 100049 Beijing, China

J. Sun et al. (eds.), *China Satellite Navigation Conference (CSNC) 2012 Proceedings*, Lecture Notes in Electrical Engineering 160, DOI: 10.1007/978-3-642-29175-3_39,

mission [1]. As a significant kind of measurement method, the differential GNSS can provide precise relative position and velocity [2, 3]. For example, in the ETS-VII mission, Carrier-Phase Differential GPS(CDGPS) was used to perform rendezvous involved two satellites, whose relative position estimation accuracy were better than 10 m and relative velocity accuracy were better than 3 cm/s [4]. This work used a method of pseudo-range smoothed by double-differential carrier phase for the distance measurement to improve the measurement accuracy of relative distance. The relative dynamics was often applied to combine the differential-GNSS to eliminate the error caused by ambiguity in GNSS sensor. Instead of the traditional C–W dynamics equation, this paper applied non-singular elements as the state variable of the dynamics model, by which the accuracy was largely improved, and the singular point of the classical Kepler's elements also avoided [5]. In addition, the J2 perturbation factor was considered in the model, which is also suitable to the elliptical orbit [6].

The extended Kalman filter was established to estimate the relative position and velocity which requires the relative dynamics as state equation. Numerical simulation and experiment demonstration was provided to verify the precision of relative position and velocity.

2 Differential-Pseudo-Range Smoothed by Differential Carrier Phase

Define two double-differential as difference value of two single-differential taking two different navigation satellites as measurement, which can delete the clock error [7, 8].

$$\begin{cases} \nabla\Delta\rho^{ij}(m,\ c) = & \nabla\Delta R^{ij}(m,\ c) + \nabla\Delta d^{ij}_{rot}(m,\ c) + \nabla\Delta d^{ij}_{mul}(m,\ c) + \nabla\Delta\varepsilon_{\rho}(m,\ c) \\ \nabla\Delta\phi^{ij}(m,\ c) = & (\nabla\Delta R^{ij}(m,\ c) + \nabla\Delta d^{ij}_{rot}(m,\ c) + \nabla\Delta d^{ij}_{mul}(m,\ c))/\lambda_{\phi} \\ & + \nabla\Delta N^{ij}(m,\ c) + \nabla\Delta\varepsilon_{\phi}(m,\ c) \end{cases} \tag{2.1}$$

Mark i and j means the number of navigation satellite, $\nabla\Delta\rho^{ij}(m,\ c)$ $\nabla\Delta\phi^{ij}(m,\ c)$ represents the double-differential of pseudo-range and carrier-phase. $\nabla\Delta R^{ij}(m,\ c)$ represents the double-differential distance between GNSS receiver and navigation satellite, $\nabla\Delta d^{ij}_{rot}(m,\ c)$, $\nabla\Delta d^{ij}_{mul}(m,\ c)$ represents the double-differential distance error from earth rotation and multipath. λ_{ϕ} represents the wavelength of carrier, N represents the integer ambiguity. $\nabla\Delta\varepsilon_{\rho}(m,\ c)$, $\nabla\Delta\varepsilon_{\phi}(m,\ c)$ represents the double-differential thermal noise. Smoothing the double-differential pseudo-range by double-differential carrier phase can get the relative position information with high accuracy and low computation load.

$$\overline{\nabla\Delta\rho}(k) = \frac{1}{M}\nabla\Delta\rho(k) + \frac{M-1}{M}\left[\overline{\nabla\Delta\rho}(k-1) - \lambda_{\phi}\nabla\Delta\phi(k,\ k-1)\right] \tag{2.2}$$

M is smoothing constant, the value can be adjusted from the area of (20–100) [9].

3 Dynamics Based on Non-Singularity Orbit Elements

3.1 About Non-Singularity Orbit Elements

Define the non-singularity orbit elements defined as follow.

$$\boldsymbol{e} = [a,\ i,\ \Omega,\ \xi,\ \eta,\ \lambda] \tag{3.1}$$

a, i and Ω represents the semi-major axis, orbit inclination and right ascension of ascending node defined by Kepler, the other three elements defined as $\xi = e\cos\omega$, $\eta = -e\sin\omega$, $\lambda = f + \omega$, e, ω and f represents the eccentricity, argument of perigee and true anomaly defined by Kepler, λ represents the argument of latitude. In non-singularity orbit elements, the correlation between ω and f disappears, so the singularity point could be avoided [10, 11].

3.2 Relative Dynamics

This paper research on formation consist of two satellites—the main satellite m and the concomitant satellite c.

The relative dynamics coordinate was the LVLH of main satellite as seen in Fig. 1. $\boldsymbol{r}$ is the relative position vector of two satellites [12]. Define non-singularity orbit elements of main and concomitant satellite as e_m and $\boldsymbol{e}_c$, the relative non-singularity orbit element of the two satellites is $\delta\boldsymbol{e} = \boldsymbol{e}_m - \boldsymbol{e}_c = [\delta a,\ \delta i,\ \delta\Omega,\ \delta\xi,\ \delta\eta,\ \delta\lambda]$. The relationship between the position of satellite and non-singularity orbit elements in inertial reference was derived as in formula (3.2) and (3.3).

$$\begin{pmatrix} x_I \\ y_I \\ z_I \end{pmatrix} = \boldsymbol{A}_z(\Omega)\boldsymbol{A}_x(i)\begin{pmatrix} a\left(\sqrt{\xi^2+\eta^2}+\cos\lambda\right)/\left(1+\sqrt{\xi^2+\eta^2}\cos\lambda\right)-a\sqrt{\xi^2+\eta^2} \\ a\sqrt{1-\xi^2-\eta^2}\left(\sqrt{1-\xi^2-\eta^2}\sin\lambda\right)/\left(1+\sqrt{\xi^2+\eta^2}\cos\lambda\right) \\ 0 \end{pmatrix} \tag{3.2}$$

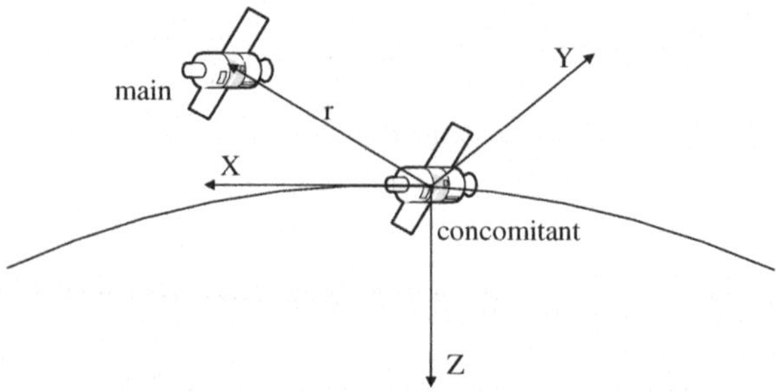

Fig. 1 Model of relative motion

$$\begin{pmatrix} \dot{x}_I \\ \dot{y}_I \\ \dot{z}_I \end{pmatrix} = \boldsymbol{A}_z(\Omega)\boldsymbol{A}_x(i) \begin{pmatrix} -\dfrac{a^2 n\left(\sqrt{1-\xi^2-\eta^2}\sin\lambda\right)/\left(1+\sqrt{\xi^2+\eta^2}\cos\lambda\right)}{a\left(1-\sqrt{\xi^2+\eta^2}\right)\left(\sqrt{\xi^2+\eta^2}+\cos\lambda\right)/\left(1+\sqrt{\xi^2+\eta^2}\cos\lambda\right)} \\ \dfrac{a^2 n\sqrt{1-\xi^2-\eta^2}\left(\sqrt{\xi^2+\eta^2}+\cos\lambda\right)/\left(1+\sqrt{\xi^2+\eta^2}\cos\lambda\right)}{a\left(1-\sqrt{\xi^2+\eta^2}\right)\left(\sqrt{\xi^2+\eta^2}+\cos\lambda\right)/\left(1+\sqrt{\xi^2+\eta^2}\cos\lambda\right)} \\ 0 \end{pmatrix} \tag{3.3}$$

Define the state vector of main satellite in inertial frame as $\boldsymbol{X}_m = [x_m y_m z_m \dot{x}_m \dot{y}_m \dot{z}_m]^T = F(\boldsymbol{e}_m)$. The relative state vector between main and concomitant satellite in LVLH of concomitant satellite can be illustrated as:

$$\boldsymbol{X} = \boldsymbol{X}_m - \boldsymbol{X}_c = \boldsymbol{A_o}[\boldsymbol{F}(\boldsymbol{e}_m) - \boldsymbol{F}(\boldsymbol{e}_c)] = \boldsymbol{F}_1(\delta\boldsymbol{e}) \tag{3.4}$$

where $\boldsymbol{A_o}$ is the transform matrix from inertial frame to LVLH of concomitant satellite [13, 14]. Derive $\boldsymbol{F}_1$ in formula (3.4) as follows:

$$\begin{cases} x = -p(\sin\lambda\delta i - \cos\lambda\sin i\delta\Omega)/(1+\xi\cos\lambda+\eta\sin\lambda) \\ y = \dfrac{V_r}{V_t(1+\xi\cos\lambda+\eta\sin\lambda)}\left(p\delta\lambda + p\delta a - 2a\xi\delta\xi - 2a\eta\delta\eta - \dfrac{p\cos\lambda\delta\xi + p\sin\lambda\delta\eta}{1+\xi\cos\lambda+\eta\sin\lambda}\right) \\ z = -p(\delta\lambda + \cos i\delta\Omega)/(1+\xi\cos\lambda+\eta\sin\lambda) \\ \dot{x} = -(V_t\cos\lambda + V_r\sin\lambda)\delta i - (V_t\sin\lambda - V_r\cos\lambda)\sin i\delta\Omega \\ \dot{y} = -\dfrac{V_r}{2a}\delta a + \dfrac{\xi\cos\lambda+\eta\sin\lambda}{p}h\delta\lambda + (V_r a\xi + h\sin\lambda)\dfrac{\delta\xi}{p} + (V_r a\xi + h\sin\lambda)\dfrac{\delta\xi}{p} \\ \dot{z} = \dfrac{3V_t}{2a}\delta a + V_r\delta\lambda - (3V_t a\xi + 2h\cos\lambda)\dfrac{\delta\xi}{p} - (3V_t a\eta + 2h\sin\lambda)\dfrac{\delta\eta}{p} + V_r\delta\Omega \end{cases} \tag{3.5}$$

where $p = a(1 - \xi^2 - \eta^2)$, $h = \sqrt{\mu p}$, V_t, V_r represents along-track velocity and radial velocity.

4 Design of EKF

4.1 State Equation

Given the initial value of non-singularity elements for main satellite as

$$\boldsymbol{e}_m(0) = [a_m(0), i_m(0), \Omega_m(0), \xi_m(0), \eta_m(0), \lambda_m(0)] \tag{4.1}$$

Considering the J2 perturbation, derive the non-singularity elements value $\boldsymbol{e}_m(k)$ at time point k as in formula (4.2).

$$\begin{cases} a_m(k)=a_m(0) \\ i_m(k)=i_m(0) \\ \Omega_m(k)=\Omega_m(0)+\dfrac{3J_2R_e^2}{2}\dfrac{\sqrt{\mu}\cos i_m(k)k}{a_m(k)^{7/2}\left(1-\xi_m(k)^2-\eta_m(k)^2\right)^2} \\ \xi_m(k)=\xi_m(0)\cos(\Lambda)-\eta_m(0)\sin(\Lambda) \\ \eta_T(k)=\xi_m(0)\sin(\Lambda)+\eta_m(0)\cos(\Lambda) \\ \lambda_m(k)=\lambda_m(0)+[1+\Delta(\theta)]\left[\dfrac{\sqrt{\mu}}{a_m(k)^{3/2}}+\dfrac{3J_2R_e^2}{4}\dfrac{\sqrt{\mu}(3\cos^2 i_m(k)-1)}{a_m(k)^{7/2}(1-\xi_m(k)^2-\eta_m(k)^2)^{3/2}}+\Lambda\right]k \end{cases} \tag{4.2}$$

where $\Lambda = \frac{3J_2R_e^2}{4}\frac{\sqrt{\mu}\left(5\cos^2 i_m(k)-1\right)k}{a_m(k)^{7/2}\left(1-\xi_m(k)^2-\eta_m(k)^2\right)^2}$, R_e represents radius of earth, detailed information about $\Delta(\theta)$ was in Song's paper [15]. Similarly, the non-singularity elements value of concomitant satellite $\boldsymbol{e}_c(k)$ at time point k could be derived referred to formula (4.2). Then the relative non-singularity elements at time point k could be derived as in formula (4.3)

$$\begin{aligned} \delta\boldsymbol{e}(k) &= \boldsymbol{F}[\boldsymbol{e}_m(0), k) - \boldsymbol{F}(\boldsymbol{e}_m(0), k] = \boldsymbol{F}[\boldsymbol{e}_m(0), k] - \boldsymbol{F}[\boldsymbol{e}_c(0) + \delta\boldsymbol{e}(0), k] \\ &= \frac{\partial \boldsymbol{F}[\boldsymbol{e}_m(0), k]}{\partial \boldsymbol{F}[\boldsymbol{e}_m(0)]}\cdot\delta\boldsymbol{e}(0) = \boldsymbol{\Phi}(\delta\boldsymbol{e}(0), k)\delta\boldsymbol{e}(k) + \boldsymbol{W}(k) = \boldsymbol{\psi}(\delta\boldsymbol{e}(k)) + \boldsymbol{W}(k) \\ \boldsymbol{W}(k) &\sim N(0, Q) \end{aligned} \tag{4.3}$$

where $\boldsymbol{W}(k)$ is the system noise, series of Gauss distribution white noise, with covariance of Q. The derivation above can be applied as the state equation of EKF.

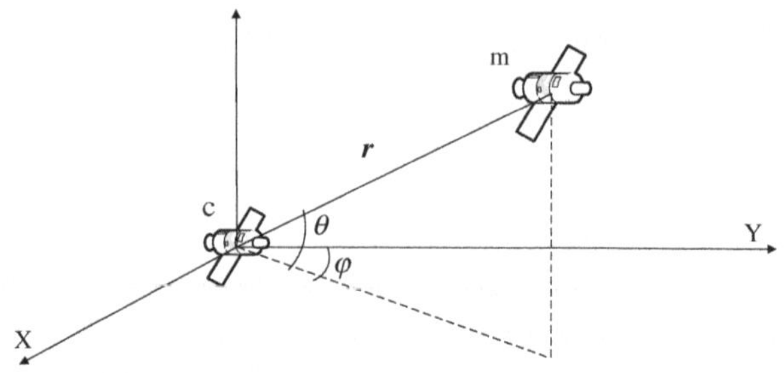

Fig. 2 Model of measurement

4.2 Measurement Equation

Take relative distance $\boldsymbol{r}$, relative azimuth φ and elevation θ as measurement vector, as seen in Fig. 2. Establish the equation between relative position, velocity and measurement vector.

$$
\begin{cases}
x(k) = r(k)\cos[\theta(k)]\cos[\varphi(k)] \\
y(k) = r(k)\cos[\theta(k)]\sin[\varphi(k)] \\
z(k) = r(k)\sin[\theta(k)] \\
\dot{x}(k) = \dot{r}(k)\cos[\varphi(k)]\cos[\theta(k)] - \dot{\varphi}(k)r(k)\sin[\varphi(k)]\cos[\theta(k)] \\
\qquad - \dot{\theta}(k)r(k)\cos[\varphi(k)]\sin[\theta(k)] \\
\dot{y}(k) = \dot{r}(k)\sin[\varphi(k)]\cos[\theta(k)] - \dot{\varphi}(k)r(k)\cos[\varphi(k)]\cos[\theta(k)] \\
\qquad - \dot{\theta}(k)r(k)\sin[\varphi(k)]\sin[\theta(k)] \\
\dot{z}(t) = r(k)\sin[\theta(k)] + \dot{\theta}(k)r(k)\cos[\theta(k)]
\end{cases}
\tag{4.4}
$$

Define measurement vector as

$$
\begin{cases}
\boldsymbol{Z}(k) = [x(k)\ y(k)\ z(k)\ \dot{x}(k)\ \dot{y}(k)\ \dot{z}(k)]^T \\
\boldsymbol{Y}(k) = \left[r(k)\ \theta(k)\ \varphi(k)\ \dot{r}(k)\ \dot{\theta}(k)\ \dot{\varphi}(k)\right]^T
\end{cases}
\tag{4.5}
$$

the measurement equation can be stated as

$$
\boldsymbol{Z}(k) = \boldsymbol{H}(\delta\boldsymbol{e}(k), k) + \boldsymbol{\Gamma}(\boldsymbol{v}(k), k) \tag{4.6}
$$

where $\boldsymbol{H}(\delta\boldsymbol{e}(k), k)$ is the relationship between relative position and non-singularity elements defined in formula (3.5). $\boldsymbol{v}(k)$ is direct measurement noise, series of Gauss distribution white noise, with covariance of R, $\boldsymbol{\Gamma}(\boldsymbol{v}(k), k)$ is the propagation noise according to formula (4.4).

$$
\boldsymbol{\Gamma}(\boldsymbol{v}(k), k) = \frac{\partial \boldsymbol{Z}(k)}{\partial \boldsymbol{Y}(k)}\boldsymbol{v}(k)\left[\frac{\partial \boldsymbol{Z}(k)}{\partial \boldsymbol{Y}(k)}\right]^T \tag{4.7}
$$

Table 1 Initial value of orbit elements for two satellites

	a /(km)	i /(°)	Ω (°)	ξ	η	λ (°)
Concomitant satellite	6768.10	42.4229	37.5477	-5.324×10^{-5}	1.500×10^{-3}	−73.0636
Main satellite	6768.05	42.4300	37.5480	-5.250×10^{-5}	1.505×10^{-3}	−73.0640

4.3 State Estimation Based on EKF

According to the state equation and measurement equation defined above, derive the equation of state estimation update as

$$\begin{cases} \delta\hat{e}(k+1,\ k) = \psi(\delta e(k)) \\ P(k+1,\ k) = \psi(k+1,\ k)P(k)\psi(k+1,\ k)^T + \Gamma(\delta\hat{e}(k))Q\Gamma^T(\delta\hat{e}(k)) \end{cases} \tag{4.8}$$

where $P(k)$ is the covariance matrix of error, and $H(\cdot)$ is the approximate function of first-order Taylor series [16]. Then derive the equation of measurement update

$$\begin{cases} \delta\hat{e}(k+1) = \delta\hat{e}(k+1,\ k) + K(k+1)[Z(k+1) - H(\delta\hat{e}(k+1,\ k))] \\ K(k+1) = P(k+1,\ k)H_{k+1}^T\left[H_{k+1}P(k+1,\ k)H_{k+1}^T + R\right]^{-1} \\ P(k+1) = [I - K(k+1)H_{k+1}]P(k+1,\ k) \end{cases} \tag{4.9}$$

where $K(k+1)$ is the filter gain, and $\psi(\cdot)$ is the approximate function of first-order Taylor series. The initial value of state and covariance matrix of error were

$$\begin{cases} \delta\hat{e}(0) = E[\delta e(0)] \\ P(0) = E\left[(\delta e(0) - \delta\hat{e}(0))(\delta e(0) - \delta\hat{e}(0))^T\right] \end{cases} \tag{4.10}$$

5 Numerical Simulation

The scenario consisted of two satellites called main satellite and concomitant satellite. The initial value of orbit elements as seen in Table 1. The perturbation included body gravity from solar and lunar, the solar radiation pressure and atmosphere drag. Simulation epoch started from December 1, 2011, 00:00:00, simulation process lasted two orbit periods. The accuracy of distance by differential-GNSS and accuracy of relative position and velocity estimation based on EKF were simulated respectively as follow.

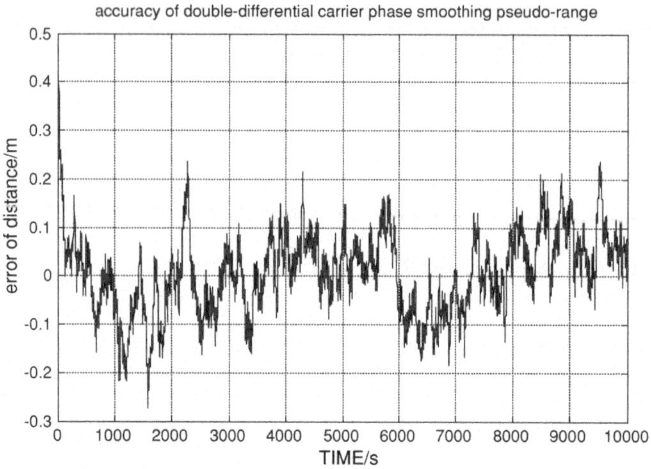

Fig. 3 Error of distance by differential-GNSS

5.1 Accuracy of Double-Differential Carrier Phase Smoothing Double-Differential Pseudo-Range

Given the measurement noise of double-differential pseudo-range and double-differential carrier-phase are series of Gauss distribution white noise, in which the accuracy of double-differential pseudo-range is $\sigma_\rho = 1\,\mathrm{m}$, while the accuracy of double-differential carrier-phase is $\sigma_\varphi = 0.02\,\mathrm{Hz}$. Design the simulation scenario based on formula (2.2), and the result was shown in Fig. 3. The final statistic data showed that the mean square error was 0.1171 m.

5.2 Accuracy of Relative Position and Velocity Estimation Based on EKF

The accuracy of distance is $\sigma_\rho = 0.1171\,\mathrm{m}(1\sigma)$, simulated in Sect. 5.1, and the accuracy of velocity as $\sigma_{\dot\rho} = 0.01\,\mathrm{m/s}(1\sigma)$, while the accuracy of angle as $\sigma_\varphi = \sigma_\theta = 0.01^\circ(1\sigma)$ and the accuracy of angular rate as $\sigma_{\dot\varphi} = \sigma_{\dot\theta} = 0.01^\circ/\mathrm{s}(1\sigma)$. Initial value of state vector was designed according to Table 1. Initial value of system noise covariance matrix was $\hat{\boldsymbol{P}}_0 = diag\left[10^{-4},\ 10^{-4},\ 10^{-6},\ 10^{-6},\ 10^{-8}, 10^{-8}\right]$, while the initial value of measurement noise covariance matrix was $\boldsymbol{R} = diag\left[\sigma_\rho^2,\ \sigma_\varphi^2,\ \sigma_\theta^2,\ \sigma_{\dot\rho}^2,\ \sigma_{\dot\varphi}^2,\ \sigma_{\dot\theta}^2\right]$. The simulation results presented the error of relative position and velocity in the x, y and z component of the main satellite LVLH frame.

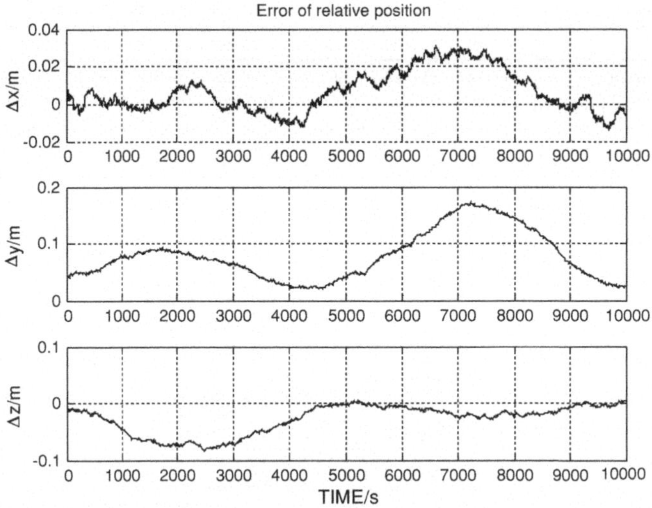

Fig. 4 Error of relative position

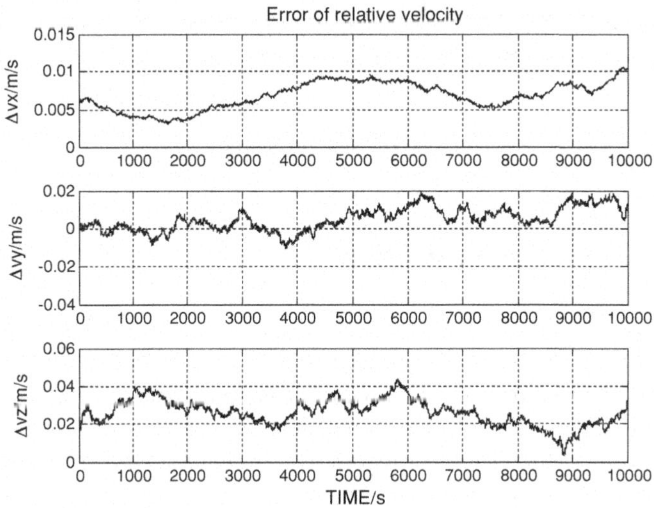

Fig. 5 Error of relative velocity

As shown in Figs. 4 and 5, the estimation algorithm based on non-singularity elements and EKF could further improve the accuracy of relative position and velocity measurement. The mean square error of position and velocity were stated in Table 2, with centimeter level of relative position estimation and millimeter level of relative velocity estimation.

Table 2 Statistics results of relative position and velocity by numerical simulation

	x (m)	y (m)	z (m)	vx (m/s)	vy (m/s)	vz (m/s)
Mean square error	0.0111	0.0424	0.0254	0.0018	6.0654×10^{-3}	6.9159×10^{-3}
Total accuracy of position	0.0507			0.0020		

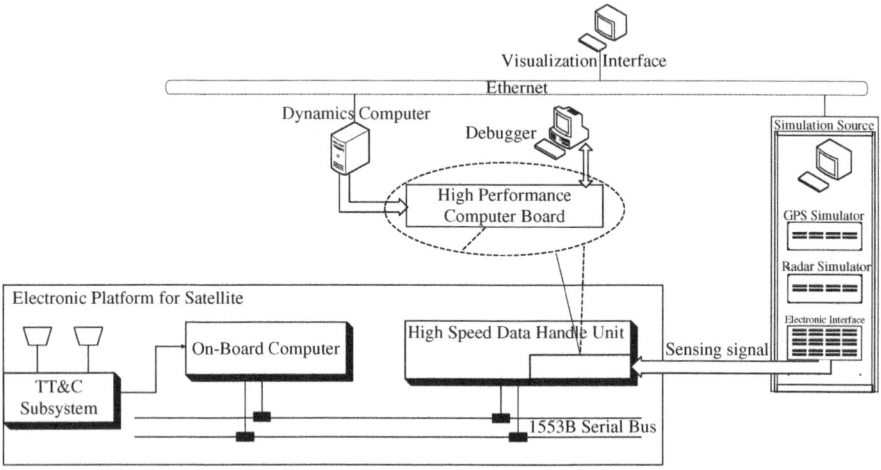

Fig. 6 Hardware-in-loop simulation experiment environment

6 Experimental Demonstration

The hardware experimental environment include an electronic platform for a simulate satellite and a simulation signal source, as seen in Fig. 6. The electronic platform consist of the Data and Command Subsystem with the 1553B as data bus, the On-board Computer is the bus controller, the High Speed Handle Unit, as a remote terminate of the data bus, provides the running platform for the algorithm of this paper. The simulation signal source creates the GPS signals, azimuth and elevation signals which would be observed by GPS receivers and radars. The signal noise could also be generated as experienced on orbit environment such as multipath, troposphere and ionosphere. The electronic interface sends the simulation signal into the High Speed Handle Unit, by which all of the simulated signal would be transformed into standard electronic signals specified by satellite platform. The dynamic computer generates a scenario of formation flying, in which the orbital environment effect would be considered such as the perturbation of J2 and air drag, the orbit information would be used to verify the accuracy of algorithm.

Table 3 Statistics results of relative position and velocity by hardware experimental

	x /(m)	y /(m)	z /(m)	vx /(m/s)	vy /(m/s)	vz /(m/s)
Mean square error	0.0120	0.0518	0.0361	0.0021	6.1857×10^{-3}	7.0132×10^{-3}
Total accuracy of position	0.0643			0.0036		

This experimental scenario was designed the same as the scenario defined in Table 1. The testing target also focused on the accuracy of relative position and velocity. Table 3 listed the statistic results, which is approximate with the number in Table 2. The hardware experimental proved the validity of the method proposed by this paper.

7 Conclusion

This paper proposed a method of relative navigation for formation satellites based on differential-GNSS, the measure information included pseudo-range and carrier-phase, combined with relative azimuth and elevation from other sensors. The estimation for relative position and velocity was designed based on EKF, with the relative dynamics as state equation. The non-singularity orbit elements were applied in relative dynamics instead of Kepler's elements, which can avoid the singular point in orbit determination. Finally, corresponding numerical simulation and hardware-in-loop experimental was provided, which showed that the method of double-differential carrier phase smoothing the double-differential pseudo range can get the distance with the accuracy of 0.1 m, and the relative position and velocity estimation based on non-singularity element and EKF algorithm can get the relative position with the accuracy of centimeter level and relative velocity with the accuracy of millimeter/s level. Results of numerical simulation and hardware experimental proved that the proposal of this paper can be applied to engineering design.

References

1. Alonso, R., Crassdis, J. L., & Junkins, J. L. (2003). *Vision-based relative navigation for formation flying of spacecraft rendezvous*. (AIAA-2003-5355)
2. Fan, L., Tang, G., & Wu, J. (2007). Application of differential carrier phase relative positioning for spacecrafts docking. *Journal of Chinese Inertial Technology, 15*(4), 435–437.
3. Parkinson, B. W. et al. (1996). *Global positioning system: Theory and application*. Washington, DC: American Institute of Aeronautics & Ast.
4. Franz, D. B., & Jonathan, P. H. (2002). Real-time experimental demonstration of precise decentralized relative navigation for formation flying space-craft. *AIAA Guidance Navigation, and Control Conference and Exhibit, Monterey*, California, USA, August 5–8.

5. Schaub, H., Vadali, S. R., Junkins, J. L., et al. (2000). Spacecraft formation flying using mean elements. *Journal of Astronautical Science, 48*(1), 69–87.
6. Wang, G., Zheng, W., & Tang, G. (2011). Application of UKF in spacecraft relative navigation based on quasi-mean element differences. *Journal of Astronautics, 32*(5), 1047–1053.
7. Fan, G., Wang, W., & Hao, X. (2005). Formation constellation states determination with carrier phase differential GPS and intra-satellite relative measurement. *Journal of Space Science, 25*(3), 218–223.
8. Fan, L., Wu, J., & Tang, G. (2008). Researches on DGPS for spacecraft autonomous rendezvous and docking. *Journal of Spacecraft TT&C Technology, 27*(3), 78–81.
9. Xie, G. (2009). *Principal of GPS and receiver design*. Beijing, China: Publishing House of Electronic Industry.
10. Yin, J., He, Q., & Han, C. (2011). Collision analysis of spacecraft relative motion based on the relative orbit elements. *Acta Aeronautica et Astronautica Sinica, 32*(1), 311–320.
11. Liu, L., Wu, B., & Yang, P. (2005). *Orbit precision determination & self-calibration technique of spacecraft*. Beijing, China: National Defence Industry Press.
12. Grassidis, J. L., & Junkins, J. L. (2004). *Optimal estimation of dynamic systems*. New York: Chapman and Hall CRC.
13. Liu, Y., & Xu, S. (2008). Polynomial interpolation filtering for relative navigation of spacecraft on elliptical orbit. *Chinese Space Science and Technology, 21*(3), 37–44.
14. Marcel, J. S. (1997). *Spacecraft dynamics and control: A practical engineering approach*. Cambridge, UK: Cambridge University Press.
15. Song, C. (2008). *Study on autonomous determination of relative orbit for satellite formation flying*. Dissertation, BUAA, Beijing, China.
16. Fu, M., Deng, Z., & Yan, L. (2010). *Kalman filter theory and its application for navigation system*. Beijing, China: Science Press.

Precise Orbit Determination of GEO Satellite Based on Helmert Variance Component Estimation Method

Shan Wu, YanYu Liu, Li Liu, Rui Guo, Feng He, XiaoJie Li and Hua Huang

Abstract Precise orbit determination of GEO satellite is a difficult point in area of satellite orbit determination. In this essay, Helmert variance component estimation method is proposed to be applied in the determination of the weights of observations in combined orbit determination. Through making χ^2 test of the results after united adjustment, it is found that when combining transfer data and pseudo-range data in orbit determination, their weights cannot be taken as equal, and thus Helmert variance component estimation is applied in the POD of GEO satellite. Numerical experiment proves that if each station determines its weight respectively, then too many groups will deteriorate the stability of computation and cause failure of orbit determination, while determining the weights by assorting data according to station type is a grouping strategy used in the orbit determination of GEO satellites. After re-determining the weights, the results pass through the χ^2 test and distinct improvement appeared in orbit accuracy via laser assessment, which indicated that the Helmert variance component estimation method is applicable for the combined orbit determination of GEO satellites using different observations.

1 Introduction

For the high orbit and geostationary characteristics of the GEO satellites, they are widely used in communication, meteorological, reconnaissance, timing, warning, tracking, data relay and scientific research in aerospace engineering. In addition, they are also becoming increasingly popular in satellite navigation. As known to all, GEO satellites play a significant role in China's regional satellite navigation system.

S. Wu (✉) · Y. Liu · L. Liu · R. Guo · F. He · X. Li · H. Huang
Beijing Global Information Application and Development Center, 100094 Beijing, China
e-mail: 1343782007@qq.com

J. Sun et al. (eds.), *China Satellite Navigation Conference (CSNC) 2012 Proceedings*, Lecture Notes in Electrical Engineering 160, DOI: 10.1007/978-3-642-29175-3_40,

However, orbit determination of GEO satellite is a difficult issue in the precise orbit determination domain. The main factors that affect the precision of GEO orbit determination are the geometric tracking condition of high orbit satellite depending on the regional tracking net, tracking geometric condition remaining almost unchanged due to the geostationary characteristic and the weak dynamic constrict of the ground station. Differing from the other navigation systems, our compass satellite navigation system provides not only abundant pseudo-range data, but also C wave forwarding measuring data. That is to say, it could take advantage of any single measurement technology to realize the precise orbit determination of GEO satellite, which could do great help to the construction and service promotion of the system [1].

GEO orbit determination processing methods based on pseudo-range and forwarding measuring data are studied in detail here, especially the weight assignment of the two kinds of data.

2 Unequal Weights of GEO Satellite Observation Measurements and Helmert Variance Component Estimation

Orbit determination based on pseudo-range, phase and self-sent and self-receive measurement of C wave is studied here. To obtain a best orbit from these data, weigh assignment must be taken into account. Actually, weight assignment appears in orbit determination through both a single technology and multi-technologies. As known to all, there are different instruments with different measurement precision on various stations while each orbit determination technology differs from the other when using multi-technologies.

Post precise orbit determination is often made through statistical methods. Taking $P = \sigma_0^2\Sigma^{-1}$ as weights of the measurements, where, Σ is the variance matrix of the measurements, σ_0^2 is an arbitrary positive number, known as unit weight variance while P donates the weight matrix [3].

Generally, measurements variance R is unknown. Thus, measurements weights could not be known precisely before data processing which is always obtained through former data. For instance, in the calculation of satellite orbit and earth rotation parameters with SLR datum in Shanghai astronomical platform, different standard bias are given according to the platform grade with the evaluation classification provided by the international organization [2–4]. Then, Helmert covariance component estimation method is introduced here to make post estimation of the residuals of the precise orbit determination results to obtain the proper weight of the measurements.

In the calculation of the precise orbit determination, iteration must be made to obtain a converge parameter solution due to the data error. If the weight elements and other parameters are processed together, it will affect the weight determination of the measurements. Besides, this will increase the number of the estimated parameters and change the condition number of the function which will reduce the stability of the solution. Thus, variance component estimation could be made independent of the orbit determination to make the calculation more controllable.

Suppose there are two independent observation groups named L_1, L_2, then, the error function could written as [5, 6]

$$\begin{bmatrix} V_1 \\ V_2 \end{bmatrix} = \begin{bmatrix} A_1 \\ A_2 \end{bmatrix} \hat{X} - \begin{bmatrix} L_1 \\ L_2 \end{bmatrix} \quad \Sigma_\Delta = \begin{bmatrix} \sigma_{01}^2 Q_1 & \sigma_{012} Q_{12} \\ \sigma_{012} Q_{12} & \sigma_{02}^2 Q_2 \end{bmatrix}$$

where, Q_1, Q_{12}, Q_2 are already known while $\sigma_{01}^2, \sigma_{02}^2, \sigma_{012}$ is not.

The purpose of the variance covariance component estimation is to obtain the estimation of $\sigma_{01}^2, \sigma_{02}^2, \sigma_{012}$ through adjusting the correction parameters V_i and $V_i^T P_i V_i$. Let $\sigma^2 = [\sigma_{01}^2 \quad \sigma_{012} \quad \sigma_{02}^2]^T = [\sigma_1^2 \quad \sigma_2 \quad \sigma_3^2]^T$. Take $\sigma^2(0) = [1 \quad 1 \quad 1]^T$ as the initial value of σ^2, then, the initial covariance and weight matrix of the observations are:

$$\Sigma_{\Delta_0} = \begin{bmatrix} Q_1 & Q_{12} \\ Q_{12} & Q_2 \end{bmatrix}, \quad P_{\Delta_0} = \begin{bmatrix} P_1 & P_{12} \\ P_{12} & P_2 \end{bmatrix}$$

Let

$$R_1 = \begin{bmatrix} P_1 & 0 \\ 0 & 0 \end{bmatrix}, \quad R_2 = \begin{bmatrix} 0 & P_{12} \\ P_{12} & 0 \end{bmatrix}$$

$$T_1 = \begin{bmatrix} Q_1 & 0 \\ 0 & 0 \end{bmatrix}, {}_2 = \begin{bmatrix} 0 & Q_{12} \\ Q_{12} & 0 \end{bmatrix}, \quad T_3 = \begin{bmatrix} 0 & 0 \\ 0 & Q_2 \end{bmatrix}$$

Then

$$\Sigma_\Delta = \sum_{i=1}^{3} T_i \sigma_i^2, \quad \Sigma_{\Delta_0} = \sum_{i=1}^{3} T_i$$

$$P_{\Delta_0} = \sum_{i=1}^{3} R_i$$

Through parameter adjustment, one can get

$$V = (AN^{-1}A^T P - I)L = GL$$

where, $N = A^T PA$ and $G = AN^{-1}A^T P - I$. Then, let

$$E(V) = GE(L) = 0$$

one can get

$$E(V^T R_i V) = E(\Delta^T G^T R_i G \Delta) = tr(G^T R_i G \Sigma)$$

with

$$E(V^T R_i V) = tr\left(\sum_{j=1}^{3} G^T R_i G T \sigma_j^2\right)$$
$$= \sum_{j=1}^{3} tr(G^T R_i G T)\sigma_j^2$$

followed, where, $i = 1, 2, 3$. Let $V = \begin{bmatrix} V_1 \\ V_2 \end{bmatrix}$, and $\hat{\sigma}^2$ be the estimation of σ^2, one can obtain

$$\begin{bmatrix} V^T R_1 V \\ V^T R_2 V \\ V^T R_3 V \end{bmatrix} = \begin{bmatrix} V_1^T P_1 V_1 \\ 2V_1^T P_{12} V_2 \\ V_2^T P_2 V_2 \end{bmatrix}$$

from which the variance covariance could be calculated.

As seen from the formulae above, the more estimated parameters the more amount of computation. In order to reduce the amount of computation, the post variance component estimation is obtained through Helmert, Ebner and Förstner approximate formula, which is proposed by Bäumker in 1984, as used in the third adjustment stage of the Europe triangle net mentioned in Ref. [5]:

$$\hat{\sigma}_i^2 = \frac{V_i^T P V_i}{f_i}$$

where

$$f_i = m_i - \frac{m_i t}{m}$$

m_i is observation number of the ith region, m mean the observation number of the whole region, t is the number of the unknown parameters. Besides, the calculation is completed through cycle algorithm.

$$\begin{bmatrix} \hat{\sigma}_1^2 \\ \hat{\sigma}_2^2 \\ \hat{\sigma}_3^2 \end{bmatrix} = \begin{bmatrix} tr(G^T R_1 G U_1) & tr(G^T R_1 G U_2) & tr(G^T R_1 G U_3) \\ tr(G^T R_2 G U_1) & tr(G^T R_2 G U_2) & tr(G^T R_2 G U_3) \\ tr(G^T R_3 G U_1) & tr(G^T R_3 G U_2) & tr(G^T R_3 G U_3) \end{bmatrix}^{-1} \begin{bmatrix} V^T R_1 V \\ V^T R_2 V \\ V^T R_3 V \end{bmatrix}$$

3 Pre-analysis of the Variance Component Estimation

As shown in today's references, whether the iteration should continue depends on the variance factor which may be the same or not. Strictly speaking, variance factor is also estimation, and it varies based on the observation swatch. So, if the random model is correct, variance component estimation can adjust the weight assignment to some degree. And the result differs from the solution of traditional optimum linear unbiased least squares. From this standpoint, it is necessary to make pre-analysis of the variance component estimation. In other words, it is necessary to judge whether the random model is correct or the weight assignment is unsuitable. Furthermore, pre-analysis of the variance component estimation can help decrease the extra amount of computation in iteration.

If the two kinds of observational data have been adjusted, of which the extra numbers are r_1, r_2, separately and the test statistics parameters are [7]

$$\chi_i^2 = \frac{V_i^T P_i V_i}{\sigma_{0i}^2} \sim \chi^2(ri), \quad i = 1, 2$$

where, σ_{0i}^2 is the pre-test variance factor. The method is called the whole test method or variance consistency-checking method.

If the two observation models are basically correct with the two statistic parameters satisfying the χ^2 test. Then, F test could be used to verify the weight assignment of the two kinds of observation data:

$$F = \frac{\chi_1^2/r_1}{\chi_2^2/r_2} = \frac{V_1^T P_1 V_1/(r_1\sigma_{01}^2)}{V_2^T P_2 V_2/(r_2\sigma_{02}^2)} \sim F(r_1, r_2)$$

which could be replaced by the χ^2 test of the whole adjustment results:

$$\chi^2 = \frac{V^T PV}{\sigma_0^2} \sim \chi^2(r)$$

where the parameters are the same to that in the whole adjustment.

The weight assignment could be taken as the right one with the above test satisfied. Then, the variance component estimation is unneeded. Or the function and random models should be tested to find out the factor which results in the above analysis. As there may exits errors of the function model in the errors of the random model, the variance component estimation results are more reliable with the correct function model.

Taking the GEO satellite orbit determination data from measurement data of September 2010 as the true value, the orbit determination period is one day. The numbers of forwarding measuring tracking station and the pseudo-range tracking station are both three. The solved parameters are the initial state of the satellite, solar pressure, and combined clock offset of each satellite and ground pair in every tracking station for the whole calculation period.

Here, the numbers of the three forwarding measuring data are 1710, 1710 and 1668 separately, while that of the three pseudo-range data are 8550, 7184 and 5582 separately. Then, the solution parameters are 9 clock offsets, 2 solar pressure parameters and 6 initial orbit status parameters. The accepted region with 95% confidence is $\left[\chi_{0.975}^2(26617) \quad \chi_{0.025}^2(26617)\right] = [26166.68 \quad 27071.10]$ with the whole adjustment results $\frac{V^T PV}{\sigma^2} = 34612.27$. From the above results, it could be found that the orbit determination result does not satisfy the χ^2 test, which reveals that the equal weight treatment of the two types of data used here is unsuitable. It means random model error will be introduced here. Therefore, a proper weight assignment scheme must be taken in order to obtain better results.

4 Orbit Determination Results and Accuracy Analysis

Verification method is applicable to the GEO satellite data combined adjustment weighted Helmert variance component estimation using the measured data validation, and design of the two options, Option I: all stations are solving their own weight; Option II: stations types of data are two types of the right.

The calculation is as follows. The first orbit determination that the right observation, the initial weights of the various stations are set to 1, each subsequent orbit determination Helmert variance component estimation the weights are corrected, repeat orbit determination, until the various types of unit weight variances.

Contain errors in the stochastic model will affect the parameter estimation accuracy of the results and statistical properties [8, 9], and ultimately affect the results of variance component estimation; due to the inevitable existence of errors in the adjustment model, it cannot be fully the right than the right to converge to the variance of unit weight equal. If forced to bring them into convergence may lead to the calculation of divergence. Chen et al. [2] set to stop when the unit weight variance ratio converges to the level of iteration, this paper set the standard with the same variance ratio of iteration stopping when the unit weight.

Following adopted in September 2010 and October 2010, the measured data of the GEO satellite joint orbit determination experiments ranging tracking station transmitting the three tracking stations in Beijing, Hainan, Sichuan, pseudo range tracking station in Beijing, Hainan and Xinjiang3 a class of stations. Orbit determination arc length for one day, the estimated parameters of the satellite initial orbit, the whole arc of a set of solar pressure parameters at the same time the estimated pseudo range tracking station for L-band combination of satellite and the ground clock parameters a0/a1/a2 (Table 1).

Joint orbit determination, the observational data in October 2010 when the program solver 10 iterations unconverged to use the laser to assess the results of orbit determination, orbit gap of several kilometers, the reason should be the observational data the classification is unreasonable, the classification of too much lead to the stability of the calculation to be destroyed, so the program is not a reasonable Helmert variance component estimation grouping scheme.

Program calculated forwarded ranging data and pseudo range observations the right ratio of 8.76 in September 2010, the program calculated both the right than 13.93 in October 2010, indicating the quality of observational data is still not very stable.

Table 1 GEO satellite joint orbit determination accuracy

	SLR assess the character depending on the accuracy	Option I		Option II	
		σ_i	SLR assess the character depending on the accuracy	σ_i	SLR assess the character depending on the accuracy
Experiment I	0.315	4011 0.257 4042 0.279 4090 0.206 1021 0.421 1041 2.125 1062 1.049	0.085	4011 0.410 4042 0.410 4090 0.410 1021 1.213 1041 1.213 1062 1.213	0.069
Experiment II	0.970	–	–	4011 0.367 4042 0.367 4090 0.367 1021 1.370 1041 1.370 1062 1.370	0.724

Helmert variance component estimation in September 2010 after the results of the assessment of satellite orbit laser increased from 0.315 m for the results of the assessment of the October 0.069, 2010 satellite Orbiter Laser increased from 0.970 to 0.724 m, two days are different degrees of accuracy, indicating such as the two observational data, the right to handle is not an optimal orbit determination processing strategy, Helmert variance component estimation method is applicable to different data in the GEO satellite joint orbit determination.

From the χ^2 test of computation results of Option Two in September, the correction analysis of weight ratio of observations after Helmert variance components estimation is obtained. If the accept domain of 95% is

$$\left[\chi^2_{0.975}(26617) \quad \chi^2_{0.025}(26617)\right] = [26166.68 \quad 27071.10]$$

The result of united adjustment in Option II is

$$\frac{V^T PV}{\sigma^2} = 26483.46$$

It shows that the result from united adjustment of this orbit determination this time passes through the χ^2 verification and the observation model after Helmert variance component estimation is basically correct.

5 Conclusion

In this paper, the test results after the orbit determination of the overall adjustment, noting that the joint orbit determination of transmitting data and pseudo orange data can not simply be such as the right to deal with, and then Helmert variance component estimation used in the precise set of GEO satellite track, results show that, if each station were solving their own weight, may be the stability of the calculation because packet too much damage, resulting in the orbit determination of failure, data classification according to the type of station, the right is a grouping for GEO satellite orbit determination strategy, re-given right to the calculation results by examining the orbital laser assess the accuracy is obviously improved. The test proved Helmert variance component estimation methods for different data in the GEO satellite joint orbit determination.

References

1. Han, C., Liu, L., & Zhao, J. (2009). Conception, definition and evaluation of Pseudo-range measurement. *Journal of Astronautic, 30*(6), 2421–2425.
2. Chen, L., et al. (2008). Evaluation methods of tracking error effect on precision of satellite navigation Pseudo-range measurement. *Hydrographic Surveying and Charting, 28*(5), 35–38.

3. Marvin, M., et al. (1990). Evaluation of GPS receiver ranging accuracy, position location and navigation symposium. *The1990's—A Decade of Excellence in the Navigation Sciences' , IEEE Plans' 90 IEEE,* pp. 314–321.
4. Irsigler, M., Avila-Rodriguez, J. A., Hein, G. W. (2005). *Criteria for GNSS multipath performance assessment* (pp. 2166–2177). Ion GNSS: Long Beach, CA.
5. Feng, X., Jin, G., Fan, J. (2010). *Experimentation and analysis of multipath in code-ranged by GNSS receiver*. CNSC 2010: Beijing.
6. Gary, A., Charles, M., Ryan, S. Y. (2005). *Assessment of GPS anti-jam system Pseudo-range and carrier phase measurement error effects* (pp. 603–617). Ion GNSS: Long Beach, CA.
7. Zhou, K., Yang, Y., & Wang, G. (2005). Influence of jamming to Pseudo-range measurement precision of GPS receiver. *Electronics Countermeasures Technology, 20*(3), 25–28.
8. Kaplan, E. D. (2006). *Understanding GPS principles and applications* (2nd ed., pp. 302–304). Norwood: Artech House, INC.
9. Xu, X., Wang, F., & Zhuang, Z. (2003). The analysis of Doppler performance of the satellite navigation signal. *Journal of National University of Defense Technology, 25*(5), 1–8.

Processing Method and Verification of Local Correlation for Spacecraft DOR Signals

Lue Chen, Geshi Tang, Songtao Han, Mei Wang and Fei Fan

Abstract Differential One-way Ranging (DOR) is an interferometry method to obtain high precision delay in deep space exploration. Two stations track the same spacecraft to obtain high precision deep space spacecraft's angle information. The process of DOR signal local correlation processing shows as follows, the model of DOR local signal is firstly constructed, then this model is used to do local correlation processing with the received spacecraft's radio signal, to obtain high precision delay. Compared with the traditional DOR method, DOR local correlation processing method can be more effective in low SNR conditions to get correct measurement information. Firstly, prior delay model is calculated by orbit prediction, and the spacecraft's emission frequency is calculated through effective frequency evaluation method, then the prior delay model and emission frequency are used to construct the DOR local model signal, then correlation phase is extracted by cross-correlation processing, finally, DOR group delay is calculated. This paper introduces the mathematical model of DOR local correlation processing in detail, and especially analyzes the technical difficulty in DOR local correlation processing. Based on the theoretical analysis, the algorithm of DOR local correlation processing is executed, and this method is verified by simulation, with the

L. Chen (✉) · G. Tang · S. Han · M. Wang
Science and Technology on Aerospace Flight Dynamics Laboratory,
100094 Beijing, China
e-mail: luechen0912@yahoo.com.cn

L. Chen · G. Tang · S. Han · M. Wang
Beijing Aerospace Control Center, 100094 Beijing, China

F. Fan
Beijing Institute of Tracking and Telecommunication Technology,
100094 Beijing, China

J. Sun et al. (eds.), *China Satellite Navigation Conference (CSNC) 2012 Proceedings*, Lecture Notes in Electrical Engineering 160, DOI: 10.1007/978-3-642-29175-3_41,

simulation measurement precision is about 0.1 ns, this verifies the effectiveness of DOR local correlation processing, and provides an effective technology for precise spacecraft orbit measurement.

1 Introduction

With the developing of spacecraft measurement technology, interferometry technology has become one of the most important navigation tools in deep space exploration missions [1], it is sensitive to tracking orbit variation in vertical direction [2]. Interferometry technology has the performance of high precision measuring angle. Two stations simultaneously receive the same source signal, and this technology measures the delay of the two tracking stations by means of radio interference. The high precision delay which is obtained from signal correlation processing can reversely calculate the accurate angular position of the signal source [3], thus, interferometry technology has been successfully applied to satellite and deep space spacecraft tracking observation and location. Since interferometry measurement technology appeared, a variety of interferometry signal processing methods have been developed, such as Very Long Baseline interferometry (VLBI), Differential One-way Range/Doppler (ΔDOR/ΔDOD), Delta Differential One-way Range/ Doppler (DOR/DOD), Same Beam Interferometry (SBI), Connected Element Interferometry (CEI), these methods have become the vital navigation tools which support the deep space exploration mission [4–6].

DOR is the most widely used in deep space exploration mission in all interferometry technologies. This technology shows as follows, the two stations receive the same one spacecraft's downlink DOR tones signal for correlation processing to obtain the geometric delay, which reflects the location relationship between spacecraft and ground stations. There are two types of DOR technologies, one is the traditional DOR method, the other one is the DOR local correlation processing method. The traditional DOR method uses the received downlink signals in two stations to directly calculate the group delay in correlation way. And DOR local correlation method constructs one local reference signal model which matches the received downlink signal, then the local reference signal and the received spacecraft downlink signal are processed in correlation way to obtain correlation phase, then do differential calculation and calculate the phase cycle ambiguity, finally the precise group delay could be obtained. DOR local correlation method has some obvious advantages compared with the traditional DOR method, for example, DOR local correlation method has more sensitive noise adaptability than the traditional DOR method, DOR local correlation method generally reduces the size of the transmit data stream between ground stations and interferometry signal processing center. It should be said that DOR local correlation method also has some difficulties, for example, the reference model parameters precisely evaluation is difficult, and the time synchronization problem should be considered carefully

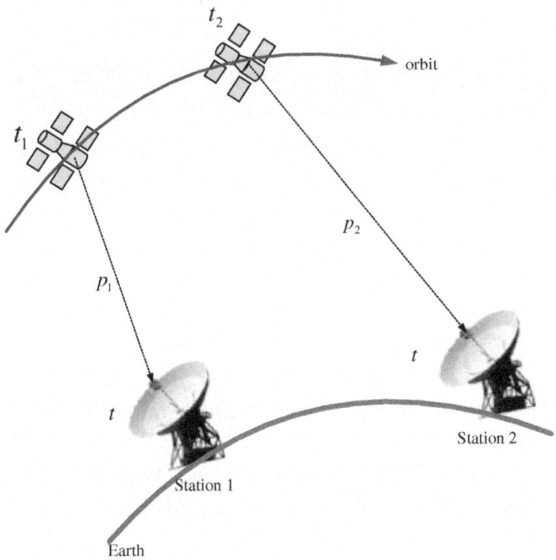

Fig. 1 Spacecraft DOR measurement geometric schematic diagram

among ground stations. However, since the potential application value, DOR local correlation method has become one of the most important developing directions in all interferometry technologies.

This paper derives DOR local correlation method in mathematical analysis way, the performance of this method is analyzed by simulation, and the effectiveness of DOR local correlation is verified.

2 DOR Local Correlation Processing Mathematical Model

Assume the light time of the tracking spacecraft downlink signal arriving Station 1 is p_1, the light time of the tracking spacecraft downlink signal arriving Station 2 is p_2, and the spacecraft DOR measurement geometric schematic diagram is shown in Fig. 1.

The relationship of spacecraft signal emission time t_1 and the station received time t shows in Formula (1) and Formula (2).

$$t_1 = t - p_1 \tag{1}$$

$$t_2 = t - p_2 \tag{2}$$

Suppose the expression of the spacecraft emission DOR tone signal shows in Formula (3) and Formula (4).

$$x(t_1) = e^{j(2\pi f_i t_1 + \phi_{oi})} \tag{3}$$

$$x(t_2) = e^{j(2\pi f_i t_2 + \phi i)} \tag{4}$$

where f_i is the ith point frequency of DOR tone signal, $i = 1, 2, \ldots, N$. N is the numbers of DOR tones, and DOR tone signal frequency can be considered that it keep stable from time to time; ϕ_{oi} is the initial phase of the DOR tone signal which contains phase noise. The received signals of the two stations can be expressed in Formula (5) and Formula (6).

$$x_{r1}(t) = e^{j(2\pi f_i (t-p_1) + \phi_{o1})} \tag{5}$$

$$x_{r2}(t) = e^{j(2\pi f_i (t-p_2) + \phi_{o2})} \tag{6}$$

Then DOR local correlation signal and the corresponding received downlink signal do correlation calculation, and the correlation phase and amplitude are extracted.

According to Formula (1) and Formula (2), the travel time p_1 of downlink signal arrives at Station 1 is as follows,

$$p_1 = \frac{|\bar{r}(t - p_1) - \bar{R}_1(t)|}{c} \tag{7}$$

where $\bar{R}_1(t)$ expresses the location vector of Station 1 when the emission signal arrives at Station1, $\bar{r}(t - p_1)$ expresses the location vector of the tracking spacecraft when the downlink signal is emit. All the above location vectors are all in J2000 inertial reference frame. Simultaneously, the travel time p_2 of downlink signal arrives at Station 2 is as follows,

$$p_2 = \frac{|\bar{r}(t - p_2) - \bar{R}_2(t)|}{c} \tag{8}$$

Only the continuous time model from discrete time model can assure continuous data processing, the simple way is using least-squares polynomial. Assume that

$$t_s < t < t_e = t_s + T_c \tag{9}$$

$$\Delta t = t - t_s \tag{10}$$

where t_s, t_e express the start time of tracking and the end time of tracking. T_c expresses the total tracking time from start time to end time. And

$$0 \le \Delta t \le T_c \tag{11}$$

So the travel time model of the 5th polynomials in (t_s, t_e) can be expressed in Formula (12)

$$p_m = \sum_{i=0}^{5} a_i \Delta t^i \tag{12}$$

According to Formula (6) and Formula (7), the received radio signal is translated into baseband signal after down conversion, and equipment phase delay is added into baseband signal, clock error also exists. Thus, the real phases of the record signal in two stations express in Formula (13) and Formula (14).

$$x_{r1\_f_i}(t) = e^{j(2\pi(f_i - f_{0i})(t-\tau_{1c}) - 2\pi f_i p_1 + \phi_{o1} + \phi_{1i})} \tag{13}$$

$$x_{r2\_f_i}(t) = e^{j(2\pi(f_i - f_{0i})(t-\tau_{2c}) - 2\pi f_i p_2 + \phi_{o2} + \phi_{2i})} \tag{14}$$

where f_{0i} is sky frequency of the ith DOR tone frequency; τ_{1c} and τ_{2c} are clock errors in two stations; ϕ_{1i} and ϕ_{2i} are the added phase delay corresponding to the ith DOR tone frequency, and the added phase delay are caused by equipment. Thus, the two time models of local correlation are respectively expressed in Formula (15) and Formula (16).

$$x_{m1\_f_i}(t) = e^{j\left(2\pi\left(f_i^m - f_{0i}\right)t - 2\pi f_i p_1^m\right)} \tag{15}$$

$$x_{m2\_f_i}(t) = e^{j\left(2\pi\left(f_i^m - f_{0i}\right)t - 2\pi f_i p_2^m\right)} \tag{16}$$

where f_i^m is the estimated emission frequency; p_1^m and p_2^m are the two travel time model values.

Correlation calculation is done by the received downlink signal $x_1(t)$ and local correlation model signal, and the cross-correlation function is obtained in Formula (17).

$$C_{10} = \frac{1}{2} A_i \cdot e^{(-j \cdot \varphi_{10})} \tag{17}$$

where A_i is the amplitude of the received DOR tone signal, and φ_{10} is the residual phase after correlation calculation.

The correlation phase of the two DOR tone signal in Station 1 is obtained as follows.

$$\begin{aligned} \phi^i_{dif1} &= 2\pi(f_i - f_{0i})(t - \tau_{1c}) - 2\pi f_i p_1 + \phi_{o1} + \phi_{1i} \\ &\quad - \left[2\pi\left(f_i^m - f_{0i}\right)t - 2\pi f_i p_1^m\right] \\ &= 2\pi\left(f_i - f_i^m\right)t + 2\pi f_{0i}\tau_{1c} - 2\pi f_i \tau_{1c} \\ &\quad - 2\pi f_i\left(p_1 - p_1^m\right) + \phi_{o1} + \phi_{1i} \end{aligned} \tag{18}$$

where the clock error is not considered, that is to say the initial phase noise and equipment phase noise are not considered, so

$$\phi^i_{dif1} = 2\pi\left(f_i - f_i^m\right)t - 2\pi f_i\left(p_1 - p_1^m\right) \tag{19}$$

Simultaneously, the correlation phase corresponding to the two point frequency signal in Station 2 is as follows,

$$\begin{aligned}\phi^{i}_{dif2} &= 2\pi(f_i - f_{0i})(t - \tau_{2c}) - 2\pi f_i p_2 + \phi_{o2} \\ &\quad + \phi_{2i} - \left[2\pi\left(f_i^m - f_{0i}\right)t - 2\pi f_i p_2^m\right] \\ &= 2\pi\left(f_i - f_i^m\right)t + 2\pi f_{0i}\tau_{2c} - 2\pi f_i \tau_{2c} \\ &\quad - 2\pi f_i\left(p_2 - p_2^m\right) + \phi_{o1} + \phi_{2i}\end{aligned} \tag{20}$$

where the clock error is not considered, thus

$$\phi^{i}_{dif2} = 2\pi\left(f_i - f_i^m\right)t - 2\pi f_i\left(p_2 - p_2^m\right) \tag{21}$$

The differential phase of the two stations corresponding to emission frequency f_1 is shown in Formula (22).

$$\begin{aligned}\phi^{f_1}_{cor} &= 2\pi f_1\left(p_2 - p_2^m\right) - 2\pi f_1\left(p_1 - p_1^m\right) \\ &= 2\pi f_1\left(p_2 - p_1\right) + 2\pi f_1\left(p_1^m - p_2^m\right)\end{aligned} \tag{22}$$

Thus

$$\begin{aligned}\phi^{f_2}_{cor} &= 2\pi f_2\left(p_2 - p_2^m\right) - 2\pi f_2\left(p_1 - p_1^m\right) \\ &= 2\pi f_2(p_2 - p_1) + 2\pi f_2\left(p_1^m - p_2^m\right)\end{aligned} \tag{23}$$

From Formula (22) and (23), we can obtain the following formula.

$$2\pi(f_1 - f_2)(p_2 - p_1) = \phi^{f_1}_{cor} - \phi^{f_2}_{cor} + 2\pi\left(p_1^m - p_2^m\right)(f_2 - f_1) \tag{24}$$

Thus the final DOR local correlation measurement expression is shown in Formula (25).

$$(p_2 - p_1) = \frac{\phi^{f_1}_{cor} - \phi^{f_2}_{cor}}{2\pi(f_1 - f_2)} - \left(p_1^m - p_2^m\right) \tag{25}$$

3 Simulation Verification

This section verifies the correctness of DOR local correlation signal processing method by simulation. Firstly, four received DOR tone signals in two stations are constructed by Simulink toolbox in Matlab. The simulation figure of the constructing DOR tone signals received in two stations is shown in Fig. 2. 'sig11' expresses the 1th DOR tone signal received in Station 1, 'sig21' expresses the 1th DOR tone signal received in Station 2, 'sig12' expresses the 2nd DOR tone signal received in Station 1, 'sig22' expresses the 2nd DOR tone signal received in Station 2. The simulation parameters is set as follows, the 1th DOR tone signal

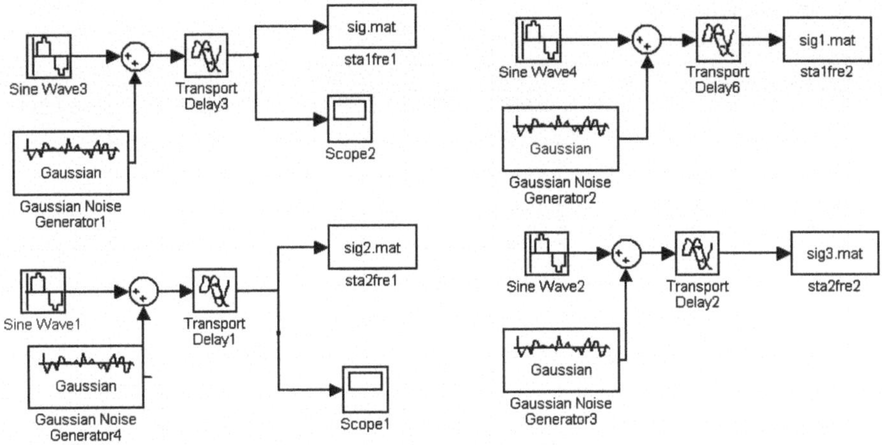

Fig. 2 Simulink DOR tone signals creation simulation figure

frequency in spacecraft is 8,464 MHz, the 2nd DOR tone signal frequency in spacecraft is 8,468 MHz, and the downlink carrier frequency is 8,460 MHz, that is to say DOR tone frequencies are respectively 4 and 8 MHz. The local frequency of the two stations are both 8,460 MHz. Assume that the geometric delay corresponding to the distance between the spacecraft and Station 1 is (2 s + 600 ns), the geometric delay corresponding to the distance between the spacecraft and Station 2 is (2 s + 1,600 ns), thus the differential delay caused by the differential distance between the same spacecraft and the two different ground stations is 1,000 ns.

From the theoretical analysis in the above section, we know that, in order to construct DOR local correlation model, constructing delay model is one of the most important things. The delay model is generally determined by orbit prediction, but the orbit prediction generally contains prediction error, which is obtained from the difference of the real delay and the orbit prediction delay. In simulation, assuming the orbit prediction value corresponding to the distance between the spacecraft and Station 1 is (2 s + 620 ns), and the orbit prediction value corresponding to the distance between the spacecraft and Station 2 is (2 s + 1,630 ns). Although the frequency model and the real spacecraft frequency exists difference, this difference keep the level of mHz is possible by modern frequency estimation technology, thus assume that the 1th received DOR tone signal frequency in Station 1 is $f_1 + 0.3$ Hz, the 2nd received DOR tone signal frequency in Station 1 is $f_2 + 0.1$ Hz, the 1th received DOR tone signal frequency in Station 2 is $f_1 + 0.1$ Hz, the 2nd received DOR tone signal frequency in Station 2 is $f_2 - 0.2$ Hz. Gauss noise with 10 dB SNR (signal noise ratio) is added in all the simulation DOR tone signals, many times experiments are done in the same conditions, the simulation delay estimation result is shown in Fig. 3, and the estimation error between DOR local correlation delay and the real delay is shown is Fig. 4.

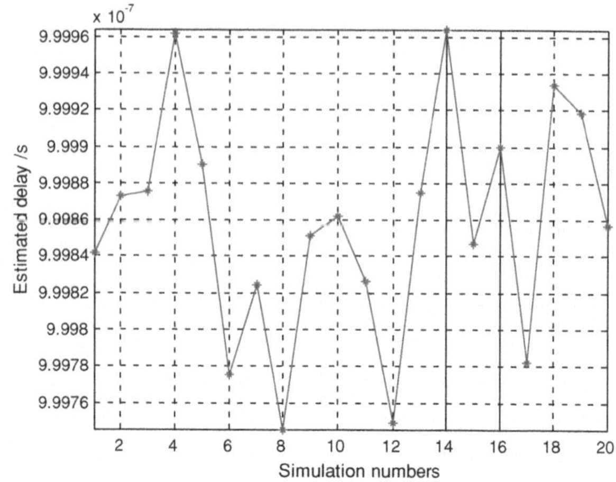

Fig. 3 The result of DOR local correlation estimation delay

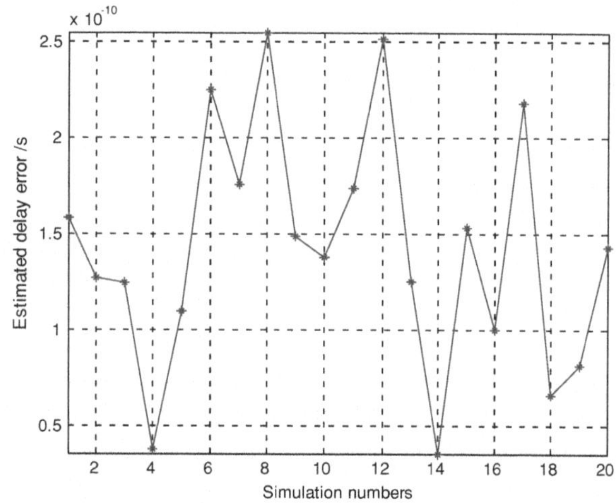

Fig. 4 The error of DOR local correlation estimation delay

The mean value of 20 times simulation results is 999.86 ns, the error between this mean value and the real differential delay (1,000 ns) is 0.14244 ns, and the standard deviation of the 20 times is 0.062786 ns. It shows that this simulation effectively verifies the correctness of DOR local correlation algorithm, meanwhile, DOR local correlation method reaches a high precision level.

When the orbit prediction error is large, phase cycle ambiguity problem will appear, and delay error caused by 2π phase ambiguity will appear too. To explain this phenomenon, the other experiments are done. The delay model corresponding to the distance between the spacecraft and the Station 2 is set with (2 s + 1,900 ns), that is to say, the error between the orbit prediction delay value and the real geometric delay is 300 ns, all the other parameters are as same as the above simulation. The calculation result is shown is Fig. 5, it shows that when the

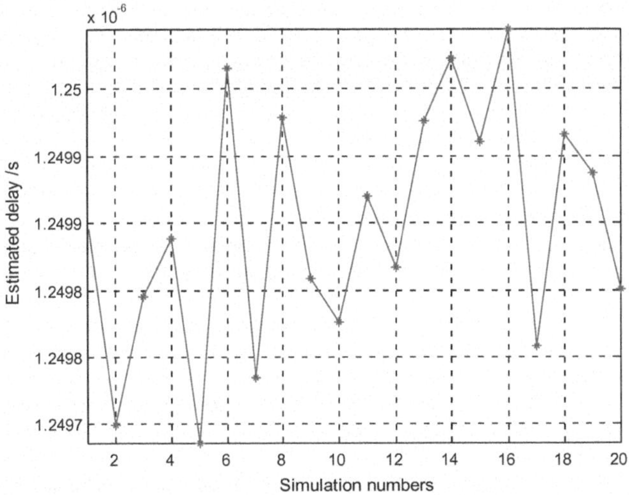

Fig. 5 DOR local correlation estimation delay when model delay error is large

model error of orbit prediction is larger, the estimation error of DOR local correlation is larger. This larger estimation is caused by phase cycle ambiguity. In Fig. 5, the mean value of estimation delay in 20 times independent experiments is 1249.85 ns, the standard deviation in 20 times independent experiments is 0.0906 ns, the error between the mean value and the real delay is 249.85 ns. The delay error caused by phase cycle ambiguity is $\frac{2\pi}{2\pi(f_2 - f_1)} = 1/(8e6 - 4e6) = 250\,\text{ns}$ Compared with 249.85 ns and 250 ns, the large error is caused by phase cycle ambiguity of the inaccuracy of delay model. Thus, in order to do DOR local correlation calculation, delay model should meet some constraint requirement to correctly calculate the group delay of DOR local correlation.

4 Conclusion

This paper constructs mathematical model of DOR local correlation signal processing, verifies the effectiveness of DOR local correlation algorithm by simulation, and obtains the result of the error smaller than 0.2 ns and the precision better than 0.1 ns. In our future work, this DOR local correlation signal processing method should be researched and verified by real spacecraft tracking signal to test the performance of this method. We believe that this DOR local correlation signal processing method can be effectively used in deep space navigation mission in China to meet the high precision measurement requirement.

Acknowledgments This work is supported by National Nature Science Foundation of China (No.11173005 and No. 61104190), *the Opening Project of Shanghai Key Laboratory of Space Navigation and Position Technology (201101)*, and China's Lunar Exploration Foundation.

References

1. Chen, L., Tang, G., Wang, M., et al. (2011). Interferometry measurement broadband correlation processing algorithm and verification. *Journal of Space TT&C Technology, 30*(6), 77–81.
2. Thornton, C. L., & Border, J. S. (2000). *Radiometric tracking techniques for deep-space navigation, deep space communication and navigation systems center of excellence (DESCANSO)*. Jet Propulsion Laboratory, Chap 3 and 5.
3. Chen, L., Tang, G., Wang, M., et al. (2011). Connected interferometry measurement signal processing and experiment and experiment verification. *Journal of Telemetry, Tracking and Command, 32*(6), 28–32.
4. Madde, R., Morley, T., & Abello, R. (2006). Delta-DOR a new technique for ESA's deep space navigation. *ESA Bulletin, 128*, 69–74.
5. Kikuchi, F., Liu, Q., Hanada, H., et al. (2009). Pico-second accuracy VLBI of the two sub-satellites of SELENE (KAGUYA) using multi-frequency and same beam methods. *Radio Science, 44*, 1–7.
6. Chen, L., Tang, G., Fan, F., et al. (2011) China's CE-2 lunar satellite experiment based on short baseline interferometry. In 62nd international astronautical congress, Cape Town.

A Modified Extend Kalman Particle Filter with Application to Relative Navigation

Xiaoliang Wang, Lixin Zhang, Xiaoping Qian, Qibing Xu, Yansong Meng and Zhe Su

Abstract To improve the accuracy of relative navigation for spacecraft formation flying in eccentric orbits, precise relative motion equation was used and Carrier-phase Difference GPS technology was adopted for relative information measurement. A modified version of Extended Kalman Particle Filter algorithm, called MEKPF, was used for navigation filter design. Simulation results provided later indicate the proposed navigation approach can provide an accuracy and consistent relative navigation output for spacecraft formation flying.

1 Introduction

Spacecrafts formation flying has attract much attention and actually has been used recently since it can be implemented as long baseline interferometry, stereographic imaging, synthetic apertures, or, as a test programme for real autonomous rendezvous, docking and proximity operations [1, 2]. However, relative navigation accuracy directly impacts the effect and success of such missions. Tillerson [3, 4] demonstrated that fuel usage for active control of formation flying satellites is a strong function of relative navigation error, and in particular, the relative velocities. The larger the error, the higher the fuel usage will be, and the shorter the mission life.

To perform the relative navigation, the Carrier-Phase Differential GPS (CDGPS) is emerging as a very promising low cost alternative to some conventional methods, such as microwave radar or laser radar. Since CDGPS has been

X. Wang (✉) · L. Zhang · X. Qian · Q. Xu · Y. Meng · Z. Su
Institute of Satellite Navigation and Intra-Satellite-Link Technology,
Academy of Space Electronic Information Technology, Xi'an, China
e-mail: xlwang12321@gmail.com

J. Sun et al. (eds.), *China Satellite Navigation Conference (CSNC) 2012 Proceedings*,
Lecture Notes in Electrical Engineering 160, DOI: 10.1007/978-3-642-29175-3_42,

used in terrestrial applications to generate very accurate relative position estimation, it is a natural candidate for spacecraft formation instrumentation. Meanwhile, the best choice of nonlinear states estimation algorithm is another key factor of desired navigation output.

Future attention should be given to the study of relative motion for spacecraft flying in formation around a planet in an elliptic Keplerian reference orbit with long formation baseline. Such formations find use in terrestrial observation, communication and stellar interferometry. Of more recent interest is the potential of use of such formations in orbits that are highly eccentric. Examples of such missions have been presented by [5].

Precise equations that describes the relative motion of spacecraft formation are needed for navigation filter design. Traditional rendezvous problem is of short duration, and there are frequent thruster firings. Therefore, the long-term accuracy of the equations of motion is not as important in the rendezvous problem as in the formation flying problem. So the Clohessy-Wiltshire equations (CW equations also as Hill's equations) or a modified version incorporating small eccentric effects is usually sufficient [6] and [7]. However, Hill's equations are based on the assumptions that the reference orbit is circular and the target orbit is very close to the reference orbit. So the nonlinear terms in the relative motion can be neglected. The modified version of Hill's equations incorporating small eccentric effects still based on the assumptions that the target orbit is very close to the reference orbit. These assumptions can result in unacceptable errors in the long-term prediction of the relative motion of spacecraft formation flying missions. To minimize the real relative navigation errors, precise relative motion equations should be adopted and a more accurate filter algorithm for the relative motion is needed.

Precise relative navigation can be achieved though traditional nonlinear states estimation methods, such as ubiquitous Extended Kalman Filter (EKF) [8]. In order to solve the divergence problem existing in EKF when model error considered, Xia [9] proposed a suboptimal Adaptive Fading Kalman Filter (AFKF) and Dan [10] applied this estimation algorithm to the relative navigation problem. However, Dan used linear Hill equation for navigation system design, which is apparently not suitable for real application. Recently, Karlgaard [11] proposed a relative navigation architecture using Huber filter algorithm for spacecraft formation flying in elliptical orbit. It used coordinate transformation in observe equation to form linear measurement which deals with non-Gaussian measurement noise.

This paper uses a modified version of Extended Kalman Particle Filter algorithm, called MEKPF, for relative navigation of spacecraft formation flying in eccentric orbits. Simulation results given later indicate that the proposed navigation approach can provide accuracy and consistent output for formation flying of spacecraft.

The organization of this paper proceeds as follows: First, relative coordinate systems and positional equations of motion were given. Then, GPS measurement models for the relative navigation were provided. Next, MEKPF algorithm for navigation states estimation were introduced. Finally, simulation results and conclusions were presented.

2 Navigation System Model

2.1 Relative Orbital Motion Equations

The equations describe the relative motion of a spacecraft, often referred as the "deputy," with respect to another spacecraft, the "chief". The relative orbit position vector, ρ is expressed in components by $\rho = [x\ y\ z]^T$ in the direction of Radial In-Track and Cross-Track respectively. The detailed derivation of the relative equations of motion for eccentric orbits can be found in [12] and [13]. The equations of motion are given by

$$\begin{aligned}
\ddot{x} &= 2\dot{\theta}\dot{y} + \ddot{\theta}y + \dot{\theta}^2 x + \frac{\mu}{r^2} - \frac{\mu(r+x)}{\left[(r+x)^2 + y^2 + z^2\right]^{\frac{3}{2}}} + w_x \\
\ddot{y} &= -2\dot{\theta}\dot{x} - \ddot{\theta}x + \dot{\theta}^2 y - \frac{\mu y}{\left[(r+x)^2 + y^2 + z^2\right]^{\frac{3}{2}}} + w_y \\
\ddot{z} &= -\frac{\mu z}{\left[(r+x)^2 + y^2 + z^2\right]^{\frac{3}{2}}} + w_z
\end{aligned} \tag{1}$$

where r is the chief orbit radius, θ is true anomaly of the chief spacecraft, μ is the gravitational parameter of the central body. w_x, w_y, w_z are acceleration disturbances in space environment in the direction of Radial, In-Track and Cross-Track, respectively which modeled as zero-mean Gaussian white-noise processes. The true anomaly acceleration and chief orbit-radius acceleration are given by

$$\begin{aligned}
\ddot{\theta} &= -2\frac{\dot{r}}{r}\dot{\theta} \\
\ddot{r} &= r\dot{\theta}^2 - \frac{\mu}{r^2}
\end{aligned} \tag{2}$$

Equations (1) and (2) is a perfect description of relative motion in space scenarios. It can be simplified if some condition satisfied. If the relative orbit coordinates are small compared to the chief orbit radius, the Eqs. (1) and (2) for the relative motion are given by

$$\begin{aligned}
\ddot{x} &= 2\dot{\theta}\dot{y} + \ddot{\theta}y + \dot{\theta}^2 x + 2\frac{\mu}{r^3}x + w_x \\
\ddot{y} &= -2\dot{\theta}\dot{x} - \ddot{\theta}x + \dot{\theta}^2 y - \frac{\mu}{r^3}y + w_y \\
\ddot{z} &= -\frac{\mu}{r^3}z + w_z
\end{aligned} \tag{3}$$

If the chief spacecraft orbit is assumed to be circular so that $\dot{r} = 0$ and $\ddot{\theta} = 0$ in Eqs. (1) and (2), the relative equation of relative motion reduce to the simple form known as the CW equations [6] and [7] (with disturbances added here):

$$
\begin{aligned}
\ddot{x} &= 2n\dot{y} + 3n^2x + w_x \\
\ddot{y} &= -2n\dot{x} + w_y \\
\ddot{z} &= -n^2z + w_z
\end{aligned}
\tag{4}
$$

where $n = \dot{\theta}$ is the mean motion.

2.2 GPS Measurement Model

Carrier-phase Difference GPS for relative orbit determination is emerging as a very promising low cost method. For the purpose of relative navigation, single differences measurement can be obtained by differencing between the antennas located in chief and deputy spacecrafts. The general single differences measured carrier phase model can be described as follow:

$$
\triangle\Phi_{ij}^k + \lambda N_{ij}^k = D_{ij}^k + c\tau_i - c\tau_j + v_{ij}^k \tag{5}
$$

where $\triangle\Phi_{ij}^k$ is the measured carrier phase of antenna i and j from GPS k. λ is the wave length of GPS carrier phase (19 cm for L1, 24 cm for L2). N_{ij}^k is the single differences integer which can be determined by integer search algorithm that has to be performed at startup before. τ_i, τ_j are the clock error of receiver i and j. v_{ij}^k is the measurement noise. D_{ij}^k is the projection of the baseline vector $\overrightarrow{ij}$ onto the Line-Of-Sight (LOS) vector $\overrightarrow{ik}$ which contains the information of relative position.

The clock errors τ can be modeled as two-order markov process as follow:

$$
\begin{aligned}
\dot{\tau} &= f + w_1,\ w_1 \sim N(0, S_\tau) \\
\dot{f} &= w_2,\ w_2 \sim N\left(0, S_f\right)
\end{aligned}
\tag{6}
$$

where f is the frequency drift.

2.3 Relative Position Estimation

System description can be given as

$$
\begin{aligned}
\dot{x} &= f(x, w) \\
y_k &= h(x_k, v_k)
\end{aligned}
\tag{7}
$$

where $x \in \mathbb{R}^x$ denotes the states, $y \in \mathbb{R}^y$ denotes the observations, $w \in \mathbb{R}^w$ denotes the process noise, $v \in \mathbb{R}^v$ denotes the measurement noise. The mappings f and h represent the deterministic process and measurement models.

Based on the relative orbit motion equation and GPS measurement model, the total states for the relative position estimation can be modeled as follow:

$$X = \left[\rho \; \dot{\rho} \; r_c \; \dot{r}_c \; \theta \; \dot{\theta} \; \tau_c \; f_c \; \tau_d \; f_d\right]^T \tag{8}$$

where *c, d* represent chief and deputy spacecraft. The nonlinear state-space model can be easily derived by combination of Eqs. (1), (2) and (6).

In this nonlinear model, the chief radius and true anomaly, as well as their respective derivation, are estimated. If this information is assumed known a prior, then these states can be removed and their respective measured values can be added as process noise in the state model. The clock errors and frequency drift of chief and deputy spacecraft are also extended as states variants.

3 Modified Extended Kalman Particle Filter

The Particle Filter is a success states estimation algorithm that has been applied in many regions as statistical and signal processing communities. The PF algorithm containing three main steps as follow:

Prediction: The particles $\{\xi_{k-1}^1, \ldots, \xi_{k-1}^N\}$, representing the posterior $f_{x_{k-1}|z_{1:k-1}}(x_{k-1}|z_{1:k-1})$, are propagated in time by sampling from the dynamical model $f_{x_k|x_{k-1}}(x_k|\xi_{k-1}^s)$ for $s = 1, \ldots, \mathrm{N}$. The propagated particles, $\left\{\tilde{\xi}_k^1, \ldots, \tilde{\xi}_k^N\right\}$, represent the prior $f_{x_k|z_{1:k-1}}(x_k|z_{1:k-1})$ at time *k*.

Observation: Given the new observation z_k, each propagated particle $\tilde{\xi}_k^s$ is assigned a weight $\omega_k^s \propto f_{z_k|x_k}\left(z_k|\tilde{\xi}_k^s\right)$. the weights are thereafter normalized to sum to one.

Resampling: Now, N new particles are sampled from the set of particles with attached weights which given by $\left\{\left(\tilde{\xi}_k^1, \omega_k^1\right), \ldots, \left(\tilde{\xi}_k^N, \omega_k^N\right)\right\}$. The frequency with which each particle is proportional to the weight (Monte Carlo sampling). The result is a particle set with equal weights, $\{\xi_k^1, \ldots, \xi_k^N\}$, representing the posterior distribution at time *k*.

The success of PF algorithm depends on the validity of the following underlying assumptions: **Monte Carlo (MC) assumption** and **Importance sampling (IS) assumption**. The IS approximation depends on how close the proposal distribution is to the posterior distribution. Gordon [14] proposed a **prior editing** method which is an ad-hoc acceptance test for proposing particles in regions of high likelihood. **Rejection methods** suppose it is possible to sample from the optimal importance distribution using an accept/reject procedure if the likelihood is bounded. Pitt [15] introduced the **auxiliary particle filter** which draw samples from the proposed auxiliary joint distribution.

The three methods presented above for designing better proposal distributions are numerous inefficiency which were not suitable for navigation filter design. Pitt [15] introduced a popular method, called **Extended Kalman Particle Filter-EKPF**, for devising proposal distributions that approximate the optimal importance distribution that relies on the first order Taylor series expansions of the likelihood and transition prior, as well as a Gaussian assumption on all the random variables. This paper using a modified version of EKPF, we called as MEKPF, for navigation filter design. The idea of MEKPF came form Adaptive Fading Kalman Filter-AFKF proposed by Xia [9] which introduce a fading factor in traditional Kalman Filter for exponential fading past observe data. Detail information of AFKF can be given in [9] which was not provided here.

The pseudo-code for the modified extended kalman particle filter follows [16].

- Initialization: $k = 0$

For $i = 1, \ldots, N$, draw the states (particles) $x_0^{(i)}$ from the prior $p(x_0)$.

- For $k = 1, 2, \ldots$

1. Importance sampling step

For $i = 1, \ldots, N$:

Compute the Jacobians $F_k^{(i)}$ & $G_k^{(i)}$ and $H_k^{(i)}$ & $U_k^{(i)}$ of the process and measurement models.

Update the particles with the EKF:

$$\bar{x}_{k|k-1}^{(i)} = f\left(x_{k-1}^{(i)}\right)$$

calculate residual $Z = y_k - h\left(\bar{x}_{k|k-1}^{(i)}\right)$ and fading factor $\lambda(k)$

$$P_{k|k-1}^{(i)} = \lambda(k) F_k^{(i)} P_{k-1}^{(i)} F_k^{T(i)} + G_k^{(i)} Q_k G_k^{T(i)}$$

$$K_k = P_{k|k-1}^{(i)} H_k^{T(i)} \left[U_k^{(i)} R_k U_k^{T(i)} + H_k^{(i)} P_{k|k-1}^{(i)} H_k^{T(i)}\right]^{-1}$$

$$\bar{x}_k^{(i)} = \bar{x}_{k|k-1}^{(i)} + K_k Z$$

$$\hat{P}_k^{(i)} = P_{k|k-1}^{(i)} - K_k H_k^{(i)} P_{k|k-1}^{(i)}$$

Sample $\hat{x}_k^{(i)} \sim q\left(x_k^{(i)} | x_{0:k-1}^{(i)}, y_{1:k}\right) = N\left(\bar{x}_k^{(i)}, \hat{P}_k^{(i)}\right)$

Set $\hat{x}_{0:k}^{(i)} \triangleq \left(x_{0:k-1}^{(i)}, \hat{x}_k^{(i)}\right)$ and $\hat{P}_{0:k}^{(i)} \triangleq \left(P_{0:k-1}^{(i)}, \hat{P}_k^{(i)}\right)$

For $i = 1, \ldots, N$, evaluate the importance weights up to a normalizing constant:

$$\omega_k^{(i)} \propto \frac{p\left(y_k | \hat{x}_k^{(i)}\right) p\left(\hat{x}_k^{(i)} | x_{k-1}^{(i)}\right)}{q\left(\hat{x}_k^{(i)} | x_{0:k-1}^{(i)}, y_{1:k}\right)}$$

For $i = 1, \ldots, N$, normalize the importance weights.

Table 1 Formation orbit parameters

Items	Chief	Deputy
Radial (km)	7178	7178
Eccentricity	0.001	0.00107
Inclination (°)	45	45
Argument of perigee (°)	0	0.00798
RAAN (°)	0	−0.01129
Mean anomaly (°)	0	0.00798

2. Selection step

Multiply/Suppress particles $\left(\hat{x}_{0:k}^{(i)}, \hat{P}_{0:k}^{(i)}\right)$ with high/low importance weights $\omega_k^{(i)}$, respectively, to obtain N random particles $\left(\tilde{x}_{0:k}^{(i)}, \tilde{P}_{0:k}^{(i)}\right)$

3. Output

The same manner as for the generic particle filter.

4 Simulation Results

Simulation results are presented in this section that show the relative navigation performance using modified extended kalman particle filter. The real relative motion parameter was not given by propagation of Eq. (1). Here, we use the relative orbit geometry description though orbit element differences proposed by Schaub [17] as real formation configuration for navigation filter comparison. Detail information can be found in reference [17] which was not given here. Formation flying in the distance of about 1 km in elliptical orbit are adopted with orbit parameters given in Table 1.

The spectral densities of the process noise components w_x, w_y, and w_z in Eq (1) are each given by $\sqrt{5} \times 10^{-11}$ m/s$^{3/2}$ which comply with the data from reference [8]. The initial clock error and frequency drift are set to 4×10^{-4} s and 3×10^{-11}/s with $S_\tau = 10^{-4}(\mathrm{s})^2$ and $S_f = 10^{-11}(/\mathrm{s})^2$ for two spacecrafts. GPS data used in this paper came from U.S. Coast Guard Navigation Center (URL: http://www.navcen.uscg.gov/gps/default.htm 2001).

The simulation time for relative motion of the two spacecrafts is 1000 s with time interval 1 s. Carrier phase measurements noise for both chief and deputy is set to 0.5 mm.

Figures 1 and 2 show the final relative position and velocity errors in this simulation. Relative position knowledge is within 0.1 m for each LVLH direction while the relative velocity knowledge is within 0.01 m per time step. The estimation of true anomaly, true anomaly rate and radius, radius rate of chief

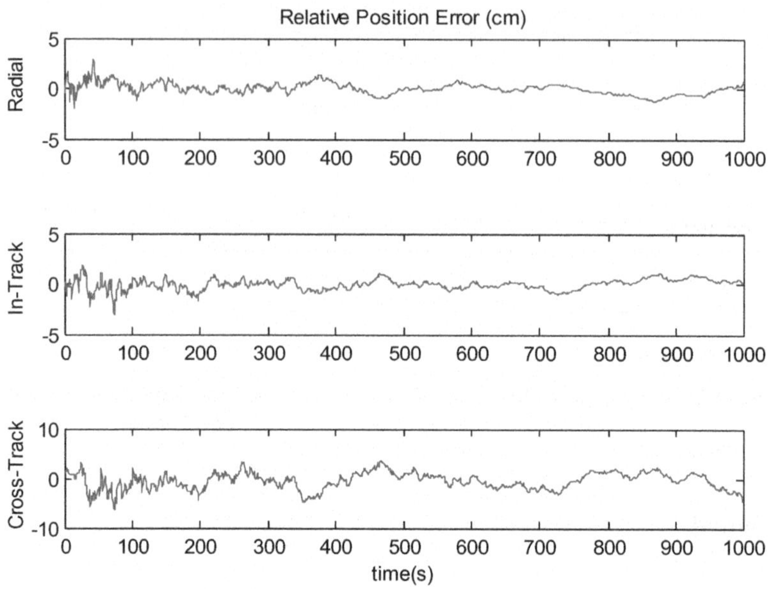

Fig. 1 Relative position error

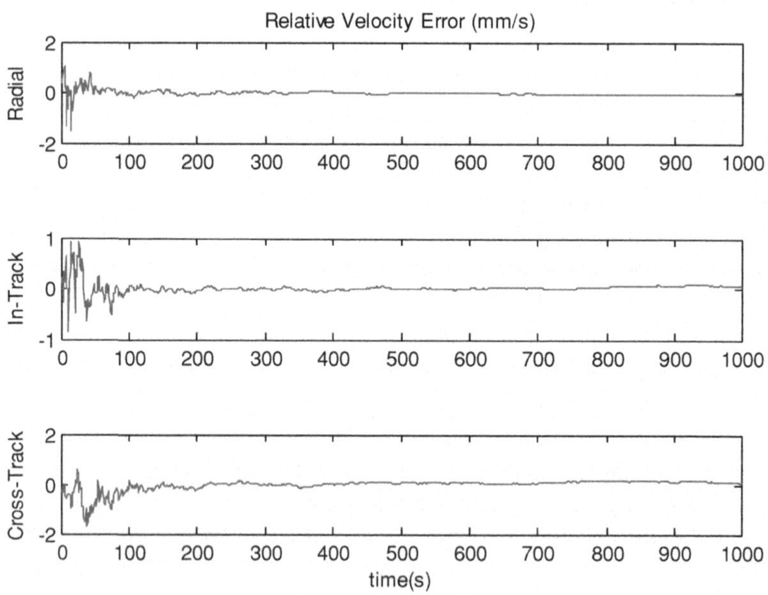

Fig. 2 Relative velocity error

spacecraft are shown in Figs. 3 and 4. The estimate error of true anomaly is less than 10^{-3} rad and 10^{-7} rad/s for true anomaly rate. The chief radius error is within 0.1 km and 10^{-3} km/s for radius rate.

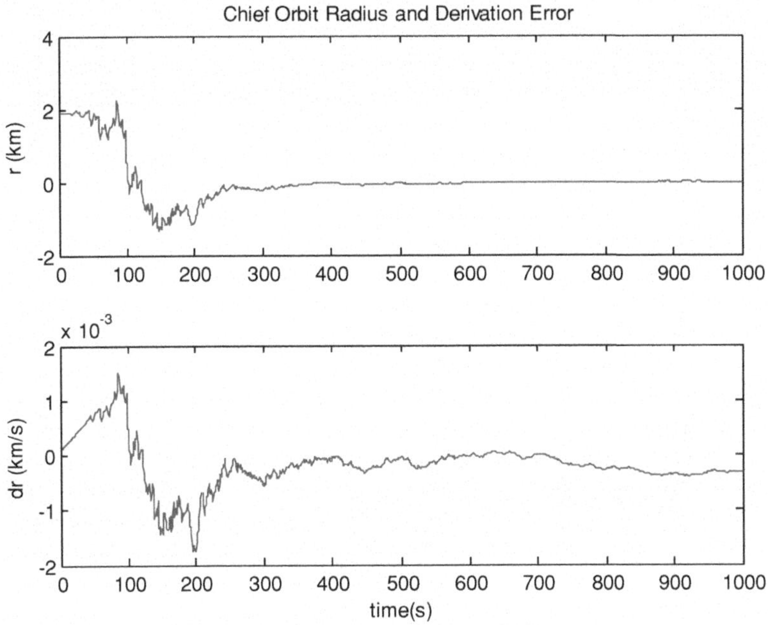

Fig. 3 Chief true anomaly and derivation error

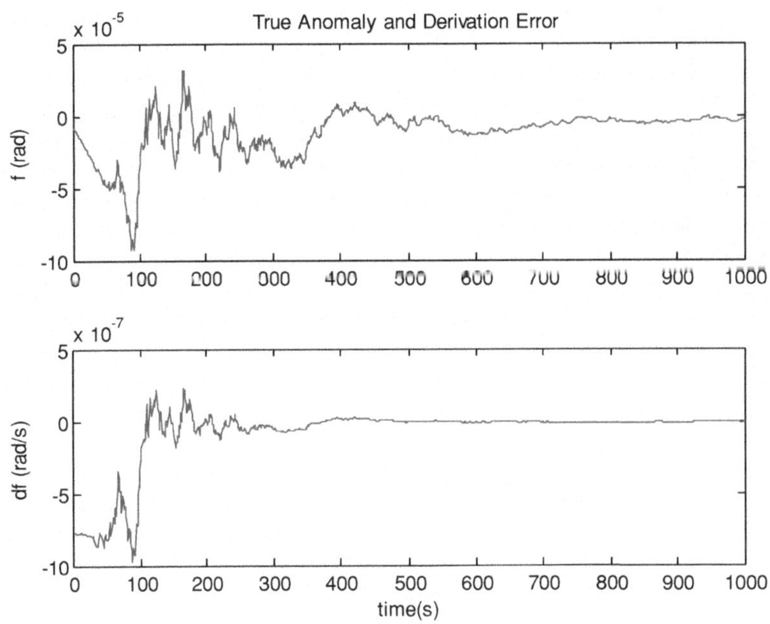

Fig. 4 Chief radius and derivation error

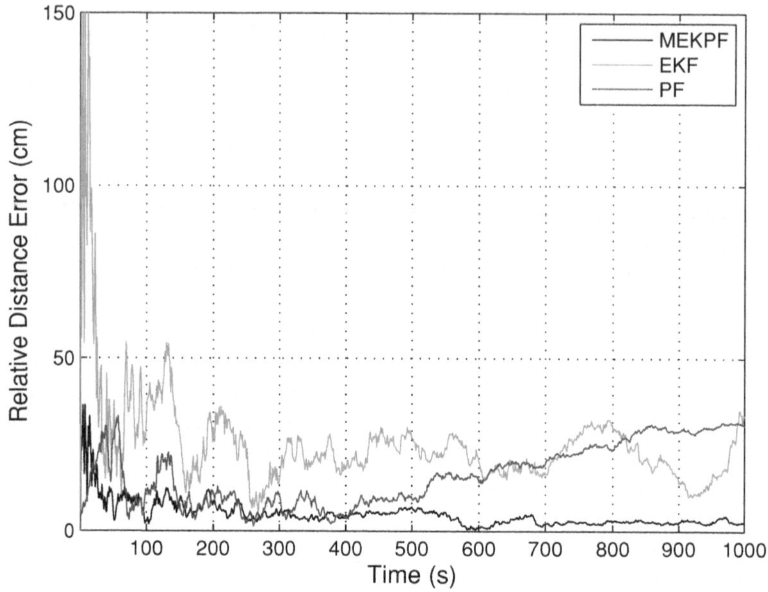

Fig. 5 Comparison of relative position error

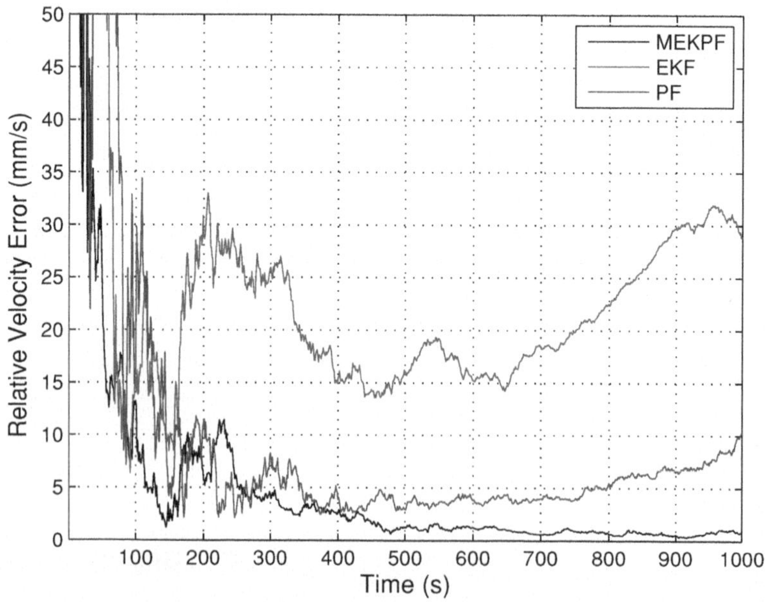

Fig. 6 Comparison of relative velocity error

In addition, filter can also provide estimation of clock error and frequency drift for both chief and deputy spacecraft which are not shown in this section.

Estimation comparison of different types of filter is provided here. The results of simulation using MEKPF and traditional EKF, also with PF are shown in Figs. 5 and 6. The estimation error for relative distance, velocity (the norm of relative position and velocity in both Radial, In-Track and Cross-Track direction) using those three methods are drawn in black, red and blue, respectively. The EKF results indicate large errors while PF perform diverge as times goes by. The errors of EKF cased by the deficient of first order Taylor series expansions for nonlinear system and the divergence of PF cased by the algorithm lack of using information of observation in update step. At the same time, MEKPF proposed in this paper can provide accuracy and consistent relative navigation output during simulation time.

5 Conclusion

A precise relative navigation system design was proposed for spacecraft formation flying in eccentric orbits. Precise relative motion equation was used and Carrier-phase Difference GPS technology was adopted for relative information measurement. Modified Extended Kalman Particle Filter was used for navigation filter design. Simulation results provided later indicate that the proposed navigation approach can provide an accuracy and consistent output for formation flying of spacecraft. The estimation RMS of relative distance is less than 14 cm while the relative velocity is better than 0.5 mm/s.

References

1. Madison, R. W. (1999). *Micro-satellite based, on-orbit servicing work at the Air Force Research Laboratory*. Kirtland AFB, NM: Air Force Research Laboratory.
2. Weismuller, T., & Leinz, M. (2006). *GN&C technology demonstrated by the orbit express autonomous rendezvous and capture sensor system*. 29th Annual AAS Guidance and Control Conference, Breckenridge, Colorado, USA.
3. How, J. P., & Tillerson, M. (2001). Analysis of the impact of sensor noise on formation flying control. *Proceedings of the American Control Conference, Arlington, VA* (pp. 3986–3991).
4. Tillerson, M. (2002). *Coordination and control of multiple spacecraft using convex optimization techniques*. S.M. Thesis, Department of Aeronautics and Astronautics, MIT.
5. Carpenter, J. R., Leitner, J. A., Folta, D. C., & Burns, R. D. (2003). *Benchmark problems for spacecraft formation flight missions*. AIAA Paper 2003-5364.
6. Clohessy, W. H., & Wiltshire, R. S. (1960). Terminal guidance for satellite rendezvous. *Journal of Aerospace Sciences, 27*, 653.
7. Hill, G. (1878). Researches in the lunar theory. *American Journal of Mathematics, 1*, 5–26.
8. Kim, S. G., Crassidis, J. L., Yang, C., & Fosbury, A. M. (2007). Kalman filtering for relative spacecraft attitude and position estimation. *Journal of Guidance, Control, and Dynamics, 30*(1), 133–143.

9. Xia, Q., Rao, M., Ying, Y., & Shen, X. (1994). Adaptive fading Kalman filter with an application. *Automatica, 30*(8), 1333–1338.
10. Dan, X., & Cao, X. (2006). Relative navigation with maneuvers using a suboptimal fading extended Kalman filter. *Proceedings of the Fifth International Conference on Machine Learning and Cybernetics, Dalian, August 13–16.*
11. Karlgaard, C. D. (2006). Robust rendezvous navigation in elliptical orbit. *Journal of Guidance, Control, and Dynamics, 29*(2), 495–499.
12. Schaub, H., & Junkins, J. L. (2003). *Analytical mechanics of aerospace systems*. New York: American Institute of Aeronautics and Astronautics, Inc.
13. Xing, G. Q., & Shabbir, A. P. (1999). Relative attitude kinematics & dynamics and its applications to spacecraft attitude state capture and tracking in large angle slewing maneuvers. *Proceedings of the 1999 Space Control Conference, MIT Lincoln Laboratories.*
14. Gorden, N. J., Salmond, D. J., & Simth, A. F. M. (1993). Novel approach to nonlinear/non-Gaussian Bayesian states estimation. *IEE Proceeding, 140*(2), 107–113.
15. Pitt, M. K., & Shephart, N. (1999). Filtering via simulation: Auxiliary particle filters. *Journal of the American Statistical Association, 94*(446), 590–599.
16. Roudolph, M., Arnaud, D., Nando, F., & Eric Wan (2000). *The unscented particle filter.* Technical Report, Cambrige University Engineering Department.
17. Schaub, H. (2002). *Spacecraft relative orbit geometry description through orbit element differences.* 14th U.S. National Congress of Theoretical and Applied Mechanics Blacksburg, VA.

Part III
Atomic Clock Technique and Time-Frequency System

Design of a Miniaturized Cavity for Space Hydrogen Masers

R. F. Yang, T. Z. Zhou and L. S. Gao

Abstract Due to large bulk and mass, the hydrogen maser in our laboratory at present stage is not suitable for space applications. The reduction of the volume of the microwave cavity is extremely important since it occupies almost 3/4 of the total volume of the hydrogen maser. In this paper, we present recent progress in designing of the sapphire-loaded cavity. Parameters were optimized for minimization of the size and maximization of the quality factor. Simulation by using the finite element method and related experiments were performed. A miniaturized sapphire microwave cavity with the total volume of about 3 dm^3 and the quality factor of 67500 is obtained. The experimental results were well consistent with those of calculation.

1 Introduction

The hydrogen maser atomic clocks operates at the hyperfine transition line of the ground state of hydrogen atoms with frequency of 1.42 GHz or wavelength of 21 cm. Hydrogen atoms with very low density in a bulb located inside a TE011 mode microwave cavity are exposed to the standing wave field in the cavity, thus the transition is stimulated. Hydrogen masers have excellent performance in short- and middle-term frequency stability, and the stability performance has not been surpassed by any other conventional frequency standards [1, 2]. One of the

R. F. Yang (✉) · T. Z. Zhou · L. S. Gao
Science and Technology on Metrology and Calibration Laboratory, 100096 Beijing, China
e-mail: yangrenfu@163.com

R. F. Yang · T. Z. Zhou · L. S. Gao
Beijing Institute of Radio Metrology and Measurement, 100096 Beijing, China

J. Sun et al. (eds.), *China Satellite Navigation Conference (CSNC) 2012 Proceedings*, Lecture Notes in Electrical Engineering 160, DOI: 10.1007/978-3-642-29175-3_43,

important applications of the hydrogen maser is to provide a precise timing reference for the space station [3].

Since the microwave cavity occupies almost 3/4 volume of a whole hydrogen atomic clock, reduction in mass and bulk of the microwave cavity is extremely important to the hydrogen maser for space applications, though these parameters are not very critical on Earth. The sapphire-loaded microwave cavity is a good choice for effectively reducing the dimension of the hydrogen maser, since sapphire can provide higher permittivity than vacuum therefore causes much better energy confinement, and the loss tangent of sapphire at microwave frequencies is among the lowest ones of all known dielectric materials, making it very beneficial to be used in a miniaturized microwave cavity [4]. The use of sapphire-loaded microwave cavity can greatly reduce the total size and weight of a hydrogen atomic clock, without distinctly degrading its stability.

References [5] and [6] reported some helpful researches of sapphire microwave cavity in the Beijing Institute of Radio Metrology and Measurement (BIRMM). The cavity is of a titanium metal cylindrical structure enclosing a sapphire tube, inside which a gas bulb containing hydrogen atoms is located. The total volume and weight of the atomic clock are 0.095 m^3 and about 55 kg, respectively, much smaller than common hydrogen maser atomic clocks [7]. Unfortunately this kind of sapphire hydrogen maser atomic clock is not suitable for space applications, because its bulk and mass are still too large.

Recently we performed a new design of the TE011-mode sapphire-loaded cavity. Main purposes are to minimize the total volume and to maximize the quality factor of the cavity. The performance of the cavity was compared with two other designs, one is made by the National Institute of Information and Communications Technology (NICT, Tokyo, Japan) [8, 9] and the other one is made by the Microwave and Optical Communication Research Institute at the University of limoges, France [10, 11].

2 Structure of the Cavity and Simulation Results

A low-loss TE011-mode microwave cavity was designed, as shown in Fig. 1. Aluminum was used for the outside cylinder due to its nonmagnetic property and low thermal expansion coefficient. A cylindrical sapphire tube was placed in the center of the microwave cavity. The tube is of single-crystal sapphire with crystallographic direction [0001] parallel to the cylinder axis of the cavity.

Figure 2 shows the result of the theoretical calculation based on the finite element simulation. Anisotropic permittivity, and dielectric loss coefficients of sapphire given in [12, 13] were used in the simulation. As shown in Fig. 2, the resonant frequencies decrease nearly linearly with the decrease of the diameter and height of the cavity, while the quality factors increases as the cavity size increases. No sharp change or jump occurs in the whole curves. When the radius of the cavity changes 1 mm, the resonant frequency shifts about 4 MHz correspondingly. When

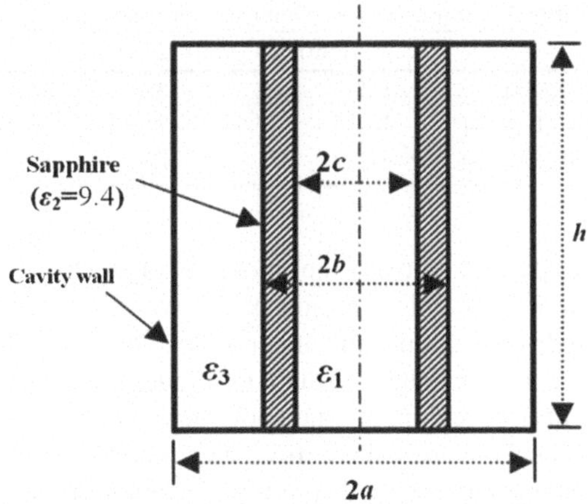

Fig. 1 Cross sectional schematic of a TE011 mode sapphire microwave cavity

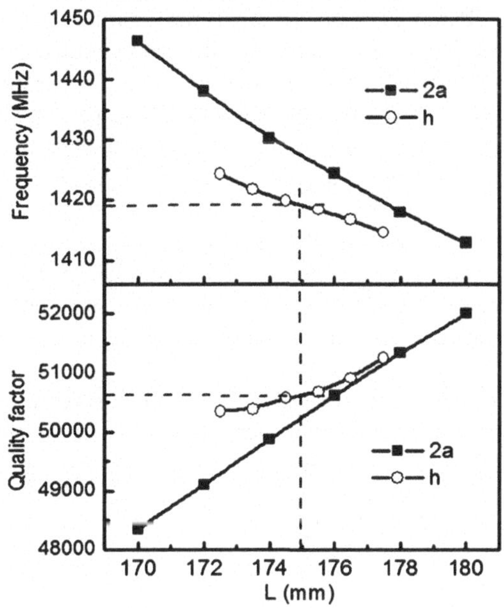

Fig. 2 Resonant frequency and quality factor as function of radius and height of the cavity

the length of cavity is fixed at 175 mm, the resonant frequency is about 1.419 GHz and the quality factor is 50600, which is consistent with the results of our previous designs [5].

For the miniaturized sapphire cavity, both inner and outer radius of sapphire tube must be finely optimized for reducing the cavity size as much as possible [14]. In our previous compact hydrogen maser [5], the thickness of sapphire tube is 6.25 mm and the corresponding resonant frequency is 1.433 GHz. The resonant frequency remains unchanged as long as we reduce the size of the microwave

Table 1 Comparison of parameters of existing three small sapphire microwave cavities

	$2a$ (mm)	$2b$ (mm)	$2c$ (mm)	h (mm)	$d = b - c$ (mm)	V (mm^3)	Q
BIRMM	150	62	38	172	12	3037950	67500
NICT [8, 9]	161.9	87.42	71.82	161.9	7.8	3332960	53380
IRCOM [10, 11]	425.9		25.3	389.3	21	55433122	506000

cavity and increase the thickness of the sapphire tube simultaneously. Because in the hollow center region of the sapphire bulb there is a sapphire bulb containing hydrogen atoms, the inner radius of the sapphire tube cannot be decreased so much. The height of the Al cylinder was chosen to be $h = 172$ mm to ensure the uniform region of magnetic field is the same as that of our previous design. According to a theoretical calculation, when the inner diameter of 38 mm and outer diameter of 62 mm of the sapphire tube, the inner diameter of 150 mm of the microwave cavity are chosen, a resonant frequency of 1.42 GHz can be obtained. The calculated quality factor of the cavity is 67500.

Parameters of our cavity are compared with those of cavities made by NICT and IRCOM in Table 1. The outer diameter of the sapphire tube of our cavity is 62 mm, which is smaller than NICT's design [8]. The diameter and height of cavity are chosen as 150 and 172 mm, respectively, close to the design by NICT [8, 9] but much smaller than the design by IRCOM [11]. Compared with our previous one, the new design realized a 28.6% reduce in cavity volume and a 29.7% increase in quality factor [5, 6]. We believe that the cavity that our new cavity is the smallest among smallest sapphire cavities for hydrogen masers.

The experimental modal of the cavity has been made. The sapphire tube was made of single-crystal material, and was put into a cylindrical aluminum cavity. The two ends of the sapphire tube were sealed by the same material as the tube, with its c-axis aligned to the cylinder axis. The cavity height was 172 mm. The sizes of sapphire and Al cavity are ensured by precise machining process. A test of the cavity has been carried out. When the inner radius of the sapphire tube is 7 mm, the test resonant frequency and quality factor are 1425.115 MHz and 64485, respectively, which are in excellent agreement with the theoretical simulation.

3 Conclusion

In conclusion, a miniaturized TE011 mode sapphire cavity for space borne hydrogen masers was designed. The total size and mass of hydrogen maser atomic clocks can be greatly reduced by using the cavity. Additional advantage of the small cavity is that the single-crystal sapphire for the cavity is very easy in fabrication and is cheap in cost.

References

1. Vessot, R. F. C. (2005). The atomic hydrogen maser oscillator. *Metrologia, 42*, s80–s89.
2. Vessot, R. F. C. (2005). The atomic hydrogen maser oscillator (erratum). *Metrologia, 42*, 468–470.
3. Schweda, H., Zivanov, H. S., Perruchoud, G., Weber, C., Thieme, B., & Baister, G. B. (2007) Performance demonstration of the on-board active hydrogen maser for the ACEs space mission of ESA. In *Proceedings of the IEEE International Frequency Control Symposium* (pp. 1116–1121).
4. Hartnett, J. G., Tobar, M. E., & Ivanov, E. N. (2006). Room temperature measurement of the anisotropic loss tangent of sapphire using the whispering-gallery-mode technique. *IEEE Transactions on Ultrasonics, Ferroelectrics, and Frequency Control, 53*(1), 34–38.
5. Zhang, D. J., Zhang, J. H., Li, C. F., & Wei, J.-l. (2002). Sapphire compact hydrogen frequency standard. *Journal of* Astronautic *Metrology and Measurement*, *22*(3), 1–8 (in Chinese).
6. Wang, N. R., Yang, R. F., Zhou, T. Z., & Gao, L. S. (2008). Frequency-temperature compensated sapphire loaded cavity for compact hydrogen masers. *Metrologia, 45*(5), 534–538.
7. Chuang, C. H., & Jair, T. C. (1980). Atomic time and frequency standards development at Shanghai observatory, China. *IEEE Transactions on Instrumentation and Measurement, 29*(3), 158–163.
8. Morikawa, T., Takahei, K., Uehara, M., Mori, K., & Tsuda, M. (2000). Design analysis of a sapphire loaded cavity for a spaceborne hydrogen maser. In *Proceedings* of *Conference* on *Precision Electromagnetic Measurements Digest, Sydney, Australia* (pp. 50—51).
9. Morikawa, T., Takahei, K., Uehara, M., Mori, K., & Tsuda, M. (2000). Design optimization of a sapphire loaded cavity for a spaceborne hy-drogen maser. In *Proceedings of 14th European Frequency and Time Forum, Torino, Italy* (pp. 462—465).
10. Hartnett, J. G., Tobar, M. E., Cros, D., Krupka, J., & Guillon, P. (2002). High q-factor Bragg-reflection sapphire-loaded cavity TE01δ mode resonators. *IEEE Transactions on Ultrasonics, Ferroelectrics and Frequency Control, 49*(12), 1628–1634.
11. Hartnett, J. G., Tobar, M. E., Stanwix, P. l., Morikawa, T., Cros, D., & Piquet, O. (2005). Designs of a microwave TE011 mode cavity for a space borne h-maser. *IEEE Transactions on Ultrasonics, Ferroelectrics, and Frequency Control, 52*(10), 1638–1643.
12. Krupka, J., Derzakowski, K., Abramowicz, A., Tobar, M. E., & Geyer, R. G. (1999). Use of whispering-gallery modes for complex permittivity determinations of ultra-low-loss dielectric materials. *IEEE Transactions on Microwave Theory and Techniques, 47*(6), 752–759.
13. Tobar, M. E., Krupka, J., Ivanov, E. N., & Woode, R. A. (1998). Anisotropic complex permittivity measurements of monocrystalline rutile between 10 and 300 K. *Journal of Applied Physics, 83*(3), 1604–1609.
14. Hartnett, J. G., Tobar, M. E., Stanwix, P., Morikawa, T., Cros, D., & Piquet, O. (2004). Cavity designs for a space hydrogen maser. In *IEEE International Ultrasonics, Ferroelectrics, and Frequency Control Joint 50th Anniversary Conference* (pp. 608–612).

The Unsymmetrical Delay Compensation in WDM Time Transmission Using Optical Fiber

Xiaofeng Li, Shuangyou Liang, Faxi Chen, Kan Zhao and Shougang Zhang

Abstract The transmission of optical wave in fiber is analyzed, and the asymmetry of transmission delay which results from phenomena of chromatic dispersion is computed under the condition of density WDM (DWDM) and sparse WDM (SWDM). Then we proposed a scheme of DWDM transmission delay compensation in time transmission using optical fiber. The theoretical calculation shows that when calculated on 50 km transmission range, the time difference when using two waves of wavelength difference of 1 nm is nearly three orders of magnitude less than when using two waves of wavelength of 1550 and 1310 nm.

1 Introduction

The technology of high precision and long distance time transmission is very important to elementary physics and metrology. For example, the time synchronization precision of femto-second is needed in the X-ray source of linearity accelerator [1]; the precision of nano-second is also needed between stations using high precision short-baseline interference method; the precision has to reach to

X. Li (✉) · S. Liang · F. Chen · K. Zhao · S. Zhang
National Time Service Center, Chinese Academy of Sciences, Xi'an, 710600 Shaanxi, China
e-mail: lixiaofeng@ntsc.ac.cn

X. Li · S. Liang · F. Chen · K. Zhao · S. Zhang
Key Laboratory of Time and Frequency Primary Standards, National Time Service Center, Chinese Academy of Sciences, Xi'an, 710600 Shaanxi, China

X. Li
Graduate University of Chinese Academy of Sciences, 100049 Beijing, China

J. Sun et al. (eds.), *China Satellite Navigation Conference (CSNC) 2012 Proceedings*, Lecture Notes in Electrical Engineering 160, DOI: 10.1007/978-3-642-29175-3_44,

0.1 ns when the ultra-high precision position is realized using radar network. Otherwise, satellite navigation network, weapon guidance network and long baseline interferometry radio telescope array etc. need high precision and long distance time synchronization. Digital communication technology which is in rapid development also needs this.

The frequency stability is very good when frequency is transmitted in optical fiber, especially short-time stability. Time transmission precision of optical fiber can be the orders of femto-second theoretically. Obviously, other method of time transmission is worse than it. By some papers, to get better effect, time and frequency transmission can be compensated in delay and phase. Some researches in other countries show that frequency stability will be better after compensation. Although there is no research on time compensation, yet by the research on the frequency, the stability of time transmission can also be far better after compensation.

The other advantage of time synchronization using optical fiber is the use of existing optical fiber network, such as internet [2] and SDH network [3]. On the premise of that the normal operation are not effected, it is reasonable to make time synchronization using these optical fiber network. In this case, the cost of optical fiber laying will be very low and the precision is also very high.

2 The Analysis to the Light-Wave Transmission

When two light waves of different wavelength pass through the fiber, their transmission time is different due to the existence of chromatic dispersion. The chromatic dispersion of the normal single mode optical fiber contains material dispersion, waveguide dispersion, and profile dispersion. The profile dispersion can be neglected because the refractive index is constant [4].

Material dispersion, Δdm, can be expressed in formula (1).

$$\Delta dm = \frac{L}{c}\lambda\frac{\partial^2 n(\lambda)}{\partial\lambda^2} \tag{1}$$

where, n is the profile of the optical fiber, λ is the wavelength.

Waveguide dispersion, $\Delta d\omega$, can be expressed in formula (2).

$$\Delta d\omega = -\frac{n_2 \Delta V}{c\lambda}\frac{\partial^2 Vb}{\partial\lambda^2} \tag{2}$$

where, n_2 is Fiber cladding refractive index, b is Standardized propagation coefficient, V is Standardized frequency.

The chromatic dispersion of single mode fiber is the sum of material dispersion and waveguide dispersion. It is shown in Fig. 1.

Obviously, the area which is enclosed by λ_1, λ_2, the horizontal axis and the curve of chromatic dispersion is that the time difference of light-wave λ_1 and λ_2

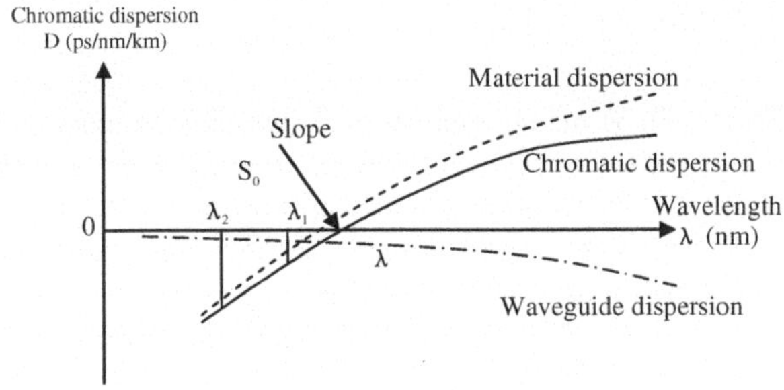

Fig. 1 The relations between wavelength and dispersion of single mode fiber

through the same optical fiber. In the case of approximate calculation, this area can be considered as trapezoid. When the slope near zero dispersion point is S_0, we will get the formula (3) which describes the above-mentioned time difference that is expressed by τ.

$$\tau = \frac{1}{2} L S_0 \left[(\lambda_2 - \lambda_0)^2 - (\lambda_1 - \lambda_0)^2 \right] (\text{ns}) \tag{3}$$

where, L is the length of the optical fiber.

At the same time, because of the influence of temperature, the time difference τ is in change. We can assume that S_0 will not change with the variation of temperature and that the wavelength of zero dispersion point changes only with temperature. So the variation, $\Delta\tau$, of the time difference τ can be described by formula (4).

$$\Delta\tau = L S_0 (\lambda_1 - \lambda_2) \frac{d\lambda_0}{dT} \Delta T (\text{ns}) \tag{4}$$

where, ΔT is the variation of temperature, $d\lambda_0/dT$ is the variation rate of zero dispersion point corresponds to the temperature optical fiber.

In the formulas (3) and (4), L and ΔT are constant for the specific application. And for single mode optical fiber, S_0 is about +0.07 ps/nm^2/km typically, $d\lambda_0/dT$ is about +0.03 nm/°C. Consequently, the choice of wavelength λ_1 and λ_2 is very important. When they are chosen properly, the time difference will be small, and the precision of time synchronization will be high.

We can get formula (5) from formula (3).

$$\tau = \frac{L S_0}{2} (\lambda_2 - \lambda_1)(\lambda_2 + \lambda_1 - 2\lambda_0) (\text{ns}) \tag{5}$$

And for formula (4), L, ΔT, $d\lambda_0/dt$ and S_0 are all constant, so $\Delta\tau$ can be described by formula (6).

$$\Delta\tau = K(\lambda_2 - \lambda_1)(\text{ns}) \tag{6}$$

where, K is constant.

By the formulas (5) and (6), we can come to the conclusion that time difference and its variation have a close relation with the difference of wavelength in a practical application.

2.1 Sparse WDM

Spare WDM means that the wavelength difference of the light-waves spreading in optical fiber is relatively large. Usually, we use the wavelength of 1550 and 1310 nm to make time synchronization.

By the formula (5), when $\lambda_1 = 1550$ nm and $\lambda_2 = 1310$ nm; if the distance of time transmission is 50 km, that is to say, $L = 50$ km; when the temperature has a variation of 10°C, $\Delta T = 10$°C. These data are used in formula (5).

$$\tau = \frac{LS_0}{2}(\lambda_2 - \lambda_1)(\lambda_2 + \lambda_1 - 2\lambda_0) = 100.8(\text{ns})$$

That is to say, when these light-wave through optical fiber, a time difference of 100.8 ns will be produced in 50 km.

Otherwise, these data are used in formula (4).

$$\Delta\tau = S_0(\lambda_1 - \lambda_2)\frac{d\lambda_0}{dT}\Delta T \times L = 0.252(\text{ns})$$

The variation of time difference is about 0.252 ns.

If the distance is 100 km, the time difference is 201.6 ns, and its variation is 0.504 ns. It is a very large error.

2.2 Density WDM

Density DWM means that the wavelength difference of the light-waves transferring in optical fiber is small. For example, when $\lambda_1 = 1549$ nm, $\lambda_2 = 1550$ nm; and the other data is the same with Sect. 2.1, we can get the time difference from the formula (5).

$$\tau = \frac{LS_0}{2}(\lambda_2 - \lambda_1)(\lambda_2 + \lambda_1 - 2\lambda_0) = 0.838(\text{ns})$$

We can also get the variation of time difference from formula (4)

$$\Delta\tau = S_0(\lambda_1 - \lambda_2)\frac{d\lambda_0}{dT}\Delta T \times L = 1.05(\text{ps})$$

And if the distance is 100 km, the time difference is 1.676 ns, its variation is 2.1 ps.

3 Time Synchronization Method

We analyze the time transmission by sparse DWM and density DWM above. Obviously, the time transmission precision depends on the method selected. If the method is suitable, the time transmission complexity would be low and the reliability would be high. We show the chart diagram of time transmission using WDM which is used by us in (Fig. 2).

Fig. 2 The Chart diagram of time Transmission using WDM

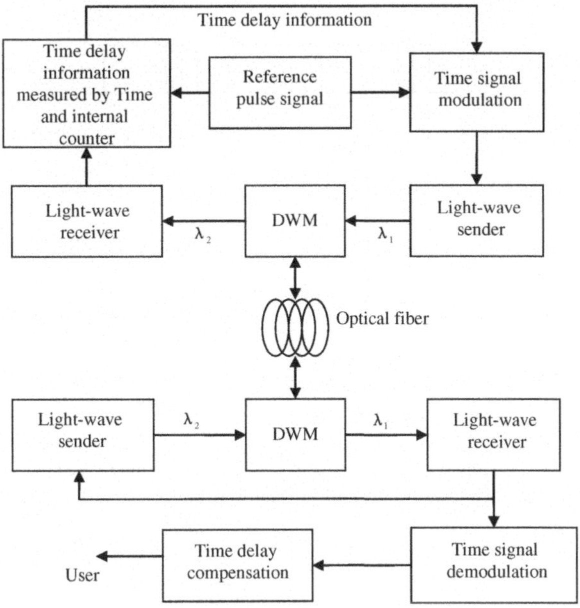

In sender, reference signal containing time information is replicated. One is the reference signal of time and internal counter which is used to measure the time difference between the reference signal and feedback signal; then time and internal counter produce a result of time delay information and it is send to time signal modulation mode. The other is send to time signal modulation mode to transfer the time information with time delay information by light-wave sender and DWM.

In receiver, the DWM gets the light-wave and send it to light-wave receiver. The light-wave receiver converts the light signal to electric signal and sends the electric signal to light-wave sender and time signal demodulation. The output of light-wave sender is transfer to sender by DWM as feedback signal. The output of time signal demodulation mode contains time information and time delay information. At last, the time delay compensation mode send the time signal to user after compensation by the time delay compensation.

The flow chart of time delay compensation is shown in Fig. 3 [5].

4 Conclusion

The precision of time transmission depends on the aim of user. By the result above, in the case of sparse DWM, two light-waves through the same optical fiber will produce the time difference of (100.8 ± 0.252) ns in 50 km and (201.6 ± 0.504) ns in 100 km; while in the case of density DWM, the time difference is (0.838 ± 0.001) ns in 50 km and (1.676 ± 0.002) ns in 100 km.

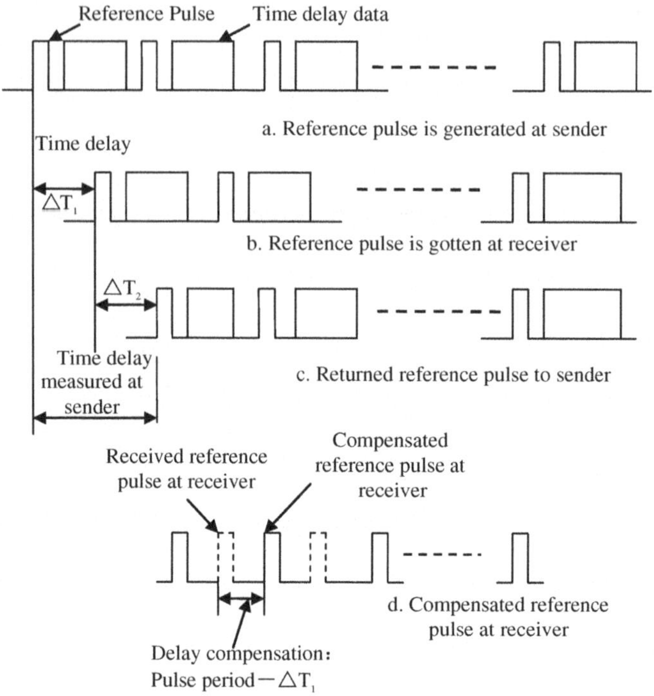

Fig. 3 The flow chart of time delay compensation

Therefore, in the short distance, the precision of time synchronization will be sub-nanosecond using density DWM, and the chromatic dispersion needn't be considered. In the long distance, the demand of the precision of time synchronization is not high, density DWM is enough.

Next, we will carry out experimental work to verify the theoretical derivation and the difference of the actual situation.

References

1. Czuba, K., Eints, F., et al. (2006). *First generation of optical fiber phase reference distribution system for TESLA*. CARE-Report-2006-037-SRF: 1–66.
2. Lopez, O., Haboucha, A., et al. (2010, 2 August). Cascaded multiplexed optical link on a telecommunication network for frequency dissemination. *Optics Express*, *18*(16).
3. Imaoka, A., & Kihara, M. (1997, February). Time signal distribution in communication networks based on synchronous digital hierarchy. *IEEE Transactions on Communications, 45*(2).
4. Tang, L. (1999). *The communication of optical fiber* (pp. 13–15). Shanghai: Fudan University Press.
5. Kihara, M., & Imaoka, A. (1996). *System configuration for standardizing SDH-based time and frequency transfer*. European Frequency Time Forum, 5–7 March, 1996, Conference Publication No. 418.

Time Scales and Time Transformations Among Satellite Navigation Systems

Pengfei Zhang, Chengdong Xu, Chunsheng Hu and Ye Chen

Abstract The definitions of time scales and the time transformations are very important for Global Navigation Satellite Systems (GNSSs) research. Different GNSSs use different internal reference time systems which include GPS Time, GLONASS Time, GALILEO System Time and BeiDou System Time. With the development of GNSSs, it is significant to set up the relationships of time scales among different GNSSs to improve the compatibility and interoperability. This paper concludes the definitions of different time scales and describes the time systems in different GNSSs firstly. Secondly, the relationships among different time scales are analyzed according to the data published by Bureau International des Poids et Mesures (BIPM) in recent years, and time transformations among different time systems are presented. Finally, these transformations are tested and validated, and the result shows that they can satisfy the accuracy requirements for most users.

1 Introduction

Navigation satellites move around the earth in high speed, for example, the speed of a GPS satellite is about 3.9 km/s. The observation time error should be less than 2.6 μs when the position error of the satellite is suggested to be <1 cm at that

P. Zhang (✉) · C. Xu · C. Hu
School of Aerospace Engineering, Beijing Institute of Technology, No.5 South Zhongguancun Street, Haidian District, Beijing 100081, China
e-mail: successful.2008@163.com

Y. Chen
School of Information and Communication Engineering,
North University of China, No.3 Xueyuan Road, Taiyuan, 030051 Shanxi, China

J. Sun et al. (eds.), *China Satellite Navigation Conference (CSNC) 2012 Proceedings*,
Lecture Notes in Electrical Engineering 160, DOI: 10.1007/978-3-642-29175-3_45,

moment [1]. 1 μs propagation time error of the signal will cause 300 m measurement distance error. If the distance needs to be measured in meter level precision, the time should be measured in nanosecond level precision [2]. The satellite signal transmitting time takes satellite navigation system time as a reference; however, the time of the receiver takes Coordinated Universal Time (UTC) as a reference. So the two time scales should be unified to meet high accuracy measurement requirements of signal propagation time. In addition, in the same satellite navigation system, it is also very important to achieve time synchronization among all the satellites for precision positioning [3]. With the development of Global Navigation Satellite Systems (GNSSs), positioning using different GNSSs has become an inevitable trend. Different GNSSs use their own internal reference time systems, so time transformations among different time systems should also be considered. This paper introduces the definitions of different time scales including astronomical time, atomic time and UTC, and it presents the relationships among them firstly. Secondly, it describes the time systems in different GNSSs including GPS Time (GPST), GLONASS Time (GLST), GALILEO System Time (GST) and BeiDou System Time (BDT), and then it shows the relationships between each time system and UTC. Finally, it derives time transformations which include both the integral part and the fractional part of the difference among different time systems.

2 Time Scales

The establishment of a time scale should take both a time starting point and a time span for consideration. The time span can make use of a periodic motion phenomenon which is repeatable observational, continuous and stable as a reference, such as the pendulum swinging, the earth rotation and the crystal oscillation [4].

2.1 Astronomical Time

Astronomical time includes Sidereal Time (ST), Solar Time (SOT), Universal Time (UT) and Ephemeris Time (ET).

2.1.1 Sidereal Time and Solar Time

Both ST and SOT are time scales based on the earth rotation. An apparent solar day is defined as the interval between two successive returns of the sun to the local meridian. Since the orbit which the earth rotates around the sun is not a circle, and the earth's axis is not strictly vertical with the rotating orbit plane, mean SOT is introduced. An apparent sidereal day is defined as the interval between two successive returns of a fixed star which is far from the earth to the local meridian. In fact, an apparent sidereal day is non-constant, so mean ST is also introduced. Mean ST and mean SOT has the following relationships [1]:

$$\begin{aligned} 1\, mean\, solar\, day &= 24\, mean\, solar\, hours = 86400\, mean\, solar\, seconds \\ &\approx 1.002737\, mean\, sidereal\, days \end{aligned} \tag{1}$$

$$\begin{aligned} 1\, mean\, sidereal\, day &\approx 0.997270\, mean\, solar\, day \\ &\approx 86164.095563\, mean\, solar\, seconds \end{aligned} \tag{2}$$

Both ST and SOT depend on the longitude of the observers. Mean SOT of Greenwich meridian is shorted for GMT. The earth was divided into 12 standard time zones. The time of each zone is equal to the mean SOT of the central meridian in this zone, the time difference between each zone and GMT can be measured by integral number of hours.

Julian Day (JD) is used to record mean solar day, which starts at 12 h January 1, 4713 BC. In order to operate the data conveniently, Modified Julian Day (MJD) was applied, the starting point is 0 h November 17, 1858 GMT [4].

2.1.2 Universal Time

UT has the same scale as mean SOT. According to the conditions whether the earth polar motion is modified or the seasonal change of the earth rotation rate is corrected, UT can be divided into UT0, UT1 and UT2 [3].

UT0: Mean SOT of Greenwich meridian which is achieved by several observatories.

UT1: UT0 adds the observation influence $\Delta\lambda$ (up to 0.06 s) which is caused by the earth polar motion.

UT2: UT1 adds the seasonal correction of the earth rotation rate ΔTs.

2.1.3 Ephemeris Time

ET is a time scale based on the earth revolution around the sun, and it is not influenced by the earth polar motion and the seasonal change of the earth rotation rate which are unpredictable. An ephemeris second is defined as 1/31556925.9747 of the tropical year at 1900 [3].

$$ET = UT2 + \Delta T \tag{3}$$

ΔT can be only decided by observation, and it can not be deduced by any formula. Obviously, this definition of the time scale is fussy and difficult to operate. In 1976, the International Astronomical Union decided to replace ET with

Table 1 The definition progress of one second

Year	One second definition
Before 1960	1/86400 mean solar day
1960–1967	One ephemeris second
Since 1967	One international atomic second

Terrestrial Dynamical Time (TDT) and Barycentric Dynamical Time (TDB). In 1991, TDT was renamed as terrestrial time.

2.2 Atomic Time

Atomic time is a time scale base on the resonant frequency of Cesium atoms. In 1967, an atomic second was defined as the duration of 9, 192, 631, 770 periods of the radiation corresponding to the transition between two hyperfine levels of the ground state of the Cesium133 atom [2]. The time scale which derives from the definition is called International Atomic Time (TAI). It is more stable and easier to operate than ET, so the ephemeris second is replaced by the atomic second as the basic unit of time measurement. The starting point of the TAI which is established by the United States Naval Observatory (USNO) is 0 h January 1, 1958 UT2. The starting point of TAI which is established by Bureau International de l'Heure (BIH) is 34 ms earlier than that of USNO [1].

The definition progress of one second is shown in Table 1.

2.3 Coordinated Universal Time

TAI has nothing to do with the earth rotation. However, it is necessary for satellite navigation to link the time with the earth rotation. Therefore, a compromise time scale—UTC is proposed. UTC is a time scale based on atomic second, and it is as close as possible to UT1. It has been adopted since 1972. When the time difference between UTC and UT1 exceeds ±0.9 s, one leap second (positive or negative) will be added or subtracted to UTC such that UTC is the closest to UT1. By now, the leap seconds which have been introduced are all positive.

$$|UT1 - UTC| < 0.9\,s \tag{4}$$

$$UTC - TAI = 1\,s \times n \tag{5}$$

where n is an adjusted parameter which is equal to the total leap seconds. The International earth rotation and reference system service (IERS) are responsible for definitively bulletining n and the difference between UT1 and UTC [5]. The difference between UT1 and UTC from year 1991 to 2011 is shown in Fig. 1.

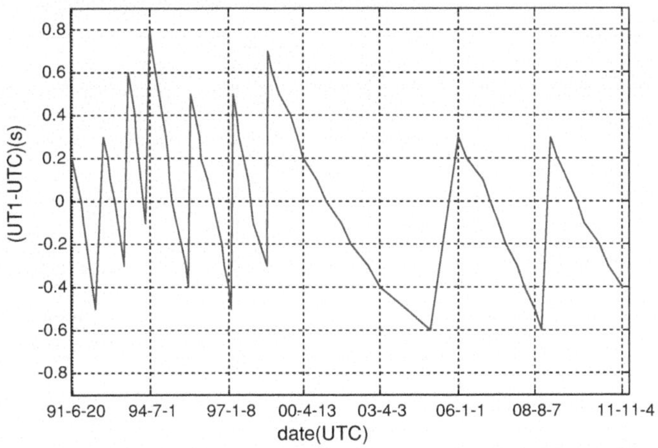

Fig. 1 Curve: [UT1-UTC] from year 1991 to 2011

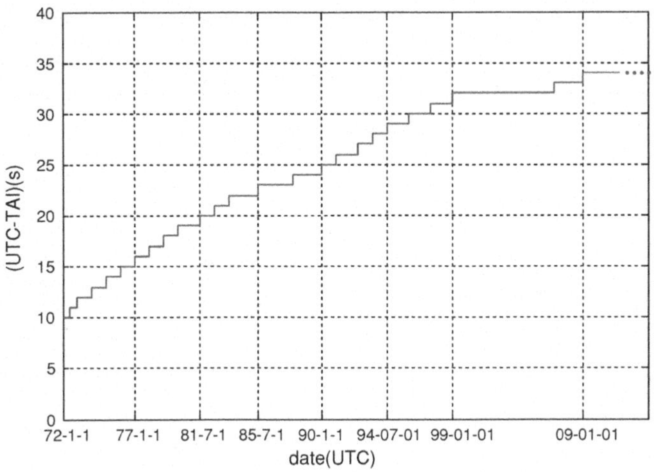

Fig. 2 Curve: leap seconds of UTC

It can be seen that the value of [UT1-UTC] is discontinuous, because a leap second was introduced in the date when a inflection point appears in the curve so that the difference between UTC and UT1 is no more than ±0.9 s.

The latest time when the leap second was loaded is 23 h 59 min 59 s December 31, 2008 UTC. It is to say the next two seconds of UTC are ordered as follows: 23 h 59 min 60 s December 31, 2008 UTC and 0 h 0 min 0 s January 1, 2009 UTC. So far, n has been equal to 34 [5]. Figure 2 shows the leap seconds of UTC since 1972.

UTC is a ‘paper’ time scale, but it is approximated by local physical representations UTC (k) through about 250 cesium atomic clocks and hydrogen

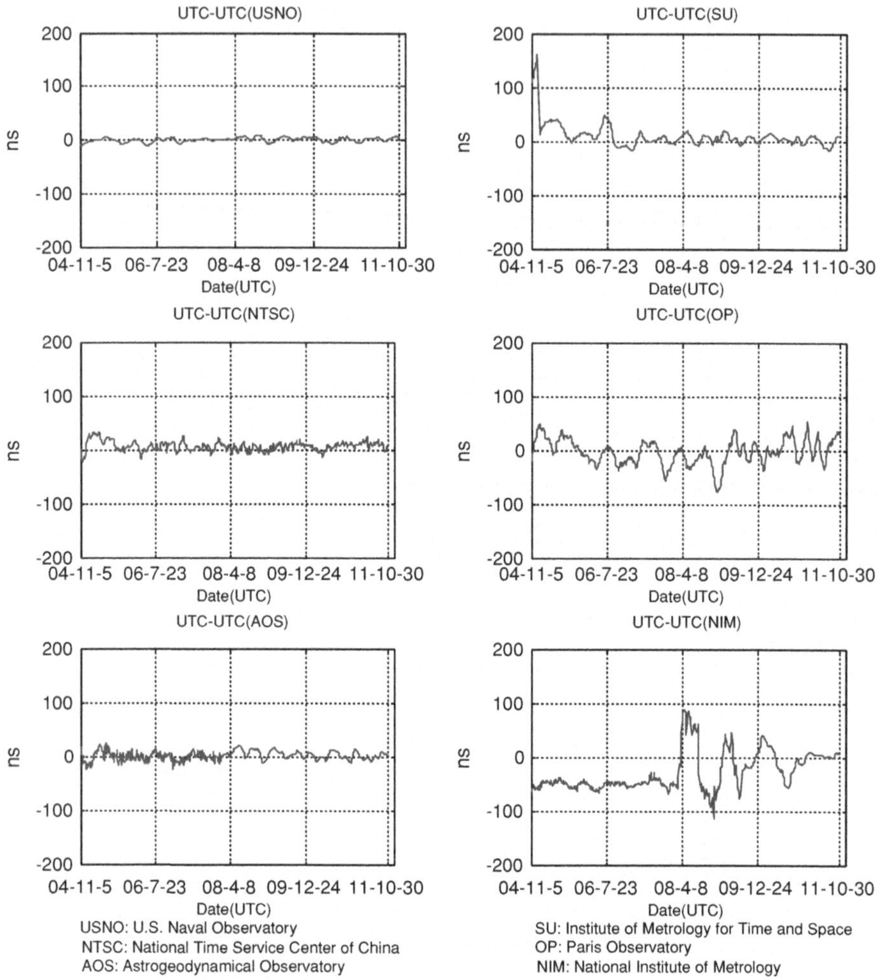

Fig. 3 Curve: [UTC–UTC (*k*)] from year 2004 to 2011

microwave blaster devices in national metrology laboratories and observatories that contribute to the formation of the international time scales at the Bureau international des poids et mesures (BIPM) [6]. The difference between UTC and UTC (*k*) is provided by the publication in the monthly BIPM Circular *T*.

Figure 3 shows some examples about the difference between UTC and UTC(*k*).

Among these local physical representations UTC (*k*), UTC (USNO) is the most closest to UTC. The difference between UTC (USNO) and UTC is kept within ± 10 ns. GNSSs rely on UTC (*k*) which is provided by local metrology laboratories and observatories, so atomic clocks technology with high accuracy is researched to be more closer to UTC by all the local metrology laboratories and observatories.

3 GNSS Time

In order to meet the needs of precision positioning and navigation, each GNSS creates its own internal reference time system. All these systems, which include GPST, GLST, GST and BDT, should be ensured as close as possible to UTC, even the leap second of UTC.

3.1 GPS Time

The length of a second in GPST, which is measured by the GPS satellite atomic clocks and the ground station clocks, is consistent with that of TAI. GPST is a continuous time system, so it does not have to be adjusted by leap second. The starting point of GPST is consistent with 0 h 0 min 0 s January 6, 1980 UTC (USNO). An epoch in GPST is distinguished by the number of seconds that have elapsed since saturday/sunday midnight and the GPS week number. The GPS number will return to zero when the week's number are up to 1024. The total number of seconds in a week is 604,800 [1]. The conversion formula between GPST and UTC is as follows:

$$\begin{aligned} GPST &= UTC(USNO) + t_1 - t_2 = UTC - t_3 + t_1 - t_2 \\ &= UTC + t_1 - (t_2 + t_3) = UTC + t_1 - \Delta t \end{aligned} \tag{6}$$

t_1 ($t_1 = 1s \times n - 19s$) which is 15 by now is the integral part of difference in seconds between GPST and UTC (USNO), it is caused by the continuity of GPST and the leap second of UTC. t_2 is the fractional part of difference in seconds between GPST and UTC (USNO). t_3 is the fractional part of difference in seconds between UTC and UTC (USNO). The change of t_3 is described in Fig. 3. $\Delta t(\Delta t = t_2 + t_3)$ which is the fractional part of difference in seconds between UTC and GPST is broadcasted by the GPS satellite navigation message and published afterwards by BIPM [7]. The change of Δt from 2009 to 2010 is described in Fig. 4.

From the above curve, the Δt from 2009 to 2010 is controlled within ±20 ns.

3.2 GLONASS Time

The length of a second in GLST, which is based on UTC (SU) and is 3 h earlier than UTC (SU) because of the specific characteristic of GLONASS, is consistent with that of TAI. The GLST is not a continuous time system, so it has to be adjusted by leap second. And the navigation capability of GLONASS will be affected because of the discontinuous time caused by leap second. The starting

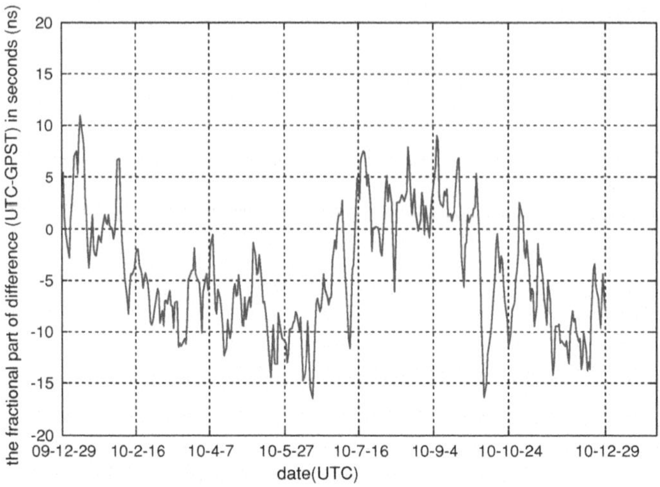

Fig. 4 Curve: the fractional part difference [UTC-GPST] in seconds from year 2009 to 2010

point of GLST is 0 h 0 min 0 s January 1, 1996 UTC (SU). An epoch in GLST, which starts from the last leap year, is described by accumulated days and the number of seconds in less than 1 day. The maximum number of accumulated days in one cycle is 1461, and the number of the total seconds which is started from the midnight between the previous day and the very day is 86400. The conversion formula between GLST and UTC is as follows:

$$\begin{aligned} GLST &= UTC(SU) + 3\text{h} - t_1' = UTC - t_2' + 3\text{h} - t_1' \\ &= UTC + 3\text{h} - (t_1' + t_2') = UTC + 3\text{h} - \Delta t' \end{aligned} \tag{7}$$

t_1' is the fractional part of difference in seconds between GLST and UTC (SU). Because GLST and UTC (SU) use the same timing method including its leap seconds, so there is no integral part of difference in seconds between them. t_2' is the fractional part of difference in seconds between UTC and UTC (SU). The change of t_2' is described in Fig. 3 $\Delta t'(\Delta t' = t_1' + t_2')$ which is the fractional part of difference in seconds between UTC and GLST is broadcasted by the GLONASS satellite navigation message and published afterwards by BIPM [7]. The change of $\Delta t'$ from 2009 to 2010 is described in Fig. 5.

From the above curve, $\Delta t'$ from 2009 to 2010 reaches hundreds of ns. Compared with Fig. 4, the magnitude of $\Delta t'$ is significantly larger than the magnitude of Δt.

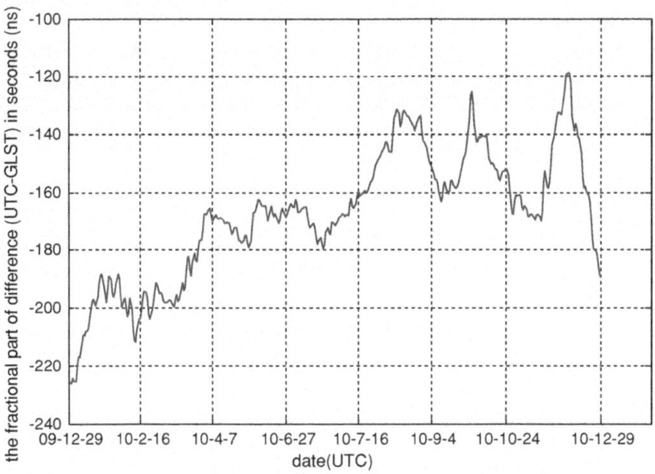

Fig. 5 Curve: the fractional part difference [UTC-GLST] in seconds from year 2009 to 2010

3.3 GALILEO System Time

The length of a second in GST which is a continuous time system is consistent with that of TAI. The deviation of GST and TAI are controlled within ±50 ns in 95% time in one year [8]. The differences between GST and TAI, GST and UTC will be broadcasted by GALILEO satellite navigation message.

To improve compatibility and interoperability with the GPS, the starting point of GST is set to 0 h 0 min 0 s January 6, 1980 UTC (USNO). An epoch in GST is similar to that of GPST, but the number of accumulated weeks will return to zero when the weeks number up to 4,096. The difference between GST and UTC includes integral part and fractional part, as shown in formula (8):

$$GST = UTC + t_1 - \Delta t'' \tag{8}$$

t_1 which equals to the t_1 in formula (6) is 15 s by now, and it is the integral part of difference in seconds between GST and UTC. $\Delta t''$ which is the fractional part of difference in seconds between GST and UTC will be broadcasted by the GALILEO satellite navigation message. By now, there is no public institutions bulletin these values.

3.4 BeiDou System Time

The length of a second in BDT which is a continuous time system is consistent with that of TAI, and there is integral part of difference in seconds between BDT and UTC. The starting point of BDT is set to 0 h 0 min 0 s January 1, 2006 UTC

(NTSC) which maintained by NTSC. An epoch in BDT is the same as an epoch in GPST [9]. The conversion formula between BDT and UTC is as follows:

$$\begin{aligned} BDT &= UTC(NTSC) + t_1''' - t_2''' = UTC - t_3''' + t_1''' - t_2''' \\ &= UTC + t_1''' - (t_2''' + t_3''') = UTC + t_1''' - \Delta t''' \end{aligned} \tag{9}$$

$t_1'''(t_1''' - 1\text{s} \times n - 33\text{s})$ which is 1(n = 34) by now is the integral part of difference in seconds between BDT and UTC. $\Delta t'''$, which is generated from the time and frequency system in master station and be published by NTSC, is the fractional part of difference in seconds between BDT and UTC. $\Delta t'''$ is controlled within ±100 ns. The change of $\Delta t'''$ is given from June 12, 2010 to August 26, 2010 in the reference [9].

4 Time Transformations Among GNSSs

With the development of GNSSs, it is significant to set up the relationships of time scales among different GNSSs to improve the compatibility and interoperability [10]. According to formulas (6)–(9) about the relationships between every GNSS time and UTC, the relationships among four GNSSs can be calculated as follows.

(1) The conversion formula between GPST and GLST:

$$GPST - GLST = t_1 - 3\text{h} - \Delta t + \Delta t'. \tag{10}$$

(2) The conversion formula between GPST and GST:

$$GPST - GST = \Delta t'' - \Delta t. \tag{11}$$

(3) The conversion formula between GPST and BDT:

$$GPST - BDT = t_1 - t_1''' - \Delta t + \Delta t'''. \tag{12}$$

(4) The conversion formula between GLST and GST:

$$GLST - GST = 3\text{h} - t_1 - \Delta t' + \Delta t''. \tag{13}$$

(5) The conversion formula between GLST and BDT:

$$GLST - BDT = 3\text{h} - t_1''' - \Delta t' + \Delta t'''. \tag{14}$$

(6) The conversion formula between GST and BDT:

$$GST - BDT = t_1 - t_1''' - \Delta t'' + \Delta t'''. \tag{15}$$

Table 2 The integral part of the difference in seconds among four GNSS times

	GPST	GLST	GST	BDT
GPST	0			
GLST	15 s–3 h	0		
GST	0	3 h–15 s	0	
BDT	14 s	3 h–1 s	14 s	0

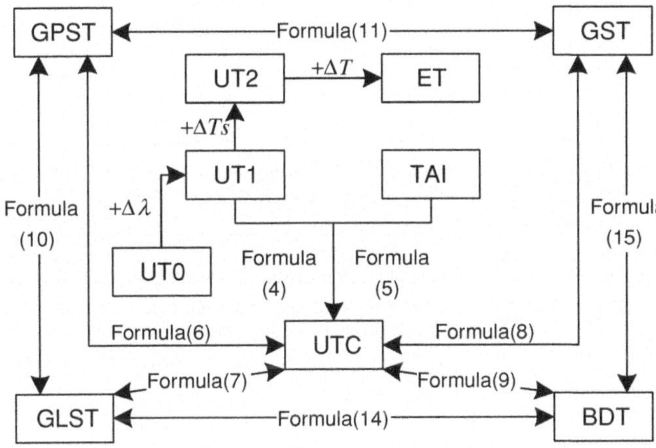

Fig. 6 The transformations among time scales and GNSS times

The difference in seconds among these GNSS times consists of integral part difference and fractional part difference. The integral part difference will change only while the leap second is adjusted. The fractional part difference is broadcasted by navigation messages of GNSSs or bulletined by related time service centers. So far, the integral part of the difference in seconds among four GNSS times is shown in Table 2.

According to the formulas which are derived above, it is convenient for time transformations among the four GNSS times and GNSSs simulation.

5 Conclusion

This paper introduces the time scales which are commonly used in GNSSs and the relationships among them. In addition, this paper describes the internal reference time systems of the four GNSSs and derives the relationships among them. Figure 6 shows the transformations among time scales and GNSS times.

GNSS times can convert to one another accurately according to the fractional part difference of seconds which is broadcasted by navigation messages and the formulas which are derived in this paper. Thus, it is convenient for processing the data of GNSSs. These formulas have been applied in GNSSs simulation, and the accuracy of them can meet the requirements.

Acknowledgments This work was supported by the National High-Tech. R&D Program, China (No.2011AA120505) and the National Natural Science Foundation, China (No.61173077).

References

1. Misra, P., & Enge, P. (2006). *Global positioning system, signals, measurements, and performance*. (2nd edn, pp. 81–89). London: Artech House Publisher.
2. Arias, E. F., Panfilo, G., & Petit, G. (2011). Timescales at the BIPM. *Metrologia, 48*, S145–S153.
3. Lewandowski, W., & Arias, E. F. (2011). GNSS times and UTC. *Metrologia, 48*, S219–S224.
4. Wang., A. (2010) *GNSS measurement data processing*. (pp. 23–30). Xuzhou: China University of Mining and Technology Press(In Chinese).
5. IERS Bulletin D [DB/OL]. Retrieved from http://www.iers.org/nn_10970/IERS/EN/Publications/Bulletins/bulletins.html?__nnn=true.
6. BIPM Circular T [DB/OL]. Retrieved from http://www.bipm.org/jsp/en/TimeFtp.jsp?TypePub=publication.
7. BIPM [TAI–GPS time] and [UTC–GPS time], [TAI–GLONASS time] and [UTC–GLONASS time] [DB/OL]. Retrieved from http://www.bipm.org/jsp/en/TimeFtp.jsp?TypePub=scale.
8. Chen, X., Fang, Y., Yin, J., Zhang, H. (2005). *Galileo satellite navigation system*. (p. 45). Beijing: Peking University Press (In Chinese).
9. Han, C., Yang, Y., & Cai, Z. (2011). BeiDou navigation satellite system and its time scales. *Metrologia, 48*, S213–S218.
10. Liu, Q., Bao, H., Wang, H., Wang, Q. (2008). Time transformation and coordinates transformation among GPS, GLONASS and GALILEO. *Journal of Science of Surveying and Mapping*, 13–15 (In Chinese).

Comparison of Short-Term Stability Estimation Methods of GNSS On-Board Clock

Hang Gong, Wenke Yang, Yong Wang, Xiangwei Zhu and Feixue Wang

Abstract The short-term frequency stability of satellite on-board clock is of great significance for GNSS applications, such as satellite clock modeling, simulation and clock offset prediction. In this paper, four short-term frequency stability estimation methods are analyzed, which need observation data from only one station in contrast with ODTS needing continuous observation of large ground monitoring networks. Performance and accuracy of all these four methods are analyzed and compared using GPS and COMPASS observation data, and their advantages and disadvantages are discussed correspondingly. For COMPASS and Galileo systems which are currently in the early stage of construction, this paper proposes some meaningful suggestions of on-board clock short-term frequency stability estimation.

1 Introduction

In satellite navigation system, performance estimation of the on-board atomic clocks will affect the accuracy of satellite clock offset modeling and prediction, and then impact user positioning and timing accuracy. On-board clock performance evaluation is of great significance for the control segment. In the application such as satellite clock offset modeling, simulation and prediction, not only characteristics determined by the medium- and long-term stability need to be known, short-term stability parameters of satellite clock also need to be obtained.

H. Gong (✉) · W. Yang · Y. Wang · X. Zhu · F. Wang
Satellite Navigation R&D Center, School of Electronic Science and Engineering,
National University of Defense Technology, Changsha, 410073 Hunan, China
e-mail: gonghang@nudt.edu.cn

J. Sun et al. (eds.), *China Satellite Navigation Conference (CSNC) 2012 Proceedings*,
Lecture Notes in Electrical Engineering 160, DOI: 10.1007/978-3-642-29175-3_46,

Similar to satellite orbit determination, satellite clock offset solution is very complex. It is usually carried out by the Orbit Determination and Time Synchronization (ODTS) method, which needs continuous observation of large ground monitoring networks and complex algorithms. Precise orbit and clock products provided by IGS analysis centers are derived from post-processing using ODTS method [1].

GPS satellite clock stability estimation is usually implemented using IGS precise clock data at present. The sampling interval of 30 s typically, even the sampling interval of 5 s provided by the Analysis Center for Orbit Determination in Europe (COD), can not meet the need of shorter sampling intervals (e.g. 1 s) frequency stability estimation. On the other hand, IGS products are only available for GPS and GLONASS system currently. COMPASS and Galileo system are in their early stage of construction, and there are no globally distributed stations for COMPASS to obtain continuous satellite tracking. In addition, only a few agencies have the measurement data of regional network due to access permission. For COMPASS and Galileo system, there are certain difficulties for common users to obtain precise satellite clock parameters presently.

In this paper, four simple short-term stability estimation methods are discussed. These methods can be implemented using a single station's measurement rather than network observation, and it is only required that the stability of the monitoring receiver clock is better than satellite clock. Compared with IGS clock products, these methods are validated using GPS data, and their estimation accuracy and application conditions are discussed at the end.

2 Single Station Estimation Methods of Satellite Clock Stability

For satellite j, the pseudorange and carrier phase observation equations of receiver k are as follows:

$$\rho_k^j = d_k^j + c\left(\Delta t_S^j - \Delta t_R^k\right) + T_k^j + s_k^j + t_j + r_k + m_k^j + \varepsilon_\rho^j \tag{2.1}$$

$$\varphi_k^i = d_k^j + c\left(\Delta t_S^j - \Delta t_R^k\right) + T_k^j + N_k^j\lambda + \varphi_0^j + s_k^j + t_j + r_k + m_k^j + \varepsilon_\varphi^j \tag{2.2}$$

where ρ_k^j is pseudorange; φ_k^i is carrier phase observation; d_k^j is geometric distance of satellite and receiver; Δt_S^j is satellite clock offset; Δt_R^k is receiver clock offset; T_k^j is atmospheric propagation delay, including the ionosphere, troposphere path delay, etc.; $N_k^j\lambda$ and φ_0^j are integral and fractional ambiguity of carrier phase; s_k^j is error caused by the Sagnac effect; t_j is transmission delay of satellite; r_k is receive delay of receiver; m_k^j is multipath error; ε_ρ^j and ε_φ^j are measurement noise of pseudorange and carrier phase.

Like one-way time synchronization method, Eqs. (2.1) and (2.2) show that, the following variables should be corrected before the computation of satellite clock offset:

1. Geometric distance d_k^j between satellite and receiver;
2. Receiver clock offset Δt_R^k;
3. Atmospheric propagation delay T_k^j , Sagnac effect s_k^j, and multipath error m_k^j;
4. Measurement noise ε_ρ^j and ε_φ^j.

Among these variables, tropospheric delay and Sagnac effect can be determined according to respective models; ionospheric delay can be determined by model or dual-frequency combination method; multipath error can be reduced by improving receiver antenna environment and multipath rejection algorithm; measurement noise is high frequency noise, and it will affect the short-term stability estimation accuracy. Considering that short-term stability computation only require relative variety of clock offset rather than its true value, carrier phase observation with lower measurement noise is preferred, and Eq. (2.2) is used.

Because only relative clock offset is needed to estimate satellite clock short-term stability, we can determine the on-board clock offset with respect to receiver clock. In this case, satellite clock stability can be obtained as long as the stability of receiver clock is better than that of satellite clock. It can be achieved easily, such as a receiver with active hydrogen maser reference. If it works in the other way, when the satellite clock stability is better than receiver, this method also provides a receiver clock stability estimation approach.

After all the above variables are corrected, the satellite on-board clock offset with respect to receiver clock is as follows:

$$\Delta t_k^j = \frac{1}{c}\left(\varphi_k^j - d_k^j\right) + \Delta\varepsilon_\varphi^j \tag{2.3}$$

Where $\Delta\varepsilon_\varphi^j$ is residual error. Receiver can obtain carrier phase observation φ_k^i with high sampling rate. According to Eq. (2.3), satellite clock offset with the same sampling rate can be obtained, as long as geometric distance of satellite and receiver is determined beforehand. The short-term stability of on-board clock can be estimated consequently.

Since receiver position can be accurately calibrated, the main problem to determine the geometric distance is how to obtain precise satellite position. There are several ways to determine the satellite position. In the following parts of this paper, the principles of several methods are described, and then the performance of these methods are analyzed and compared.

3 Comparison of Several Estimation Methods

There are three primary approaches to reduce calculation errors of geometric distance of satellite and receiver: firstly, we can obtain accurate satellite position through IGS precise ephemeris products or corrected broadcast ephemeris, etc.; secondly, we can accurately determine the geometric distance, such as using satellite laser ranging; thirdly, we can try to eliminate the effect of geometric distance error by two-way ranging or observation fitting, etc. These methods are described and compared with each other in the following five parts.

3.1 Estimation Method Based on the Precise Ephemeris

For GPS and GLONASS systems, the accurate satellite position can be obtained directly from IGS precise ephemeris product. IGS final ephemeris has the accuracy of 1–2 cm with sampling interval of 15 min. Interpolation is required to obtain satellite position with higher sampling rate. Using windowed trigonometric and polynomial interpolation algorithm [2], the interpolation error can be less than 1 mm. Its impact on short-term stability estimation can be ignored.

The estimation method based on precise ephemeris is carried out as follows [3, 4]:

1. The transmission time of satellite for each observation is calculated;
2. Satellite position is calculated by precise ephemeris interpolation;
3. Geometric distance of satellite and receiver for each observation is calculated;
4. Carrier cycle slips of carrier phase observations, atmospheric propagation delay, Sagnac effect, and multipath errors are corrected;
5. The satellite on-board clock offset with respect to receiver clock is calculated using Eq. (2.3), and its short-term stability is computed consequently.

We use carrier phase observations of the USNO Stations to verify this method. This station provides GPS measurements at the rate of 1 Hz, and its receiver clock refers to a hydrogen maser which is disciplined by UTC (USNO). Because UTC (USNO) is very close to GPS system time (less than 2 ns during the experiment), the satellite position error introduced by the error of satellite transmission time is minimized. Figure 1 shows the stability of GPS PRN14 on-board clock stability obtained by this method and compared to the results of COD 5 s sampling interval precision clock product in the same period. The experiment period is from GPST 2010-9-10 06:00:00 to GPST 2010-9-10 12:00:00.

The above result shows that, the short-term stability estimation based on precise ephemeris is consistent with IGS precise clock product within the sampling interval of 5–1000 s, and the calculated results from L1, L2 and ionosphere free combination of measurements are basically the same. Ionosphere free combination result has a slight deterioration within the sampling interval of less than 10 s due to measurement noise magnified. For the sake of simplicity, the following discussions are base on L1 carrier measurements.

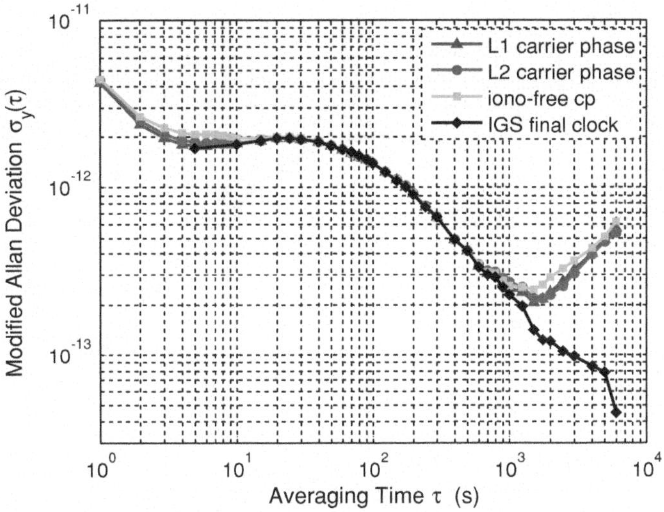

Fig. 1 Frequency stability of GPS PRN14 based on the precise ephemeris

3.2 Estimation Method Based on Smoothed Broadcast Ephemeris

As to COMPASS and Galileo, there is no precise ephemeris available presently but broadcast ephemeris. Broadcast ephemeris is predicted results of the post-processed satellite position data using ODTS method. The prediction error of the satellite orbit both in cross-track and along-track direction normally is 5–10 m [5]. But the prediction error in radial direction which changes relatively slow is much smaller for it is fully observed, with only 1/3 to 1/4 of the error in cross-track and along-track directions [5], which does not affect the estimation of the short-term stability of the satellite clock. Therefore, without precise ephemeris, we can try to use broadcast ephemeris to estimate the stability.

Taking GPS as an example, the broadcast ephemeris is updated every 2 h. Each group of ephemeris parameters is available for 4 h centered at Time of Ephemeris (TOE). The accuracy is highest at TOE, and gradually reduces with the aging of ephemeris. Each group of ephemeris parameters is a combination of forecasted fitting results using data of a specific time range. Thus the calculated satellite orbit parameters according to each group of ephemeris parameters jump at the time of ephemeris parameters update. In order to reduce the impact of the jump on the short-term stability estimation, it is needed to smooth the broadcast ephemeris.

For each epoch, there are two groups of ephemeris parameters available. It is possible to use the adjacent two groups of broadcast ephemeris parameters to calculate the satellite positions in order to avoid satellite orbit jumps. As both groups of ephemeris parameters are available at the same time, such smoothing does not introduce additional error. As the predicted error of each group of

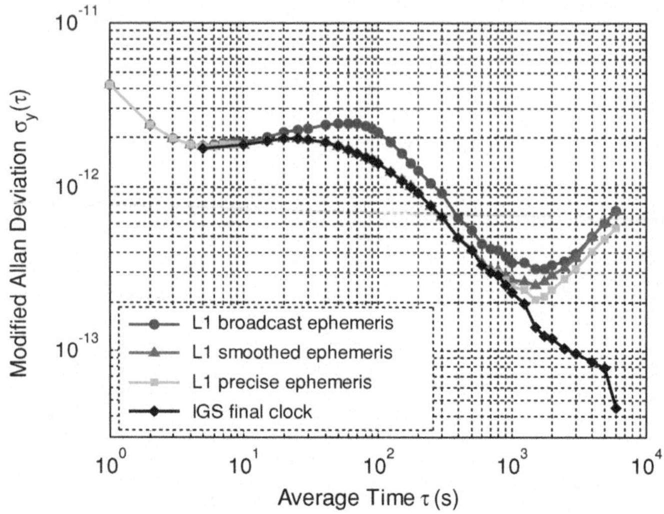

Fig. 2 Frequency stability of GPS PRN14 based on smoothed broadcast ephemeris

broadcasted ephemeris is roughly the same, it is suggested here to smooth broadcast ephemeris using linear weighted method according to the ephemeris age, namely:

$$p = \frac{t_f - t}{t_f - t_b} \cdot p_b + \frac{t - t_b}{t_f - t_b} \cdot p_f \tag{3.1}$$

where t is the observation epoch, t_b and t_f are TOE for the two adjacent groups of broadcast ephemeris respectively, and P_b and P_f are satellite positions calculated from the above two groups of broadcast ephemeris for t. Experiments show that the linear weighted smooth method eliminates the jumps of calculated satellite orbit, which helps to improve the estimation accuracy of short-term stability.

Based on the same observation data of GPS L1 carrier phase, the short-term stability of GPS PRN14 on-board clock is estimated using smoothed broadcast ephemeris with linear weighted method as in Fig. 2.

As to COMPASS, there is no precise clock available to validate the estimation results. The observation data obtained from two monitoring stations (Beijing and Sanya) within the same observation period are used to estimate the short-term stability of satellite clock. The estimation results of Beijing and Sanya are compared to validate themselves one another. The receivers of the two stations refer to hydrogen masers synchronized with COMPASS system time (BDT).

The observation period is from BDT 2011-10-09 06:00:00 to BDT 2011-10-09 18:00:00. The B3 carrier phase observations and broadcast ephemeris are used to estimate the short-term stability of SV09 on-board clock. The estimation results are shown in Fig. 3.

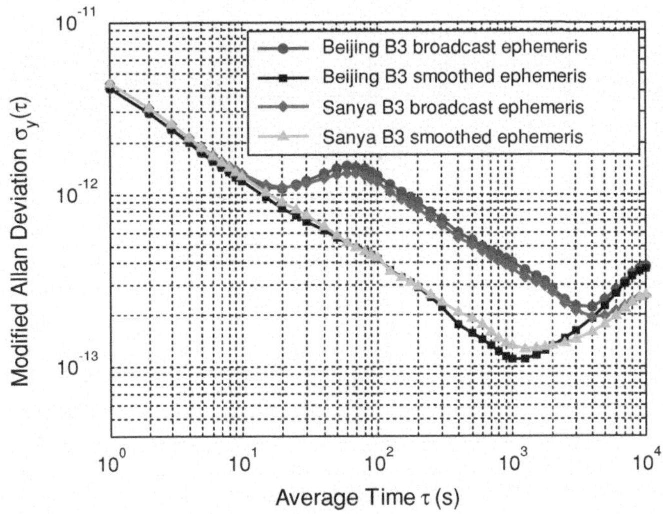

Fig. 3 Frequency stability of COMPASS SV09 based on smoothed broadcast ephemeris

3.3 Estimation Method Based on Satellite Radio Ranging

References [6, 7] have researched on estimating the geometry distance of the satellite and receiver using Satellite Laser Ranging (SLR). In GNSS systems, laser ranging is generally used to validate the estimated satellite orbit parameters. Due to the constraint of weather and other factors, laser ranging is unable to achieve continuous observation, which is not adaptive to evaluate the short-term stability. The proposed methods in [6, 7] are not suitable to estimate the short-term stability.

GEO satellite broadcast ephemeris is not precise enough to estimate the short-term stability due to the difficulty in orbit determination of GEO satellite. In this case, the smoothed broadcast ephemeris method is not available any more. However, COMPASS GEO satellites can be observed continuously by the time synchronization monitoring stations. COMPASS GEO satellite has a C-band transponder on it to achieve time synchronization between monitoring stations. Similar to SLR, Satellite Ratio Ranging (SRR) works in the way that the monitoring station receives the satellite transferred signal transmitted by the station itself [8]. By measuring the travel delay from the station to the satellite and back, the geometry distance between the satellite and station can be determined. If the GNSS receiver and the monitoring station have the same position and refer to the same clock, then the satellite on-board clock offset can be calculated by Eq. (2.3).

Several errors should be corrected before SRR and GNSS carrier phase observations are used to calculate the satellite clock offset: one is the time difference between the observation epochs of the GNSS receiver and the SRR observation station; the other is the distance difference between the antenna centers of the GNSS receiver and the SRR observation station.

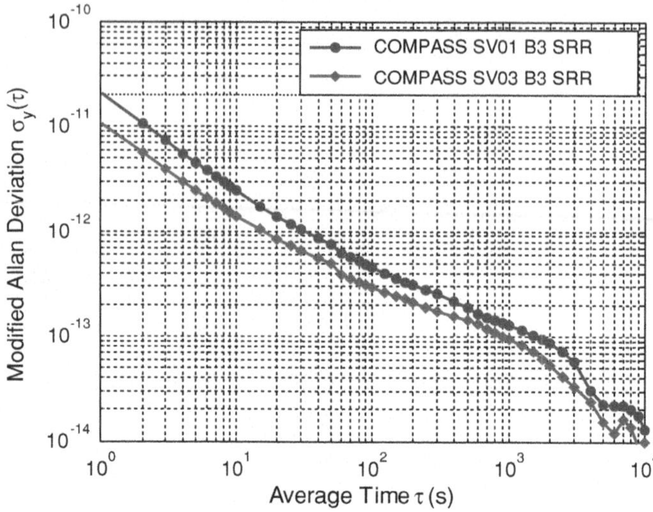

Fig. 4 Frequency stability of COMPASS SV03 based on satellite radio ranging

Using SRR and GNSS receiver observation data of Sanya station, the short-term stability of SV01 and SV03 on-board clocks are estimated. The observation period is from BDT 2011-10-09 06:00:00 to BDT 2011-10-09 18:00:00. The stability results are shown in Fig. 4.

3.4 Estimation Method Based on Carrier Phase Fitting

According to Eq. (2.1), besides the geometry distance between satellite and receiver, the pseudorange contains satellite clock offset with respect to receiver clock, signal propagation delay, the Sagnac effect, multi-path and observation noises. Most of the variables mentioned above change slowly except clock offset and observation noises. Based on this, Ref. [3] proposed a polynomial fitting method to estimate clock offset without geometric distance determination. Carrier phase observation is fitted with high order polynomial and the residue is obtained as the estimation of the satellite and receiver clock difference.

Figure 5 shows the results using the polynomial fitting and IGS final clock offset of GPS PRN14. The corresponding result of COMPASS SV09 is also shown in Fig. 5, compared to the smoothed broadcast ephemeris method.

From Fig. 5, it is observed that the estimated short-term stability based on carrier phase fitting is usually a little optimistic than the one obtained from IGS final clock product. It may be explained as below:

The clock offset can be regarded as the sum of two parts: nonrandom part and random part. The nonrandom part, which includes the initial clock offset and errors

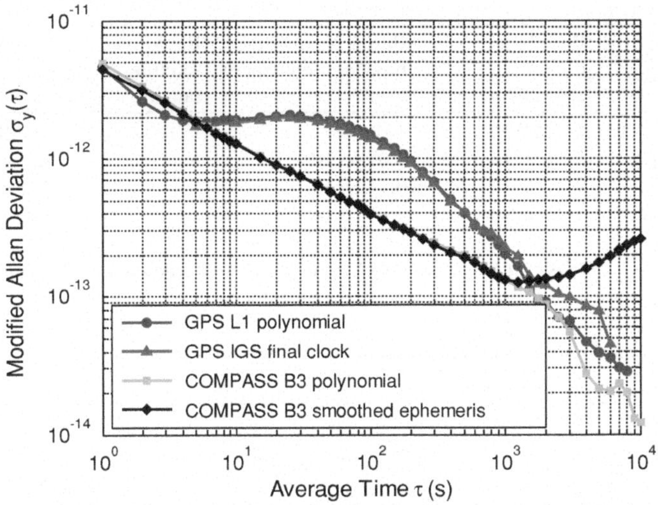

Fig. 5 Frequency stability of GPS PRN14 and COMPASS SV09 based on carrier phase fitting

introduced by clock frequency bias and frequency drift, can be modeled as two order polynomial with observation epoch as the independent variable. The random part is mainly affected by clock phase noise.

The polynomial fitting mentioned here is usually in the order of much higher than 2, typically 24 [3]. As a result, the residue of the high order polynomial fitting of carrier phase observation contains only part of the random part of the clock error and the observation noises. So this method may obtain results slightly optimistic for short-term stability estimation.

3.5 Comparison of Estimation Methods

Four methods have been discussed above to estimate the short-term stability of GNSS on-board clock. They are compared with each other in Table 1 in four aspects: applied sampling intervals, estimation accuracy, applied GNSS systems and satellites. The estimation accuracy is evaluated as the maximal bias between estimated stability and the corresponding IGS final clock product for sampling intervals 5–1000 s. There is no precise clock offset available for COMPASS satellite on-board clock, so estimation accuracy for COMPASS is not presented.

From Table 1, we can draw the following conclusions:

1. The short-term stability estimation by the above three methods except SRR method agrees well with the IGS clock data for sampling interval 1–1000 s.
2. The dual-frequency iono-free combination increases the observation noise, making stability for sampling intervals within 10 s is worse than the stability estimated using single-frequency.

Table 1 Estimation accuracy of these methods

	Based on the precise ephemeris	Based on smoothed broadcast ephemeris	Based on satellite radio ranging	Based on carrier phase fitting
Sampling interval	1–1000 s	1–1000 s	1–1000 s	1–1000 s
Accuracy of L1	5.27×10^{-14}	6.67×10^{-14}	–	7.23×10^{-14}
Accuracy of L2	8.93×10^{-14}	9.73×10^{-14}	–	7.16×10^{-14}
Accuracy of iono-free	2.54×10^{-13}	2.54×10^{-13}		7.90×10^{-14}
Availability for GNSS systems at this stage	GPS/ GLONASS	GPS/GLONASS/ COMPASS/ Galileo	COMPASS	GPS/GLONASS/ COMPASS/ Galileo
Availability for satellites	MEO/GEO/IGSO	MEO/GEO/IGSO	GEO	MEO/GEO/IGSO

3. The stability estimated based on precise ephemeris with single frequency is bias from IGS precise clock product by less than 1×10^{-13}, for the sampling interval of 5–1000 s.
4. The stability estimated based on smoothed broadcast ephemeris agrees well with precise ephemeris, and is bias from IGS precise clock product by less than 1×10^{-13} for the sampling interval of 5–1000 s, thanks to the elimination of jumps of calculated satellite orbit.
5. The estimation method based on SRR is limited to GEO on-board clock and the GNSS monitoring stations that have the ability to transmit signal to the satellite.
6. The stability estimated based on carrier phase fitting is usually a little optimistic than the one estimated using precise ephemeris due to high order fitting.
7. For the constructing GNSS systems such as COMPASS and Galileo, common users can obtain the short-term stability of on-board clock using estimation method based on smoothed broadcast ephemeris or carrier phase fitting.

4 Conclusion

Short-term frequency stability estimation of satellite on-board clock is of great significance for GNSS application such as satellite clock modeling and simulation. The ODTS method, which is well-known and commonly used, needs continuous observation from large monitoring networks and complex algorithms. COMPASS and Galileo are in their early stage of construction and COMPASS does not have worldwide observation networks to track the satellite continuously. In addition, limited access permission has also restricted common users from obtaining precise on-board clock parameters.

Four estimation methods of the short-term stability of GNSS on-board clock are proposed in this paper, which only need observation from one single station. All the four methods are validated and compared using GPS or COMPASS observation. Compared to IGS final clock product, the estimation accuracy of these methods is presented for GPS. The advantages and disadvantages of these four methods are pointed out, which is meaningful for the users of COMPASS and Galileo in their early stage of construction.

References

1. Dow, J. M., Neilan, R. E., & Rizos, C. (2009). The international GNSS service in a changing landscape of global navigation satellite systems. *Journal of Geodesy, 83*, 191–198.
2. Horemuz, M., & Andersson, J. V. (2006). Polynomial interpolation of GPS satellite coordinates. *GPS Solutions, 10*, 67–72.
3. Delporte, J., Boulanger, C., & Mercier, F. (2011). Straightforward estimations of GNSS on-board clocks. In *2011 Joint Conference of the IEEE International Frequency Control Symposium and European Frequency and Time Forum*, San Francisco, CA, USA.
4. Gonzalez, F., Waller, P. (2007). GNSS clock performance analysis using one-way carrier phase and network methods. In *Proceedings of the 39th Precise Time and Time Interval (PTTI) Applications and Planning Meeting*, Long Beach, CA, USA.
5. Roulston, A., Talbot, N., & Zhang, K. (2000). An evaluation of various GPS satellite ephemerides. In *Proceedings of the 13th International Technical Meeting of the Satellite Division of The Institute of Navigation (ION GPS 2000).*
6. Qin, X., Yang, Y.-X., Jiao, W., & Wang, G. (2004). Determination of navigation satellite clock bias using SLR and pseudorange data. *Acta Geodaetica et Cartographica Sinica, 33*, 205–209 (in Chinese).
7. Lu, X. F., Jia, X. L., & Yang, Z. Q. (2008). Determination of navigation satellite clock bias using SLR and GPS dual frequency phase-smoothed pseudo-range data. *Geomatics and Information Science of Wuhan University, 33*, 237–240 (in Chinese).
8. Yang, X., Ma, L., Sun, B., Lei, H., Sun, L., Li, Z., Zhou, W., et al. (2011). The method of time synchronization based on the combination of COMPASS GEO pseudo-range and two-way data. In *2011 Joint Conference of the IEEE International Frequency Control Symposium and European Frequency and Time Forum*, San Francisco, CA, USA.

An Integrity Monitoring Algorithm for Satellite Clock Based on Test Statistics

Xinming Huang, Hang Gong, Wenke Yang, Xiangwei Zhu and Gang Ou

Abstract As a core part of satellite payload, satellite clock has a direct influence on GNSS service performance. Therefore, it is necessary to monitor its integrity. In this paper, real-time test statistics for satellite clock phase and frequency anomaly are built, and an integrity monitoring algorithm for satellite clock based on these statistics is proposed. It is composed of two parts, including phase anomaly monitoring and frequency anomaly monitoring, which mainly aims at three anomalies comprising occasional bad or outlier points, anomaly jumps that later return to stable values and phase or frequency steps. The new algorithm can not only detect phase and frequency anomalies, but also identify and deal with clock anomaly instantly, including eliminating and replacing the anomaly data. Satellite clock anomalies are simulated using the IGS observation clock data and performance of the new algorithm is verified. The results show that the new algorithm has a perfect performance of detecting the satellite clock anomaly including both phase anomaly and frequency anomaly. At the same time, it can also eliminate and replace the anomaly data, and the replaced error is negligible. The integrity monitoring algorithm for satellite clock proposed in this paper is helpful to GNSS satellite clock autonomous integrity monitoring.

1 Introduction

As the frequency and time reference of space satellites, satellite clock has a direct influence on GNSS service performance. Therefore, it is necessary to monitor its integrity. In order to meet the real-time requirements of special services such as

X. Huang (✉) · H. Gong · W. Yang · X. Zhu · G. Ou
Satellite Navigation R&D Center, National University of Defense Technology, 410073 Changsha, China
e-mail: huangxinming_2007@163.com

J. Sun et al. (eds.), *China Satellite Navigation Conference (CSNC) 2012 Proceedings*, Lecture Notes in Electrical Engineering 160, DOI: 10.1007/978-3-642-29175-3_47,

aviation and safety of life services, Rodriguez et al. [1] for the first time proposes the idea of satellite autonomous on-board clock monitoring, and related algorithms are given, including the dynamic ADEV and moving average filtering. Algorithm based on the dynamic ADEV is easy to identify a variety of satellite clock anomalies, but it is not real-time. The moving average filtering method is real-time, but its detection performance is unsatisfactory. Lorenzo and Tavella [2] presents a frequency jump detector for space clocks, which can detect the satellite clock frequency jump, but it is also not real-time. Clearly, an algorithm for satellite clock autonomous integrity monitoring is useful.

It can be known from Weiss et al. [3] that satellite clock anomalies mainly consist of phase anomaly and frequency anomaly. Existing satellite clock integrity monitoring algorithms usually aim at single anomaly, such as phase anomaly or frequency anomaly. However, satellite clock anomalies often include both phase anomaly and frequency anomaly. In this case, a better approach is to consider both phase and frequency anomalies.

In this paper, real-time test statistics for phase and frequency anomaly of satellite clock are built, and an integrity monitoring algorithm for satellite clock based on these statistics is proposed. It is composed of two parts, including phase anomaly monitoring and frequency anomaly monitoring. The new algorithm can not only detect phase anomaly and frequency anomaly, but also identify and deal with clock anomaly instantly, including eliminating and replacing the anomaly data. Satellite clock anomalies are simulated using the IGS observation clock data and performance of the new algorithm is verified.

2 Anomaly Test Statistics of Satellite Clock

The satellite clock anomalies comprise phase anomaly and frequency anomaly, which differ with each other and need integrity monitoring separately. In this section, real-time test statistics for phase and frequency anomalies of satellite clock are built, which are used to detect phase and frequency anomalies.

There are two types of satellite onboard atomic clocks: rubidium and cesium clock. Cesium clock can be modeled as a linear model, while rubidium clock, with short-term linear frequency drift, can be modeled as a quadratic polynomial model [4]. In this paper, test statistic is constructed based on rubidium clock. For satellite cesium clock, the construction of anomaly test statistic is similar to satellite rubidium clock.

2.1 Phase Anomaly Test Statistic

It can be known that the rubidium clock model can be expressed as a quadratic model [5], which is computed following

$$x(t) = a_0 + a_1 \cdot t + a_2 \cdot t^2 + \varepsilon_x(t) \tag{1.1}$$

where a_0 is initial atomic clock phase deviation, a_1 is initial frequency deviation, $a_2/2$ is initial frequency drift in atomic clocks; $\varepsilon_x(t)$ is atomic clock phase noise, which is often assumed as Gaussian white noise. Equation (1.1) can be also expressed following

$$X = HA + \varepsilon \tag{1.2}$$

where $H = \begin{bmatrix} 1 & t_1 - t_0 & (t_1 - t_0)^2 \\ \vdots & \vdots & \vdots \\ 1 & t_n - t_0 & (t_n - t_0)^2 \end{bmatrix}, A = \begin{bmatrix} a_0 \\ a_1 \\ a_2 \end{bmatrix}, X = \begin{bmatrix} x_1 \\ \vdots \\ x_n \end{bmatrix}, \varepsilon = \begin{bmatrix} \varepsilon_1 \\ \vdots \\ \varepsilon_n \end{bmatrix}.$

In order to achieve real-time detection of phase anomaly, a new phase anomaly test statistic is constructed using the Time Interval Error (TIE) by taking difference between the measured and predicted phase offsets. Phase offset of rubidium clock can be predicted by the quadratic model. According to Guo [5], satellite clock parameters can be computed following

$$\hat{A} = \begin{bmatrix} \hat{a}_0 \\ \hat{a}_1 \\ \hat{a}_2 \end{bmatrix} = (H^T H)^{-1} H^T X \tag{1.3}$$

The one-step predicted phase offset is given as

$$\hat{x}_{n+1} = \begin{bmatrix} 1 & t_{n+1} - t_0 & (t_{n+1} - t_0)^2 \end{bmatrix} \cdot \begin{bmatrix} \hat{a}_0 \\ \hat{a}_1 \\ \hat{a}_2 \end{bmatrix} = T_{n+1} (H^T H)^{-1} H^T X \tag{1.4}$$

where $T_{n+1} = \begin{bmatrix} 1 & t_{n+1} - t_0 & (t_{n+1} - t_0)^2 \end{bmatrix}$.

We will construct the new phase anomaly test statistic by discussing the Probability Density Function (PDF) of $z = x_{n+1} - \hat{x}_{n+1}$.

As ε_i follows a Gaussian distribution, clock phase offset also follows a Gaussian distribution, as $x_i \sim N(T_i A, \sigma^2)$

It can be deduced from Eq. (1.4) that $\hat{x}_i$ follows a Gaussian distribution, as $\hat{x}_{n+1} \sim N(T_{n+1} A, (T_{n+1} (H^T H)^{-1} T_{n+1}^T) \sigma^2)$.

Statistic $z = x_{n+1} - \hat{x}_{n+1} = (x_{n+1} - T_{n+1} A) - (\hat{x}_{n+1} - T_{n+1} A)$ consists of two independent statistics following Gaussian distribution, so it also follows a Gaussian distribution. Its mean and variance are computed following

$$E(z) = E((T_{n+1} A + \varepsilon_{n+1}) - T_{n+1} (H^T H)^{-1} H^T (HA + \varepsilon)) = 0 \tag{1.5}$$

$$\begin{aligned} \operatorname{var}(z) &= E((x_{n+1} - \hat{x}_{n+1})(x_{n+1} - \hat{x}_{n+1})^T) \\ &= (1 + T_{n+1} (H^T H)^{-1} T_{n+1}^T) \cdot \sigma^2 \end{aligned} \tag{1.6}$$

Therefore, $z \sim N(0, (1 + T_{n+1} (H^T H)^{-1} T_{n+1}^T) \sigma^2)$

In order to improve detection performance of phase anomaly, statistic $Z = (z/\sigma_z)^2 = ((x_{n+1} - \hat{x}_{n+1})/\sigma_z)^2$ is adopted. Z is the phase anomaly test statistic chose in this paper, which follows a $\chi^2(1)$ distribution.

According to equation $P(\chi^2 > \chi^2_\alpha) = \alpha$, with the confidence level of $1 - \alpha$ the detection threshold value should be set as $x^2_\alpha(1)$.

2.2 Frequency Anomaly Test Statistic

Frequency offset of rubidium clock can be expressed following [5]

$$y(t) = \frac{dx(t)}{dt} = a_1 + a_2 \cdot t + \varepsilon_y(t) \tag{1.7}$$

where a_1 is initial frequency deviation, a_2 is initial frequency drift in atomic clocks; $\varepsilon_y(t)$is atomic clock phase noise, which is often assumed as Gaussian white noise. Equation (1.7) can be also expressed following

$$Y = HA + \varepsilon \tag{1.8}$$

Where $H = \begin{bmatrix} 1 & t_1 - t_0 \\ \vdots & \vdots \\ 1 & t_n - t_0 \end{bmatrix}, A = \begin{bmatrix} a_1 \\ a_2 \end{bmatrix}, Y = \begin{bmatrix} y_1 \\ \vdots \\ y_n \end{bmatrix}, \varepsilon = \begin{bmatrix} \varepsilon_1 \\ \vdots \\ \varepsilon_n \end{bmatrix}$.

In order to achieve real-time detection of frequency anomaly, a new frequency anomaly test statistic is constructed by taking difference between the measured and predicted frequency offsets. Predicted frequency offset can be achieved by the linear prediction model of frequency offsets.

As construction of frequency anomaly test statistic is similar to phase anomaly test statistic, we only gives the new test statistic here.

In this paper, the new test statistic adopted is expressed following

$$Z = ((y_{n+1} - \hat{y}_{n+1})/((1 + T_{n+1}(H^T H)^{-1} T^T_{n+1}) \cdot \sigma^2))^2 \tag{1.9}$$

Where $T_{n+1} = [\, 1 \; t_{n+1} - t_0 \,]$.

Z is the frequency anomaly test statistic chose, which follows a $\chi^2(1)$ distribution.

According to equation $P(\chi^2 > \chi^2_\alpha) = \alpha$, with the confidence level of $1 - \alpha$ the detection threshold value should be set as $x^2_\alpha(1)$.

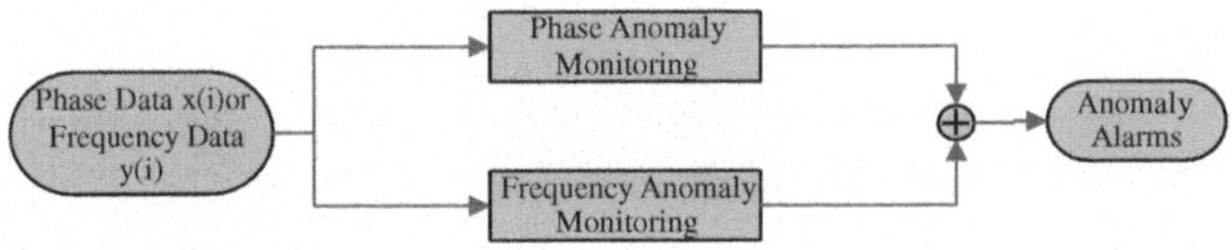

Fig. 1 Architecture of satellite clock integrity monitoring algorithm

3 Integrity Monitoring Algorithms

Figure 1 shows the architecture of satellite clock integrity monitoring algorithm. The algorithm consists of two parts, namely, phase anomaly monitoring and frequency anomaly monitoring.

The basic flow of satellite clock integrity monitoring algorithm is as follows:

- Acquiring clock data from inter-satellite ranging observations or on-board measurement and comparing equipments, including clock phase offset data and frequency data.
- Anomaly monitoring phase and frequency of acquired clock data.
- Combining results of phase anomaly monitoring and that of frequency anomaly monitoring to determine whether satellite clock is in good state.

We will discuss the principles of phase anomaly monitoring and frequency anomaly monitoring detailedly in the following.

3.1 Phase Anomaly Monitoring

Figure 2 illustrates the block diagram of phase anomaly monitoring. Phase anomaly monitoring is achieved on the basis of phase anomaly statistic.

The basic flow of phase anomaly monitoring is as follows:

- Acquiring predicted satellite clock phase offset by linear model, and constructing phase anomaly test statistic Z by taking difference between the measured and predicted phase offsets.
  Where $Z = (z/\sigma_z)^2 = ((x_{n+1} - \hat{x}_{n+1})/\sigma_z)^2$.
- Comparing detection threshold with statistic Z. The detection threshold value is set as $x_\alpha^2(1)$ with the confidence level of $1 - \alpha$.
- If value of statistic Z is smaller than detection threshold, clock can be considered in good state. Otherwise, clock can be considered anomaly. In such instance, on one hand, anomaly type will be identified and alarm set. On the other hand, clock anomaly will be dealt instantly, including eliminating and replacing the anomaly data.

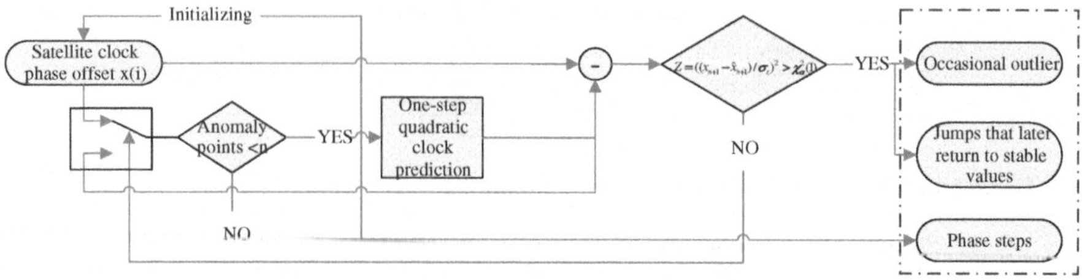

Fig. 2 Block Diagram of phase anomaly monitoring

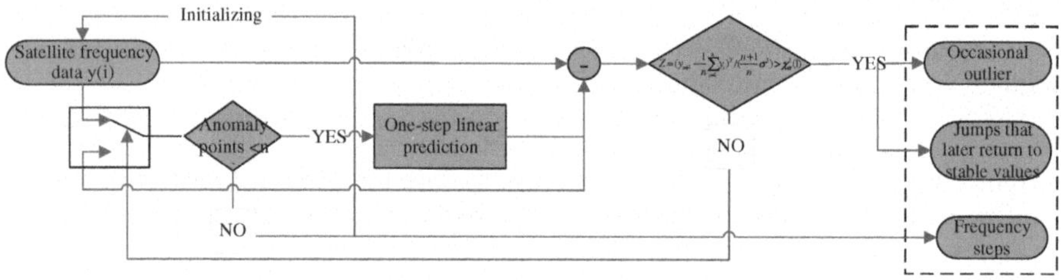

Fig. 3 Block diagram of frequency anomaly monitoring

The algorithm can not only detect phase anomaly and frequency anomaly, but also identify and deal with clock anomaly instantly. Thus the monitoring algorithm can achieve automatic detection and processing of anomaly.

3.2 Frequency Anomaly Monitoring

Figure 3 illustrates the block diagram of frequency anomaly monitoring. Frequency anomaly monitoring is achieved on the basis of frequency anomaly statistic.

As frequency anomaly monitoring is similar to phase anomaly monitoring, we will not discuss it in detail here.

4 Algorithm Verification

To validate performance of the proposed algorithm, satellite clock anomalies are simulated using the IGS observation clock data. The selected satellite clock is Rubidium clock, whose PRN number is 2. Time period of selected IGS data range from 29 to 30 August 2010 and its sampling interval is 15 min.

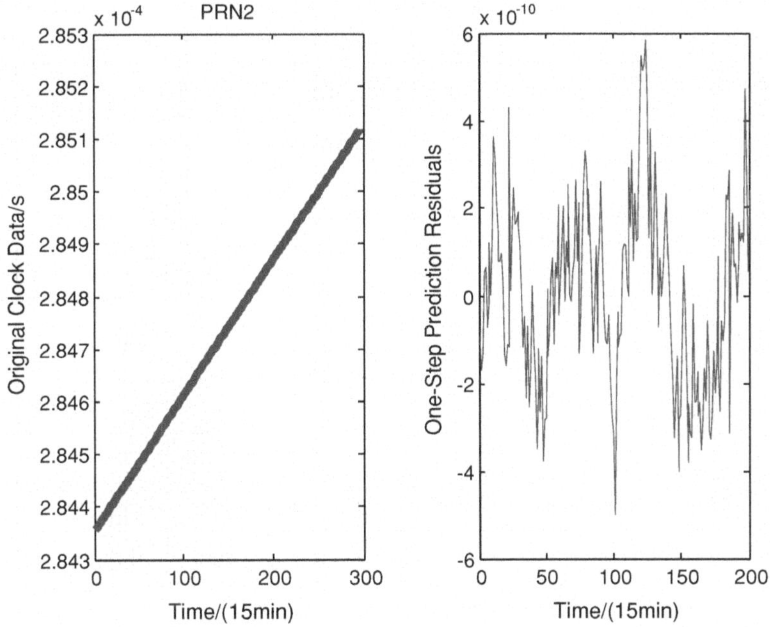

Fig. 4 Original clock data and one-step prediction residuals

In the following sections the satellite clock model will be verified at first, and then IGS data will be processed statistically to verify validity of test statistic proposed. Afterwards performance of the new algorithm will be verified using the simulated anomaly satellite clock data.

4.1 Verification of Clock Model and Test Statistic

The accuracy of IGS clock data is 0.1 ns, so that it can be regarded as undistorted clock data, which can represent natural state of satellite clock.

Figure 4 shows original clock data and one-step prediction errors acquired from a 96-point sliding window. It can be seen that one-step prediction residuals contain obvious periodic fluctuation. Therefore, the quadratic clock model ought to include periodic terms, which should be extracted and removed.

In this paper, the periodic terms will be extracted and removed by spectral analysis. Norden et al. [6] and Fei [7] give specific approaches to determine periodic terms. So there will be no longer detailed description of these methods, and only extraction results will be given.

Figure 5 shows one-step prediction residuals with seven relatively obvious periodic terms removed and its histogram.

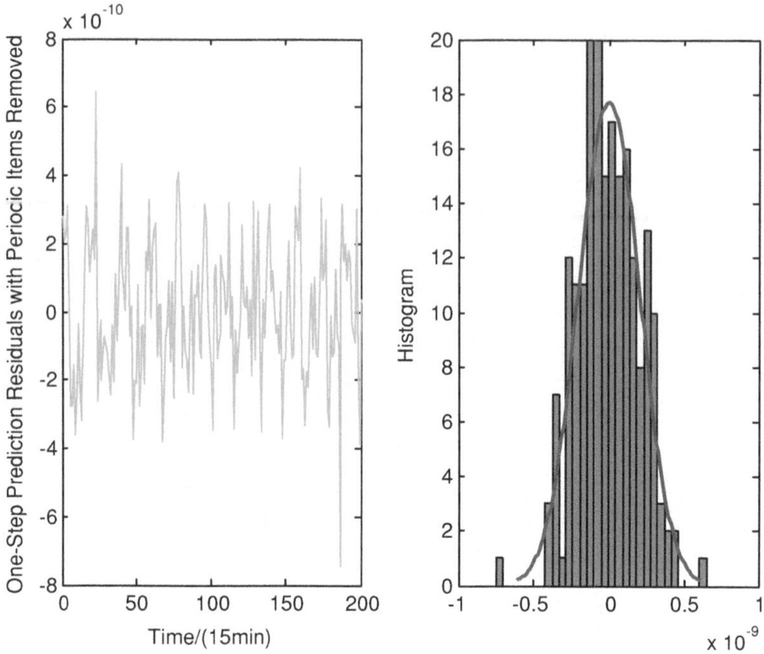

Fig. 5 One-step prediction residuals with periodic items removed and its histogram

Fig. 6 Histogram for phase anomaly test statistic

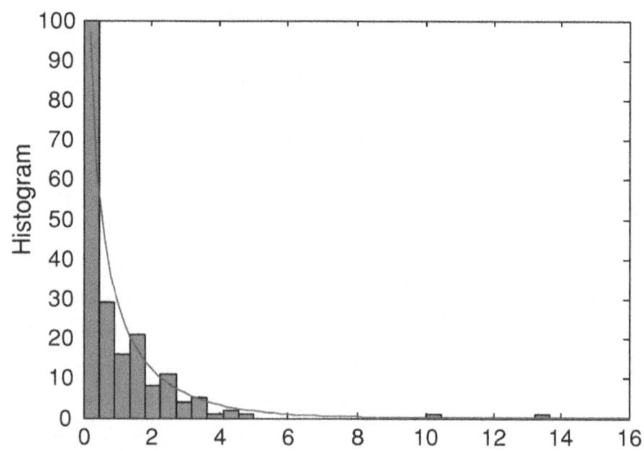

As observed in Fig. 5, one-step prediction residuals with obvious periodic fluctuation removed follow a Gaussian distribution approximately. Therefore, the quadratic model could be considered as a valid clock model with periodic terms removed.

Figures 6 and 7 show histograms of test statistics proposed in this paper, including phase anomaly test statistic and frequency anomaly test statistic.

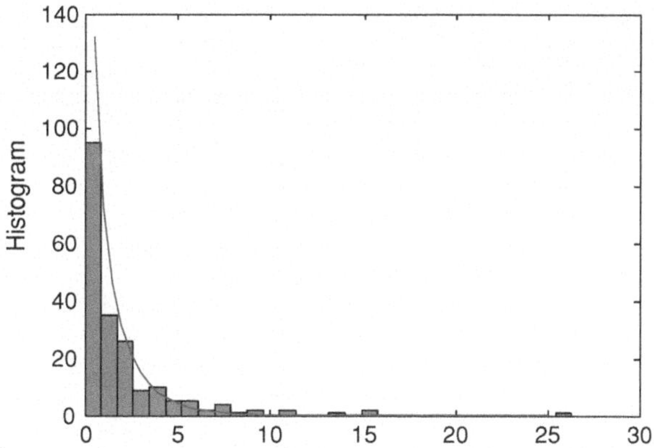

Fig. 7 Histogram for frequency anomaly test statistic

It can be seen from the figures that the proposed phase and frequency anomaly test statistics obey the Chi-square distribution approximately.

4.2 Phase Anomaly

Clock model and anomaly test statistics have been verified to be valid in previous section. In this section, we will validate performance of the new algorithm using the simulated anomaly phase data on the basis of previous section.

As satellite clock is very stable, making it difficult to get anomaly data, we will simulate satellite clock anomalies by adding abnormal data to IGS observation clock data similar to Ref. [8]. In this section, original IGS clock data from 100 to 108 points plus 8 ns to simulate phase anomaly.

Figure 8 shows simulated abnormal satellite clock data, and Fig. 9 shows detection values obtained from phase anomaly test statistic.

As phase anomaly test statistic Z follows a $\chi^2(1)$ distribution, with the confidence level of $1 - \alpha = 0.995$ the detection threshold value should be set as $x_{\alpha}^{2}(1) = 7.789$. It can be seen from Fig. 9 that simulated clock data from 100 to 108 points are abnormal.

The algorithm can also eliminate and replace the anomaly data. Table 1 shows replaced errors of anomaly data. It can be seen that the algorithm can deal with abnormal data perfectly, and the replaced error is sub-nanosecond.

4.3 Frequency Anomaly

Verification of the algorithm proposed under the circumstance of frequency anomaly is similar to that of phase anomaly. We will also simulate satellite clock

Fig. 8 Simulated phase anomaly data

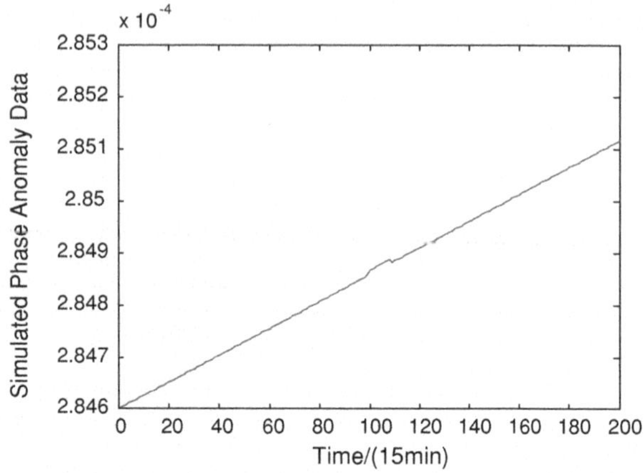

Fig. 9 Detection value of phase anomaly test statistic

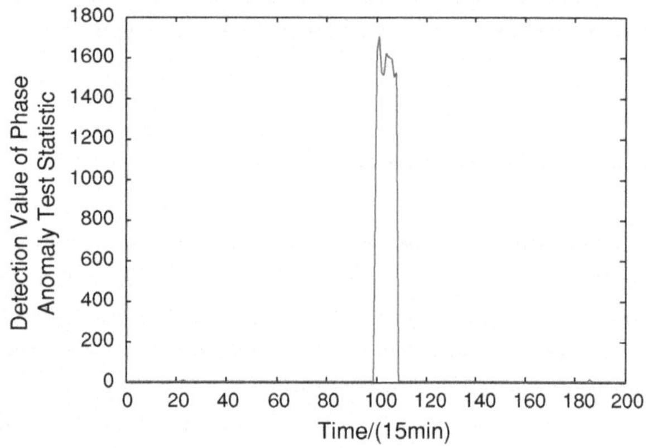

Table 1 Replaced error of anomaly points

Anomaly points	Replaced errors (s)
100	1.631202703293588e-10
101	3.467576195257359e-10
102	1.030683462839230e-10
103	1.382399600607237e-10
104	1.377543100205755e-10
105	8.744623710039978e-11
106	5.904497382561505e-11
107	1.640259615957006e-10
108	1.101885686643697e-10

anomalies by adding abnormal data to IGS observation clock data. In this section, original IGS clock frequency offset data from 100 to 108 points plus 2e-12 to simulate frequency anomaly.

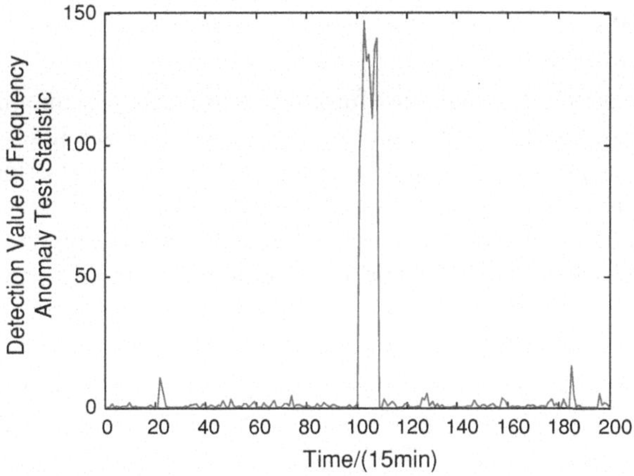

Fig. 10 Detection value of frequency anomaly test statistic

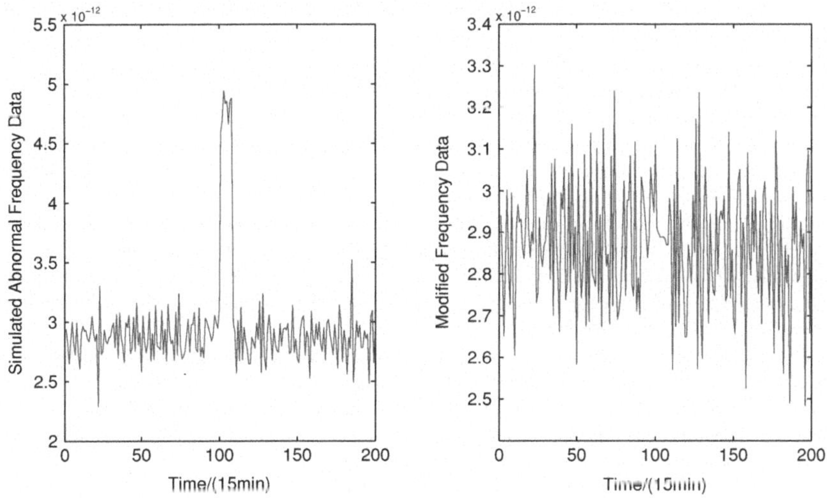

Fig. 11 Simulated frequency anomaly data and modified frequency data

As frequency anomaly test statistic Z follows a $\chi^2(1)$ distribution, with the confidence level of $1 - \alpha = 0.995$ the detection threshold value should be set as $x_{\alpha}^2(1) = 7.789$.

Figure 10 shows detection values obtained from phase anomaly test statistic, it can be seen that simulated frequency data from 100 to 108 points are abnormal. Figure 11 shows simulated frequency anomaly data and modified frequency data. It can be seen that the algorithm can deal with abnormal data perfectly.

5 Conclusion

In this paper, real-time test statistics for phase and frequency anomaly of satellite clock are built, and an integrity monitoring algorithm for satellite clock based on these statistics is proposed, which is composed of two parts, including phase anomaly monitoring and frequency anomaly monitoring. The new algorithm can not only detect satellite clock phase and frequency anomalies, but also identify and deal with clock anomaly instantly, including eliminating and replacing the abnormal data.

References

1. Rodríguez, I., García, C., & Catalán, C. et al. (2009, September). Satellite autonomous integrity monitoring (SAIM) for GNSS systems. *ION GNSS 2009*.
2. Galleani, L., & Tavella, P. (2010). An algorithm for the detection of frequency jumps in space clocks. *In Proceedings of the 42th Annual Precise Time and Time Interval (PTTI) Applications and Planning Meeting*.
3. Weiss, M., Shonme, P., & Beard, R. (2010). On-board GPS clock monitoring for signal integrity. *In Proceedings of the 42th Annual Precise Time and Time Interval (PTTI) Applications and Planning Meeting*.
4. Allan, D. W. (1987, November). Time and frequency (time-domain) characterization, estimation and prediction of precision clocks and oscillators. *IEEE Transactions on Ultrasonics, Ferroelectrics, and Frequency Control*. UFFC-34(6).
5. Guo, H.-R. (2006). Study on the analysis theories and algorithms of the time and frequency characterization for atomic clocks of navigation satellites. Information Engineering University.
6. Zhang, B., Ou, J., Yuan, Y., & Hao, M. (2007). *Fitting method for GPS satellite clock errors using wavelet and spectrum analysis*. Geomatics and Information Science of Wuhan University, 3 (28), pp. 715–718.
7. Huang, N. E., Shen, Z., et al. (1998). The empirical mode decomposition and the Hilbert spectrum for nonlinear and non-stationary time series analysis. *Proceedings of the Royal Society of London, Series A: Mathematical and Physical Sciences, 454*, 903–995.
8. Fei, N. (2008). *Theory and technique on GNSS integrity augment*. Dissertation submitted to PLA information engineering university, Zheng Zhou.

Satellite Clock Parameter Short-Term Prediction Using Piece-Wise Adaptive Filter with State Noise Compensation

Li Liu, Lan Du, LingFeng Zhu, ChunHao Han, GuiFen Tang and Xin Shi

Abstract Because prediction update of satellite clock correction parameters is considerable frequent, there are a large number of transition points in satellite clock predictions. The conventional Batch for fitted clock parameters usually ignores the relevance between adjacent fit subsections; consequently, there might appear clock leaps at dense prediction update points. In addition, it is very desirable to further improve the precision of shore-term prediction. We introduce a piece-wise adaptive EKF algorithm for near real-time clock estimation and shore-term prediction. Dynamically varied state variance factor is used to compensate equivalently the inaccurate adopted value of observation variance factor and thus improve the subsection-specific estimation precision, whereas the prediction precision is improved by using ultra-short prediction of one hour to estimate the best state variance factor. The algorithm is experimentally validated with GEO clock observations. The comparison with the Batch results shows that the short-term prediction precision is about 0.8 ns and is close to 50% enhancement with respect to that of the Batch approach. Most of all, the new approach smoothes the frequent clock leaps at the transition points of consecutive subsections dramatically, where the sampled leaps almost restrain with 1 ns. As a result, more steady short-term prediction and real-time evaluation for GEO/IGSO/MEO satellite clocks of Compass constellations become achievable.

L. Liu (✉) · L. Zhu · C. Han · G. Tang · X. Shi
Beijing Global Information Center of Application and Exploration,
100094 Beijing, China
e-mail: liuli_cy@sohu.com

L. Du
Information Engineering University of PLA, 450052 Zhengzhou, China

J. Sun et al. (eds.), *China Satellite Navigation Conference (CSNC) 2012 Proceedings*,
Lecture Notes in Electrical Engineering 160, DOI: 10.1007/978-3-642-29175-3_48,

1 Introduction

In order to provide satellite navigation services, precise satellite clock and orbit have to be estimated and broadcasted to users in real-time by navigation message. Precise clock parameters, usually linear or quadratic polynomial coefficients, are estimated from the satellite clock offset series measured from clock comparison [1]. Therefore, high precision time synchronization between on-board and ground clocks is one of key missions for navigation satellite system [2]. There are two meanings for high precision time synchronization: one is accurate clock comparison, which is the base of time synchronization; the other is precise clock prediction, which is the realization of time synchronization [2].

At present, batch algorithm is usually performed to obtain satellite clock parameters in the master control center of Compass satellite navigation system. Estimation period of about 2 hours combined with update rate of 1 hour is used in the clock parameter generation sufficiently allowing for the actual stability of the on-board clock. This rapid update prediction strategy yields a leap in the predicted clock offsets every hour at the transition point between two adjacent piecewise clock parameters. Because regular batch algorithm merely considered data quality control within subsection instead of the variations over adjacent subsections, the clock leap can reach even 1–2 ns, which restrains clock correction precision and consistency.

With the current performance of clock comparison, the estimation and prediction strategy of satellite clock parameters according to the clock behavior might be more desirable in order to provide reliable positioning and timing information for users.

Many scholars have studied some clock correction models for satellite clocks and obtained a lot of good results [5–7]. But the short-term fit and prediction procedure also limit some other available clock correction model such as multinomial and grey models due to too many model parameters needed. Adaptive filter has the merit of simultaneous process as well as observation noises [8–11]. In practice, each of the two stochastic models uses a fixed empirical covariance and then an adaptive factor is induced to balance the contribution from the two kind of stochastic information [12, 13]. Conventionally, in order to ensure the accuracy of the filter parameters, the adaptive factor is adjusted for each of the observable data set. If the adjusting of adaptive factor is calculated in short period, the process will be cumbersome. Even more important, overall smoothing among consecutive piece-wise prediction can not be obtained by means of adaptive filter without considering the data relevancy between adjacent subsections.

We introduce an improved piecewise adaptive filter with state noise compensation (SNC) for clock parameter estimation. Instead of adaptive factor, the adaptive state noise factor of process noise is used in adaptive filter. For clock parameter estimation, the best state noise factor is then fixed to equivalently compensate the inaccurate value of observation noise variance. Therefore, the state noise factor only needs to be estimated once for every subsection, but it can adaptively adjust for different subsections. The algorithm is implemented with the real observations of Compass GEO satellite clock for validation. Based on conventional Batch and EKF and the new approach, a total of sixty-six consecutive subsections of 3 day data are used to estimate and predict the clock offsets respectively. Comparison within the three algorithms shows that the new method has a comparable enhancement in estimation precision and prediction precision for subsection. Moreover, the clock leaps of prediction update points are reduced dramatically. In general, the new approach demonstrates a promise to generate short-term satellite clock parameters in near real-time for GEO/IGSO/MEO satellites of Compass constellations.

2 The New Approach

Short-term prediction of satellite clock normally uses piecewise linear model, where a two-parameter state vector consists of clock bias and clock frequency bias $\boldsymbol{x} = (a_0, a_1)^{\mathrm{T}}$. A filter data set is satellite clock offset series computed from time transfer between satellite and ground station. The procedure includes time update and measurement update [14].

2.1 Time Update

Assuming high and fixed sampling rate is used, only short-term stability of satellite clock needs to be considered. Incorporating process noise to compensate the change of frequency bias, the state propagation equation can be expressed as

$$\dot{\boldsymbol{x}} = \boldsymbol{A}\boldsymbol{x} + \boldsymbol{B}u \tag{1}$$

where $\boldsymbol{A}$ and $\boldsymbol{B}$ are state propagation matrix and process noise mapping matrix, $u(t)$ is process white noise. They are identified as

$$\begin{cases} \boldsymbol{A} = \begin{pmatrix} 0 & 1 \\ 0 & 0 \end{pmatrix}, \quad \boldsymbol{B} = \begin{pmatrix} 0 \\ 1 \end{pmatrix} \\ \begin{cases} E(u) = 0 \\ \sigma^2(u) = \sigma_u^2 \delta(t - \tau) \end{cases} \end{cases} \tag{2}$$

where σ_u^2 is variance coefficient of process noise, t is time and τ is the measurement interval.

Under an assumption, $u(t)$ is white noise with known covariance, the algorithm is known as State Noise Compensation.

Integrating Eq. (1) from state estimation $\hat{\boldsymbol{x}}_{k-1}$ and its covariance matrix $\hat{\boldsymbol{P}}_{k-1}$ at epoch t_{k-1}, we have one-step state prediction $\bar{\boldsymbol{x}}_k$ and its covariance matrix $\bar{\boldsymbol{P}}_k$ at epoch t_k. The discrete time update equations are

$$\begin{cases} \bar{\boldsymbol{x}}_k = \Phi(t_k, t_{k-1})\hat{\boldsymbol{x}}_{k-1} \\ \bar{\boldsymbol{P}}_k = \Phi\hat{\boldsymbol{P}}_{k-1}\Phi^T + \boldsymbol{Q}_\eta \end{cases} \tag{3}$$

where $\Phi(t_k, t_{k-1})$ and $\boldsymbol{Q}_\eta(t_k, t_{k-1})$ are state transition matrix and process noise covariance matrix, and they are

$$\Phi(t_k, t_{k-1}) = \begin{pmatrix} 1 & \Delta t \\ 0 & 1 \end{pmatrix} \tag{4}$$

$$\begin{aligned} \boldsymbol{Q}_\eta(t_k, t_{k-1}) &= \sigma_u^2 \int_{t_{k-1}}^{t_k} \Phi(t,\tau)\boldsymbol{B}\boldsymbol{B}^T\Phi^T(t,\tau)d\tau \\ &= \sigma_u^2 \begin{pmatrix} \frac{1}{3}\Delta t^3 & \frac{1}{2}\Delta t^2 \\ \frac{1}{2}\Delta t^2 & \Delta t \end{pmatrix} \end{aligned} \tag{5}$$

where $\Delta t = t_k - t_{k-1}$ is the measurement update interval. It is noted that with $\boldsymbol{Q}_\eta$ involved, the original deterministic variable of clock frequency bias is modified by including a random component that is a constant-diffusion Brownian motion process.

2.2 Measurement Update

Assuming the clock offset observations are corrupted only by uncorrelated Gaussian noise, the measurement equation is as follows

$$Y_k = \begin{pmatrix} 1 & 0 \end{pmatrix}\begin{pmatrix} a_0 \\ a_1 \end{pmatrix} + \varepsilon_k = \boldsymbol{H}\boldsymbol{x}_k + \varepsilon_k \tag{6}$$

Where Y_k and ε_k are clock offset observation and its pre-fit residual at epoch t_k, a_0 is the variable of clock bias, a_1 is the variable of clock frequency bias, $\boldsymbol{H}$ is the observation/state mapping matrix.

From (3) and (6), the gain matrix $\boldsymbol{K}$ and the predicted residual y_k at epoch t_k are

$$\begin{cases} \boldsymbol{K} = \bar{\boldsymbol{P}}_k\boldsymbol{H}^T(H\bar{\boldsymbol{P}}_k\boldsymbol{H}^T + R_k)^{-1} \\ y_k = Y_k - \boldsymbol{H}\bar{\boldsymbol{x}}_k \end{cases} \tag{7}$$

where R_k is the variance of clock offset observation, and assuming observation noise and process noise are uncorrelated.

The measurement update equations are

$$\begin{cases} \hat{\boldsymbol{x}}_k = \bar{\boldsymbol{x}}_k + \boldsymbol{K} y_k \\ \boldsymbol{P}_k = (\boldsymbol{I} - \boldsymbol{K}\boldsymbol{H})\bar{\boldsymbol{P}}_k \end{cases} \tag{8}$$

2.3 Equivalent Compensation of Stochastic Model Parameters

In the Kalman filter with SNC, the estimated results depend to some extent on the accuracy of stochastic model parameters, $\boldsymbol{Q}_\eta$ and R_k, or σ_u^2 and σ_ε^2 [6]. The adopted values of the two model parameters should be conformed corresponding to their own model noise level. The fact is, the clock frequency bias over a few hours should be relatively stable without any operations such as frequency modulation or phase modulation [15]. Therefore, σ_u^2 should be chosen as an empirical value of second stability due to the one second measurement update interval. On the other hand, the clock observations from time transfer between satellite and ground station show some trend of variation in σ_ε^2 over a few hours.

Actually, these two stochastic parameters effect on the measurement update just by the gain matrix $\boldsymbol{K}$. Eq. (7) yields the gain factor of clock bias parameter a_0

$$K_1 = \frac{\sigma_u^2}{\sigma_u^2 + \sigma_\varepsilon^2} \tag{9}$$

It is obvious that the gain K_1 indicates a weight-balance relationship between the measurement and process noises. In practice, we usually treat short-term data as stochastically identical in precision, that is, assigning σ_ε^2 with a fixed value. But the empirical value might not be correct enough in the data processing. Eq. (9) implies that we can adjust state noise variance σ_u^2 herein to compensate the inaccuracy of adopted value of σ_ε^2, which still ensure the equivalent value of the gain K_1.

With the equivalent adjustment to the two parameters of σ_u^2 and σ_ε^2 in Eq. (9), the gain K_1 should be able to reflect the overall variations of σ_ε^2 during filter and short-term prediction period. That is, when measurement quality declines from sub-section to sub-section, the state noise variance σ_u^2 should adaptively enlarge in order to restrain the function of measurement update; when measurement quality becomes higher, σ_u^2 should be adaptively decrease in order to boost the function of measurement update. Obviously, simply fixing certain empirical value for σ_u^2 can not demonstrate the adaptive effect mentioned above.

Another advantage of using the process compensated filter lies in the fact that it can balance the decrease in the covariance matrix that occurs at the observation point in Eq. (8). Under this situation, the estimation procedure will always be sensitive to new observations [14].

3 Adaptive Adjust Method of State Noise Variance

It usually uses the nearest 2 hours of data set as sub-section measurements to fit the filter and moves forward to the next sub-section per hour. Therefore, the adjacent sub-sections always have 1 hour of data set in common. With the specific data overlap rate as high as 50% between the adjacent sub-sections, we introduce an adaptive approach for the best estimation of state noise variance $\hat{\sigma}_u^2$. The last 3 hours of data is used in the estimation process. With the first 2 hours of data, we estimate the clock parameters of the third sub-section. Then we do an ultra-short prediction of 1 hour with the estimated parameters to the end of current sub-section. Compared with the real observation at the last data point, the best value of $\hat{\sigma}_u^2$ can be obtained and fixed in the current filter. The algorithm can be summarized as follows,

(1) Based on the conventional batch algorithm, fitting the clock parameters with the first 2 hours of data (or directly fetch the last estimation results). Then eliminating coarse and putting the clear data into the Kalman filter.
(2) Initializing the results from batch algorithm, varying the value of σ_u^2 within [1e-8, 1e-22], starting the filter with SNC and predicting the end of the 3 h, the biases da_0 between the predicted value and the observations for every value of σ_u^2 are recorded, and then the function of da_0 varied with σ_u^2 can be obtained.
(3) Finding the value of σ_u^2 corresponding to the minimum da_0 and taking it as the best adaptive $\hat{\sigma}_u^2$ for the current subsection.

4 Validation and Results

The new proposed algorithm has been tested with the real observations of Compass GEO satellite clock offset data of 3 days. The clock offset data is divided consecutively into 66 subsections of 2 hours as fit period, which has 1 hour data overlap between any two adjacent subsections. Three algorithms, Batch, EKF1 with fixed σ_u^2 and EKF2 with adaptively adjust σ_u^2, are used to estimate and predict clock error. And the initial values of the later two algorithms adopt Batch results.

4.1 Overall Variation in Data Quality

Figure 4.1 shows sample time series of the RMS of observation residuals resulted from Batch algorithm. With respect to the commonly used data noise level, the overall variation in data quality from subsection to subsection is notable.

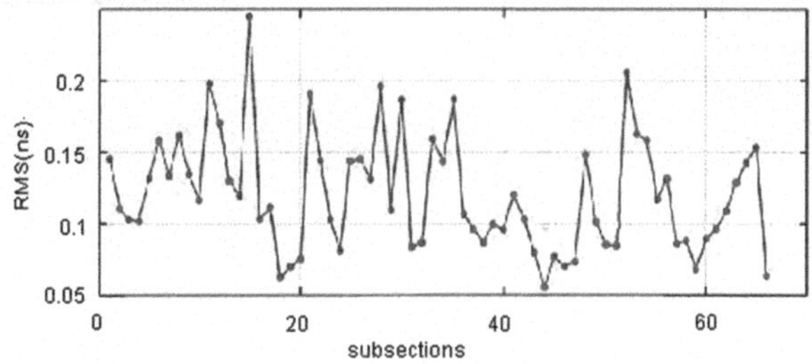

Fig. 4.1 RMS of observation residuals resulted from Batch algorithm

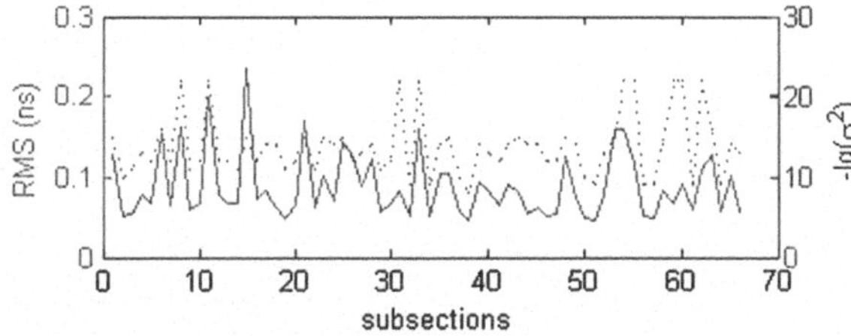

Fig. 4.2 RMS of predicted residuals from EKF2 (*real*) versus $\hat{\sigma}_u^2$ (*dashed*)

The larger RMS denotes the worst consistency of clock offset, which directly influence the estimated clock parameters and the predicting precision.

Figure 4.2 shows the clock prediction residuals (real line) and the best process noise variance σ_u^2 (dashed line) of EKF2. We can see that the changing range of dashed line is from 1e-8 to 1e-22, and the variation trend of the two random parameters is consistent. If data quality of the subsection become lower, the clock prediction residuals will be larger, and the process noise variance will be smaller in order to restrain the measurement update, and vice versa. For example, corresponding the 0.2 ns clock prediction residuals, the estimation value of process noise variance $\hat{\sigma}_u^2$ reduce to 1e-22.

Illumination: horizontal axis is the subsection numbers of clock prediction, left vertical axis is the rms of clock prediction, and right vertical axis is the logarithm of process noise variance σ_u^2.

4.2 Short-Term Prediction Results

Prediction results from three algorithms are agree for normal subsections, where σ_u^2 for EKF1 is fixed as 1e-22. It also validates the algorithm correction of the two process compensated filters. But from the overall sampling subsections, the proposed EKF2 with adaptive σ_u^2 demonstrates the best smoothing not only within the subsections but also inter-sections.

(1) **Maximum predicted clock bias**

Figure 4.3 shows sample time series of the maximum predicted clock bias for Batch and EKF2 algorithms.

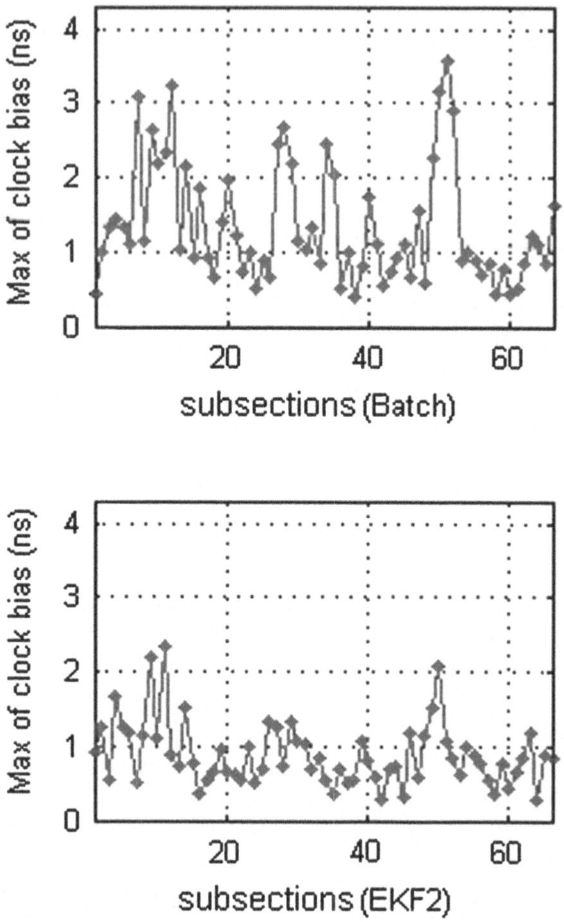

Fig. 4.3 Max. predicted clock bias for EKF2 and batch

The upside figure indicates the averaged maximum predicted clock bias from Batch algorithm is bout 1.5 ns, but certain parts can even reach 3 ns. However, according to EKF2 algorithm, shown in the downside figure, the averaged level is 0.8 ns and it remains less than 2 ns basically throughout the sections.

Figure 4.4 shows estimated and predicted residuals of Batch and EKF2 for subsection fifty. The relatively flat trend in EKF2 indicates it can simultaneously improve both the estimation precision and prediction precision, while the maximum bias is reduced significantly from 3.6 ns to 2.05 ns.

Figure 4.5 shows improvement statistics in maximum clock bias for two filters with SNC compared with that of Batch. The upside figure of EKF2 indicates most of the corrected values are positive, whereas a few values are negative but still less

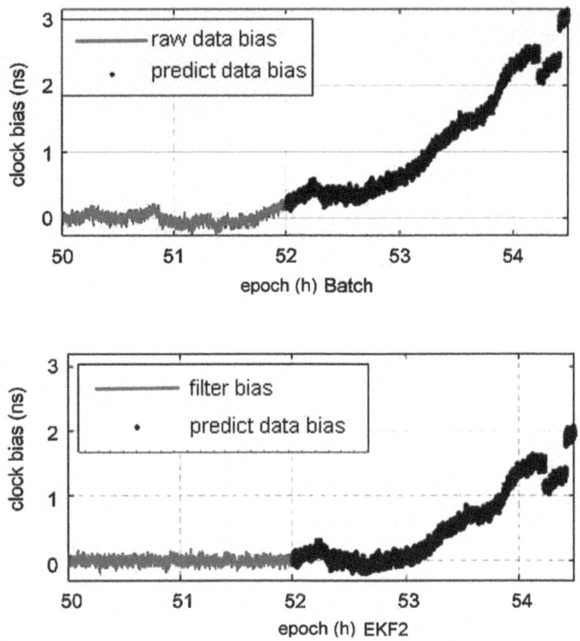

Fig. 4.4 RMS of fit and predicted residuals of 50th subsection for EKF2 and batch

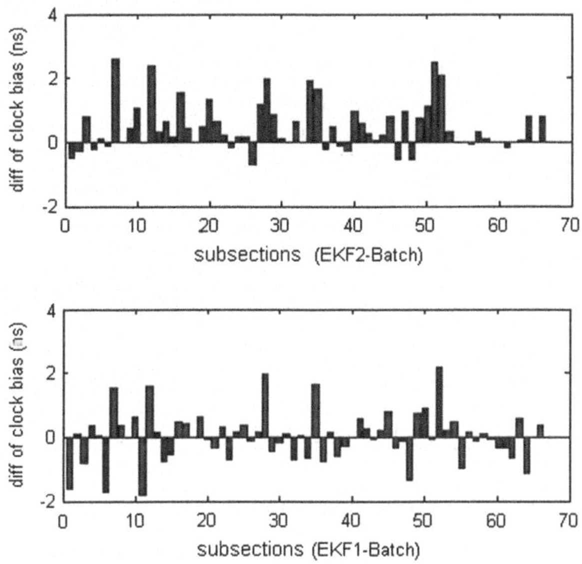

Fig. 4.5 Max. predict clock bias for EKF1-Batch and EKF2-batch

than 1 ns. Further check reveals those worse ones appearing at where the maximum predict biases are negligible. The downside figure displays no improvement of EKF1 with respect to Batch, where σ_u^2 uses 1e-14 as fixed value for all subsections.

Fig. 4.6 Clock bias leaps from EKF2 and batch

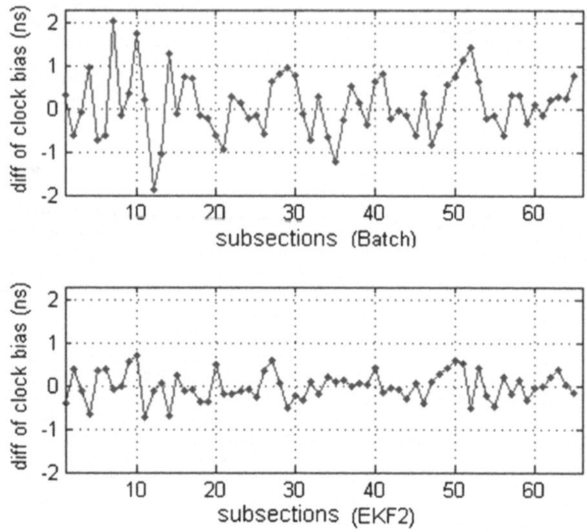

Fig. 4.7 Leap differences for EKF1 and EKF2 compared with batch

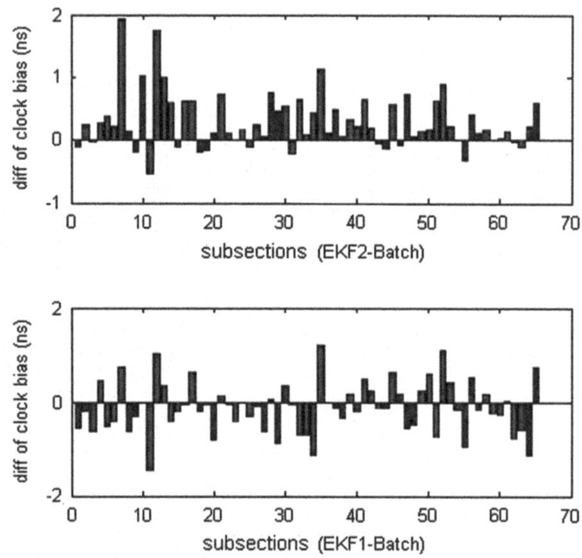

(2) **Clock leaps of transition points**

The very high update rate of short-term prediction yields dense transition points among subsections. If the piece-wise clock parameter estimations change comparably, the predicted satellite clock offset will exhibit some big leaps. Figure 4.6 shows the contrast of clock leaps at transition points between EKF2 and Batch. Because the conventional Batch merely focuses on the data set within subsection, a certain part of leap is larger than 1 ns, whereas some worse ones reach up to 2 ns.

Contrastively, EKF2 utilizes half data of previous subsection along with all of current data to search the best value for σ_u^2, namely, short prediction. This method makes the best of data relevancy between adjacent subsections, so the whole of sampled leaps have been restrained within 1 ns.

Figure 4.7 shows improvement statistics in clock leaps for two filters with SNC. The upside figure strengthens the merits of EKF2 algorithm by smoothing considerable transition leaps in Batch. Likewise, still some negative values indicates a little bigger than that of Batch, but they correspond to negligible predict biases. For example, the worst leap difference of -1 ns appears in subsection eleven, but the absolute leap for Batch is close to zero (see Fig. 4.6, downside). Again, the downside figure displays no improvement of EKF1 with respect to Batch, where σ_u^2 uses 1e-14 as fixed value for all subsections.

5 Conclusions

We have developed a new approach for short-term satellite clocks estimation and prediction. The new approach sufficiently makes use of data relevancy of adjacent subsections instead of treating subsection separately. The advantage of the new approach is: the best adaptive state variance factor is estimated by utilizing the peculiar specification in data processing of Compass, namely, 50% of data overlap between any two adjacent subsections. Therefore, the new approach can provide precise and steady short-term clocks prediction from a long run point of view.

The new approach is validated by comparing its clock estimates and short-term predictions with the values obtained using Batch and EKF with a fixed state variance factor from 66 consecutive subsections of GEO clock offset observations. The short-term prediction precision is about 0.8 ns and is close to 50% enhancement with respect to that of the Batch approach. The major differences occur when prediction biases of Batch are small and negligible. The cause is most likely abnormal data-quality variations appearing in adjacent subsections and resulting in inaccurate estimates for state variance factor during the ultra-short prediction. More investigation will be carried out in order to obtain a consistent result.

Most of all, the new approach smoothes the frequent clock leaps at the transition points of consecutive subsections dramatically. The sampled leaps almost restrain with 1 ns. In addition, the short prediction can demonstrate another advantage of inspecting and evaluating the prediction results as good as 1 hour in real-time, which is rather practical and effective in fault-positioning of the routine operation and management of Compass.

The new approach overcomes the major drawback of the conventional Batch approach, where the dense update of clock parameters of 1 hour is ignored and brings certain big leaps at transition points, and it retains the stable use of satellite clock corrections for users. The effective adoption of the process noise compensation and adaptively estimating the state variance factor makes the clock parameters more stable before generating navigation message. The significantly improved prediction performance makes steady short-term clock prediction at a high update rate possible.

Acknowledgments Funding information: National natural science foundation of China (41174027), open foundation from key laboratory of dynamic geodesy of CAS (L09-04).

References

1. Xu, Q. F. (1994). *GPS satellite navigation and precise positioning*. Beijing: PLA Publishing House (in Chinese).
2. Liu, L. (2004, April). *Relativity time comparison and high-precision time synchronization technology*. The PLA Information Engineering University.
3. Liu, L., Zhu, L. F., Han, C. H., et al. (2009). The model and experimental analysis of two-way satellite and ground time transfer. *Acta Astronomic Sinica, 50*(2), 189–196 (in Chinese).
4. Yang, F.-M., Li, X., et al. (2003). *Progress of laser time transfer technology*. Annual meeting of national time and frequency.
5. Zheng, Z., Chen, Y., & Lu, X. (2008). An improved grey model for the prediction of real-time GPS satellite clock bias. *Acta Astronomica Sinica, 49*(3), 306–320.
6. Zheng, Z., Dang, Y., Lu, X., & Xu, W. (2010). Prediction Model with periodic item and its application to the prediction of GPS satellite clock bias. *Acta Astronomica Sinica, 51*(1), 95–102.
7. Guo, C., & Teng, Y. (2010). Performance analysis of satellite clock bias based on wavelet analysis and neural network. *Acta Astronomica Sinica, 51*(4), 395–403.
8. Yang, Y. (2003). Properties of the adaptive filtering for kinematic positioning. *Acta Geodaetica Et Acrtographica Sinica, 32*(3), 189–192.
9. Yang, Y., & Gao, W. (2005). Influence comparison of adaptive factors on navigation results. *Journal of Navigation, 58*(3), 471–478.
10. Yang, Y. (2006). *Adaptive navigation and kinematic positioning*. Beijing: Publishing House of Surveying and Mapping.
11. Yang, Y., & Gao, W. (2006). An optimal adaptive Kalman filter. *Journal of Geodesy, 80*, 177–183.
12. Yang, Y., & Cui, X. (2008). Adaptively robust filter with multi adaptive factors. *Survey Review, 40*(309), 260–270.
13. Huang, G., Yang, Y., & Zhang, Q. (2011). Estimate and predict satellite clock error using adaptively robust sequential adjustment with classified adaptive factors based on opening windows. *Acta Geodaetica et Cartographica Sinica, 40*(1), 15–21.
14. Montenbruck, E. G. (2000). *Satellite orbits—models, methods, and applications* (pp. 293–302). New York: Springer-Verlag.
15. Guo, H., & Yang, Y. (2009). Analyses of main error sources on time-domain frequency stability for atomic clocks of navigation satellites. *Geomatics and Information Science of Wuhan University, 34*(2), 218–221.

Scalar Weighed Least Square Combination Model for Clock Offset Prediction

Chao Song and Jinming Hao

Abstract The clock offset prediction plays an important role in time and frequency community. Because the conventional prediction models have both advantages and disadvantages, the results of these models combine with synthesizing the characteristics of various models. In order to solve the problem, a scalar weighed least square combination model is proposed. To demonstrate the efficiency of this method, the clock offset data of GPS satellites are chosen and comparation is made among three methods. Gray model, polynomial model and,the new combination model. The result shows that reliability and the precision of the clock offset prediction with the new method is improved.

1 Introduction

It is much important for the Global Navigation Satellite System to possess high precision satellite clock offset. Scholars both in domestic and abroad have studied the prediction of satellite clock offset for a long time. Cui [1] and his team have studied the application of gray model in the prediction while Guo [2] has done a lot of detailed work in the prediction methods. Besides, neutral network and wavelet analysis have been applied successfully in the prediction of satellite clock offset.

Each method has both advantages and disadvantages. In this paper, two methods are combined together by scalar weighed least square combination model to predict the clock offset.

C. Song (✉) · J. Hao
Institute of Surveying and Mapping, Information Engineering University,
Zhengzhou, People's Republic of China
e-mail: 979290126@qq.com

J. Sun et al. (eds.), *China Satellite Navigation Conference (CSNC) 2012 Proceedings*, Lecture Notes in Electrical Engineering 160, DOI: 10.1007/978-3-642-29175-3_49,

2 Two Clock Offset Prediction Models

Two models, gray model and polynomial model, are discussed here.

2.1 *Polynomial Model*

The most popular prediction model is quadratic polynomial model [3]:

$$\Delta t = a_0 + a_1(t - t_0) + a_2(t - t_0)^2 + \int_{t_0}^{t} y(t)dt \tag{2.1}$$

where, a_0 is the satellite clock offset, $a_1$is speed, and a_2 is drift at t_0. $\int_{t_0}^{t} y(t)dt$ is a random item with statistical characters of the stability of the clock.

With more than 3 clocks offset data, the coefficients a_0, a_1 and a_2 could be estimated by means of the least square method. Then, the clock offset at t_i could be calculated through Eq. (2.1).

The nature of polynomial model is to fit the clock offset series with power functions.

2.2 *Gray Model*

In gray model, a new number series is obtained by accumulating or regressing original number series and used in the discrete prediction model.

The nature of the gray model is to fit the clock offset series with exponential functions. The model construction is a simple differential equation, of which the solution is

$$f(x) = c_1 e^{-qx} + c_2 \tag{2.2}$$

where c_1 and c_2 are constant; in which, when $q > 0, f(x)$ infinitely decreases and approaches c_2 while $f(x)$ infinitely increases if $q < 0$. Obviously, it is unrealistic for clock offset increasing and decreasing infinitely. Thus, exponential functions are not suitable for fitting clock offset data.

3 Scalar Weighed Least Square Combination Model

Scalar weighed least square combination model originates from the scalar weighed linear minimum variance best combination estimation algorithm in the combined navigation. Supposing that the outputs and its corresponding covariance of n models at t_k are $\widetilde{X}_k^1, \widetilde{X}_k^2, \ldots, \widetilde{X}_k^n$ and $\sum \widetilde{X}_k^i (i = 1, 2, \ldots, n)$, then, the final combination solution is

$$\widehat{X}_k = \sum_{i=1}^{n} w_i \widetilde{X}_k^i \tag{3.1}$$

w_i represents weighed coefficients, and it satisfies

$$\sum_{i=1}^{n} w_i = 1 \tag{3.2}$$

to ensure the unbiased.

Assuming $\widetilde{X}_k^1, \widetilde{X}_k^2, \ldots, \widetilde{X}_k^n$ are independent from each other. To obtain minimum posterior covariance matrix, the right side of Eq. (3.3)

$$\sum\nolimits_{\widehat{X}_k} = \sum_{i=1}^{n} w_i^2 \sum\nolimits_{\widetilde{X}_k^i} \tag{3.3}$$

should be the least, which equals to

$$f = tr(\sum\nolimits_{\widehat{X}_k}) = \min \tag{3.4}$$

The condition extreme of Eq. (3.4) is

$$f = \sum_{i=1}^{n} w_i^2 \sum\nolimits_{\widetilde{X}_k^i} + 2\lambda (\sum_{i=1}^{n} w_i - 1) = \min \tag{3.5}$$

Let $\partial f / \partial w_i = 0$, and

$$\frac{\partial f}{\partial w_i} = 2 w_i tr(\sum\nolimits_{\widetilde{X}_k^i}) + 2\lambda = 0 \tag{3.6}$$

then

$$w_i = -\frac{\lambda}{tr(\sum_{\widetilde{X}_k^i})} \tag{3.7}$$

Based on (3.2)and (3.7), one attains

$$-\lambda \sum_{i=1}^{n} \frac{1}{tr(\sum_{\widetilde{X}_k^i})} = 1 \tag{3.8}$$

If

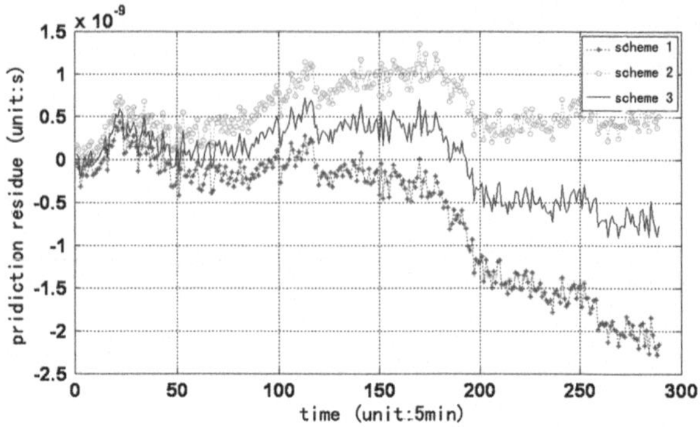

Fig. 1 PRN05 prediction residue time series

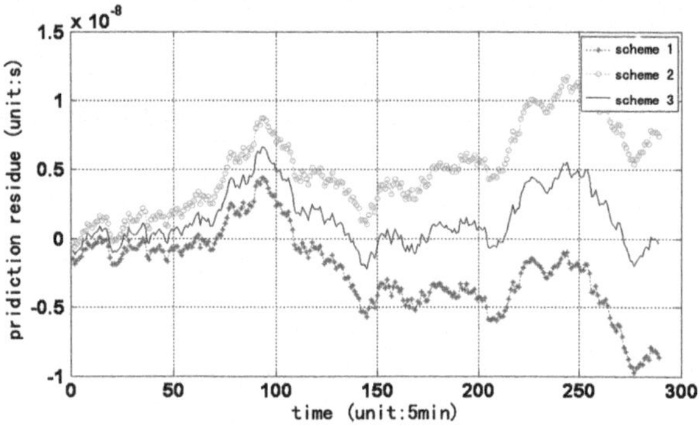

Fig. 2 PRN09 prediction residue time series

$$p_i = \frac{1}{tr(\sum_{\tilde{X}_k^i})} \tag{3.9}$$

then

$$\lambda = -\frac{1}{\sum_{i=1}^{n} p_i} \tag{3.10}$$

Combining (3.7), (3.9) and (3.10), one obtains

$$w_i = \frac{p_i}{\sum_{i=1}^{n} p_i} \tag{3.11}$$

Finally, the combination solution is [4]

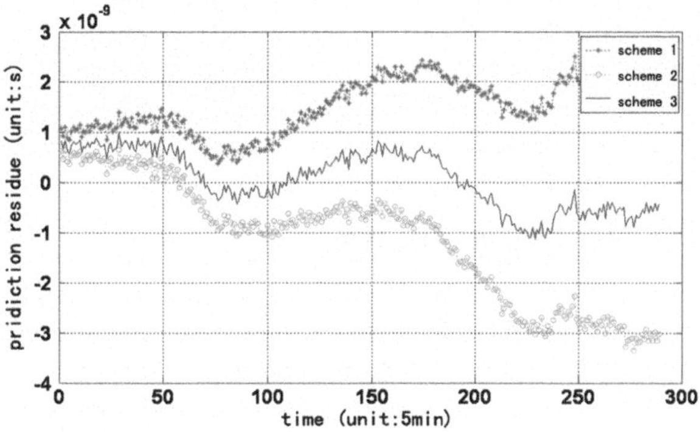

Fig. 3 PRN11 prediction residue time series

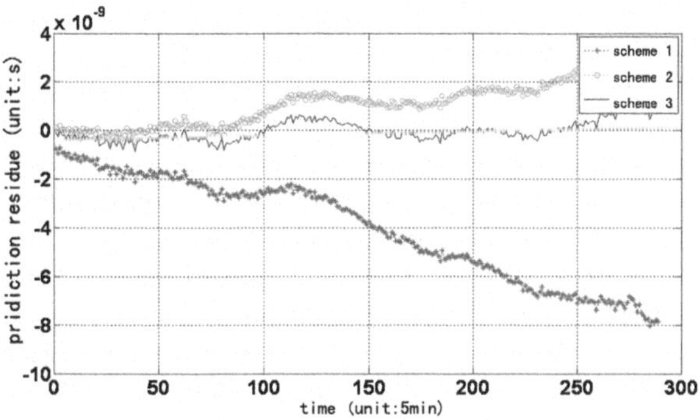

Fig. 4 PRN12 prediction residue time series

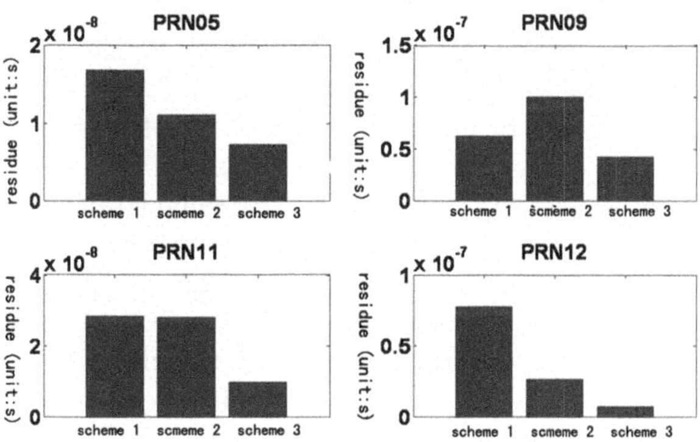

Fig. 5 Prediction RMS

$$\widehat{X}_k = \left(\sum_{i=1}^{n} p_i\right)^{-1}\left(\sum_{i=1}^{n} p_i \widetilde{X}_k^i\right) \tag{3.12}$$

4 Calculation Analysis

The precise GPS clock offset data used here are provided by IGS with sampling interval at 5 min. Three schemes are taken into account here, gray model, quadratic polynomial model and scalar weighed least square combination model considering both the former two models.

Precise clock offset data of the first day of the 1635th week are used in the fitting model to predict that of the next day. Taking the precise clock offset of IGS as the true value, the prediction RMS of the residuals is

$$RMS = \sqrt{\frac{1}{N}\sum_{i=1}^{N}(t_i - T_i)^2} \tag{4.1}$$

where t_i is the predicted clock offset, T_i is the precise clock offset of IGS, N is the epoch number of the predicted clock offset.

Under the three schemes, four GPS satellites are chosen in the calculation. The results are shown in the following figures.

According to Fig. 5, the combined model is better than any of the single one.

Besides, from Figs. 1, 2, 3 and 4, it could be drawn that the precision of the combined model depend on that of the single model. As the precision of single model of PRN09 (Fig. 2) is lower than that of the other three figures, the precision of the combined model is correspondingly lower.

5 Conclusions

The application of scalar weighed least square combination model in the prediction of the satellite clock prediction is studied here. Calculation results show that the precision of the combined model is better than that of its single model and depend on the precision of its single model.

However, only two models are combined in the new model. In addition, due to the single model with low precision, the precision of the combined model is not high. Therefore, various methods such as spectrum analysis, wavelet analysis, neutral network and so on are able to involved in the combined model to enhance the precision of prediction of satellite clock offset. Further research will be made to discuss how to obtain better combined model with the methods mentioned above.

References

1. Cui, X., & Jiao, W. (2005). Grey system model for the satellite clock error. *Predicting Geomatics and Information Science of Wuhan University*, *30*(3), 447–449.
2. Guo, H. (2006). *Study on the analysis theories and algorithms of the time and frequency characterization for atomic clocks of navgation satellites*. Dissertation, Information Engineering University.
3. Li Z, & Huang, J. (2005). GPS surveying and data processing. Wuhan: Wuhan University Press.
4. Yang, Y. (2006). *Adaptive navigation and kinematic postioning*. Beijing: Surveying and Mapping Press.

Study of Main Techniques for Space Passive Hydrogen Masers

Yonghui Xie, Jiayu Dia, Wenxing Chen, Yong Zhang, Jiayang Liu, Jixing Peng, Tiexin Liu and Chuanfu Lin

Abstract Development of space passive hydrogen maser in Shanghai Astronomical Observatory was briefly reported. Main techniques, including novel cavity-bulb assembly, absorption pump, nickel tube purifier and time-separated control scheme of electronics package, were described. By using the techniques, performances and life time of the prototype were improved. Engineering modification is under way.

1 Introduction

Passive hydrogen maser (PHM) has better performance than the widely used rubidium clock in frequency stability, frequency drift. Using the PHM as the satellite clock instead of the rubidium clock, the positioning precision of a global navigation satellite system (GNSS) will be significantly improved. The small frequency drift of the PHM, typically in 5E-15/day order, is also helpful for autonomous navigation of the GNSS. Recently the size and the weight of PHM have been greatly reduced for GNSS application. For example, the engineering model of the space PHM for Galileo has realized a size of 210 mm × 250 mm × 500 mm and a weight of 18 kg [1], and has been used in Galileo satellites.

Shanghai Astronomical Observatory has a long history in research of active hydrogen masers, and has been developing the space PHM since 2005. Main techniques have been investigated, key assemblies and a test system of PHM have

Y. Xie (✉) · J. Dia · W. Chen · Y. Zhang · J. Liu · J. Peng · T. Liu · C. Lin
Shanghai Astronomical Observatory, Chinese Academy of Sciences, Shanghai, China
e-mail: xyh@shao.ac.cn

J. Sun et al. (eds.), *China Satellite Navigation Conference (CSNC) 2012 Proceedings*,
Lecture Notes in Electrical Engineering 160, DOI: 10.1007/978-3-642-29175-3_50,

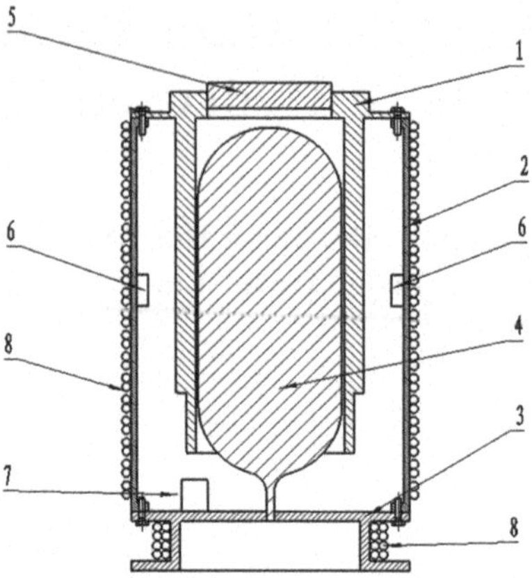

Fig. 1 structure of an electrode cavity for PHM. *1* electrode tube; *2* cylinder cavity; *3* base plate; *4* storage bulb; *5* frequency adjust knob; *6* coupling loop; *7* varactor diode loop; *8* thermal control coil

been made, providing a good foundation for further construction of a space PHM prototype. In this paper we report main features of design of the system.

2 Physics Package

2.1 Novel Kind of Electrode Cavity

A novel electrode microwave cavity for physics package was employed for its high cavity quality factor and filling factor. The structure of the cavity as shown in Fig. 1.

Electrode tube, as shown in Fig. 2, is the core component of the cavity. Several electrodes are firmly fixed on the top flange, and gaps occur between electrodes. Electrodes and gaps form an induction-capacitance structure for microwave resonance. The structure of electrode tube is comparatively simple and has less disturbance to the microwave field. So the cavity has higher Q factor than the well known magnetron cavity [1, 2]. A simulation showed that the filling factor of the cavity is about 0.5.

2.2 Absorption Pump

We developed absorption pump to meet the needs of long operation time and light weight of the PHM. In the design a non-evaporable getter was employed, and vacuum interface, exterior and interior structures of the pump were optimized. Activation of gas absorber and high temperature protection were taken into account. Several absorption pumps have been built and tested. Absorption ability test showed that the pump can absorb up to 40 bar*l of hydrogen gas, and the vacuum can be kept at 10^{-5} to 10^{-4} Pa after gas absorption. The absorption pump

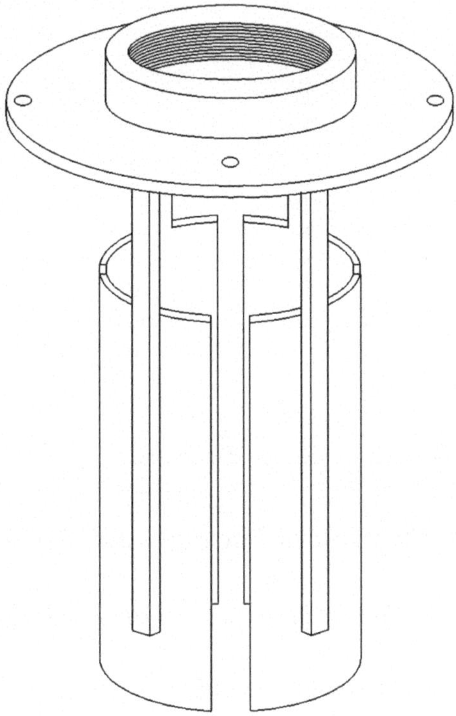

Fig. 2 Structure of the electrode tube

has been applied in a passive hydrogen maser. The vacuum pressure of the system has met our design requirement, and atomic gain is up to acceptable 3 dB. Life time and reliability of the pump are still under test.

2.3 Nickel Purifier

The nickel purifier is used to purify hydrogen gas from the hydrogen source. Key component of the nickel purifier is nickel tube. The operation principle of the nickel purifier is that hydrogen atoms can penetrate nickel tube, while other atoms can not.

Stability of hydrogen atom flux is essential to steady operation of a PHM. In our design atom flux is stabilized by controlling heating power of the nickel purifier. Working temperature of the nickel tube is chosen to be nearly 500°C. The temperature is much higher than that used for a palladium purifier, but the atom flux shows still a low sensitivity to temperature. The final flux fluctuation is controlled to be within 1% by the nickel purifier, acceptable for realization of the long term stability of a PHM. Atom flux produced by the nickel purifier at room temperature is nearly zero, which means the hydrogen wastage at stand-by state is very low.

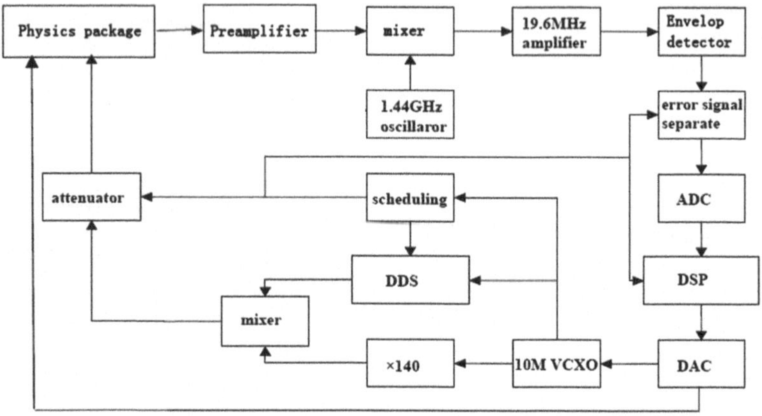

Fig. 3 Block-diagram of the PHM

3 Electronics Package and Performance Test

3.1 Time-Separated Control Method

Block-diagram of the PHM is shown in Fig. 3. Electronics package contains two frequency control loops, i.e. VCXO control loop and cavity control loop. The former realizes the lock of VCXO frequency to the hydrogen atomic transition line and the latter realizes the lock of cavity frequency to 1.420405 GHz. A scheduling module was used to determine working state of the two loops. The 20.405 MHz signal produced by DDS is frequency modulated and mixed with the 1400 MHz signal produced by multiplication of the 10 MHz VCXO signal, thus the 1420.405 MHz is obtained as the interrogation signal for hydrogen atomic transition. The two control loops have the same carrier wave frequency but different modulation frequency and modulation index for interrogation signal, and these two loops work time-separately. Corresponding error signals are induced in the physics package and time separately acquired. After computation of DSP the error signals form control signals for VCXO and varactor respectively.

3.2 Performance Test

Physics package and electrics package have been integrated into a test system of PHM, and working parameters of the system have been optimized. A preliminary test has been performed.

Frequency stability is tested by using the frequency comparator VCH-314, and the reference clock is two active hydrogen masers. Result shows that the frequency stability of the PHM system meets application demand.

4 Conclusion

Techniques for space passive hydrogen maser have been studied at Shanghai Astronomical Observatory. The techniques include a novel electrode cavity, absorption pump, nickel purifier, and time separated control loop. Several test systems of PHM have been developed, and the space prototype of PHM is in progress.

References

1. Wang, Q., Mosset, P., & Droz, F. (2006). Verification and optimization of the physics parameters of the onboard Galileo passive hydrogen maser. *38th PTTI Meeting*, Washington.
2. Demidov, N. A., Belyaev, A. A., Sakharov, B. A., & Uljanov, A. A. (1993). Passive hydrogen maser frequency stability and accuracy investigation. *7th EFTF Meeting*, Neuchatel.

Generation of Broadband Frequency Entangled Biphotons for Quantum Clock Synchronization

Run-ai Quan, Rui-fang Dong, Fei-yan Hou, Yun Bai, Yu Zhang, Tao Liu and Shou-gang Zhang

Abstract: Frequency entangled biphotons based on spontaneous parametric down conversion (SPDC) of nonlinear crystal are widely used in Quantum clock synchronization protocols. The time-correlation width of the entangled source determines the accuracy of the attainable synchronization, which relies on the spectral bandwidth of the generated biphotons. We theoretically investigate the generation of frequency anti-correlated biphotons from chirped periodically-poled potassium titanyl phosphate (C-PPKTP). It is demonstrated that an ultra-broad-band entangled biphoton source with a width of 857 nm was obtained, by using a 10 mm-long crystal with a chirping of 9.7×10^{-6} μm^{-2}, and a cw pumping source with a wavelength of 792 nm. The corresponding time correlation width is only 3.5 fs, which implies feasible clock synchronization accuracy in femtosecond scale. We further demonstrate that the narrowing of the time-correlation width slows down dramatically by increasing the chirping and the length of the nonlinear crystal, which provides a theoretical instruction for us to trade off between the complexity of the crystal fabrication and the sufficiently narrow time-correlation width.

R. Quan · R. Dong (✉) · F. Hou · Y. Bai · Y. Zhang · T. Liu · S. Zhang
Key Laboratory of Time and Frequency Primary Standards, National Time Service Center, Chinese Academy of Science, 710600 Xian, China
e-mail: dongruifang@ntsc.ac.cn

Y. Bai · Y. Zhang
Graduate University of Chinese Academy of Science, 100039 Beijing, China

J. Sun et al. (eds.), *China Satellite Navigation Conference (CSNC) 2012 Proceedings*, Lecture Notes in Electrical Engineering 160, DOI: 10.1007/978-3-642-29175-3_51,

1 Introduction

Quantum clock synchronization is a new clock synchronization technology, which is based on quantum mechanical principles to break through the shot noise limit on the classical synchronization precision. Since proposed one decade ago [1], it posed a bright future on improving the distant clock synchronizing accuracy dramatically. Therefore, researches on how to accomplish an advanced quantum clock synchronization system have been extensively carried out [2−4]. In present quantum clock synchronization protocols, frequency-entangled biphoton sources are the most widely used quantum resources [3, 5], the time correlation width of which realized by the quantum interference measurement determines the achievable synchronization accuracy. It is reported that [3] the clock synchronizing accuracy is increased with the decrement of time-correlation width, which is dependent on the broadening of the spectral bandwidth of the biphoton sources. Therefore, investigation on generation of broadband, frequency entanglement has been attracting particular attentions, and a large number of theoretical and experimental works have been carried out [6–13].

The frequency entangled biphotons are commonly prepared by spontaneous parametric down conversion process in a nonlinear crystal, in which lower-frequency biphotons were produced by an intensive pump with higher frequency. When the pump is monochromatic, the frequency anti-correlated biphotons will be obtained. On the other hand, if the pump is pulsed, with the nonlinear crystal satisfying the extended phase matching condition and being sufficiently long, the frequency correlated biphotons will be obtained [14, 15]. Up to now, many attempts have been done on the generation of broadband frequency entangled biphotons. For example, Shimizu et al. [9] realized a frequency anti-correlated biphotons with a bandwidth of 35 nm by the use of periodically poled MgO-doped stoichiometric lithium tantalite (PPMgSLT) crystal. Nasr et al. [10] achieved pairs of photons with a bandwidth of 300 nm using chirped PPSLT. Periodically poled lithium niobate (PPLN) [11] and beta barium metaborate (BBO) [12] were also reported as the nonlinear medium for generation of broadband frequency entangled biphotons. Recently, $Al_xGa_{1-x}As$ material-based Bragg reflection waveguides [13] were used to obtain frequency correlated biphotons with tunable bandwidth from 1 nm to 450 nm. Compared with the above nonlinear crystals, potassium titanyl phosphate (KTP) crystal has many advantages, such as high conversion efficiency, high damage threshold, and weaker photon refractive effect and so on. Furthermore, it has a relatively low coercive field, which is in great help for the periodic poling fabrication of the nonlinear crystals to realize the period reversal with a depth of more than 1 mm [16]. All of these characteristics determine KTP to be an excellent candidate in the SPDC application.

In this paper, based on spontaneous parametric down conversion (SPDC) of KTP crystal, we theoretically studied the generation of frequency entangled biphoton state. By using a chirped periodically-poled KTP (C-PPKTP) crystal pumped by a quasi monochromatic light, it is demonstrated that an ultra-broadband frequency anti-correlated entangled biphoton source can be obtained. Consider a 10 mm-long C-PPKTP crystal with a chirp parameter 9.7×10^{-6} μm^{-2} and a cw pumping source with a wavelength of 792 nm, the biphoton spectrum will reach 857 nm, the relevant time correlation width is only 3.5 fs, which corresponds to a feasible clock synchronization with accuracy in femtosecond scale. We further analyzed the dependence of the time-correlation width of the generated broadband biphoton state on the chirping parameter as well as the length of C-PPKTP crystal. The result shows that, in contrast to the continuous broadening feature of the biphoton spectral bandwidth, the narrowing of the time-correlation width slows down dramatically by increasing the chirping and the length of the nonlinear crystal.

2 Mechanism of Generating Frequency Anti-Correlated Biphoton State

The output of a parametric down conversion crystal can be expressed as follows [14]:

$$|\psi\rangle = \iint d\omega_s d\omega_i A(\omega_s, \omega_i) a_s^+(\omega_s) a_i^+(\omega_i)|0\rangle, \tag{1}$$

where $a_s^+(\omega_s)$ and $a_i^+(\omega_i)$ are the creation operators for signal s and idler i modes, respectively, and $|0\rangle$ is the vacuum state. $A(\omega_s, \omega_i)$ is the joint spectral amplitude of the biphoton state. If we assume collinear plane-wave propagation and neglect inconsequential normalizing factors, $A(\omega_s, \omega_i)$ can be written as [17]:

$$A(\omega_s, \omega_i) = \sigma(\omega_s, \omega_i)\Phi(\omega_s, \omega_i), \tag{2}$$

where

$$\sigma(\omega_s, \omega_i) \propto E_p(\omega_s + \omega_i = \omega_p), \tag{3}$$

$$\Phi(\omega_s, \omega_i) = \int_0^L dz \exp[i\Delta k(\omega_s, \omega_i)z], \tag{4}$$

The quantity $\sigma(\omega_s, \omega_i)$ stems from the intense pump and is determined by the pump spectrum. $\Phi(\omega_s, \omega_i)$ is the phase-matching function, which depends on the nonlinear crystal length L and on the phase mismatch wave vector. For the first-order quasi phase matching (QPM) condition

$$\Delta k(\omega_s, \omega_i) = k_s(\omega_s) + k_i(\omega_i) - k_p(\omega_s + \omega_i) - \frac{2\pi}{\Lambda}, \tag{5}$$

where $k_{p,s,i} = \omega \cdot n_{p,s,i}(\omega)/c$ denote the wave numbers of the pump, signal and idler photons, respectively. Λ is the poling period of the nonlinear crystal.

By using a quasi monochromatic cw light of frequency $2\omega_0$ as the pump, $\sigma(\omega_s, \omega_i) \to \delta(\omega_s + \omega_i - 2\omega_0)$. When the SPDC crystal is operated at frequency degeneracy, the signal and idler spectra are both centered around half the pump frequency, i.e. ω_0. The biphoton state can then be reduced to

$$|\psi\rangle = \iint d\Omega \phi(\Omega) a_s^+(\omega_0 + \Omega) a_i^+(\omega_0 - \Omega)|0\rangle, \tag{6}$$

where Ω denotes the frequency derivation from the center frequency ω_0. The frequency spectral amplitude of the biphoton state is determined by the phase matching function: $\phi(\Omega) \propto \Phi_L(\omega_0 - \Omega, \omega_0 + \Omega)$, which features the frequency anti-correlation between the signal and idler photons. Therefore, the bandwidth of the biphoton spectrum is determined by the bandwidth of the phase-matching function.

Introducing spatial chirp into the quasi phase matching condition, the phase-mismatch wave vector can be rewritten as

$$\Delta k(\Omega) = k_s(\omega_0 + \Omega) + k_i(\omega_0 - \Omega) - k_p(2\omega_0) - \mathrm{K_g(z)}, \tag{7}$$

where $\mathrm{K_g(z)} = 2\pi/\Lambda = \mathrm{K_0} - \alpha \mathrm{z}$ is the inverse grating vector, which is linearly dependent on the coordinate z. α is defined as the chirp parameter, and $\mathrm{K_0}$ corresponds to the phase matching condition at the center frequency of the SPDC: $\mathrm{K_0} = k_s(\omega_0) + k_i(\omega_0) - k_p(2\omega_0)$. For type II phase-matched SPDC crystals, $k_s(\omega) \neq k_i(\omega)$.

To simplify the analysis, we expand the wave vector in Taylor series about frequency ω_0 and only maintain until the first derivative term

$$\begin{aligned} k_j(\omega) &= \frac{n_j(\omega)\omega}{c} \approx k_{j,0}^0 + (\omega - \omega_0)k_{j,0}', \\ k_{j,0}' &= \frac{dk_j(\omega)}{d\omega}\bigg|_{\omega=\omega_0}, j = p, s, i. \end{aligned} \tag{8}$$

Then formula (7) can be rewritten as

$$\Delta k(\Omega) = \alpha z - (k_{i,0}' - k_{s,0}')\Omega. \tag{9}$$

The frequency anti-correlated biphoton state is given by:

$$\begin{aligned} \phi(\Omega) &= \int_0^L dz \exp[i(\alpha z + k_{i,0}' - k_{s,0}')\Omega)z] \\ &\propto \exp[-\frac{i(k_{i,0}' - k_{s,0}')^2\Omega^2}{4\alpha}]\{erf[\frac{(-1)^{3/4}(k_{i,0}' - k_{s,0}')\Omega}{2\sqrt{\alpha}}] \\ &- erf[\frac{(-1)^{3/4}((k_{i,0}' - k_{s,0}')\Omega + 2\alpha L)}{2\sqrt{\alpha}}]\}. \end{aligned} \tag{10}$$

The time-correlation width of the biphoton state is determined by the HOM interferometric measurement [18], in which the coincidence count rate is given by

$$P(\tau) \propto \int_T dt_1 \int_T dt_2 \langle \psi | E_1^-(t_1) E_2^-(t_2) E_2^+(t_2) E_1^+(t_1) | \psi \rangle, \tag{11}$$

where the subscripts 1, 2 denote the photodetectors at the two outputs of the HOM interferometer, T is the integration time window of the detectors. $E_j^{\pm}$ denote the positive- and negative-frequency field operators at detector j, when omitting the irrelevant normalization constants, they are defined by

$$\begin{aligned} E_j^+(t_j) &= \int d\omega_j \hat{a}_j(\omega_j) e^{-i\omega_j t_j}, \\ E_j^-(t_j) &= \int d\omega_j \hat{a}_j^+(\omega_j) e^{+i\omega_j t_j} \quad \text{for } j = 1, 2. \end{aligned} \tag{12}$$

By substituting (12) into (11) and taking the limit $T \to \infty$, the coincidence rate will be expressed as:

$$P(\tau) \propto \int d\Omega \{ |\phi(\Omega)|^2 [1 - \cos(2\Omega\tau)] \} = 1 - \tilde{F}(2\tau), \tag{13}$$

where τ denotes the time delay between the signal and idler photons, $\tilde{F}(2\tau)$ is the Fourier transform of the joint spectrum function $|\phi(\Omega)|^2$. It is featured by a dip at $\tau = 0$, the width of which represents the time-correlation width.

3 Theoretical Analyses of Frequency Anti-Correlated Biphotons in C-PPKTP

The refractive indices variation of the KTP crystal is described by the modified Sellmeier equation [19, 20]. By using a quasi monochromatic cw light of 792 nm as the pump, for frequency degeneracy of down conversion, the poling period of the PPKTP crystal is calculated to be about $\Lambda_0 = 46.1$ μm. In the case of the spatial chirping of the poling period, we take the inverse grating vector to be $K_g(z) = 2\pi/\Lambda_0 - \alpha z$ with $z \in [0, L]$. Fixing the crystal length to be 10 mm, we study the characteristics of generated frequency anti-correlated biphoton state at two specific chirping parameter values: $\alpha = 9.2 \times 10^{-7}$ μm^{-2} and $\alpha = 9.7 \times 10^{-6}$ μm^{-2}. The spectrum distribution $|\phi(\Omega)|^2$ and the coincidence interferograms are plotted in Fig. 1. It can be seen that, increasing α from 9.2×10^{-7} μm^{-2} to 9.7×10^{-6} μm^{-2} broadens the biphoton spectrum from 71 to 857 nm, while the time-correlated width of the biphoton state, represented by the full width at half maximum of the HOM dip, is depressed from 37.3 to 3.5 fs.

Theoretical simulations of the SPDC biphoton spectrum and the coincidence interferogram are further implemented as a function of the chirping parameter α.

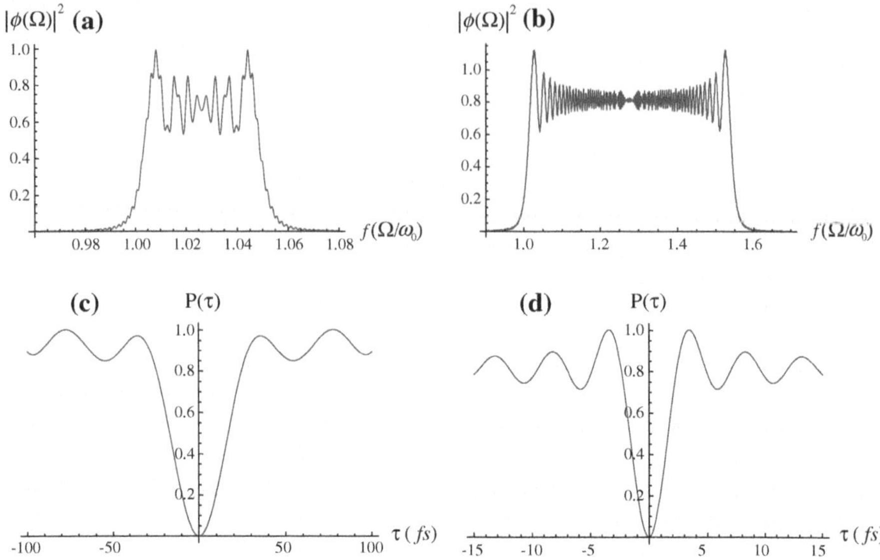

Fig. 1 The calculated spectrum distribution $|\phi(\Omega)|^2$ of the biphoton state are plotted in **a** for $\alpha = 9.2 \times 10^{-7} \mu\text{m}^{-2}$ and **b** for $\alpha = 9.7 \times 10^{-6}\ \mu\text{m}^{-2}$. The normalized coincidence interferogram of biphotons are plotted in **c** for $\alpha = 9.2 \times 10^{-7}\ \mu\text{m}^{-2}$ and **d** for $\alpha = 9.7 \times 10^{-6}\ \mu\text{m}^{-2}$. The time-correlation width of the frequency anti-correlated biphotons is denoted by the FWHM of the HOM dip

As shown in Fig. 2, the increased chirping leads to broadening of the spectrum bandwidth of the biphotons, and narrowing of the time-correlation width. It is seen from Fig. 2a that, no theoretical upper bound exists on the attainable bandwidth. Nevertheless, the depression of the dip width slows down dramatically as the chirping parameter increases. On the other hand, higher chirping means more complicated fabrication of the crystal. Therefore, a proper chirping parameter α should be chosen to achieve a sufficiently narrow time-correlation width. In our case, a chirping of 6.4×10^{-6} suffices to achieve a time-correlation width of 5 fs, with a corresponding joint spectrum bandwidth of 552.8 nm.

As crystal length affects the generated biphoton characteristics as well, we fix the chirping parameter to be $\alpha = 9.2 \times 10^{-7}\ \mu\text{m}^{-2}$ and investigate the joint spectrum and HOM dip for different crystal lengths. The plots are shown in Fig. 3. It can be seen that, as the crystal length increases, the biphoton spectrum bandwidth is broadened almost linearly, correspondingly the time correlated width is suppressed. However, the suppression gets asymptotic with the lengthening of the crystal. For example, when the crystal length is increased from 60 to 70 mm, the time-correlation width only varies from 5.96 to 5.36 fs. This dependence is similar with that on the chirping parameter. Therefore, there is an optimal crystal length to trade off between the complexity of the crystal fabrication and the sufficiently narrow time-correlation width.

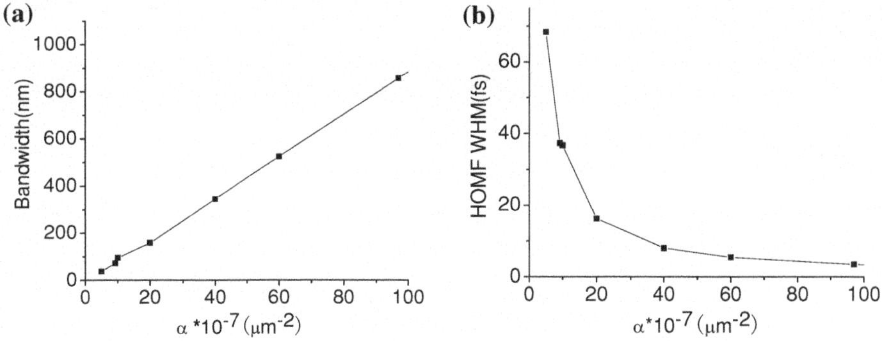

Fig. 2 Plots of the spectrum bandwidth **a** and time-correlation width **b** of the biphoton state for the chirping parameter α increasing from $5 \times 10^{-7}\ \mu m^{-2}$ to $9.7 \times 10^{-6}\ \mu m^{-2}$

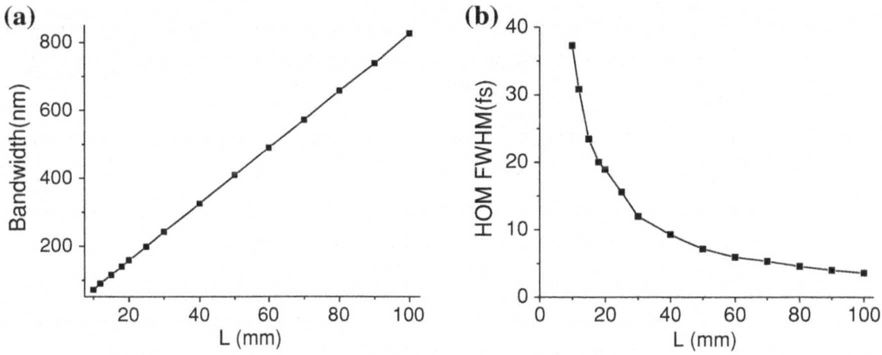

Fig. 3 Plots of the spectrum bandwidth **a** and time-correlation width **b** of the biphoton state as a function of the C-PPKTP crystal length

4 Conclusions

In summary, we theoretically analyzed the generation of frequency anti-correlated biphotons in C-PPKTP crystal by using a cw light with a wavelength of 792 nm as the pump. By increasing the chirping and the length of the nonlinear crystal, the biphoton spectral bandwidth is continuously broadened; no theoretical upper bound exists on the broadening. On the other hand, the time-correlation width of the corresponding biphoton state slows down dramatically with the increment of the chirping and the crystal length. For application in quantum clock synchronization systems, in which the time-correlation width plays an important role, a compromise between the synchronization accuracy and the nonlinear crystal fabrication can be done by such analysis.

Acknowledgments The work has been supported by the National Natural Science Foundation of China (Grant No.11174282).

References

1. Giovannetti, V., Lloyd, S., & Maccone, L. (2001). Quantum-enhanced positioning and clock synchronization. *Nature, 412*, 417–419.
2. Giovannetti, V., Lloyd, S., Maccone, L., & Wong, F. N. C. (2001). Clock synchronization with dispersion cancellation. *Physical Review Letters, 87*, 117902.
3. Giovannetti, V., Lloyd, S., & Maccone, L. (2002). Positioning and clock synchronization through entanglement. *Physical Review A, 65*, 022309.
4. Zhang, J., Long, G. L., Deng, Z., Liu, W., & Lu, Z. (2004). Nuclear magnetic resonance implementation of a quantum clock synchronization algorithm. *Physical Review A, 70*, 062322.
5. Valencia, A., Scarcelli, G., & Shih, Y. (2004). Distant clock synchronization using entangled photon pairs. *Applied Physics Letters, 85*, 2655–2657.
6. Tanakal, A., Okamoto, R., Lim, H. H., Subashchandran, S., Okano, M., Kurimura, S. et al. (2011). Collinear ultra-broadband parametric fluorescence generated from 10%-chirped quasi phase matched device, in CLEO/Europe and EQEC 2011 conference digest, OSA Technical digest (CD) (Optical Society of America, 2011), paper EA_P7.
7. Hendrych, M., Shi, X., Valencia, A., & Torres, J. P. (2009). Broadening the bandwidth of entangled photons: a step towards the generation of extremely short biphotons. *Physical Review A, 79*, 023817.
8. Kitaeva, G. K., Chekhova, M. V., & Shumilkina, O. A. (2009). Generation of broadband biphotons and their compression in an optical fiber. *JETP Letters, 90*, 172–176.
9. Shimizu, R., & Edamatsu, K. (2009). Controlled frequency entanglement of photons in type-II spontaneous parametric down-conversion, in nonlinear optics: Materials, fundamentals and applications, OSA technical digest (CD) (Optical Society of America, 2009), paper JTuB20
10. Nasr, M. B., Carrasco, S., Saleh, B. E. A., Sergienko, A. V., Teich, M. C., Torres, J. P., et al. (2008). Ultrabroadband biphotons generated via chirped quasi- phase-matched optical parametric downconversion. *Physical Review Letters, 100*, 183601.
11. O'Donnell, K. A., & U'Ren, A. B. (2007). Observation of ultrabroadband, beamlike parametric downconversion. *Optics Letters, 32*, 817–819.
12. Torres, J. P., Macià, F., Carrasco, S., & Torner, L. (2005). Engineering the frequency correlations of entangled two-photon states by achromatic phase matching. *Optics Letters, 30*, 314–316.
13. Abolghasem, P., Hendrych, M., Shi, X., Torres, J. P., & Helmy, A. S. (2009). Bandwidth control of paired photons generated in monolithic Bragg reflection waveguides. *Optics Letters, 34*, 2000–2002.
14. Giovannetti, V., Maccone, L., Shapiro, J. H., & Wong, F. N. C. (2002). Extended phase-matching conditions for improved entanglement generation. *Physical Review A, 66*, 043813.
15. Kuzucu, O., Fiorentino, M., Albota, M. A., Wong, F. N., & Kärtner, F. X. (2005). Two-photon coincident-frequency entanglement via extended phase matching. *Physical Review Letters, 94*, 083601.
16. Sang, M. (2003). *Quasi-phase matched PPKTP fabrication and relative application research*. Doctoral thesis of Tianjin University.
17. Giovannetti, V., Maccone, L., Shapiro, J. H., & Wong, F. N. C. (2002). Generating entangled two-photon states with coincident frequencies. *Physical Review Letters, 88*, 183602.
18. Hong, C. K., Ou, Z. Y., & Mandel, L. (1987). Measurement of subpicosecond time intervals between two photons by interference. *Physical Review Letters, 59*, 2044–2046.

19. Fradkin, K., Arie, A., Skliar, A., & Rosenman, G. (1999). Tunable midinfrared source by difference frequency generation in bulk periodically poled KTiOPO4. *Applied Physics Letters, 74*, 914–916.
20. Fan, T. Y., Huang, C. E., Hu, B. Q., Eckardt, R. C., Fan, Y. X., Byer, R. L., et al. (1987). Second harmonic generation and accurate index of refraction measurements in flux-grown $KTiOPO_4$. *Applied Optics, 26*, 2390–2394.

Study of the Physics Package for High Performance Rubidium Frequency Standards

Songbai Kang, Wenbing Li, Pengfei Wang, Feng Zhao, Feng Qi, Fang Wang, Gang Ming, Baihua Xia, Shaofeng An, Da Zhong and Ganghua Mei

Abstract Frequency stability of an atomic frequency standard depends mainly on the Signal to Noise Ratio (SNR) of atomic discrimination signal. For a Rubidium Atomic Frequency Standard (RAFS), the SNR is closely related to the characteristics of the cavity-cell assembly in physics package. The RAFS in our laboratory is designed based on the slotted tube cavity, and a typical stability of $1.5 \times 10^{-12}\ \tau^{-1/2}$ (1–1000 s) has been achieved. The current cavity used in our design is of a disadvantage of field inhomogeneity. To improve further the stability of the RAFS, a modification of the cavity has been carried out. With this modification, the field homogeneity has been greatly improved, and the cavity size was increased also, enabling to hold a larger absorption cell. According to a theoretical evaluation, the SNR of physics package was enhanced by nearly two times, meeting the need to design a RAFS with a stability better than $1.0 \times 10^{-12}\ \tau^{-1/2}$.

S. Kang (✉) · W. Li · P. Wang · F. Zhao · F. Qi · F. Wang · G. Ming · S. An · D. Zhong · G. Mei
Key Laboratory of Atomic Frequency Standards,
Wuhan Institute of Physics and Mathematics,
Chinese Academy of Sciences, 43007 Wuhan, China
e-mail: kangsongbai@wipm.ac.cn

G. Mei
e-mail: mei@wipm.ac.cn

B. Xia
National Defense Information Academy,
430012 Wuhan, China

W. Li · P. Wang · G. Ming
Graduate School, Chinese Academy of Sciences,
430071 Wuhan, China

J. Sun et al. (eds.), *China Satellite Navigation Conference (CSNC) 2012 Proceedings*, Lecture Notes in Electrical Engineering 160, DOI: 10.1007/978-3-642-29175-3_52,

1 Introduction

Application in Global Satellite Navigation System (GNSS) is one of the most significant motivations for development of atomic frequency standards. The satellite onboard atomic frequency standard is of great influence on the positioning precision of the GNSS. Up to now the onboard atomic frequency standards used in navigation satellites are rubidium atomic frequency standards (RAFS), cesium beam frequency standards and passive hydrogen frequency standards. Among these three traditional atomic frequency standards, the RAFS obtained the widest applications due to its small volume, low power consumption, low production cost and high reliability. So far the RAFS has been used in satellites of American GPS, European Galileo and Chinese Compass systems. In recent decades, the performance of onboard RAFS has been greatly improved. The enhanced rubidium atomic frequency standard [1] produced by PerkinElmer for GPS-IIF and future GPS-III has realized a stability better than $1.0 \times 10^{-12}\,\tau^{-1/2}$ (1–10^5 s), which is recognized as the best RAFS in the world.

A rubidium atomic frequency standard normally consists of the physics package and the electronic package. The physics package is the core of a RAFS since the frequency stability is mainly determined by the Signal to Noise Ratio (SNR) of the discrimination signal supplied by the physics package. The frequency stability for the RAFS can be described by [2]

$$\sigma_y = \frac{0.16}{\frac{S}{N} \cdot Q} \tau^{-1/2}, \tag{1}$$

where Q is the quality factor of the rubidium clock transition line and S/N is the SNR. Main contributions to the SNR come from the cavity-cell assembly in the physics package. The cavity-cell assembly normally consists of a microwave cavity, a rubidium absorption cell and a rubidium filter cell, and the absorption cell is inside the cavity, and the filter cell can be located inside or outside the cavity. There are two main factors affecting the physics package's SNR. The first one is the microwave field distribution inside the cavity should be suitable for stimulating the rubidium clock transition. According to quantum mechanics, the rubidium clock transition is a magnetic dipole transition, therefore the magnetic lines of stimulating field should be parallel to the direction of the quantum axis, and densely distributed in central area in the cavity, where the absorption cell is located. This is the reason why a TE_{011} mode cavity is more suitable to be used in the RAFS than a TE_{111} mode one. The other factor is the volume of the cavity. A large cavity enables to use a large absorption cell inside, in this case more rubidium atoms could be involved in transition process, so a high S/N could be expected. In our RAFS design a slotted tube cavity was used [3]. Aiming to build a RAFS with a stability better than $1.0 \times 10^{-12}\,\tau^{-1/2}$, recently we made a modified design of the cavity, by which the field distribution inside the cavity was optimized and the cavity volume was increased. In this paper we report main features of the design, and analyze briefly its potential ability in realizing stability performance of a RAFS.

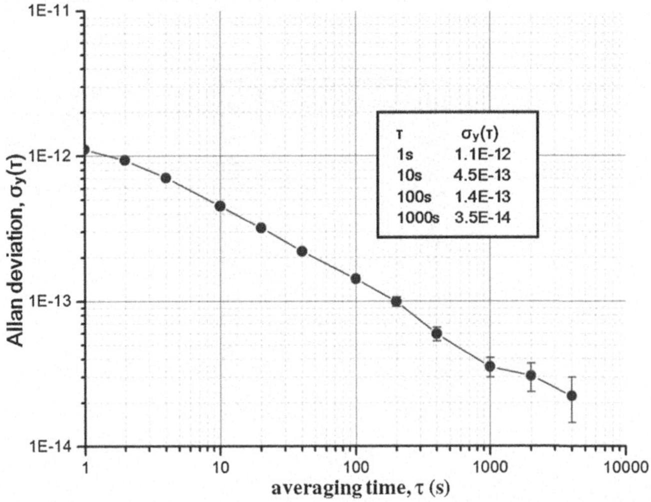

Fig. 1 Typical short term frequency stability of rubidium standards in our laboratory

2 Main Characteristics of the Current Cavity-Cell Assembly

Figure 1 shows a test result of the short-term frequency stability of the RAFS manufactured in our lab, from which a stability of about $1.5 \times 10^{-12}\tau^{-1/2}$ (1–1000 s) can be seen. According to Eq. (1), the SNR of our RAFS's physics package has to be enhanced by nearly two times to meet the requirement to design a RAFS with a stability better than $1.0 \times 10^{-12}\,\tau^{-1/2}$.

The structure of the cavity-cell assembly used in our lab is shown in Fig. 2a, which is based on the patented slotted-tube microwave cavity [3]. The assembly is mainly composed of a slotted tube, a cylindrical cavity body, a dielectric ring, and a top cap. Both the filter and absorption cells are set inside the slotted tube. The photocell is used as the photo detector, and the coupling loop is used to send microwave signal into the cavity. The C-field coil outside the cavity body creates a static and weak magnetic field, defining the quantum axis for clock transition. The magnetic shielding mounted outside the whole cavity-cell assembly is used to cancel the outer magnetic field interference to the atomic system. The slotted tube, as shown in Fig. 2b, is the core of the cavity. The resonance frequency and field distribution are mainly determined by the structural parameters of the slotted tube.

The inner diameter of the slotted tube for current design is 14 mm, enabling to contain an absorption cell with diameter of 14 mm. The field distribution in the cavity has been theoretically simulated and experimentally tested [4]. The obtained magnetic lines in two sections of the cavity are shown in Fig. 3. From Fig. 3a, one can see that the magnetic lines in the place for absorption cell have large components parallel to the quantum axis, this is the main reason why our current RAFS presents a good performance. But Fig. 3a shows also that the microwave field is

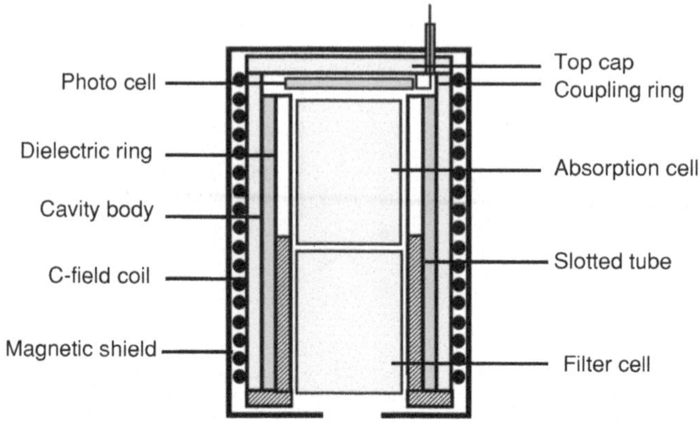

(a)

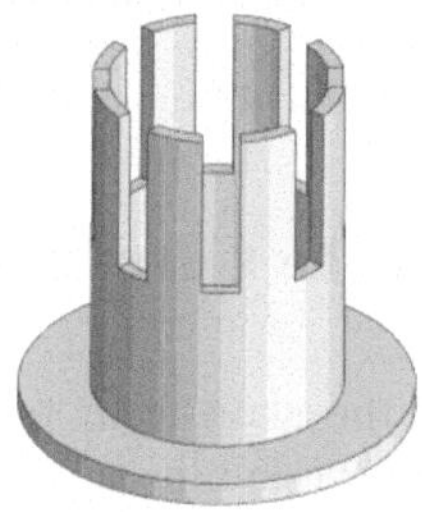

(b)

Fig. 2 **a** Structure of the cavity-cell assembly based on the slotted tube cavity. **b** Structure of the slotted tube

Fig. 3 Magnetic field lines in **a** side view section and **b** top view section of the current slotted tube cavity. *Arrows* and *colors* of the lines represent direction and amplitude of magnetic field (*red*: strong; *green*: weak). The quantum axis is parallel with the cavity axis

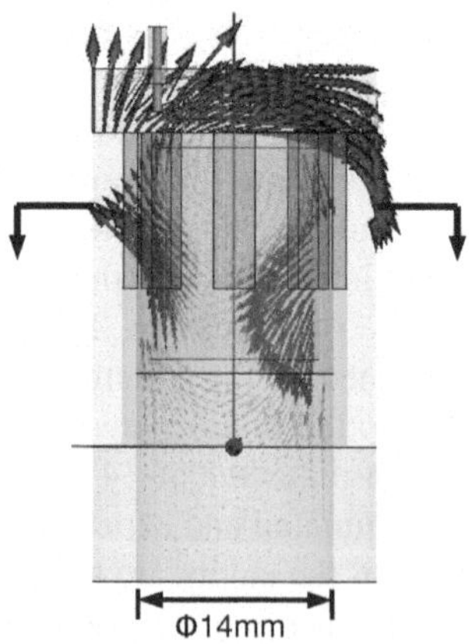

(a)

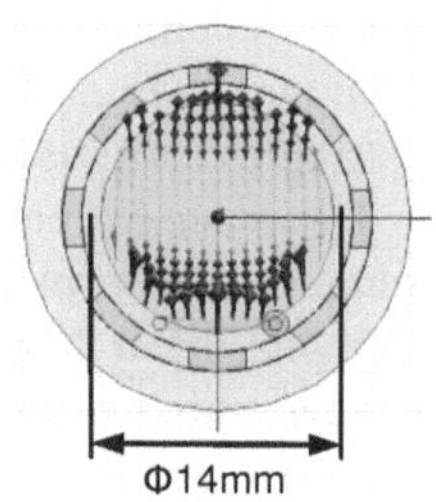

(b)

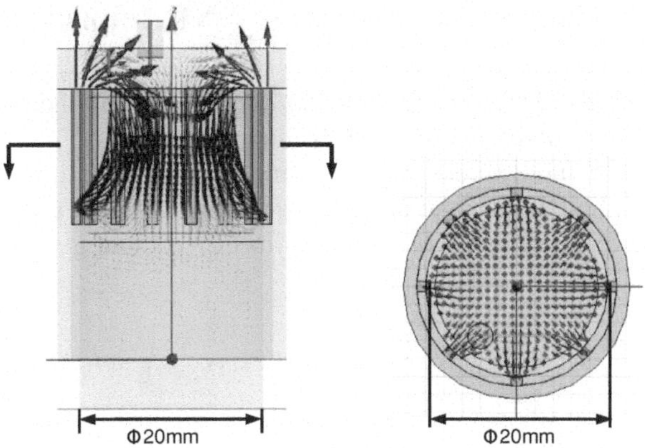

Fig. 4 Magnetic field lines in **a** side view section and **b** the top view section of the modified slotted tube cavity. *Arrows* and *colors* of the lines represent direction and amplitude of magnetic field (*red*: strong; *green*: weak). The quantum axis is parallel with the cavity axis

inhomogeneous in radial direction. The magnetic field is strong in the region near to slots or electrodes of the slotted tube but weak in the central region, much like the standard TE_{111} mode. The inhomogeneity of the microwave field can be also seen from a top view of the field distribution, as shown in Fig. 3b. There exist obviously two strong field regions in the cavity, and in other regions the field is comparatively weak. Since the clock transition of rubidium atoms is induced by microwave field, and the transition probability is proportional to the field intensity, an inhomogeneous field will lead to a limitation for the SNR of the atomic signal.

3 Modified Design of the Cavity-Cell Assembly

The result obtained by theoretical simulation revealed the design defect of the current slotted tube cavity, but mean while pointed out the way to modify the design. The modification includes two aspects. One is to make the field distribution inside the cavity more homogeneous to maximize atomic resonance transition signal in the whole absorption cell region. The other is to increase the size of the slotted tube so that a big size absorption cell can be used. The theoretical simulation design was made with the High Frequency Structure Simulation software (HFSS). Through adjusting the structure parameters of the slotted tube and other components of the cavity, a modified cavity with a slotted tube of 20 mm inner diameter has been designed. The microwave field distribution in the cavity is shown in Fig. 4, from which one can see that the magnetic lines are parallel to the quantum axis to a great extent, and are uniformly distributed in almost the whole region inside the slotted tube. The magnetic field distribution is close to the standard TE_{011} mode. At the same time, the increase in inner diameter of the slotted tube enable to hold a 20 mm diameter absorption cell instead of previous 14 mm diameter one. Both improvement in field homogeneity and increase in cavity size should have positive influence to enhance the atomic transition signal.

4 Theoretical Analysis of SNR Improvement

In this section we analyze the improvement in the SNR possibly resulted from the cavity modification described above.

A more direct way is to measure the discrimination signal i_0 and the background light signal I_0 for physics packages with and without the modification, then calculate the SNRs according to the formula given by Vanier [5]:

$$\frac{S}{N} = \sqrt{\frac{i_0^2}{4eI_0}}. \tag{2}$$

But we cannot perform such measurement since the modified physics package has not be built yet at present time.

Here we use the method given by Tremblay et al. [6]. for a theoretical analysis. By using the method, the SNR can be calculated provided the pumping light intensity, the relaxation rates of atoms and the field distribution in cavity. Tremblay et al. have given the results of RAFS's using cavity operated in mode TE_{011}, TE_{111} and TE_{101} [7]. For convenience for theoretical calculation, we assume that our previous slotted tube cavity is of mode TE_{111} with magnetic field described as

$$H_z(r, \theta, z) = A_1 J_1\left(\frac{1.84r}{R}\right)\cos\theta \sin\left(\frac{\pi}{L}z\right), \tag{3}$$

and magnetic field in the modified cavity can be described as

$$H_z(r, \theta, z) = A_2 \sin\left(\frac{\pi}{L}z\right). \tag{4}$$

In Eqs. (3) and (4) A_1 (A_2) represents the magnetic amplitude of the microwave field feed into the cavity, R is the radius of absorption cell, and L is the length of the cell. Equation (3) is well known since it describe fully the standard TE_{111} mode field distribution, while Eq. (4) can be understood as the TE_{011} mode with radial inhomogeneity removed. The approximation seems to be reasonable since the field distribution of the modified cavity, as shown in Fig. 4, is phenomenally similar to that of the standard TE_{011} mode around the cavity axis. The parameters used in our calculation, including those of pumping lights, rubidium atoms and microwave cavity, are presented in Table 1.

The theoretical signal to noise ratios in various pumping light intensities for physics packages using cavities with and without the modification are shown in Fig. 5. All the data were obtained under the optimized microwave power. From Fig. 5 it can be seen that, for both cavities, with total pumping light intensity increases, the SNR increases and then decreases. The reason for decrease of SNR in high light intensity is that the redundant pumping light leads to a saturated broadening of atomic spectral line. Therefore, to obtain the maximum SNR, both the pumping light intensity and the microwave power are needed to be optimized.

Table 1 Parameters used in the numerical calculation

Ratio of the intensities of the pumping lights D1, D2	3:1
Difference between the resonance frequency of the corresponding lines in the emission and absorption spectra, $\upsilon_{ei}-\upsilon_{ai}$	0 Hz
Linewidth of the pumping light, $\Delta\upsilon_{ei}$	1 GHz
Linewidth of the ^{87}Rb optical absorption spectrum, $\Delta\upsilon_{ai}$	1 GHz
Absorption cross sections of D1,D2 (maximum value), $\sigma 1$, $\sigma 2$	1.5×10^{-15} m^2, 3.2×10^{-15} m^2
Resonance frequency of the ^{87}Rb atoms, f_0	6.835 GHz
Number of ^{87}Rb atoms per unit volume (68°C), n_0	5.8×10^{11} cm^{-3}
Population relaxation rate, γ_1	500 Hz
Coherence relaxation rate, γ_2	350 Hz
Size of the current absorption cell	7 mm, 16 mm
Number of subdivisions, r, θ, z	7, 40, 16
Size of the absorption cell after the modification	10 mm, 16 mm
Number of subdivisions, r, θ, z	10, 1, 16

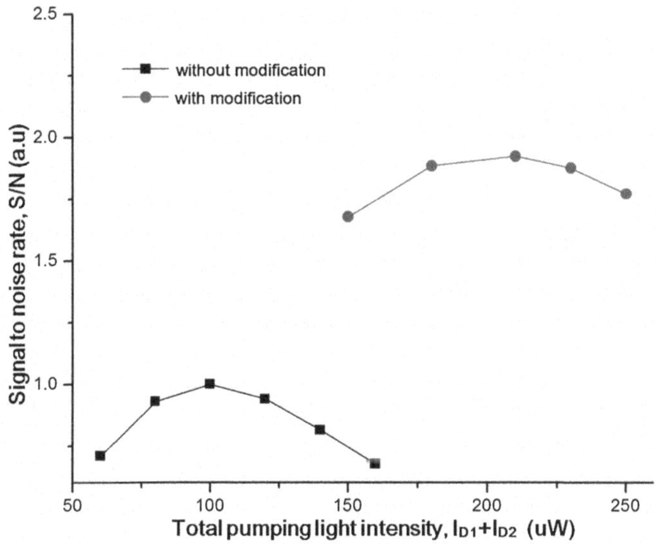

Fig. 5 Theoretical signal to noise ratios in various pumping light intensities for cavities with (*filled circle*) and without (*filled square*) modification. All the data were obtained under the optimized microwave power

Comparing two curves shown in Fig. 5, it is easy to find that the SNR with cavity modification is about twice as that of without cavity modification. According to the relationship between the frequency stability and the SNR given by Eq. (1), the frequency stability of the RAFS using the modified cavity should be better than $1.0 \times 10^{-12}\ \tau^{-1/2}$ since that of the current RAFS is $1.5 \times 10^{-12}\ \tau^{-1/2}$.

5 Summary

The RAFS using the slotted tube cavity as the core of physics package has realized a frequency stability of $1.5 \times 10^{-12}\ \tau^{-1/2}$ (1–1000 s) in our laboratory. For further performance improvement of the RAFS, the characteristic of the current slotted tube cavity was analyzed. It was found that the microwave field provided by the cavity is spatially inhomogeneous, which affects the SNR of the physics package. A modification of the cavity aiming at improvement of field homogeneity was carried out. Theoretical simulation shows that field homogeneity has been improved to great extent. In the modification the size of the cavity was increased also, so that a larger absorption cell than before can be used. A theoretical evaluation shows that cavity modification could result in a SNR enhancement by a factor of two, which implies that the frequency stability of the RAFS using the modified cavity can be better than $1.0 \times 10^{-12}\ \tau^{-1/2}$.

References

1. Vannicla, F., Beard, R., White, J., Senior, K., Kubik, T., & Wilson, D. (2009). GPS block IIF rubidium frequency standard lift test. *41st Annual Precise Time and Time Interval (PTTI) Meeting*, pp. 449–456.
2. Vanier, J., & Audoin, C. (1989). *The Quantum physics of atomic frequency standards*. Bristol: Adam Hilger.
3. Mei, G., Zhong, D., An, S., Liu, J., & Huang, X. (2001). Miniaturized microwave cavity for atomic frequency standard. US Patent, No. 6 225 870 B1.
4. Li, W., Kang, S., Ming, G., & Mei, G. (2011). Optimization of microwave field distribution of the slotted tube cavity. *China Time and Frequency Symposium*, pp. 40–43.
5. Vanier, J., (1981). On the signal-to-noise ratio and short term stability of passive rubidium frequency standards. *IEEE Transactions on Instrumentation and Measurement*, *IM-30*(4), 277–282.
6. Tremblay, P., Cyr, N., & Tetu, M. (1985). Pumping light intensity transmitted through an inhomogeneously broadened line system: Application to passive rubidium frequency standards. *Canadian Journal of Physics, 63*, 1563.
7. Tremblay, P., Cyr, N., & Tetu, M. (1984). Evaluation of the performance of passive rubidium frequency standards using cavity operated in mode TE_{011} TE_{111} and TE_{101}. *38th Annual Frequency Control Symposium*, pp. 408–415.

A Quantitative Testing Method of Quartz Resonators' Acceleration Sensitivity Based on a MEMS Sensor

Longzhe Ji, Qingxiao Shan, Qian Tang, Jun Yang and Ming Lin

Abstract Quartz resonators are widely used as the core components in navigation satellite systems because of their high Q value and low cost. But quartz resonators are sensitive to acceleration and accelerations in different directions affect resonators in different degree. On the most sensitive direction we define an acceleration sensitivity vector $\vec{\Gamma}$. In this paper, a quantitative testing method is proposed which used a MEMS three-axis sensor and the vector $\vec{\Gamma}$ can be disposed at the sensor's coordinates. A testing system which mainly consists of a MEMS sensor, an FPGA and a spectrum analyzer was also built. We studied the relationship between acceleration and resonator's frequency shift by imposing vibration experiments along sensor's each axis respectively. With the sensor getting acceleration data, the resonator's frequency offset and acceleration sensitivity of each axis could be calculated accurately.

1 Introduction

The reference frequency sources are the important part of the space electronic instruments and equipments, they play key roles in fields such as satellite navigation, exactitude timing, atomic clock, time synchronization, remote sensing imagining and space communications. Quartz resonators are widely used as the frequency standards to generate the precise frequencies and time signals, but they are very sensitive to accelerations, which deeply influence the frequency stability

L. Ji (✉) · Q. Shan · Q. Tang · J. Yang · M. Lin
College of Mechatronics Engineering and Automation, National University of Defense Technology, 410073 Changsha, China
e-mail: crystal_nudt@163.com

J. Sun et al. (eds.), *China Satellite Navigation Conference (CSNC) 2012 Proceedings*, Lecture Notes in Electrical Engineering 160, DOI: 10.1007/978-3-642-29175-3_53,

and worsen the phase noise of resonators. A great deal of serious vibrations and concussions come into being when satellites launch and move in space, the continuous exist of accelerations may degrade the steady performance of quartz resonators [1–5].

In this paper, we will have a brief look at how acceleration affects the frequency stability of the quartz resonators, then we focus on the quantitative testing method of the acceleration sensitivity based on a Micro Electro Mechanical System (MEMS) three-axis sensor SMB380 and a Xilinx Spartan-6 FPGA, the testing principle and the experiment will be described in details.

2 Acceleration Effects on Quartz Resonators

2.1 The Principle of the Acceleration Effects

Without any surrounding noise, a quartz resonator can provide a steady sinusoidal voltage signal with a single frequency, which is called the resonator's carrier frequency and expressed by f_0. When vibration or concussion occurs, the resonator's frequency output will have a tiny change, the frequency can be expressed by $f(a)$ which is denoted as follows:

$$f(a) = f_0 + f' \tag{2.1}$$

where f' is the frequency shift induced by the acceleration a.

Large numbers of experiments have proved that this kind of frequency shifts of quartz resonators are relative to the acceleration $\vec{a}$ as is shown below:

$$f' = \vec{\Gamma} \bullet \vec{a} \bullet f_0 \tag{2.2}$$

Here $\vec{\Gamma}$ is the acceleration sensitivity vector of the resonator. If the acceleration $\vec{a}$ is induced by simple harmonic motion, then it can be written as the equation:

$$\vec{a} = \vec{A} \cos(2\pi f_v t) \tag{2.3}$$

in which $\vec{A}$ stands for the peak acceleration and f_v is the vibration frequency.

Assuming that the vibration only exists in one certain direction, then we bring Eqs. (2.2) and (2.3) into Eq. (2.1) and we get a new equation:

$$f(a) = f_0 + \Delta f \cos(2\pi f_v t) \tag{2.4}$$

where Δf is the peak frequency shift due to the acceleration $\vec{a}$.

From Eq. (2.4) we figure out that the output frequency of the resonator changes in range of $f_0 \pm \Delta f$. What's more, the frequency spectrum can be analyzed as is shown in Fig. 1. A basic conclusion is that the essential of the acceleration effect is

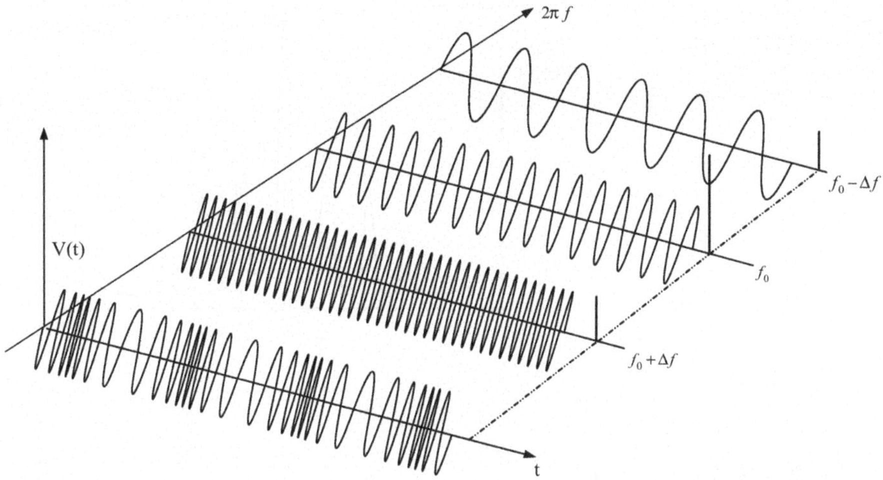

Fig. 1 The spectrum analysis under accelerations

a kind of frequency modulation which influences the resonators' frequency stability.

2.2 Analysis in Frequency Domain

When influenced by the accelerations, the output voltage of quartz resonator can be described as [6]

$$V(t) = V_0 \cos[2\pi f_0 t + (\frac{\Delta f}{f_v}) \sin(2\pi f_v t)] \tag{2.5}$$

As the frequency modulation theory and the Jacobi-Angel Equation

$$e^{jz\sin\theta} = \sum_{n=-\infty}^{+\infty} J_n(z) e^{jn\theta} \tag{2.6}$$

Eq. (2.5) can be expanded as an infinitude series [6]

$$\begin{aligned} V(t) = V_0\{&J_0(\beta)\cos(2\pi f_0 t) + J_1(\beta)\cos[2\pi(f_0 + f_v)t] \\ &- J_1(\beta)\cos[2\pi(f_0 - f_v)t] + J_2(\beta)\cos[2\pi(f_0 + 2f_v)t] \\ &+ J_2(\beta)\cos[2\pi(f_0 - 2f_v)t] + \cdots\} \end{aligned} \tag{2.7}$$

Here the modulation index is defined as $\beta = \Delta f / f_v$, the phase noise of the nth vibration-induced sideband can be donated as

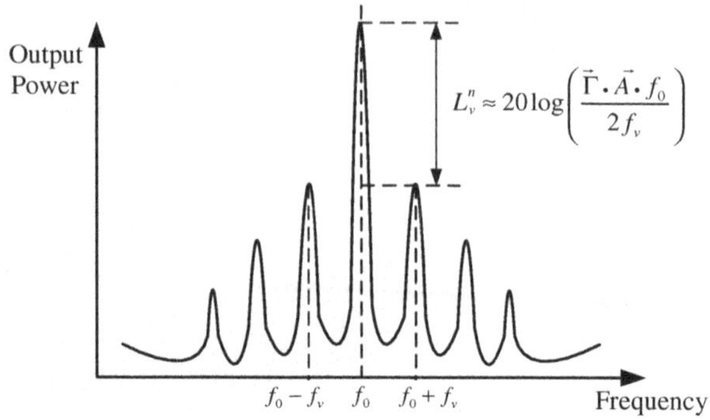

Fig. 2 Some other spectrum elements under vibration

$$L_v^n = [J_n(\beta)/J_0(\beta)]^2 \tag{2.8}$$

or another expression

$$L_v^n \approx 20\log\left(\frac{\vec{\Gamma}\cdot\vec{A}\cdot f_0}{2f_v}\right) \tag{2.9}$$

when $\beta < 0.1$ according to frequency modulation theory.

From Eqs. (2.7) and (2.9), under the simple harmonic vibration, quartz resonator's output frequency is added to some other spectrum elements which is based on the vibration frequency f_v. Because of the exists of accelerations, resonator's phase noise is weakened deeply and there comes sideband modulation effect [7–10], as is shown in Fig. 2.

The acceleration sensitivity Γ can be calculated easily from Eq. (2.9) as

$$\Gamma = \frac{2f_v}{A \times f_0} \times 10^{L_v^n/20} \tag{2.10}$$

2.3 The Acceleration Sensitivity Vector

As is discussed above, quartz resonators are sensitive to accelerations. What's more, accelerations in different directions will affect quartz resonators in different degree. On the most acceleration sensitive direction, we can define an acceleration sensitivity vector $\vec{\Gamma}$ of the quartz resonator. For a certain quartz resonator, the acceleration sensitivity vector $\vec{\Gamma}$ is the proper character, which never changes with the surrounding factors [11].

The acceleration sensitivity vector $\vec{\Gamma}$ can be analyzed into a three-dimensional reference frame. Supposing the three axes of this reference frame are X axis, Y

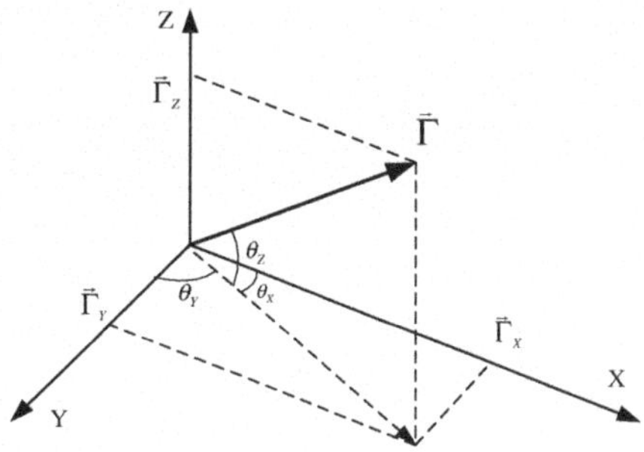

Fig. 3 The acceleration sensitivity vector in space

axis and Z axis, and the angles between $\vec{\Gamma}$ and each axis are θ_X, θ_Y and θ_z as is shown in Fig. 3, so the vector $\vec{\Gamma}$ can be described as

$$\vec{\Gamma} = \vec{\Gamma}_X \cos\theta_X + \vec{\Gamma}_Y \cos\theta_Y + \vec{\Gamma}_Z \cos\theta_Z \tag{2.11}$$

where $\vec{\Gamma}_X$, $\vec{\Gamma}_Y$ and $\vec{\Gamma}_Z$ are three acceleration sensitivity vectors in the three directions in space.

Obviously, the analyzing method for the vector $\vec{\Gamma}$ is not one and only because there are innumerable reference frames in space, but the direction and magnitude of a resonator's $\vec{\Gamma}$ is certain, only $\vec{\Gamma}_i$ and $\cos\theta_i$ are different.

3 The Quantitative Testing System

3.1 Testing Method and Principle

From Eq. (2.10) we draw the conclusion that the acceleration sensitivity of one certain direction can be calculated if we get the vibration parameters and the induced sideband. Assuming that the vibration is a simple harmonic one, a MEMS three-axis sensor SMB380 is attached onto the quartz resonator so that the acceleration amplitude and frequency in each directions can be tested exactly, called a_X, a_Y and a_Z. So the acceleration A in the most acceleration sensitive direction of the resonator can be described as

$$A = a_X \cos\theta_X + a_Y \cos\theta_Y + a_Z \cos\theta_Z \tag{3.1}$$

where θ_X, θ_Y and θ_Z are the angles of each axis. On the other hand, a digital spectrum analyzer is used to measure the sideband and phase noise L_v^n of the resonator.

We take the MEMS sensor's three axes X, Y and Z as the preference frames, put the vibration onto the resonator along each axis, so the three acceleration sensitivity elements $\vec{\Gamma}_X$, $\vec{\Gamma}_Y$ and $\vec{\Gamma}_Z$ can be calculated by the above method. Then

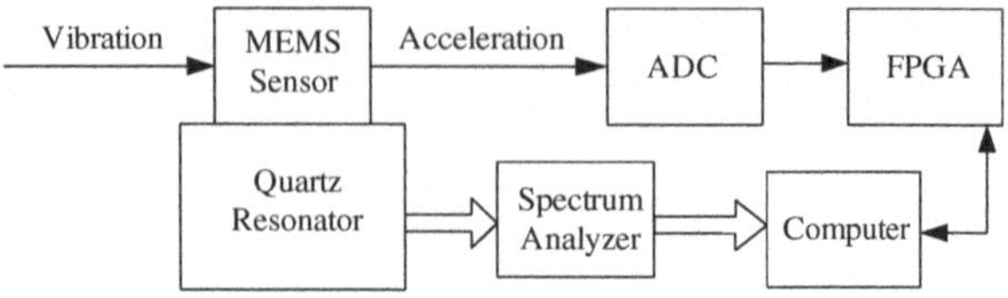

Fig. 4 The testing system

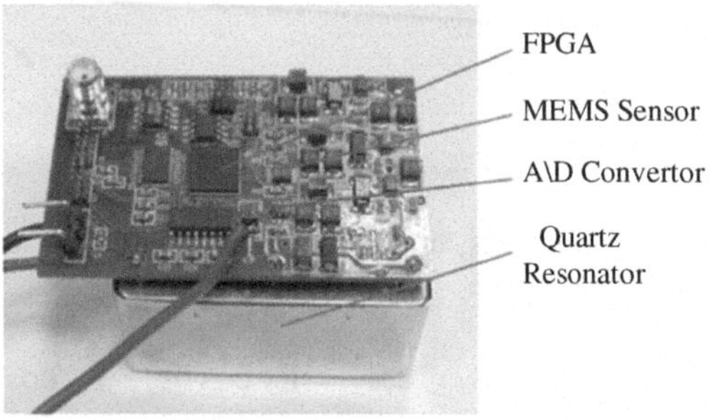

Fig. 5 Resonator with the testing system

we follow the vector algorithm principle and take use of the current known $\vec{\Gamma}_X$, $\vec{\Gamma}_Y$ and $\vec{\Gamma}_Z$, the resonator's acceleration sensitivity vector $\vec{\Gamma}$ and the three angles θ_X, θ_Y and θ_z will be computed easily.

3.2 The Testing System Design

As is described above, the quantitative testing system of the acceleration sensitivity is mainly made up of a MEMS sensor SMB380, an FPGA of Xilinx Spartan-6 series and a spectrum analyzer, Fig. 4 shows the system design.

We have manufactured some resonator samples with this testing system, as Fig. 5 shows, the circuit board is attached onto the resonator so that SMB380 is able to sense the exact accelerations of surrounding vibrations.

4 Experiment

Figure 6 shows the project of the vibration experiment equipment. It consists of the vibration platform, the controller part, the resonator part, the power-supply part and the spectrum analyzer. The actual experiment platform in our laboratory is shown in Fig. 7.

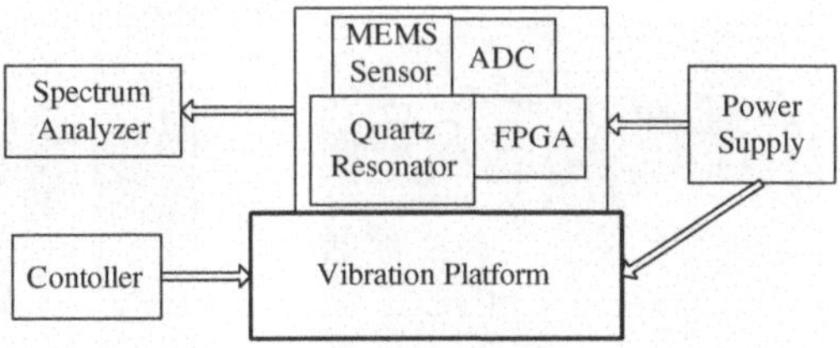

Fig. 6 The project of the vibration experiment

Fig. 7 The experiment platform

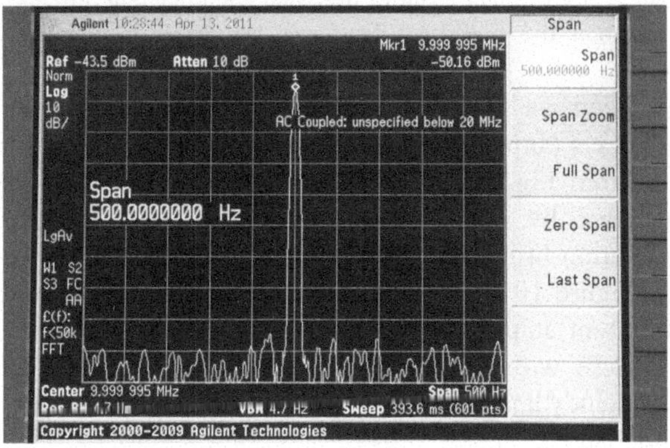

Fig. 8 Resonator's output without vibration

4.1 The Relationship Between Acceleration and Frequency Variation

After building the vibration experiment platform, the acceleration $\vec{a}$ can be tested by the MEMS sensor SMB380 and the sideband L_v^n is also known with the help of the spectrum analyzer, which makes the research of the relationship between acceleration and variation of frequency easier.

In condition without any vibration, the resonator's output is a pure carrier frequency 10 MHz, as is shown in Fig. 8.

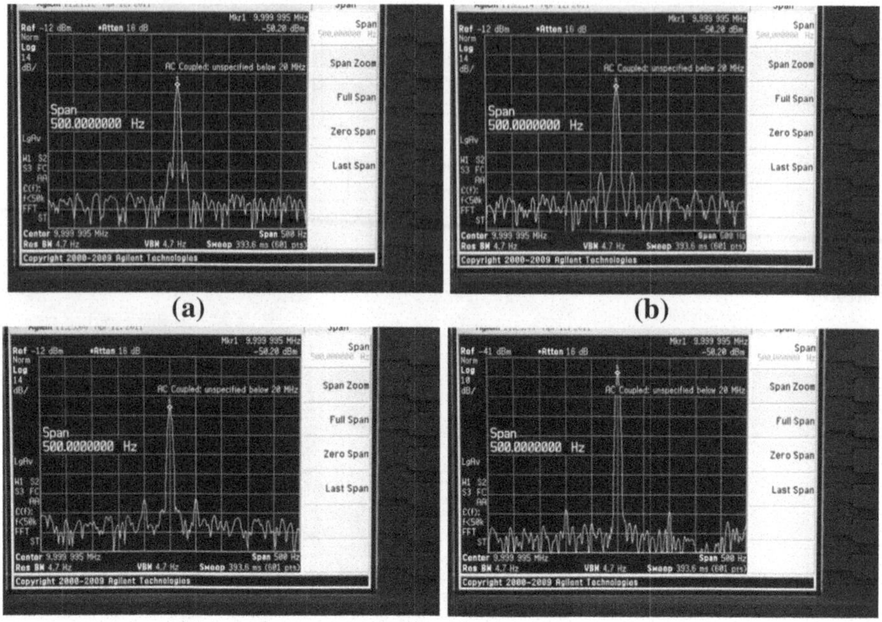

(a) (b)

(c) (d)

Fig. 9 Experiment result of the same amplitude but different frequency vibrations. **a** 2 g,15 Hz Vibration. **b** 2 g, 30 Hz Vibration. **c** 2 g, 50 Hz Vibration. **d** 2 g, 100 Hz Vibration

Then we set the vibration parameters to different values and observe the resonator's output changes. Figure 9 provides some experiment results.

From Fig. 9 we can draw the conclusion that the position of the sideband is decided by the vibration frequency, at the same amplitude, high frequency vibration brings low sideband.

Figure 10 shows the experiment result of the same vibration frequency but different amplitude accelerations.

Here we can see that the positions of the sideband represent the vibration frequency and the amplitude of the accelerations are relative with the power of the sidebands.

According to the above experiment results, when acceleration occurs to quartz resonator, the resonator's carrier frequency will be modulated and there exist some symmetrical sidebands, whose positions are decided by the vibration frequency and powers are due to the acceleration parameters.

4.2 The Quantitative Testing of the Acceleration Sensitivity

The acceleration sensitivity $\vec{\Gamma}$ of a certain resonator would not change with different surroundings. So if the acceleration sensitivity vectors of each axis

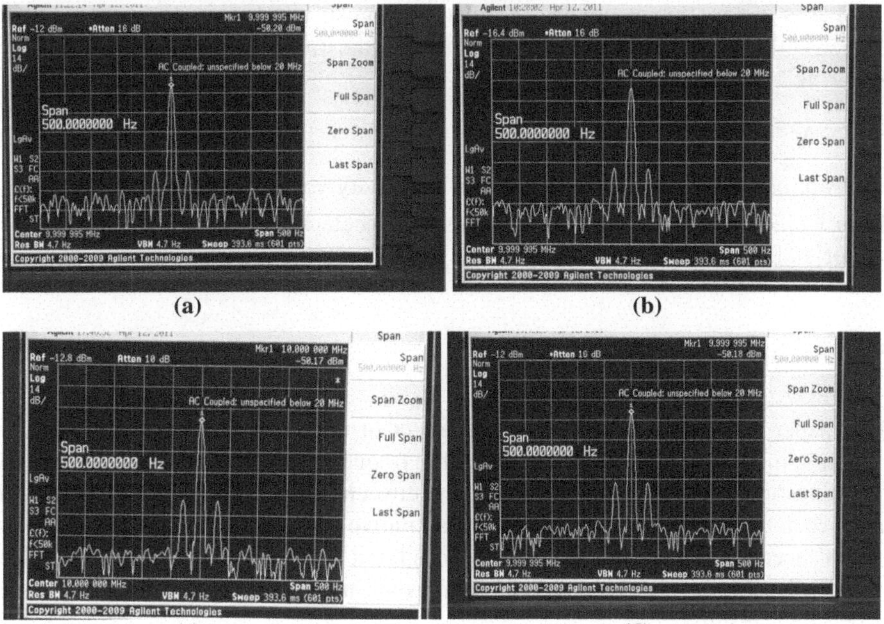

(a) (b)

(c) (d)

Fig. 10 Experiment result of the same frequency but different amplitude vibrations. **a** 2 g, 30 Hz Vibration. **b** 4 g, 30 Hz Vibration. **c** 6 g, 30 Hz Vibration. **d** 8 g, 30 Hz Vibration

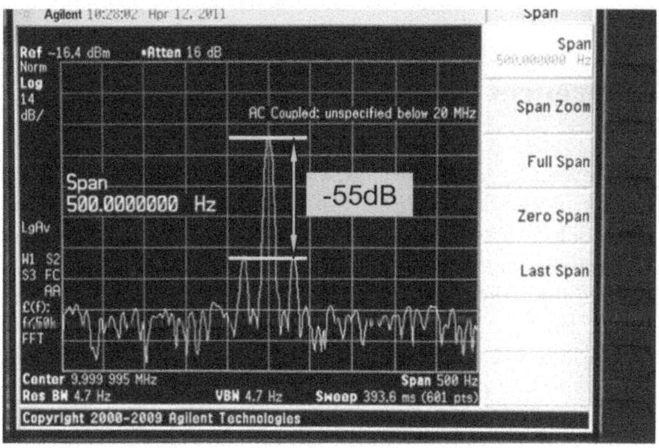

Fig. 11 Testing the acceleration sensitivity

$\vec{\Gamma}_X$, $\vec{\Gamma}_Y$ and $\vec{\Gamma}_Z$ are all known, it's not hard to calculate $\vec{\Gamma}$ with them. Here we take the condition of 4 g, 30 Hz vibration on one axis for example, as Fig. 11 shows.

By observing the output frequency spectrum shown on the spectrum analyzer, the phase noise of the 1st sideband is about -55 dB, if we bring other parameters into Eq. (2.10), then we can calculate the acceleration sensitivity as

$$\Gamma = \frac{2f_v}{A \times f_0} \times 10^{L_v^n/20} = \frac{2 \times 30}{4 \times 10 \times 10^6} \times 10^{-\frac{55}{20}} \approx 2.667 \times 10^{-9}/g \tag{4.1}$$

That is to say, this resonator's acceleration sensitivity in this direction is about 2.667×10^{-9}/g, which means the quantity of the frequency change induced by every "g" is

$$\Delta f = \Gamma \times a \times f_0 = 2.667 \times 10^{-9} \times 1 \times 10 \times 10^6 = 0.02667\text{Hz} \tag{4.2}$$

With the same method, the other two sensitivity vectors can also be calculated. Then $\vec{\Gamma}_X$, $\vec{\Gamma}_Y$ and $\vec{\Gamma}_Z$ can determine a certain acceleration sensitivity $\vec{\Gamma}$ due to the vector's algorithm principle.

5 Conclusion

This paper introduced the acceleration effect of quartz resonators, discussed the principle of how accelerations influence the frequency stability and phase noise of quartz resonators, studied the relationship between resonator's frequency variation and acceleration parameters. Then a quantitative testing method of quartz resonators' acceleration sensitivity was proposed and a testing system was also made up. Finally, the paper designed a vibration experiment platform and provided some experiment results.

How to reduce the acceleration effect on the reference frequency source is now becoming a new problem in navigation satellite applications and many other fields. The relative techniques will have nice developing potentials in future.

References

1. Steven, J., & Gregory, A. (2010). Reducing the acceleration sensitivity of AT-strip quartz crystal oscillators. *IEEE International Frequency Control Symposium* (pp. 25–30).
2. Bloch, M., Mancini, O., McClelland, T., Terracciano, L. (2009). Acceleration "g" compensated quartz crystal oscillators. *Joint Meeting of the European Frequency Forum and the IEEE International Frequency Control Symposium* (pp. 175–180).
3. Howe, A., LanFranchi, L., Cutsinger, L., Hati, A., Nelson, C. (2005). Vibration-induced PM noise in oscillators and measurements of correlation. *IEEE International Frequency Control Symposium and Exposition* (pp. 494–498).
4. Larry, D., & Errol, P. (1993). The effects of metal depositions upon acceleration induced frequency shifts for quartz crystal units. *IEEE International Frequency Control Symposium* (pp. 560–573).
5. Fred, L., & Jacques, G. (1992). Environmental sensitivity of quartz oscillators. *IEEE Transactions on Ultrasonics, Ferroelectrics and Frequency Control* (pp. 241–247).
6. Filler, R. (1988). The acceleration sensitivity of quartz crystal oscillator: A review. *IEEE Transactions on Ultrasonics and Frequency Control* (pp. 297–305).
7. He, M., LiMei, X., WeiBin, T., Lu, L. (2006). Experimental analysis of vibration-induced phase noise of crystal oscillator circuit. *IEEE International Conference on Mechatronics and Automation,* (pp. 1559–1564).
8. Filler, R. (1981). The effect of vibration on frequency standards and clocks. *35th Annual Symposium Frequency Control* (pp. 31–41).

9. Driscoll, M. (1989). Quartz crystal resonator g-sensitivity measurement methods and recent results. *43rd Annual Symposium Frequency Control* (pp. 419–426).
10. Francisco, J., Julia, C., & Juan, P. (1995). New digital compensation technique for the design of a microcomputer compensated crystal oscillator. *IEEE Transactions on Industrial Electronics* (pp. 307–315).
11. Milliren, B., Martin, D., & Emmons, D. (1988). An acceleration compensated precision quartz oscillator. *42nd Annual Symposium Frequency Control* (pp. 334–341).

The Exploration of Satellite Clock and Ephemeris Error Correction in Wide Area Differential System

ChengLin Cai, XiaoHui Li and HaiTao Wu

Abstract Real time principle of wide area differential GPS (WADGPS) uses a generally inverted positioning method of four-dimensional solution of satellite clock and ephemeris corrections. Compared with the real time single point positioning, its positioning accuracy of this method is improved greatly, but the DOP of four-dimensional calculation is up to several hundred, under the error amplification effect of DOP, 1–3 m positioning precision level of WADGPS is difficult to breakthrough. To overcome the limit of DOP, the feasibility on large and small part decomposition of satellite clock and ephemeris error is analyzed, and optimal differential performances of the exploring two smallness correction methods: Small DOP and no DOP constraint, are analyzed, and they will be the foundation for further exploration.

1 Introduction

WADGPS has two main methods: one is dynamics method for unreal time application, another is kinematics method for real time application. International GPS service (IGS) uses the dynamics method to obtain precise ephemeris, the

C. Cai (✉)
Guilin University of Electronic Technology, 541004, Guilin, China
e-mail: chengcailin@126.com

X. Li
National Time Service Center, Chinese Academy of Sciences,
710600 Lintong, China

H. Wu
Academy of Opto-Electronics, Chinese Academy of Sciences,
100094 Beijing, China

J. Sun et al. (eds.), *China Satellite Navigation Conference (CSNC) 2012 Proceedings*, Lecture Notes in Electrical Engineering 160, DOI: 10.1007/978-3-642-29175-3_54,

ephemeris error can eliminate to centimeter with this method [1]. Satellite clock and ephemeris corrections of the WAAS of American and the EGNOS of European are calculated together with the kinematics method of inverted geometry [2], as for MEO satellite, its four-dimensional DOP(4D_DOP) is high up to several hundred [3], and as for GEO satellite, its 4D_DOP is more than 1000 [4]. The precision of ephemeris correction is difficult to improve because of the limit of great DOP. At present, real-time positioning accuracy of WADGPS can only improve to 1–3 m with the 4D method [5, 6], and it is difficult to meet requirements of the first Aviation Precision Approach (CATI) [7]. Therefore, many countries are actively exploring the new wide area differential method or technique for further improving its accuracy. At first, the mechanism of novel wide area differential method previously proposed is deeply elaborated [4, 8], and then the higher accuracy differential method is explored.

2 The Decomposition Mechanism of Large and Small Part of Ephemeris Error

It is analyzed that 3D_DOP of satellite geometric orbit determination can be greatly reduced in document [4], as for the GEO satellite, it can be reduced to about 30, and as for MEO satellite, it can be reduced to about 10. If the satellite clock and orbit are separated, and the satellite clock error done firstly, 3D calculation of satellite position can be used. Due to DOP reduction, the precision of differential solution can be greatly improved, this method is named as small DOP constraint's wide area differential method, but how to realize this idea? The direct emissive satellite navigation system, the two-way time synchronization method between satellite and earth by pseudo code ranging or laser ranging must be used to realize the absolute separation of satellite clock error, especially, the laser ranging can control the synchronization error within 0.1 ns. In order to realize time synchronization with this kind of synchronization method, advanced time synchronization equipment must be equipped on satellite and in master control station. If there is no device of the two-way time synchronization, equal accuracy can be reached in a certain way?

Firstly, the distributing characteristics of ephemeris error in a certain service area is analyzed. Because navigation satellites are 20,000 or 36,000 km from the earth, the radiation angle between two different stations is very small, that is shown in Fig. 1. From Fig. 1, as for MEO satellite, we can see the radiation angle between the East and West station is less than 8.63°, and as for GEO satellite, its radiation angle is less than 5.43°. The small beam radiation angle determines that the radial ephemeris error also has small variation regularity within this area, and the large radial ephemeris error can be corrected through one station in the service area, the large part of it in every station is corrected, only a small ephemeris error is left, this is the decomposition mechanism of large and small part of ephemeris error.

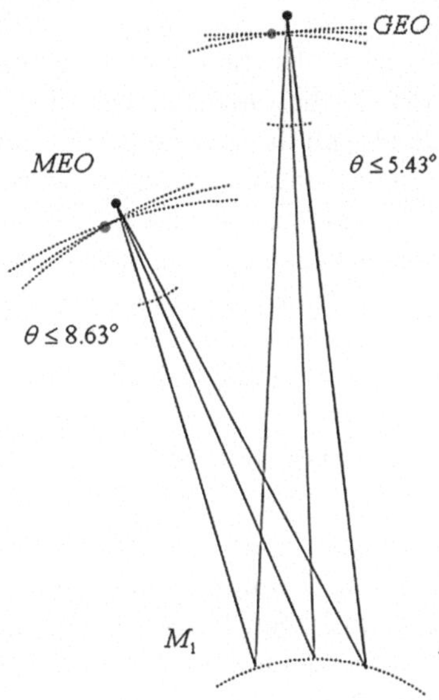

Fig. 1 The decomposition mechanism of large and small part of ephemeris error

Relative correction and absolute correction thought of ephemeris error are put forward with reverse thought in Document [8]' Supplementary explanation on the correction of this method is shown here. In fact, the pseudo-range observation value output by the receiver has no ephemeris radial error, because the navigation signal is directly from the satellite the true position, just because it contains the ionosphere and troposphere delay and the local clock error, etc., if the propagation error can be deducted, the pseudo range can be expressed as:

$$\rho = R_s + c \cdot \Delta t_s + c \cdot \Delta t_i \tag{1}$$

In the above equation, R_s is expressed as the geometric distance between the satellite true position and monitoring station, Δt_s is satellite clock error, Δt_i is the station clock deviation. As to the master station with time synchronization, $\Delta t_i \approx 0$, the pseudo-range error includes the satellite clock error if the receiver noise is not considered. But because the satellite true location 'S' do not know, the broadcast ephemeris 'E' is replaced 'S', the pseudo range can be expressed as:

$$\rho = R_E + c \cdot \Delta t_s + c \cdot \Delta t_i \tag{2}$$

In the above positioning equation, R_E is expressed as the distance between the satellite and the station, in fact, the '=' is not completely established because of the broadcast ephemeris error, this will cause positioning error, In order to make the '=' established, radial ephemeris error $\Delta\rho_E$ is added or subtracted in the right of the equation, that is shown as

$$\rho = R_E - \Delta\rho_E + c \cdot \Delta t_s + c \cdot \Delta t_i \tag{3}$$

Among them, $R_S = R_E - \Delta\rho_E$, the implication is that the pseudo range error is corrected in the direction of approaching R_S, through the correction of $\Delta\rho_E$, the Eq. (3) can be turned into the Eq. (1), so the precise satellite ephemeris can be obtained with the inverse positioning principle. Because $\rho - R_E$ is known in every monitoring station, as for the master station with the time synchronization, the satellite clock error can be separated with the two-way time synchronization method between satellite and earth, $\Delta\rho_E$ can naturally be obtained after the separation of the satellite clock error, this is the absolute separation (AS) of satellite clock and ephemeris error [8].

The inverse thought is used below, if satellite ephemeris error is not corrected, the user will directly use broadcast ephemeris for positioning calculation, its positioning equation needs to be changed as:

$$\rho + \Delta\rho_E - c \cdot \Delta t_s = R_E + c \cdot \Delta t_i \tag{4}$$

The simple transposition is done in the above formula, but the most significant consideration can be brought, it means that the direction correcting of pseudo range approaching to R_E with comparison to AS, the correction direction is reverse. Because of the master station's time synchronization, the pseudo range can limitlessly approach to R_E, and the positioning solution is very accurate. As for the secondary station, according to the decomposition principle of the large and small part error, its amendment is the same as that of the master station, which is the large part, the remaining radial ephemeris error (that is the small part) is related to spatial position, the small correction of ephemeris error, can be obtained with an inverse geometry method, and the same precise solution of master station can be obtained with small correction of ephemeris error.

In fact, in order to ensure high precise synchronization among stations, the two way satellite alignment link need be established, secondary station only uses satellite navigation signal to realize coarse synchronization and to remove the station clock deviation without this link in the same station, the tiny satellite ephemeris correction can be solved. Specific correction method is shown in Ref. [7].

3 Positioning Simulation with Small DOP Constraint

Simulation platform adopts 7 stations: Xi'an, Changchun, Shanghai, Urumqi, Kashi, Kunming, Sanya, the simulation constellation respectively uses pure GEO and MEO constellation, validated users are Wuhan and Chengdu. If ionosphere delay and troposphere delay error are eliminated, the pseudo range only contains satellite clock error and ephemeris error, the large and small decomposition and correction principle is used, the large part is corrected in the master station with time synchronization, and the large part correction is corrected with one station, and the small part is corrected with 3D calculation, the positioning results of monitoring station and validated user are shown in Table 1.

Table 1 The differential positioning (m) and timing (ns) with no measurement noise (1σ)

Reference station and user	GEO constellation		MEO constellation		
	Lon dev	Lat dev	Lon dev	Lat dev	H dev
Lintong	0.14	0.11	0.11	0.03	0.04
Chuangchun	0.12	0.06	0.26	0.02	0.17
Shanghai	0.05	0.08	0.13	0.10	0.07
Wulu	0.07	0.02	0.14	0.15	0.04
Kunming	0.08	0.01	0.11	0.02	0.05
Beijing	0.08	0.06	0.03	0.04	0.02
Kashi	0.03	0.23	0.03	0.19	0.03
Haikou	0.06	0.05	0.05	0.11	0.01
Wuhan (inside)	0.06	0.10	0.04	0.07	0.01
Chengdu (inside)	0.14	0.10	0.03	0.05	0.02

From Table 1, we can see, if the pseudo range has no measurement noise, the optimal positioning accuracy in network with small DOP constraint method is:

(1) In the GEO constellation, the longitude direction's positioning error is less than 0.15 m; and the latitudinal direction's positioning error is less than 0.25 m.
(2) In the MEO constellation, the longitude direction's positioning error is less than 0.3 m, the latitudinal direction's the positioning error is less than 0.2 m, the height direction's positioning error is less than 0.2 m.

Results in the Table 1 further indicate whether in GEO or MEO constellation, through 3D calculation combined multiple stations, their small corrections can be obtained, and differential user positioning effect of the final remaining ephemeris error in the service area can be maintained to decimeter, even centimeter level. Because the 3D_DOP is very small, it powerfully illustrates this differential method has the most accurate potential ability to the user positioning correction. As long as the receiver measurement noise (carrier phase smoothing or adaptive filtering based on time series), and other error except ephemeris error and satellite clock error can be effectively suppressed, the user positioning accuracy can reach real time decimeter level.

In comparison with WAAS, this method has the following advantages: (1) The correction thought is different from WAAS, WAAS follows the principle of satellite clock and ephemeris correction to real value, but this method follows the principle of the sum of the pseudo-range error correction close to zero, and ephemeris correction is only small correction. (2) WAAS uses 4D calculation of satellite clock and ephemeris, but our method is 3D calculation, 3D_DOP is much less than 4D_DOP, so error amplification effect on user positioning was obviously reduced. (3) Through one station correction, it can eliminate all the satellite clock error and most of radial ephemeris error, thus the final correction's effect to user positioning becomes much smaller. This small DOP solution is very suitable for differential stations not designed to be grid and not strictly synchronized.

4 Pure Pseudo-Range Correction Method with no DOP Constraint

With geometric orbit determination method, differential correction is restricted to the DOP. The small DOP constraint method can reduce DOP constraint from the two following aspects: (1) the error involved in the calculation is no longer a total error, but only the small part; (2) the error amplification effects are obviously reduced due to 3D_DOP greatly reduced. Therefore, the small DOP constraint method is a good differential method for the differential network with no grid design, user differential positioning accuracy level close to the true value can be maintained to decimeter. If the remaining radial ephemeris error correction has no DOP constraint, it is no doubt that the user pseudo-range errors can be close to 0, thereby the best differential positioning accuracy can be obtained.

From Fig. 1, we can see that the bigger beam radiation angle with the main station is, the larger remaining radial ephemeris error is, therefore, when the master station is regarded as a reference point, the spatial correlation of the remaining radial ephemeris error has certain relationship to beam radiation angle. Whether the strong correlation with space position can be shown with the beam radiation angle?

First of all, the clock deviation of each differential station is not considered, namely, we assume that the every station is strict time synchronization, and its pseudo-range error includes only the satellite clock and ephemeris error (we also assume troposphere and ionosphere error can be completely eliminated), and layout every 5° a monitoring station in longitude line and latitude line through master station, there are 11 monitoring stations in longitude line, and 15 stations in latitude line, with the multiple station four GEO satellites, the left two figure in Fig. 2 can be plotted with their pseudo-range errors, residual ephemeris pseudo-range error after the one master station correction are shown in the middle of the two sub-Fig. 2, compared with the left figure, it is the small part. With the three stations of west, the master and the East, two order polynomial model can be established, the results of the model subtracting the residual radial ephemeris error of 15 stations are in right of Fig. 2. With the three station of the north, the master, south station, the results of the model subtracting the residual radial ephemeris error of 11 stations are in the last Fig. 2. From the right two diagrams can be drawn through the model correction, the residual results: it is less than 2 mm in longitude direction, and less than 1 cm in the latitude direction. This description based on the beam radiation angle is very effective to space model for residual radial ephemeris error. The beam radiation angle based on the master station reference point can be converted to the differential between the master station and secondary station, therefore, in fact, the space distribution of residual radial ephemeris error can be modeled by the latitude and longitude.

Analysis in Fig. 2 only solves radial ephemeris error correction in longitude line and latitude line through master station, we need to adopt the rule to extend the surface line, and to solve curved surface model correction for all users

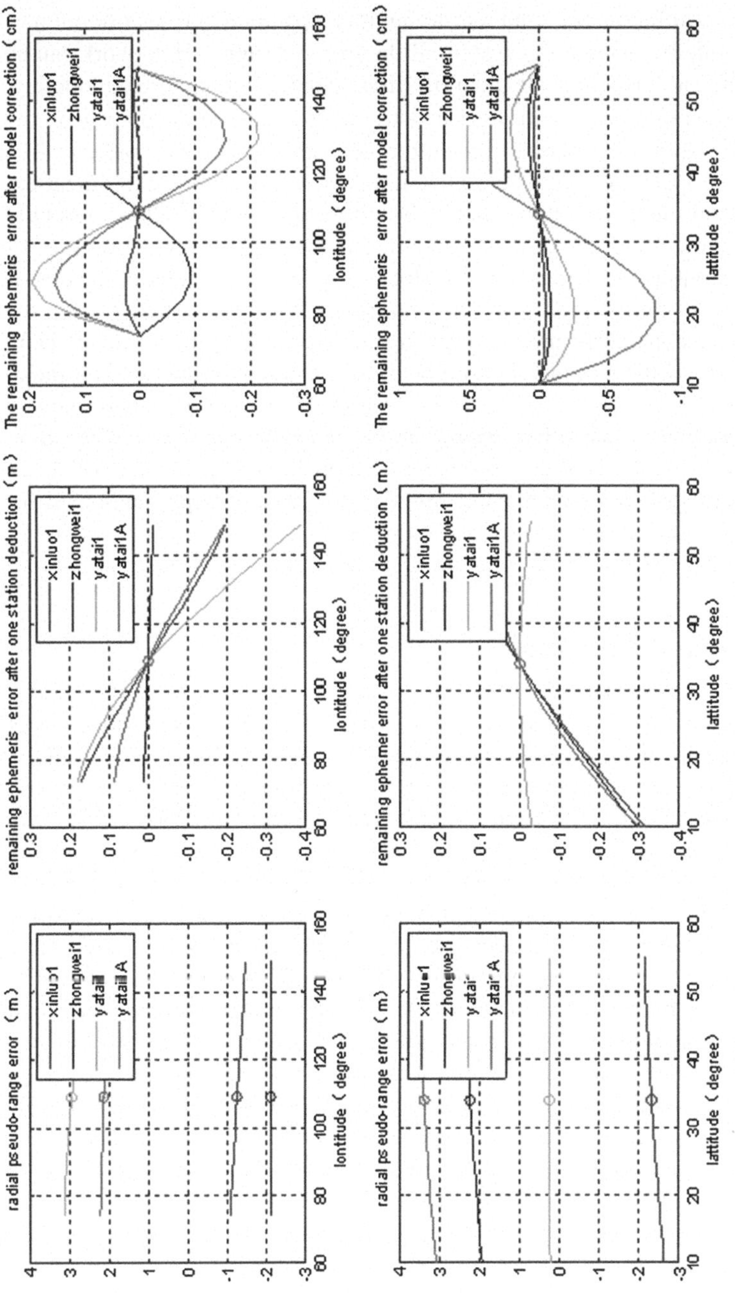

Fig. 2 Remaining radial ephemeris error be corrected by two order polynomial model

differential within the network. If the surface model can be established, the curved surface model coefficients can be directly broadcasted, which is convenient for users to use these differential coefficient to achieve ephemeris error correction.

For Chinese satellite navigation system, if the future differential network can be designed as grid, and it can provide the most direct condition for surface modeling. $5^{\circ} \times 5^{\circ}$ grid of the monitoring stations is constructed for simulation analysis in the following. In the analysis process, other reference stations except master station are only assumed to be coarse synchronization, not to be strict synchronization. Firstly, the station clock deviation should be eliminated with single differential among satellites, therefore, the data for modeling is the remaining ephemeris error after single differential (the below left subplot in Fig. 3), the simulation results as shown in Fig. 3. As can be seen from the graph: with the grid surface model, the relative remainder ephemeris error can reach millimeter. This model is not DOP constraint, as long as the parameters are given, user can obtain the residual radial ephemeris error. The user's pseudo range error after the correction is approximate to millimeter (assuming that other errors can be precisely corrected). Therefore, the differential users can obtain very high real time positioning precision. The wide area differential method of no DOP constraint can be a kind of important means for high precise and real time differential positioning.

5 Conclusion and Prospect

Because there are always some residuals after troposphere and ionosphere correction, the residual error performs as random error. Because the geometric distribution for ground differential network observation to satellite is poorer, directly cause that DOP is very large, because DOP has amplification effect, very small residual error will bring large error of ephemeris correction, this is the key reason that WAAS or other traditional differential methods are difficult to realize to 1 m accuracy of wide area differential. Our previous work aimed at real-time decimeter level of wide area differential positioning precision, so as to satisfy less than 1 m precision of wide area differential COMPASS. This paper analyzes for optimal differential performance of two wide area differential method: the small DOP constraint and without DOP constraints. The common point for the two methods lies in the one station correction of the large part, and the different point lies in the small correction, the small DOP constraints method do not require rule graphical layout of differential station, and strong flexibility, and higher differential positioning precision than WAAS, it is a relatively practical differential method; the requirement for no DOP constrain is to build a grid differential network, because the calculation of the differential component are free from DOP limit, its positioning accuracy is very high, so this method is a great potential differential method.

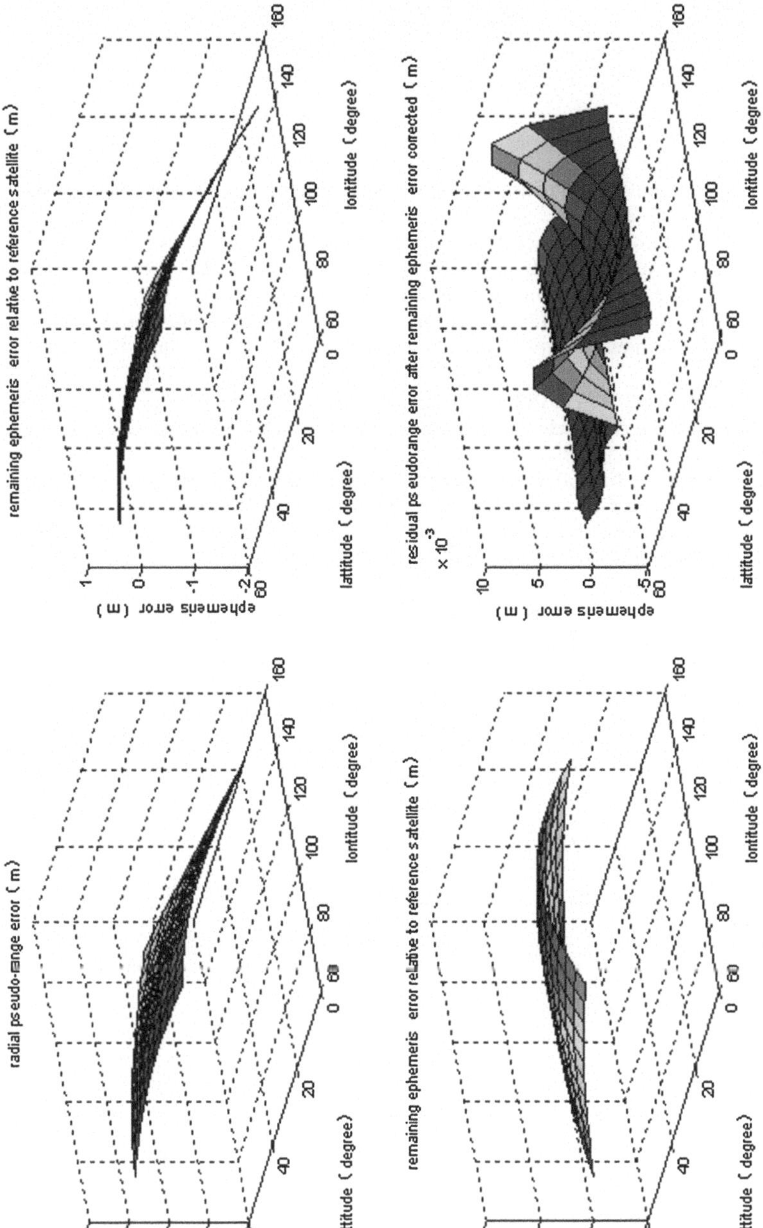

Fig. 3 5° × 5° grid surface model correction

Acknowledgments This work was supported by the National Natural Science Foundation of China (Grant No. 10778715), the National Key Basic Research Development Program of China (Grant No. 2007CB815502).

References

1. Kouba, J., & Héroux, P. (2001). Precise point positioning using IGS orbit and clock products. *GPS Solutions, 5*(2), 12–28.
2. Marcio, A. (1998). Regional approach to wide area DGPS. The University of Nottingham.
3. Yeou-Jyh, T. (1999, August). *Wide area differential operation of the global positioning system: Ephemeris and clock algorithms*. California: Stanford University.
4. Cai, C. L., Li, X. H., & Wu, H. T. (2009). An analysis of the wide area differential method of geostationary orbit satellites. *Science in China Series: Physics Mechanics and Astronomy, 2*, 310–314.
5. Aminloo, R. (1995). What is WAAS? *Air Transport World, 32*(9), 55.
6. Whitehead, M. L. (1998). A close look at Satloc's real-time WADGPS system. *GPS Solutions, 2*(2), 45–63 (Wiley).
7. Yunck, T.P., Bertiger, W. I., et al. (1995, April). A robust and efficient new approach to real time wide area differential GPS navigation for civil aviation, JPL Report D.
8. Cai, C. L., Li, X. H., & Wu, H. T. (2009). A satellite clock error correction method in wide area differential system. *Chinese Science Bulletin, 54*(24), 4616–4623.

Study on Microwave Circuit for Chip Scale Atomic Clock

Jiehua Chen, Deng Wei, Zhang Yi, Yuanchao Wang and Sihong Gu

Abstract We have studied the microwave generating circuit base on Dielectric Resonator Oscillator (DRO), the circuit base on Injection Locked Oscillator (ILO). According our study results, we consider that both schemes are not ideal candidates for Chip Scale Atomic Clock (CSAC). We have chosen a microwave synthesizing scheme for CSAC, and have realized the circuit. We have realized an atomic clock pattern with the circuit and based on the circuit, and with the clock we have experimentally studied the behavior of the circuit. Our study results show that the microwave generating scheme is a competitive candidate for CSAC.

1 Introduction

Coherent Population Trapping (CPT) atomic clocks, with many advantages such as small size, low power consumption and fast startup etc., are promising to be applied in broad areas. As the microwave cavity is not needed in a CPT atomic clock, its physics package can be fabricated into chip scale through Micro-Electro-Mechanical Systems (MEMS), combined with the Application Specific Integrated Circuit (ASIC) chips, chip scale atomic clock supported by a button cell battery can be realized [1]. In principle the CPT atomic clock is the only one which can be miniaturized to the Chip Scale Atomic Clock (CSAC). Applied in ground receiver, CSAC will enhance the functions of satellite navigation and position system.

J. Chen (✉) · D. Wei · Z. Yi · Y. Wang · S. Gu
Wuhan Institute of Physics and Mathematics, The Chinese Academy of Sciences,
430071 Wuhan, People's Republic of China
e-mail: andychen@wipm.ac.cn

S. Gu
e-mail: shgu@wipm.ac.cn

J. Sun et al. (eds.), *China Satellite Navigation Conference (CSNC) 2012 Proceedings*,
Lecture Notes in Electrical Engineering 160, DOI: 10.1007/978-3-642-29175-3_55,

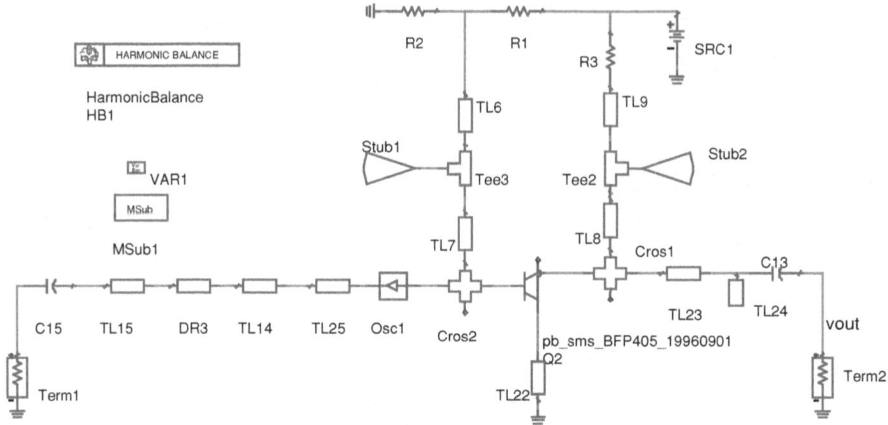

Fig. 1 ADS machine simulation schematic of the DRO based circuit

Products of CSAC by Symmetricom Inc. are commercial available now, the volume of the CSAC is 16 cm^3, power consumption is 115 mW, and its frequency stability reaches 1×10^{-11}/h [2]. In China, several groups have been carrying out researches of CPT atomic clock and CSAC. Our group has realized CPT atomic clocks, one of our developed CPT atomic clock reaches volume 160 cm^3, and frequency stability $4 \times 10^{-11}\, \tau^{-1/2}$. CSAC has also being studied in our group and certain progresses have been achieved.

In our designed CSAC scheme, ^{87}Rb is chosen as working atom, to obtain coherent bi-chromatic laser, 3417 MHz microwave is used to modulate VCSEL. In this paper we present our study results about microwave generating circuits: the study result about the Dielectric Resonator Oscillator (DRO) scheme, the Injection Locked Oscillator (ILO) scheme, and the scheme with which we designed and realized circuit for our CSAC prototype.

2 Dielectric Resonator Oscillator

NIST proposed a DRO based microwave circuit scheme [3], by which the 3417 MHz microwave is selected out from the frequencies in the oscillate circuit with a DRO, thus the circuit outputs only 3417 MHz microwave. We have experimentally studied this scheme, Fig. 1 is our obtained ADS machine simulation schematic structure of the DRO based circuit, and Fig. 2 is our ADS stimulated frequency spectrum which is also digitally presented in Table 1.

According to the stimulated result we have designed and realized the circuit, RO4003C is chosen for the circuit substrate board, the size of the realized circuit is 15 × 15 mm, and power consumption is 15 mW. Figure 3 is our measured output spectrum of the circuit, the output 3417 MHz microwave power of is −4dBm.

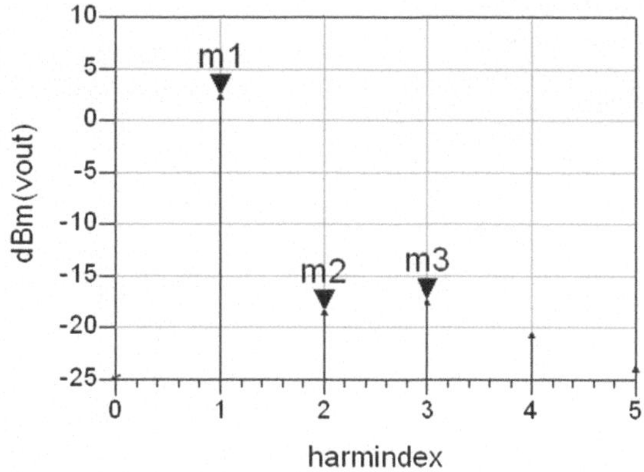

Fig. 2 ADS stimulation frequency spectrum (Power (m1) = 2.557 dBm; Power (m2) = −18.238 dBm; Power (m3) = −17.180 dBm)

Table 1 Frequency spectrum

Harmindex	0	1	2	3	4	5
Frequency (GHz)	0	3.417	6.835	10.25	13.67	17.09

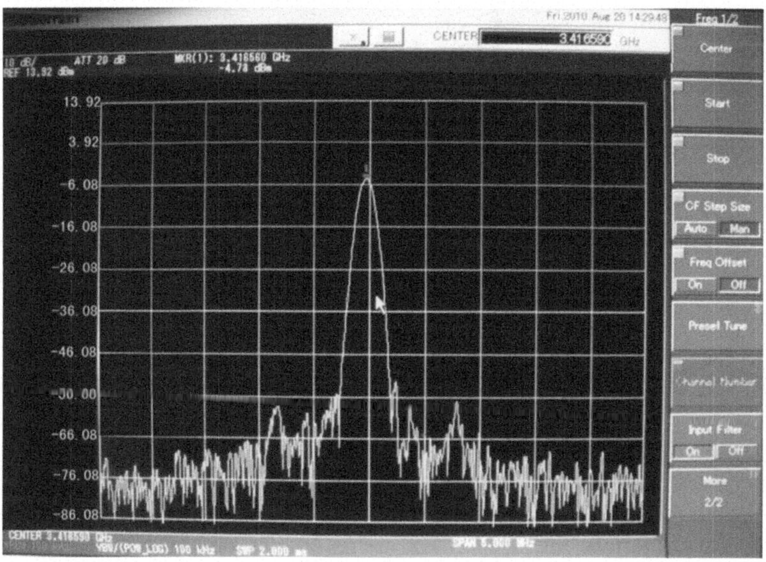

Fig. 3 The measure spectrum pattern of the DRO circuit output

Compared with other ways, the phase noise of the microwave generated by this DRS based circuit is not good. As only 3417 MHz microwave is provided, although a ^{87}Rb CPT atomic clock can be realized, whereas it will confined the application of the atomic clocks as in general application a 10 MHz output is required, whereas extra resource will if arranging converting 3417 to 10 MHz.

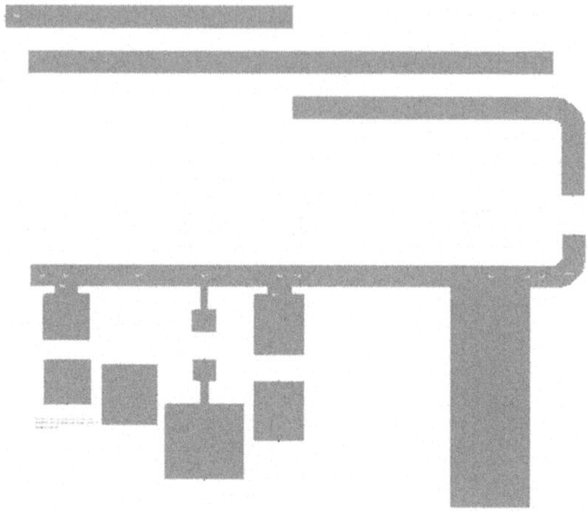

Fig. 4 ADS machine stimulating schematic of the DRO based circuit

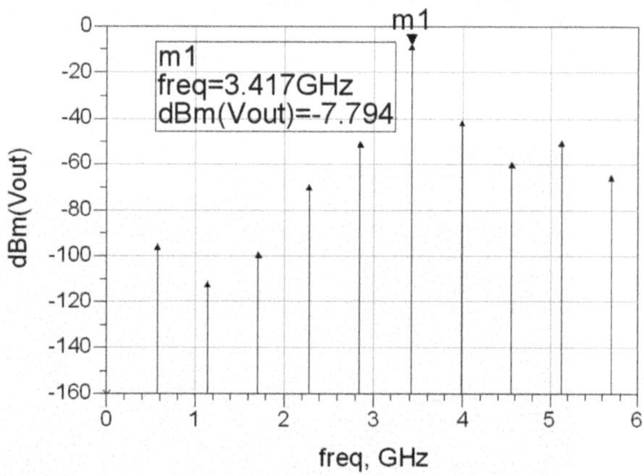

Fig. 5 ADS stimulation frequency spectrum

3 Injection Locked Oscillator

Teledyne Technologies Inc has realized a principle prototype CSAC, its volume is only 1 cm^3, power consumption 30 mW, and frequency stability reaches 1×10^{-11}/h [4]. In the CSAC the microwave circuit is realized with the ILO scheme. In the circuit the 3417 MHz microwave is obtained by directly multiplying from 569 MHz, which bring circuit the advantages of small size and low power consumption. We have also studied this scheme, Fig. 4 is the ADS machine stimulation schematic of the ILO base circuit, and Fig. 5 is our stimulated its frequency spectrum.

According to the stimulated result we have designed and realized the circuit, its size is 20 × 10 mm and power consumption is less than 10 mW. With the inputted microwave of Fig. 6 (left), Fig. 6 (right) is the measured output frequency distribution from the circuit. It is seen from Fig. 6 that the phase noise is better than

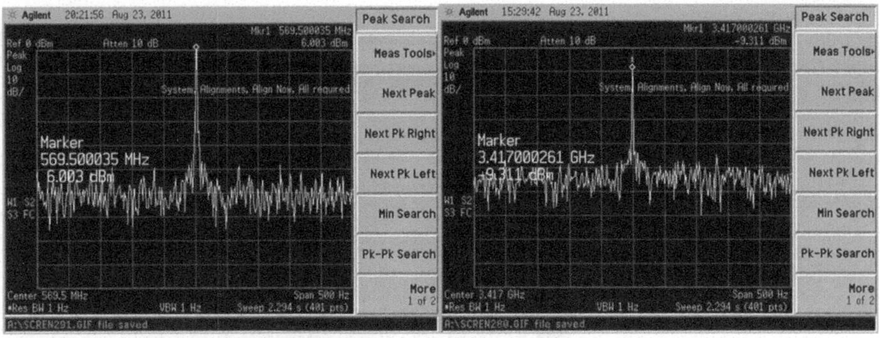

Fig. 6 The microwave spectrum pattern about the ILO circuit, the *left* is the input one and the *right* is the output microwave from the circuit

that of Fig. 3. However, an unconventional crystal oscillator has to be arranged for the circuit, and the realized atomic clock also does not provide 10 MHz standard frequency output.

We want to mention that although with the two schemes small size and low power consumption circuits can be realized, both of them don not suit for realizing the ASIC chip.

4 The Scheme We Chosen for CSAC

Aiming at applying in CSAC, we have designed a microwave generating circuit, and Fig. 7 is its stimulation result. Although by this scheme the volume and power consumption of the circuit are both slight more than the previously discussed two circuits, it suits for integrating into ASIC chip thus size and power consumption can be reduced by implementing integration. Moreover, beside 3417 GHz the circuit provides 10 MHz signal, which makes the realized atomic clock more application flexible.

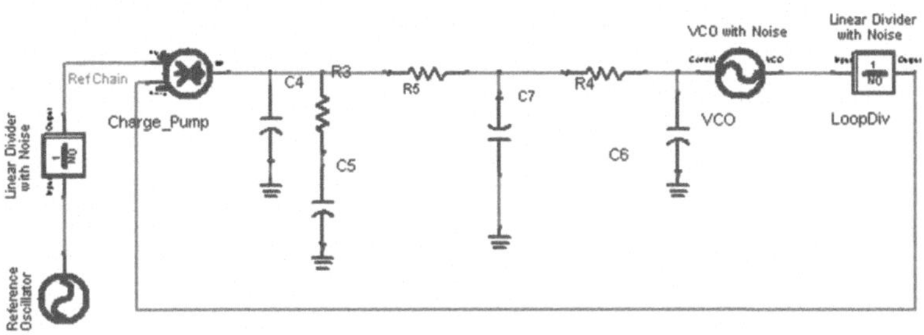

Fig. 7 ADS machine stimulation schematic of the our designed circuit

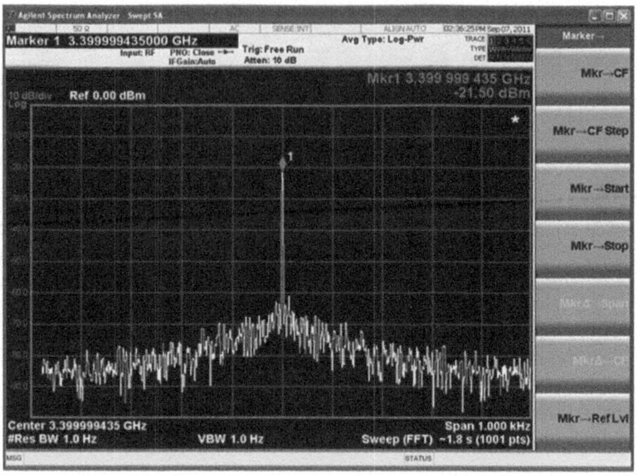

Fig. 8 The microwave spectrum of microwave output from the microwave chip

By connecting a physics package with the circuit we have realized a close loop lucked CPT atomic clock pattern, and the experimental results reveal that its functions have reached the design requirement. Through ASIC we have obtained an integrated chip, in which the main components of the realized circuit have been integrated [5]. The size of the chip is 5 × 5 mm, and Fig. 8 is the measured spectrum from the chip.

5 Conclusion

We have experimentally studied microwave generating circuit schemes based on DRO and ILO, both of them are competitive for realizing small size and low power consumption microwave generating circuit. However, they don't suit for realizing microwave generating chip. Moreover, the realized atomic clocks don't output 10 MHz standard frequency, which will considerably limit their application.

Aiming at CSAC, we have designed and realized a microwave generating circuit, and realized an atomic clock pattern with the circuit and based on the circuit we have also realized an ASIC chip. The experimental result reveals that chip basically meets the requirement of CSAC. The atomic clock realized this way will provide 10 MHz standard frequency output, which makes it more application flexible, so that we consider the microwave generating scheme a competitive candidate for CSAC.

References

1. Chen, J. (2010). Doctoral Dissertation. Chinese Academy of Sciences.
2. http://www.symmetricom.com/media/files/downloads/product-datasheets/DS_SA.45sCSAC.pdf, (2011).

3. Brannon, A., Member, S., et al. (2009). *A low-power, low phase noise local oscillator for CSAC. IEEE*. pp. 1535–1538.
4. Denatable, J. L., Borwick, R. L., et al. (2008). *Compact, low power CSAC. IEEE*. pp. 67–70.
5. Chen, J., Deng, W., Zhang, Y., et al. (2011). Micro-CPT atomic clock electronic research progress. *China Time and Frequency Symposium*. p. 240.

Analysis on Performance Relation Between Time–Frequency Architecture and Positioning Service of a Satellite Navigation System

Jun Lu, Zhi-Wu Cai and Hong-Wei Zhou

Abstract Performance improvement of time and frequency technology has played a key role in GPS URE accuracy development. The time and frequency elements related to system service performance were analyzed and concluded based on positioning theory and main error sources analysis regarding actual engineering situation, such as system time, clock, time comparison and so on. According to URE performance requirement, the performance analysis architecture about time and frequency indexes were put forward for the first time. Performance relation model between time and frequency indexes and URE clock component and its error estimation formula were studied and build. Simulations were made and reference data about time and frequency indexes and URE were given. The result may be used as technique reference to satellite navigation system research and Chinese Beidou system development.

1 Introduction

Pseudo-range measurement is the basis of modern satellite navigation system implementation. The essence of Pseudo-range measurement is time measurement. With the broad applications and performance improvement of satellite navigation

J. Lu · H.-W. Zhou
Beijing Institute of Tracking and Telecommunication Technology, 100094 Beijing, China

Z.-W. Cai
Beijing Global Information Center of Application and Exploitation, 100094 Beijing, China

Z.-W. Cai (✉)
Nanjing University of Aeronautics and Astronautics, 210000 Nanjing, China
e-mail: caizhiwu@gmail.com

J. Sun et al. (eds.), *China Satellite Navigation Conference (CSNC) 2012 Proceedings*, Lecture Notes in Electrical Engineering 160, DOI: 10.1007/978-3-642-29175-3_56,

system, performance requirements to atomic clocks, time standard, time synchronization and so on have increased step by step. More stable satellite clock and relative time and frequency precise measurement and control technologies are the key factors of GPS IIR and IIF satellite URE decrease [1–4]. The very precise pseudo-range measurements of Galileo experiment satellites benefit from satellite hydrogen clocks application. Chinese Beidou navigation system have performed services of RNSS positioning, navigation, and timing formally, which depended on the high performance satellite clock and high precise time transferring technologies advances [5].

The PNT services of satellite navigation system depend on two basic kinds of parameter in navigation messages which are satellite orbit and clock parameters. With precise orbit dynamic model constraint, satellite orbit parameters could be determined with high accuracy and stability, which precision may reaches to decimeter or even centimeter level. However, the precision of satellite clock parameters relate to more sophisticated factors, such as satellite clock, system time, time comparison between satellite and ground or satellite and satellite etc. The achievement capabilities of high precise clock difference have been the bottleneck of high precise satellite navigation service implement. It is an important and complex problem how to analysis and design time–frequency index demands to ensure the navigation system PVT service accuracy. Analysis and discussion about performance relation of time–frequency and positioning will be expanded in this paper.

2 Performance Relation Between Time–Frequency Architecture and Positioning Service

Usually, navigation system positioning accuracy could be expressed by the multiplication of User Equivalent Range Error (UERE) and Position Dilution of Precision (PDOP). PDOP mainly determined by constellation structure and user position. UERE is composed of navigation signal errors, propagation delay errors, user receiver measurement errors and expressed by radial equivalent range quantity.

According to the error source features, main error sources of UERE may be concluded into three kinds and time–frequency errors related to every kind error source involve several items. Detailed analysis is described as following.

(1) Navigation message parameter errors

Navigation message parameter errors mainly include orbit parameter errors, clock parameter errors and ionosphere parameter errors. The satellite orbit and clock parameter errors are usually expressed by User Range Error (URE). Here, the clock parameters represent time offset between local time of satellite clock and system time, which are usually acquired through time difference observing and modeling based on satellite and ground time comparison technology.

Considering Chinese actual situation, the navigation system background was designed as the following: global constellation of MEO satellites, regional monitor stations distributed in China, satellite and ground links and inter-satellite links. On the mode of system operation and control implementation based on ground stations, the system time is produced in the master station. And the time difference between satellite clock and system time standard could be got through direct high precise satellite and ground time comparison links when MEO satellites lie in visible arc of ground master stations. While MEO satellites lie in invisible arc of ground master stations, the indirect time comparison could be implemented through the combination of satellite and ground time comparison links and inter-satellite comparison links. Another way of indirect time comparison is implemented through the combination of satellite and ground time comparison links and inter-station comparison links. Therefore, satellite clock predication error is related to satellite clock stability, system time standard stability, satellite and ground time comparison accuracy, inter-satellite time comparison accuracy and inter-station time comparison accuracy etc.

On the mode of autonomous navigation, it is assumed that system time standard was established based on composition of all satellite clocks and there were not ground stations supplies. Here, satellite clock predication error is related to satellite clock stability, system time standard and inter-satellite time comparison accuracy etc.

Besides, a lot of other factors will also influence the stability and accuracy of satellite clock signal output, such as the temperature and magnetic variance of satellite clock circumstance, additional loss of other devices in satellite time–frequency subsystem.

(2) Navigation signal dithering errors

Navigation signal dithering errors mean the frequency and phase noise and difference of carrier and pseudo-code components between emitting navigation signal and normal navigation signal. Its main error sources include frequency uncertainty, frequency stability, phase noise of basic frequency signal from satellite clock, and additional phase noise within satellite time–frequency subsystem. For example, one technique in GPS selective availability (SA) policy was adding frequency dithering to basic frequency of satellite clock frequency output artificially, that leads to obvious error increase of GPS positioning service.

(3) User equipment measurement errors

User equipment measurement errors mean all errors produced in the procession from navigation signal running away from satellite antenna to pseudo-range and carrier measurement finished in receiver channels, that usually expressed as User Equipment Error (UEE). Its main error sources include ionosphere delay error, troposphere delay error, multiple path error, receiver measurement error and receiver time–frequency unit signal error etc.

After concluding and regulating the above errors, they are divided into three layers and organized in a tree structure as the following figure. PVT performances of navigation system are put in input layer, and time–frequency architecture performances related to input layer are put in middle layer, and the last determined time–frequency technique demands synthesized from above constraints are put in output layer. This figure explained the relation of performance elements between time–frequency performance and positioning service in detail, which is helpful to budget the performance demands of atomic clocks, time comparison links, and time measurement etc. according with system overall service performance (Fig. 1).

3 Analysis Model of Time–Frequency Architectures

In the service indexes of system positioning, URE index is the critical restriction index. In the point of time and frequency view, the prediction accuracy of satellite clock offset is one of the primary weights, and also synthetically indicates all kinds of time and frequency influences. Therefore, the index of satellite clock offset prediction is the main aim of analysis.

The frequently used technology of two-way satellite time and frequency transfer (TWSTFT) and clock offset prediction model of quadratic polynomial are the basis of analysis. The clock offsets of the time scale of satellite clock between ideal time scale, and system time between ideal time scale all can be denoted by quadratic time model. The time model of satellite clock is as following:

$$T_i - t = a_{i,0}^s + a_{i,1}^s (t - t_0) + \frac{1}{2} a_{i,2}^s (t - t_0)^2 + R_i^s(t) \tag{3.1}$$

In the formula, T_i means the time of satellite clock i, t is the ideal time scale, $a_{i,0}^s$ $a_{i,1}^s$ $a_{i,2}^s$ respectively represent the time offset of satellite clock i between ideal time scale, frequency accuracy (frequency offset), frequency excursion rate. $R_i^s(t)$ is the noise of satellite clock i, and reflects the frequency stability of satellite clock i.

Similarly, the time scale of navigation system can be represented by quadratic polynomial. With respect to the time scale of Beidou navigation system, it has the following formula:

$$BDT - t = a_0^{BDT} + a_1^{BDT}(t - t_0) + \frac{1}{2} a_2^{BDT}(t - t_0)^2 + R^{BDT}(t) \tag{3.2}$$

In the formula, t is the ideal time scale, a_0^{BDT} , a_1^{BDT} and a_2^{BDT} respectively represent the model parameters, and R^{BDT} is the random noise which represents frequency stability of BDT. The satellite clock offset with respect to the time scale of navigation system can also be represented by quadratic polynomial model, and the prediction model of clock offset is as following:

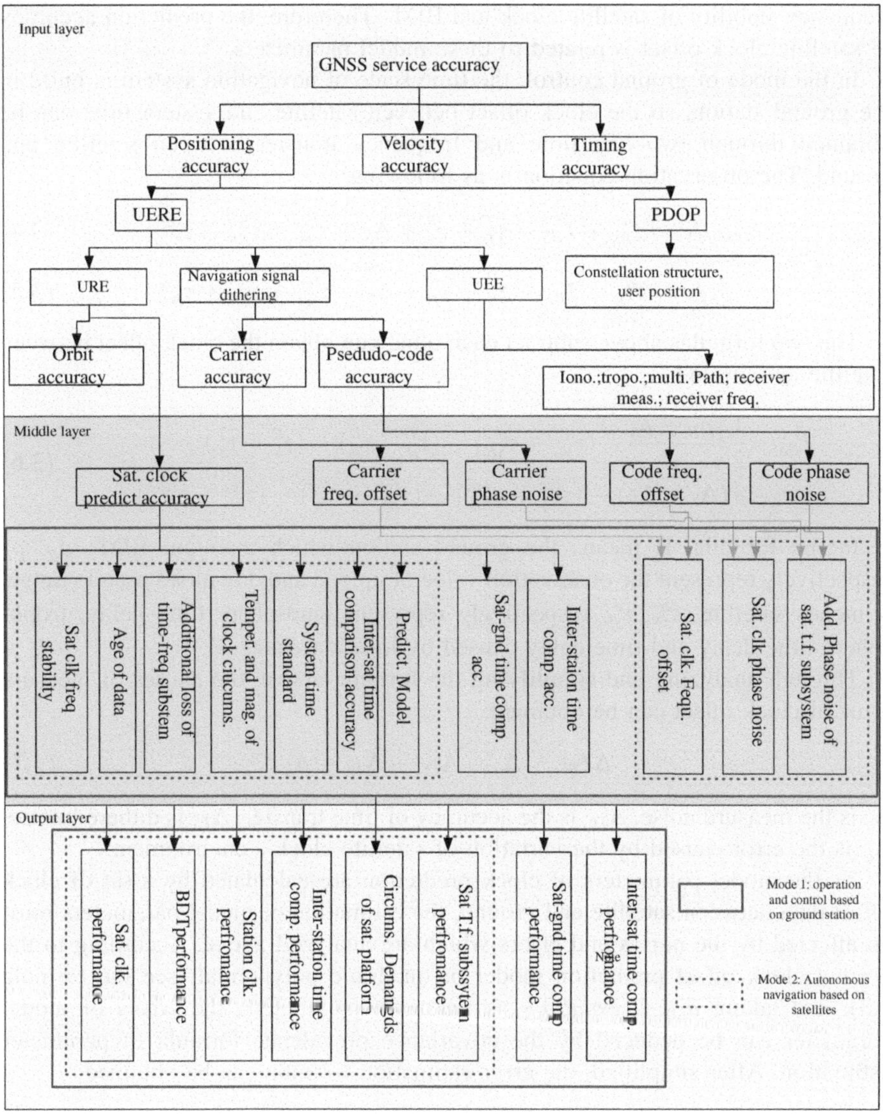

Fig. 1 Decomposition tree of positioning service and time–frequency performance

$$T_i - BDT = \tilde{a}_{i,0} + \tilde{a}_{i,1}(t - t_0) + \frac{1}{2}\tilde{a}_{i,2}(t - t_0)^2 + \tilde{R}_i(t) \tag{3.3}$$

In the formula, $\tilde{a}_{i,0}$, $\tilde{a}_{i,1}$ and $\tilde{a}_{i,2}$ respectively represent the time offset, frequency offset and frequency excursion rate at the time of t_0. $\tilde{R}_i(t)$ is synthetical noise, which is part of clock offset and cannot be modeled. It is respect to the frequency stability of satellite clock and BDT. Therefore, the prediction accuracy of satellite clock offset is related to these model parameters.

In the mode of ground control, the time scale of navigation system is build in the ground station, so the clock offset between satellite and system time can be obtained through two-way time and frequency transfer between satellite and ground. The observation equation is as following:

$$\rho_B/c = \tau_{BS}^{u} + T_{BS} + \tau_B^{T} + \tau_S^{R} + \Delta_S + \tau_{BS}^{ion} + \tau_{BS}^{tro} + \tau_S^{sta} \tag{3.4}$$

$$\rho_S/c = \tau_{SB}^{d} - T_{BS} + \tau_S^{T} + \tau_B^{R} + \Delta_B + \tau_{SB}^{ion} + \tau_{SB}^{tro} + \tau_B^{sta} \tag{3.5}$$

The two formulas above subtract each other can obtain the clock offset between satellite and ground:

$$\begin{aligned} T_{BS} =& \frac{1}{2}\Big[\frac{\rho_B - \rho_S}{c} - (\tau_{BS}^{u} - \tau_{SB}^{d}) - (\tau_B^{T} - \tau_B^{R}) - (\tau_S^{R} - \tau_S^{T}) \\ & -(\Delta_S - \Delta_B) - (\tau_{BS}^{ion} - \tau_{SB}^{ion}) - (\tau_{BS}^{tro} - \tau_{SB}^{tro}) - (\tau_S^{sta} - \tau_B^{sta})\Big] \end{aligned} \tag{3.6}$$

In the formula, B means the ground station which maintain BDT, ρ_B, ρ_S respectively represent the observation value of upload and download pseudo range. S means satellite, τ_{BS}^{u}, τ_{SB}^{d} respectively represent ionosphere time delay, troposphere time delay and time delay caused by location changes.

Through analyzing and simplifying the formula above, the causes of measure error of clock offset can be obtained:

$$\Delta T_{BS} = \Delta_n + \Delta_{TC} + \Delta_{TS} + \Delta_E \tag{3.7}$$

Δ_n is the measure noise, Δ_{TC} is the accuracy of time transfer, Δ_{TS} is dithered noise, Δ_E is the error caused by the variation of satellite clock's environment.

As the model parameters of clock prediction are calculated by a set of clock offset data between satellite and ground, the estimation of model parameters must be affected by the noises and errors which are analyzed above. According to the normal clock offset prediction model of quadratic polynomial (see the formula 3.3), and taking $\tilde{a}_{i,0}$, $\tilde{a}_{i,1}$ *and* $\tilde{a}_{i,2}$ as unknown parameters, the errors of model parameter can be deduced by the covariance prevalence formula of parameter estimation. After simplified, the error estimation formula can be obtained:

$$\sigma_{a_0} = \frac{\sigma_0}{\sqrt{n}} \tag{3.8}$$

$$\sigma_{a1} = \frac{\sigma_0}{\sqrt{\tau_0^2 \times \frac{1}{6}n(n+1)(2n+1)}} \tag{3.9}$$

$$\sigma_{a2} = \frac{\sigma_0}{\sqrt{\tau_0^4 \times \frac{1}{30}n(n+1)(2n+1)(3n^2+3n-1)}} \tag{3.10}$$

Table 1 Clock deviation predication accuracy of different models and observation intervals

Observation intervals (min)	Predication model	Maximum predication error (ns)	Minimum predication error (ns)	Mean value of predication error (ns)	Mean square error of predication (ns)	Predication error (RMS)
1	Linear	1.92	7.99e−3	7.99e−1	4.92e−1	9.38e−1
	Parabola	2.01	1.68e−2	1.00	5.14e−1	1.13
3	Linear	1.94	6.32e−3	7.95e−1	4.88e−1	9.32e−1
	Parabola	2.15	4.83e−2	9.99e−1	5.03e−1	1.12
5	Linear	2.06	1.61e−2	8.14e−1	5.05e−1	9.57e−1
	Parabola	2.36	8.36e−3	1.02	5.49e−1	1.16
10	Linear	1.89	1.25e−2	7.63e−1	5.10e−1	9.17e−1
	Parabola	2.43	9.12e−3	9.69e−1	5.92e−1	1.13
15	Linear	2.06	5.63e−2	8.08e−1	5.19e−1	9.59e−1
	Parabola	2.39	4.70e−3	1.02	6.32e−1	1.19

In the formula, σ_0 is the measure error of clock offset, τ_0 is sampling interval, n is sampling times, and observation time is $N = n \times \tau_0$.

The model relation of prediction accuracy of satellite clock offset is:

$$\Delta(T_i - BDT) = \Delta\tilde{a}_{i,0} + \Delta\tilde{a}_{i,1}(t - t_0) + \frac{1}{2}\Delta\tilde{a}_{i,2}(t - t_0)^2 + \tilde{R}_i(t) \tag{3.11}$$

In which,

$$\tilde{R}_i(t) = \sqrt{\sigma_{y,BDT}(\tau)^2 + \sigma_{y,S}(\tau)^2} \times \tau \tag{3.12}$$

Using the formula above, the prediction error of satellite clock offset can be estimated. When the linear model of clock offset prediction is used, formula 3.8 and 3.9 can also be taken to analyze the error of model parameter $\tilde{a}_{i,0}$ and $\tilde{a}_{i,1}$.

4 Simulation Data Analysis

The relationship between time frequency index and clock offset index of URE is simulated, and the main errors considered in the simulation are the following: the stability of satellite clock is $3 \times 10^{-14}/10^4$ s; the stability of system time is $1 \times 10^{-14}/10^4$ s; the error of time transfer between satellite and ground is 0.6 ns; the error of time transfer between satellites is 0.6 ns; other errors caused by environment changes such as temperature and magnetic field is 0.6 ns.

Data with different observation time intervals are analyzed by linear and quadratic polynomial model. The observation time is 12 h long, and prediction time is 2 h long. 126 groups of data are calculated every time, and the detailed results are in the Table 1.

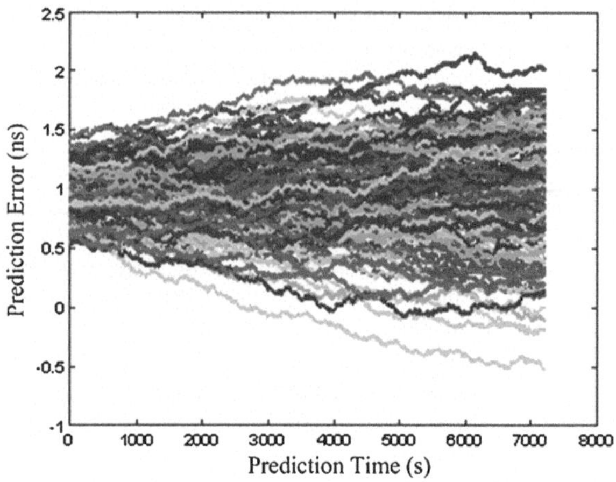

Fig. 2 Clock deviation predication accuracy of 126 groups data based on parabola models

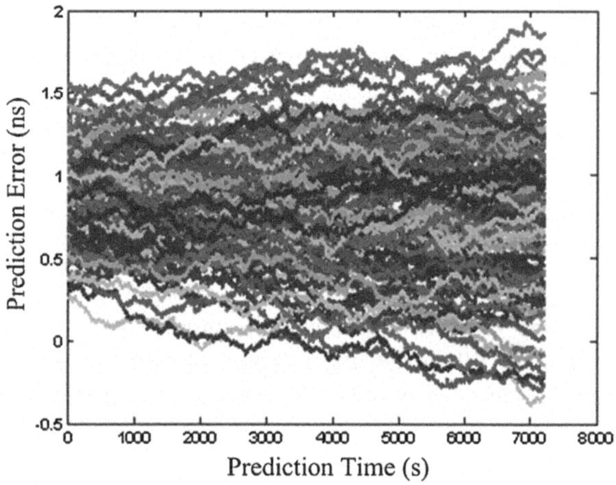

Fig. 3 Clock deviation predication accuracy of 126 groups data based on linear model

The predication error of 126 groups of data with 5 min sampling interval is showing in the Figs. 2 and 3.

Through the simulation, the predication error of clock offset with 2 h can reach the accuracy of 1–2 ns, in the condition of that: the satellite clock stability is $3 \times 10^{-14}/10^4$ s, the error of time transfer between satellite and ground is 0.6 ns, and the error of time transfer between satellites is 0.6 ns. The mean value of predication error is close to the sum of system errors of time transfer between satellite and ground, and time transfer between satellites. It cannot be reduced by data smooth or fit. In practical project, the system error of time transfer should be reduce or control by calibration onboard or other calculation methods to improve the prediction accuracy of clock offset further.

5 Conclusion and Suggestion

To implement high precision services of navigation, position and timing, all kinds of direct or indirect error should be considered adequately. URE and time–frequency performance decomposition tree structure described the basic error influence relation and could be used as error budget foundation. Specially, several key indices, which include satellite clock predication error, satellite and ground time comparison error and system time standard performance etc., should be analyzed and controlled strictly to ensure the availability of system service accuracy.

Because of the complexity and crossed influences of lots of indices, more data analysis or experiments are needed to determine the time–frequency performance influences to navigation system service accuracy.

References

1. Kaplan, E. D. (2007). *Principle and application of global position system* (2nd ed.). Publishing house of electronics industry.
2. Vittorini, L. D., & Collins, R. (1998). Cedar rapids. GPS URE/UE evolutionary improvements and end-user accuracy results. In *Position location and navigation symposium, IEEE 1998, Palm Springs, CA, USA, 20–23 April 1998*. pp. 403–411.
3. Colonel Harold "Stormy" Martin . (2011). *GPS status and modernization*, http://www.pnt.gov.
4. Boeing, W. E., & Boeing, E. W. (1997). Atomic frequency standards for the GPS IIF satellites. *29th annual precise time and time interval (PTTI) meeting, Long Beach, CA, 2–4 December*, pp. 201–212.
5. China Satellite Navigation System Management Office (2011). *Development report of beidou satellite navigation system*, http://www.beidou.gov.cn,14, November .

The manufacturer's authorised representative in the EU is Springer Nature Customer Service Centre GmbH, Europaplatz 3, 69115 Heidelberg, Germany. If you have any concerns regarding our products, please contact ProductSafety@springernature.com

Printed and bound by CPI Group (UK) Ltd, Croydon, CR0 4YY
25/09/2026
02222839-0013